NOBEL LECTURES

PHYSICS

2001–2005

NOBEL LECTURES

INCLUDING PRESENTATION SPEECHES
AND LAUREATES' BIOGRAPHIES

PHYSICS

CHEMISTRY

PHYSIOLOGY OR MEDICINE

LITERATURE

PEACE

ECONOMIC SCIENCES

NOBEL LECTURES

INCLUDING PRESENTATION SPEECHES
AND LAUREATES' BIOGRAPHIES

PHYSICS

2001–2005

EDITOR

Gösta Ekspong

Department of Physics
University of Stockholm

World Scientific

NEW JERSEY • LONDON • SINGAPORE • BEIJING • SHANGHAI • HONG KONG • TAIPEI • CHENNAI

Published by

World Scientific Publishing Co. Pte. Ltd.
5 Toh Tuck Link, Singapore 596224
USA office: 27 Warren Street, Suite 401-402, Hackensack, NJ 07601
UK office: 57 Shelton Street, Covent Garden, London WC2H 9HE

NOBEL LECTURES IN PHYSICS (2001–2005)

ISBN-13 978-981-279-447-5
ISBN-10 981-279-447-6
ISBN-13 978-981-279-448-2 (pbk)
ISBN-10 981-279-448-4 (pbk)

Printed in Singapore by Mainland Press Pte Ltd

FOREWORD

Since 1901 the Nobel Foundation has published annually *Les Prix Nobel* with reports from the Nobel award ceremonies in Stockholm and Oslo as well as the biographies and Nobel lectures of the Nobel laureates. In order to make the lectures available for people with special interests in the different fields the Nobel Foundation gave Elsevier Publishing Company the right to publish in English the lectures for 1901–1970, which were published in 1964–1972 through the following volumes:

Physics 1901–1970	4 vols.
Chemistry 1901–1970	4 vols.
Physiology or Medicine 1901–1970	4 vols.
Literature 1901–1967	1 vol.
Peace 1901–1970	3 vols.

Thereafter, and until the year 2000, the Nobel Foundation has given World Scientific Publishing Company the right to bring the series up to date and also publish the Prize Lectures in Economics from the year 1969. Between the years 2001–2005, the Nobel Foundation has given World Scientific Publishing Company a publishing right through an agreement between Nobel Media AB and World Scientific Publishing Company.

The Nobel Foundation is very pleased that the intellectual and spiritual message to the world laid down in the laureates' lectures, thanks to the efforts of World Scientific, will reach new readers all over the world.

Michael Sohlman
Executive Director
Stockholm, April 2008

PREFACE

The present volume contains the Nobel lectures in physics for the years 2001–2005. Also included are the autobiographies of the Nobel Laureates and the presentation speeches given by members of the Royal Academy of Sciences at the award ceremony in Stockholm.

The 2001 prize was given for the experiments achieving Bose-Einstein condensates of gases. They required difficult manipulation of atoms to keep them confined while cooling them to unprecedented low temperatures some millionth of a degree above absolute zero. These achievements opened a new field of research. Two fascinating stories about this are found in the lectures by Carl Wieman and Eric Cornell and by Wolfgang Ketterle, respectively.

For opening new windows to observe solar and cosmic events the 2002 prize was awarded to Raymond Davis Jr, Masatoshi Koshiba and Riccardo Giacconi. The births of neutrino astronomy and of X-ray astronomy are the subjects of these lectures. Many surprising and remarkable discoveries have been made. The X-ray window in astronomy has revealed a Universe pervaded by echoes of high-energy explosions and abrupt changes. R. Davis Jr was the first to observe neutrinos from the centre of the sun. The supernova 1987a occurred 170,000 years ago and emitted a short burst of an immense number of neutrinos. During their travel time to Earth all our human history developed. Then after 10 seconds it was all over.

The oldest Nobel Laureate ever is Vitaly Ginzburg. In his lecture he gives advice to young physicists about worthwhile problems to attack also outside his own field of research and he challenges colleagues to take more interest in certain specified areas of interest. His autobiography reveals a life full of drama. The 2003 prize to him and to Alexei Abrikosov and Anthony Leggett are about the understanding of the relation between order and the macroscopic quantum phenomena of superconductivity and superfluidity.

The nature of the strong force in action within the atomic nucleus has been sought since the 1930s when its existence was first recognized. The discovery story with its many pitfalls and winding road to a final success is well told in the three prize lectures of 2004 by David Gross, Frank Wilczek and David Politzer. Together with other contributions in the field of elementary particles their achievement led to the formulation of the Standard Model of Elementary Particle theory.

The Nobel prizes of 2005 were devoted to the field of light. Roy Glauber laid the foundation for a deep analysis of the quantum nature of light. The advent of lasers as light sources revolutionised several areas of science and technology. A recent revolution towards extreme precision in areas of spectroscopy and frequency measurements is the subject of the fascinating lectures by John Hall and Theodore Hänsch.

Usually original, scientific papers are written in a dry and impersonal style. This tradition makes it hard for outsiders to appreciate the hard work behind a discovery or invention and the enthusiasm and joy the researcher feels when the end result is in sight.

On the other hand the Nobel lectures are in most cases written in a generously personal style with detailed glimpses given of the meandering road to success thus giving the reader the possibility to share the excitement of final success. Therefore the present volume is very suitable and recommendable for university teachers and students.

Gösta Ekspong

CONTENTS

Physics 2001

ERIC A. CORNELL, WOLFGANG KETTERLE and CARL E. WIEMAN

"for the achievement of Bose-Einstein condensation in dilute gases of alkali atoms, and for early fundamental studies of the properties of the condensates"

THE NOBEL PRIZE IN PHYSICS

Speech by Professor Sune Svanberg of the Royal Swedish Academy of Sciences.
Translation of the Swedish text.

Your Majesties, Your Royal Highnesses, Honoured Nobel Laureates, Ladies and Gentlemen,

Three quarters of a century ago, Bose and Einstein made theoretical predictions that this year's Nobel Laureates in Physics succeeded in implementing 70 years later in the form of an exciting new state of matter. The so-called Bose-Einstein condensate they achieved has the same relationship to a piece of ordinary matter as a laser beam has to the light from a bulb.

The world consists of interacting particles and waves. In our ordinary perception of reality, we are fairly confident that we can distinguish easily between waves and particles, but in fact they can sometimes switch identities. Light waves may be regarded as a stream of massless particles, also called photons, and particles of matter sometimes have a wavelike nature. Generally speaking, the corresponding wavelength is extremely short, but for atoms that move slowly, it may be observable. Einstein predicted that if a gas is cooled to very low temperatures, all the atoms should gather in the lowest energy state. Matter waves of the individual atoms then merge into a single wave; indeed, they can be said to "sing in unison." Thousands of atoms behave like one big superatom. This is Bose-Einstein condensation. However, the phenomenon can only occur with a special type of particles called bosons. Electrons are an example of another type of particles. They are not at all "sociable"; they absolutely do not wish to be in the same state and behave in the same way, but instead they arrange themselves in the increasingly complex electronic shells of atoms, building up the periodic system of the elements.

To achieve Bose-Einstein condensation, a gas must be cooled to less than one millionth of a degree above absolute zero. The 1997 Nobel Laureates had developed effective laser-based methods for cooling and trapping atomic gases, but the path to Bose-Einstein condensation remained extremely difficult. Carl Wieman pointed toward a successful method: laser cooling of alkali atoms in a so-called magneto-optical atom trap and then continued reduction of speed through evaporative cooling, which is a way of systematically getting rid of the fastest atoms. After extensive efforts by the research team in Colorado, Eric Cornell solved the last remaining problem that prevented condensation. The successful experiments using rubidium atoms were reported in 1995.

Wolfgang Ketterle worked independently of Cornell and Wieman, and 4 months after them he reported large condensates of laser-cooled sodium atoms. Ketterle was able to demonstrate that the condensate actually behaved as a single coherent wave. He did an experiment similar to the one where two stones are thrown simultaneously at a calm water surface and their wave patterns roll into each other, strengthening and weakening each other in a systematic way. This is in greater contrast to what happens when uncoordinated matter, for example two fistfuls of sand, are thrown on the water surface. Ketterle was also able to extract a beam of coherent matter from the condensate, thus achieving the first atom laser. An ordinary laser yields coherent radiation, an atom laser a stream of coherent matter.

When a gas consisting of uncoordinated atoms turns into a Bose-Einstein condensate, it is like when the various instruments of an orchestra with their different tones and timbres, after warming up individually, all join in the same tone.

After the pioneering experiments of this year's Laureates, condensates have now been achieved by more than twenty additional research teams. Many fascinating applications are conceivable. Precision measurements utilizing slow atoms may offer big surprises: perhaps what we today call natural constants are not completely constant. The new control over matter that Bose-Einstein condensation provides may have far-reaching practical applications, for example in lithography and nanotechnology.

Professor Cornell, Professor Ketterle, Professor Wieman,

Your groundbreaking work on Bose-Einstein condensation has opened up a very fruitful area of research and potential applications. On behalf of the Royal Academy of Sciences I wish to congratulate you on your great accomplishments. I now ask you to step forward to receive your Nobel Prizes from the hands of His Majesty the King.

ERIC A. CORNELL

I was born in Palo Alto, California in 1961. My parents were completing graduate degrees at Stanford. Two years later we moved to Cambridge, Massachusetts, the city I consider to be my hometown. My father was a professor of civil engineering at MIT, and my mother taught high school English. The family, including my younger brother and sister, accompanied my father on sabbatical years to Berkeley, California and Lisbon, Portugal. These were wonderful experiences for me and no doubt they are in part to blame for my lifelong love of travel.

My mother taught me to read when I was still quite young, and at least in my memory I passed the majority of my childhood reading. My head was always bubbling over with facts and it seems to me this had little to do with my paying close attention in school and more to do with my voracious and omnivorous reading habits. Indeed in elementary school I often kept my desktop slightly open and affected an alert-looking pose that still allowed me to peek into the desk where I kept open my latest book, as interesting as it was irrelevant to the academic subject at hand. Every so often my hand slipped surreptitiously into the desk to turn the page. In the intervening three decades I have spent plenty of time lecturing in front of a classroom of my own, and in retrospect I realize I was seldom fooling anyone. Most of my teachers probably found I made less trouble if they let me read.

Some nights, especially in the early summer when the late evening light kept my west-facing bedroom from getting very dark, I had trouble falling asleep at my appointed bedtime. My parents probably felt that reading me a story was a little redundant, but on occasion my father would come in and suggest to me a "problem" to think about. Stewing over these problems was supposed to help me go to sleep. It never did that, but it did get me in the lifelong habit of thinking about technical issues at all sorts of random moments in my daily life, and not only (or even primarily) during scheduled "thinking time." Some of my father's bedtime problems I now recognize as classic physics brainteasers. A man driving a van full of beehives comes to a bridge. The combined weight of the truck, bees, and beehives barely exceeds the safety limit of the bridge. The driver comes up with the idea of banging on the side of the van, so that all the bees swarm out of the hive and fly around in the back of the van. Does the fact that the bees are now all airborne make the truck light enough to safely cross the bridge? Other problems were exercises in mental estimation. If you hold out your thumb, at arms length, you can just about cover the moon with your thumb. The moon is a quarter of a million miles away. How big is it?

The 1970s, the decade of my teenage years, was a transitional period in American youth culture. It was already past the peak of the era when science-minded kids built radios, model airplanes, rockets – things of that sort. But it was certainly well before the heyday of computers and video games. I was partly old-fashioned and partly modern. I certainly remember building model rockets. It was fun to watch the rocket blast into the air, suspenseful to wonder if the parachute would open to bring the rocket safely back. I didn't really enjoy the assembling the model kits very much, and usually I couldn't be bothered to paint the thing, or even to stick on the decals. A more vivid memory for me was designing a model of my own. Besides the store-bought kits, the Estes Model Rocketry company in those days also sold by mail various sizes of cardboard tubing, balsa-wood sheets, nosecones, and gun-powder rocket engines. Estes also published a terrific little booklet full of quantitative design tips. A key issue in rocket design is to make sure that the center of mass is well forward from the fins, lest the rocket be aerodynamically unstable. My father showed me how (after a candidate design was laid out on graph paper) to calculate the center of mass of the assembly based on the masses and distribution of the component parts. I designed an over-sized, under-powered, clunky sort of rocket. I didn't care how high it would go – I wanted it to rise slowly enough that I could watch to see if its orientation wobbled during the flight. On its maiden flight it lifted off the ground with all the ponderousness of a Saturn V, rising steady and true but rolling slightly about its long axis (had I glued the fins on crooked?) as it gained altitude. The engine burn completed, and then the parachute popped and my creation drifted with the wind to land on the roof of a schoolhouse. My parents suggested I go on Monday morning to ask the school's janitor to retrieve my rocket, but this I was too shy to do.

My freshman year of high school I joined the chess and math clubs. The clubs met after school in the computer-instruction classroom, under the loose supervision of a genial polymath with the unlikely name of Mr. Wisdom. Between rounds of speed chess I read enough of a programming manual to teach myself to write programs on the school's DEC mainframe in the language Basic. For several months I was really captivated with this new activity. The exercises in the Basic manual seemed pretty tedious so I invented a few projects for myself, including a program to generate word puzzles for the math club newsletter. After a semester or so, my infatuation with computers burnt out as quickly as it had begun. Not enough substance there to sustain interest, I felt. This episode is probably the basis for my lifelong distaste for "computers for computers' sake" – it's a kids' game, I think. A second legacy of my brief childhood infatuation with computers was a life-long secret preference for programming in Basic, although during my years of apprenticeship in other scientists' labs I was compelled to learn both C and Fortran. When eventually I had the opportunity to establish a lab of my own, one of my first acts as a young principal investigator was to write a program to output a precisely timed sequence of electronic pulses to control the lasers and magnetic fields in what was to become the first successful Bose-Einstein condensation apparatus. Of course, I wrote the program in Basic!

Celeste Landry, in 1984.

Some of my classes in high school were pretty interesting and I benefited from having several very intelligent and inspiring teachers. Among these were John Samp, a physics teacher, and JoAnn Walther, an English teacher. After the Nobel Prize announcement, I got back in touch with them and was delighted to learn that they are still (as of 2001) teaching at my old high school.

Just before my final year of high school, my brother, sister and I moved with my mother to San Francisco. I spent my last year of high school there, at Lowell High School. Lowell High was a so-called "magnet school," drawing academically inclined students from all over the city. My fellow students there were very smart, but the really novel thing was that they actually seemed to put a lot of effort into their school work. By the end of my first semester there, I began to get into that habit as well. Something else new at Lowell was that it was "cool" to excel at school, at least among the Asian kids with whom I mostly hung out. Without the transitional year at Lowell, my first year as an undergraduate at Stanford would have been a horrible shock.

The truth is that first year at Stanford was a shock anyway, although not for academic reasons. Everyone was beautiful, self-confident, self-satisfied. Later I moved into a student-run, co-op house and felt more at home in that "alternative" residential atmosphere. It was there I met my future wife, Celeste Landry, although our lives took us separate ways for many years and we were not to marry until more than ten years later.

My first job in physics was as a "scanner" at the Stanford Linear Accelerator Center. As a freshman I needed to earn a little money and I was looking for a way to learn about science at the same time. The advertised hourly wage was unusually high for a campus job, which should have been a danger sign. On my first day on the job, a postdoc spent 30 minutes or so showing me how to call up symbolic representations of an endless series of archived detector "events," for display on a graphics terminal. There was a particular kind of rare event I was to look for – I can't remember now exactly what it was – characterized by a certain precise number of photons, of muons, etc. The postdoc explained to me how to distinguish different sorts of particles on the basis of the amounts of energy they deposited in various sorts of detectors, spark chambers, calorimeters, what have you. When I recognized a promising event, I was to flag it by pressing a certain key on the terminal, and, "pop", another event would come up on the screen for my consideration. After my

30-minute training period was up, the educational part of the job (and incidentally the part of the job involving any human interaction) was essentially finished. I could come in whenever I wanted, work as many hours as I wanted. The money was great but towards the end of the third mind-numbing afternoon of staring at the graphics terminal I realized my sanity was at risk. I decided to quit right then and there, and wandered around the data center looking for someone to notify of my decision. There were plenty of people buzzing around the room, but no one looked familiar. It occurred to me that, after the original 30-minute training period, I had never again seen the postdoc who had taught me the tricks of the high-energy physics trade. Finally I just wandered out of the building, never to return. Over the course of my three afternoons I had worked my way through hundreds of stored events, and flagged four of them as promising candidates. Is it possible those four events eventually got my postdoc a nice assistant professor position at the University of Chicago? One can always wonder!

Meanwhile, I was taking freshman physics with Blas Cabrera, then only in his second year as a professor, and eventually I worked up the nerve to approach him after class. Did he have a position in his lab for an undergraduate? He did! I started off building some data acquisition electronics for a scanning magnetometer, sharing a lab bench with a fellow undergraduate, Charlie Marcus. For the remainder of my years at Stanford I worked afternoons and summers for low-temperature physics groups on campus. I really enjoyed this experience, and it was these jobs, more than anything else, that persuaded me to pursue a career in scientific research.

Roughly halfway through my undergraduate years, I began to worry that my future was choosing me, instead of the other way around. Time seemed to be accelerating. Had I really already completed nearly two years of college? I was taking lots of science classes, spending lots more time in physics labs, and was doing well there. In a little more than a year, the most natural thing for me to do would be to apply to physics graduate school. Doubtless I would be admitted, and then – zoom – off I would go into a pre-defined future as a scientific researcher. It seemed somehow too pat, too canned. When was it that I actually got to decide the course of my own future life? Perhaps I would be happier pursuing something a little more explicitly intellectual than physics. Maybe a return to my first love, of books, was in order. I had been studying Mandarin Chinese for a quarter or two. I took a great interest in politics. Couldn't I put together some sort of future with all that in mind? The first thing I needed was to buy a little time to think it over, lest I be out the door with a degree before I knew what had happened. A Stanford program called Volunteers in Asia seemed to offer me that time. So the summer following my second year of college, I went off to the YMCA in Taichung, Taiwan, to teach conversational English. The work was pleasant and not very hard; I had a lot of time to read and to think and to study Chinese. Six months after that, I left Taiwan, first for Hong Kong and then for mainland China, where I spent another three months studying still more Chinese and generally kicking around the country.

Travel provided many interesting experiences, but perhaps the most useful lesson I learned was that I really had no proficiency for learning the thousands of characters of the written Chinese language. It is not that my memory is generally poor. I am very good at remembering the lyrics to popular songs. A single line from a popular song probably represents about as many bits of information as a single Chinese character. If I could have displaced the one set of information with the other, I would have had no problem storing in my brain the 5000 characters necessary for advanced Chinese literacy. As it was, I realized choosing the study of Chinese literature as my life's work was probably a mistake. Conversely, I came to realize that being good at something is hardly a reason to avoid doing it.

I returned to Stanford with much more of a sense of purpose. I continued to take elective courses in such topics as poetry and political science, but I allowed myself to enjoy my physics courses and my work in the labs. My last two years at Stanford I worked for the gyroscope-based general relativity experiment of Francis Everett and co-workers, with my final year's work growing into an honors project. Everett was the titular advisor of my honors thesis, but I worked more closely with John Turneaure, a research professor. The gyroscope relativity experiment needed data on the low-temperature adsorption properties of helium on various technical materials such as OFHC copper, fused quartz and so on. I inherited a recently abandoned apparatus and was told to extend the range of temperatures and go beyond monolayer coverage. I went to see John for advice as needed, but other than that I was left to work alone. No doubt I wasted a lot of time reinventing the wheel, but I loved the sensation of "having my own lab."

For graduate school I returned to Cambridge. In the spring of 1985, shopping around for a graduate school and a research project, I met Dave Pritchard at MIT. He spun me a wonderful yarn: by very precisely measuring the mass difference between the helium-3 and tritium, one can determine the total amount of energy released in the beta decay of tritium. Combine this mass measurement with a determination (no big deal, Dave implied) of the endpoint of the beta-ray spectrum, and one has measured the rest mass of the electron neutrino! There were hints, in those days, that the neutrino might have a rest mass as large as ten eV, a value of cosmological significance. Think of it, Dave said: working with two or three other students on a bench-top experiment, one might just find the missing dark mass and close the universe! It sounded awfully good to me. It still does, as I retell it today.

Thus in the fall of 1985 I joined Dave's single-ion cyclotron resonance experiment. The idea was to trap a single ion in a Penning trap, measure its cyclotron frequency to great accuracy, then swap in a different species of ion and do a comparison measurement. The ratio of cyclotron frequencies should be just the inverse of the ratio of masses. Two graduate students, Robert Weisskoff and Bob Flanagan, and a postdoc, Greg Lafyatis, had the apparatus designed and largely assembled by the time I arrived, but we didn't succeed in trapping and detecting single ions until three years later. The work got to be pretty frustrating and when at last one morning Robert finally acquired the definitive signal from a single ion, he said "That is that." By that afternoon he had begun writing his thesis and he did not return to the ion lab again. A new graduate student Kevin Boyce had recently joined the group and the two of us spent a couple of years learning how to make precision measurements on the single ions.

It is hard to overstate how much I learned from Dave Pritchard over my five years as a graduate student. He was seldom in the lab, but he ate lunch with us students several days a week, and held regular progress meetings as well. Meeting with Dave could be a fairly overwhelming experience. He frequently was in a sort of quizmaster mode, in which he peppered his student with questions. "How big is this effect? You don't know? That's fine, but why don't you estimate it for me then? No, don't offer to go away and think about it – work it out right now, out loud, for the benefit of all of us here." His quiz sessions could be aggravating or even intimidating, but in the end I found them to be great training. Dave liked to show us how widely disparate effects in quantum and classical physics could be understood with the same basic and rather small set of ideas such as resonance, adiabaticity, stationary points, dressed states, entropy and so on. To this day I have ambitions of designing a course called "The Seven Most Useful Ideas in Physics," that would somehow condense and codify the Pritchardian wisdom. Thus it was that when my five years of grad school were over, while I had come nowhere near to finding the Universe's missing mass, I still felt enthused enough about physics research to proceed on to a postdoc.

There are relatively few experiments in atomic physics these days that don't involve the use of a laser. One major shortcoming in my graduate education in preparing me for a career in atomic physics research was that I had not learned any laser techniques. I felt my postdoctoral job had better fill in that lacuna. Looking for a postdoc job, I made the usual rounds, visiting Yale, Stanford, Bell Labs, Gaithersburg, and so on. Laser cooling was in its heyday in 1990, and as I traveled around I saw all the major programs. I was a little daunted by the size and complexity of the experiments, and worried also that maybe all the really interesting experiments had already been done. Finally, I went out to Boulder to give a talk to Dave Wineland's group in NIST labs. Dave Wineland was and is one of the towering figures in ion trapping, so I felt a little foolish, earnestly describing to his group my modest contribution, but I soldiered on through my talk. No job offer was forthcoming, but as luck would have it, in the audience was a former Wineland-group postdoc, Sarah Gilbert. Sarah called her husband, Carl Wieman, who was looking to hire a postdoc, and suggested that he invite me to make the one kilometer trek from NIST labs over to JILA, on the University of Colorado campus, to visit his lab. At this time the main focus of Carl's research was on precision measurements of parity violation in cesium, but my attention was immediately drawn to his smaller, laser cooling experiment. In contrast to the other laser cooling experiments I had seen, which took up the better part of a room, Carl's experiment could have fit on a card table. Using diode lasers instead of Ar^+-pumped dye lasers, and using a tiny little vapor cell instead of an atomic beam machine, the whole experiment seemed accessible and compact, even cute. There was just one graduate student working on the project, and this impressed me as well – if a single student could make it work, how hard could it be? (It would be almost a year later before I realized that Chris Monroe was not exactly an average graduate student!) It was clear to me that during a two-year postdoc I could learn how to make a fun little laser-cooling set up like Carl's, and, looking ahead, it also seemed to me that I could duplicate such an experiment as an assistant professor without much trouble. It would be sufficiently easy to constract that that I would have energy, time and money left over to use the cold atoms in turn to study something else; I would not be compelled to catch up with the established major AMO groups that were studying the cooling process itself.

With an offer from Carl in my pocket, I went back to Cambridge to write up my dissertation. While considering the offer, I began to think for the first time of attempting to see Bose-Einstein condensation (BEC). BEC was a natural thing for atomic physics student at MIT to think about: occupying the office next to Dave Pritchard was Dan Kleppner, co-leader (with Tom Greytak) of one of the major groups attempting to see BEC in spin-polarized hydrogen. The idea of BEC was in the air, and I had seen a number of talks on the topic. Just a year earlier the MIT BEC group had dramatically succeeded in implementing evaporative cooling out of a magnetic trap, a clever idea due to Harold Hess. The MIT hydrogen experiment was daunting in its size and complexity, whereas it seemed to me that if one took as one's starting point the relatively tractable vapor-cell, laser-cooling technology that Wieman was using, it wouldn't be so much of a stretch to imagine souping it up into an apparatus capable of evaporatively cooling to BEC. So I decided to head off to Boulder for a couple of years.

After accepting Carl's offer I postponed actually moving to Boulder for three months while my then girlfriend finished her PhD as well. In the meantime I took a very short-term postdoctoral position working with Joel Parks at the Rowland Institute, helping him design and build a Paul trap for ionized atomic clusters.

In October of 1990 I arrived in Boulder. I found working with Carl to be a very congenial experience. Carl and I share very similar tastes in what makes for an interesting physics experiment, and I was happy to assimilate a fraction of his seemingly endless bag of technological ideas. Carl taught me to decide what part of the experimental apparatus really mattered, and then to spare no effort improving that part. Conversely, Carl emphasized that one needs to recognize where "good enough" was indeed good enough, and to waste no time worrying about it. I learned from Carl's student, Chris Monroe, as well. I had always been reluctant to mess with the innards of a store-bought piece of equipment, lest I break something. Chris' ever-fearless attitude was, if that gizmo isn't doing what we need it to do now, how much worse off will we be even if we *do* break it? As my two-year postdoctoral appointment wound up, Carl, Chris and I had essentially defined what needed to be done to make BEC with the hybridized method of laser cooling followed by magnetic trapping and evaporative cooling.

During those early years in Boulder, I spent a lot of time trying to imagine what a Bose-Einstein condensate would be like, if we could ever make one. Would it be superfluid, like liquid helium? Would it be coherent, like a laser? What do "superfluid" and "coherent" really mean? I understood these words in the context of the experiments the words had been invented to describe, or at least I thought I did, but it seemed to me that to understand how these words applied to a dramatically different physical system, one had to have a much deeper understanding. Superfluidity and lasing were two of my favorite topics in physics, but each was surrounded by a vast thicket of lore and literature. It was hard to step off of the well-worn paths through these thickets, hard for a newcomer to get a fresh look at the underlying phenomena. If one could make a gas-phase condensate, one would have a less brambled system against which to test one's own physical intuition. Meditations along these lines converted me from BEC dabbler to true believer.

It was with some zealotry, then, that I took the "hybrid cooling to BEC" pitch on the road in 1992, in an effort to find a faculty job. Berkeley and MIT did not bite, but I had offers from Haverford College, University of Virginia and JILA/NIST. The environment at JILA for doing AMO research was so strong, I decided to accept their offer and remain, against the advice of several people who pointed out the potential risks of remaining in the shadow of

My research group, in 2002. Back row, from left, Peter Engels, Tetsuo Kishimoto, Eric Cornell, Paul Haljan. Middle row, Heather Lewandowski, Jeff McGuirk, Pam Leland and Dave Harber. Front row, Ian Coddington and Ying-Ju Wang.

my postdoctoral advisor. As it turned out, over the years Carl was to be extremely fair in the sharing of credit, and I have never regretted my decision to stay at JILA.

The scientific developments from 1990 to 1995 leading to BEC are discussed in the companion article. In the mid-1990s I ran a secondary research project in parallel with my BEC effort. The idea was to extend the techniques of laser cooling into solid-state systems. We never got it to work. In the end, my sunny optimism was trumped by my complete lack of training in solid-state spectroscopy. As it turned out, a group at Los Alamos National Labs has since successfully cooled a solid using a related experimental approach. Also in the mid-90s, Dana Anderson and I began a project to construct waveguides for matter waves. Our first successes were based on hollow glass fibers, but our ongoing collaboration now focuses on guiding atoms with the magnetic fields from lithographically patterned wires. The bulk of my group's research efforts over the last seven years has focused on elucidating the properties of BEC. With every passing year, BEC proves that it still has surprises left for us. Most lately my group has been pursuing studies of quantized vortices in BEC and of spin-waves in ultra-cold atoms. This latter work required us to retreat back above the BEC transition temperature! (Although we are still comfortably within a millionth of a degree of absolute zero.)

I have been very fortunate over the years in the graduate students and postdocs who have come to work in my lab. Their hard work, talent and creativity have made me look good. I have been fortunate also to live in a society that values scientific research, and is willing to support people to do it.

In 1993, Celeste Landry and I rekindled an old romance and we were married in January of 1995, in the Stanford Faculty Club. At the time of our wed-

Sophia (left) and Eliza Cornell on the bank of the Colorado River.

ding, I had upcoming professional travel to the ICOLS conference in Capri scheduled for June, and we planned to delay our honeymoon until then. Just two weeks before the ICOLS conference, the BEC experiment finally succeeded. In beautiful Capri, with lovely Celeste, I felt on top of the world.

The next year I experienced a still keener pleasure, attending the birth of our daughter, Eliza. Her younger sister, Sophia, arrived in 1998. The four of us live in an old brick house in the shade of two large silver maples in central Boulder.

Photo © Carl E. Wieman

Carl E. Wieman

CARL E. WIEMAN

I was born on March 26, 1951 in the small town of Corvallis, Oregon. A number of years earlier my newly wed parents N. Orr and Alison Wieman, like somewhat belated pioneers, had driven their decrepit car across the country to settle deep in the forests of the Oregon coastal range. My father began working in the lumber industry and during most of my childhood he worked as a sawyer in a sawmill. I was the fourth of five children. Most of my childhood was spent in the woods of Oregon where lumber was the sole industry. Probably some of my spirit of independence came from growing up far from other houses and towns. The nearest tiny store was always many miles away over unpaved mountain roads. Some of my earliest childhood memories are of the long school bus rides that my siblings and I used to take over those winding roads to go to school.

Much of my youth was spent wandering around in the forests of towering Douglas fir trees. I also spend much of my time reading and picking fruit and fir cones to earn spending money. Every Saturday my family would make a long expedition to the nearest town to do the week's worth of shopping. A stop at the public library was always part of these trips. Although I was unaware of it at the time, my parents must have made special arrangements for their children to use the library since we lived far outside the region it was supposed to serve. The librarians would also overlook the normal five-book limit and allow me to check out a large pile of books each week that I would then eagerly devour. That experience has left me with a profound appreciation for the value of public libraries. At the time I was quite envious that my friends had televisions while we did not, but in retrospect I am very grateful that I spent this time reading instead of watching TV.

I went to primary school (up to grade 6) at Kings Valley grade school. It was a tiny rural school that had expanded from one to three rooms shortly before I enrolled. For the seventh grade I had to take the much longer bus ride (almost interminable for an impatient 13 year old!) to the small town of Philomath. My young idealistic teachers in mathematics and science there had a significant influence on me. I particularly remember my science teacher, Ron Tobias, who was just starting his first teaching job. I am sure that Philomath 7th grade, with all its children of loggers and farm workers, for whom education was not a particularly high priority, must have been a very tough job for a young teacher. At times I could sense hints of his frustration. However Mr. Tobias did a great deal to kindle my interest in science with his enthusiasm and knowledge. I still remember his explanations (far better than any of the material from my college courses!) of the structures of atoms in the periodic table and how these structures determined the various chemical properties and molecular reactions.

After 7th grade my parents moved to Corvallis (home of Oregon State University) so that my siblings and I could both avoid the long bus rides and take advantage of the better school system offered by this "big city" of 25,000. It was a heady day for me when we moved into a house that had a central heating system instead of just a wood stove and had an actual paved street out front! Although I was never a very sociable child, Corvallis provided me with somewhat more comfortable companions. My intellectual interests and the liberal political attitudes of my parents were always somewhat at odds with the leanings most of my previous rural classmates, but I fit in better with the children of faculty at OSU. I became close friends with a very smart boy, Brook Firey, whose father Bill was a Professor of mathematics. One summer Bill gave Brook and I our own private course in geometry. It was a rewarding and eye-opening experience to get a glimpse of the richness of mathematics, even elementary geometry, as viewed by a true mathematician. And of course, at that age, I did not realize there was anything unusual about a University professor spending a few hours each day to provide personal instruction to two fourteen year olds.

Brook and I also spent many hours engrossed in all sorts of projects constructing and investigating things. I think that much of my talent and enjoyment at improvising solutions to experimental problems goes back to those homebuilt projects. In this regard my older brother Howard also inspired me; he was always tinkering with machines and building astonishingly elaborate toys for his younger siblings. Carrying out these individual projects also developed in me a good sense of self-reliance and a sense when a piece of improvised apparatus was likely (or unlikely) to be adequate. This sense is one that I often see missing in students whose education has been confined to formal instruction.

During high school I was a good student, but never quite at the top of the class. I mastered the material, but was usually a little too independent to do precisely what the teacher wanted, and so was never considered among the very best students. Usually the worse the teacher (at least according to me), the lower was my standing. Although always interested in science, my most memorable classes were in literature and writing. From 7th through 10th grade I was a passionate chess player, spending hours a day on it. I traveled all over Oregon and occasionally to nearby states to play in tournaments. I was highly ranked in the northwest US among my age group, but at the ripe old age of 16 decided to "retire" to spend my time in more productive activities. Those activities were studying and playing tennis.

My high school grades, although not outstanding, were good enough to get me accepted into MIT. From what I now know about college admissions, I suspect my admission was considerably helped by their being intrigued to have a student who had spent much of his life literally in the wilds of Oregon. My accomplishments as a competitive chess and tennis player may also have helped. Although it may seem surprising that a boy from the woods of

Oregon would aspire to go to MIT, my family always had a strong interest in education. Both my parents graduated from college and had come from well-educated families. My grandfather Henry Wieman was a rather well known Professor of Theology at the University of Chicago. Out of my four siblings there are two Ph.D.s, including a successful nuclear physicist, as well as a high-level software engineer.

My tendency to intensely pursue a particular activity to the exclusion of everything else was and is one of my most notable strengths and weaknesses. After "retiring" from chess, my focus turned to tennis. That continued after I went to MIT, and I played intercollegiately my freshman year. I also learned to play squash rackets and took to it so naturally that I was quickly at the top of the freshman intercollegiate squash team. My squash career was notable in that I can claim to have lost to some of the best players in the country, including one future national champion. Unfortunately my rather fierce competitive drive exceeded my limited physical capacities, and after surviving several minor injuries caused by throwing myself into walls and such, by the end of my freshman year I had seriously damaged my right elbow from excessive practice at squash and tennis. After several unsuccessful treatments, I then switched to playing left handed, and by early in my second year of college was starting to again be competitive in both sports at the intercollegiate level. At that point I developed serious elbow problems in my left arm, and reluctantly came to the conclusion that at age 19, it was time for my second "retirement". It was only then I turned my full attention to physics.

As one might imagine, going from the woods of Oregon to MIT was quite a culture shock. I did not do particularly well in classes my freshman year, but I greatly enjoyed an informal freshman seminar on physics that I had with Professor Al Hill. He was a gruff but kindly old faculty member. Although I had a general interest in physics at least since seventh grade, particularly the behavior of light and atoms, I was not totally convinced when I started at MIT that I wanted to go into physics. However, after this seminar and its casual far ranging discussions about physics, Al Hill encouraged me and suggested that I should get involved in research. I discussed this with my freshman advisor who was Daniel Kleppner, and he took me on to work in his laboratory my first summer of college. This was a dramatic change from my employment the previous summer. Then, just out of high school, I had worked in the lumber mill, "pulling on the green chain". As the green lumber came out of the mill on a large conveyor chain, my job was to pull it off and stack it in the appropriate pile. This was an exhausting job that gave me a clear taste of what real labor was like. Every now and then when I am fed up with some aspect of my job as an academic, it is useful to reflect on that summer on the green chain to remind myself how well off I am compared to all those people who spend their lives doing *real* work.

I quickly became deeply engaged in research as an undergraduate and continued to work in Dan Kleppner's research group until I left for graduate school. I found this much more interesting and educational than taking courses, and quickly adopted a philosophy of taking as few courses as possible. Since

I never did terribly well in most normal courses anyway, particularly ones that had exams, this worked out well. I was actually remarkably successful at avoiding courses, helped in large part by the events of the times. The last few weeks (and the dreaded final exams) of my freshman year were canceled because of massive protests over the Vietnam War, and during the following years there were many opportunities to participate in experiments in various sorts of alternatives to normal classes. I took full advantage of all of these alternatives, and their rather lax requirements and oversight. I also spent countless hours discussing physics with the graduate students and postdocs (notably Dave Pritchard) in Dan Kleppner's research group. My education as a physicist came largely from my work in Dan's lab and these interactions with him and his group. I also spent much time in physics discussions with an informal seminar group (the "physics family") run by Rai Weiss and Al Hill.

In spite of (or because of) this unorthodox education, I ended up far more enthusiastic about physics than most of my classmates, as well as having a much better grasp of many basic concepts such as quantum mechanics. Of course I was considerably weaker in the formal solving of problems, and I still have not learned much of the standard material of the undergraduate curriculum. However, when I needed to know some material, I was completely comfortable with going out and learning it myself in a way that I discovered was not typical for my classmates. My undergraduate experience has always left me deeply suspicious of the claims of those who say a student cannot become a physicist without being required to take courses covering a whole list of specific topics. I have had a pretty successful career in optics and atomic physics without having a course in either, for example. Some may argue that this could only work because I was an extraordinary student, and the more typical student must be required to take a formal curriculum with a large number of courses and exams. However, it might be noted that before obtaining this unusual "education" there was little to indicate that I was anything special as a physics student. So one could equally well argue that it was not me that was exceptional, but rather the education I received. Perhaps if far more students learned physics in the haphazard way that I did, many more of them might turn out as motivated and successful as I have been.

I did become extremely involved in research as an undergraduate. Through a chain of circumstances, helped out no doubt by my enthusiasm and willingness to put in long hours of work, I ended up with my own lab and my own experiment. This involved the construction and use of a tunable dye laser, which at that time was a very new and exciting device. This was the beginning of Kleppner's group moving into the use of lasers to study atomic physics. I spent my time blasting atoms with a dye laser tuned to the atomic resonance line and looking carefully at what happened. To a large extent much of my subsequent career has been variations on this basic theme. After spending many very late nights by myself taking data in the lab and showering every day at the athletic center after exercising, I started to wonder why I was paying all that money, of which I had little, to rent a dormitory room I almost never saw. So driven by my involvement in the research and a desire to save money, I actually moved into my lab. After about a half a year, living in the lab got pretty old, and so I moved into a normal apartment, but the story of my being so devoted to experimental physics that I actually lived in the lab has tended to follow me ever since.

My work with dye lasers made me aware of the exciting developments in narrowband dye lasers and their applications to atomic spectroscopy being done by Ted Hänsch. That, along with the far superior weather and more relaxed academic atmosphere, convinced me to move from MIT to Stanford for graduate school. At Stanford I resisted the natural temptation to immediately jump into laser spectroscopy again, and so I spent a year looking fairly carefully into all of the different faculty and research areas in the department. However, in the end I concluded that working on laser spectroscopy with Hänsch was the best option. I began working with his group as they were developing a very high power narrowband dye laser for exciting the 1S–2S transition in hydrogen. Ted was a new enthusiastic young professor, the technology and the experiment were new and exciting, and because of my previous background I was able to become thoroughly involved in the experiment almost immediately. It was a fun time, made more so by the fact that we soon observed the H 1S–2S transition and used it to measure the Lamb shift of the 1S state. For my thesis work I then went on to develop the technique of polarization spectroscopy and built the first single mode continuous wave dye laser at 480 nm to further improve the 1S Lamb shift measurement and greatly improve the determination of the 1S–2S isotope shift.

As I neared completion of my Ph. D., I became interested in the subject of parity violation in atoms. This was predicted by the theory of electroweak unification, but had not been seen. It seemed like the natural next step to my thesis work in that it was using precision spectroscopy of atoms to test fundamental physics. But rather than further test QED in atoms, which by then I was ready to accept as being confirmed as well as ever need be, the parity violation work was looking for new physics in atoms that went beyond QED and was far from certain. It offered the possibility of using atomic physics to do important elementary particle physics. I took a position as an assistant research scientist at University of Michigan to pursue these studies. I joined Bill Williams' ongoing experiment to measure this parity violation in atomic hydrogen using microwave spectroscopy. Shortly after I arrived at Michigan I found that the research scientist position I had taken was not the research faculty position that I had expected. It had all the disadvantages of a regular postdoc, but none of the advantages in that there was not sufficient research money in the grant to cover my salary, so that I also had to teach, and I had to be responsible for much of the administration of the research group. However, I threw myself into the experiment and worked extremely hard, and my position was converted into a regular assistant professor position after a couple of years. Shortly after this I developed a somewhat different formulation for how to describe atomic parity violation experiments. This allowed me to see clearly how to compare the sensitivities of a large variety of different experimental approaches. At that same time I was also becoming increasingly disillusioned with the hydrogen experiment, and my new formulation made it clear to me that a quite different approach, using laser spectroscopy of cesium, would have a far better chance of success.

For a variety of reasons I chose to pursue the cesium experiment on my own, after first receiving assurances from the department chair that this was a suitable activity. Unfortunately, my abandoning of the hydrogen experiment to pursue my own atomic parity violation experiment lead to considerable friction with senior faculty and general strife within the department. As a young assistant professor naïve in departmental politics I was quite vulnerable, and had a difficult time during my subsequent years at Michigan. However, during that time, Sarah Gilbert and I were able to get funding from Research Corporation and then NSF and used it to thoroughly develop a novel experimental approach for measuring atomic PV in cesium. Sarah was a graduate student that I had met soon after I arrived at Michigan. We then worked intensively with graduate students Rich Watts and Charlie Noecker to implement this difficult experiment. By 1984 we had made sufficient progress to indicate the viability of our approach, and this attracted an offer of a faculty position at the University of Colorado in Boulder. I eagerly accepted the offer.

The year 1984 was a very active one for me. First, Sarah Gilbert completed her Ph. D., and I accepted the job at Colorado. In late August we then packed up the entire lab into the back of a rental truck along with all the personal furniture of the graduate students, and then Sarah, Rich, Charlie, Charlie's girlfriend, and I set out on a modern day pioneer caravan across the Great Plains to Boulder, Colorado. After quickly unpacking the truck, Sarah and I then left to fly out to Oregon for our wedding. We had been anticipating this since shortly after we met, but we had delayed until after Sarah finished her degree. We then returned to Boulder to start our new jobs and a new lab.

In the supportive environment of JILA and the Department of Physics in Boulder, along with lots of very hard work, the four of us, Sarah, Rich, Charlie, and myself, were able to make rapid progress and in less than a year completed our first measurement of parity violation in cesium. As with all my experiments, I had started out wildly optimistic as to the difficulty and time required for this experiment, and it was both a tremendous relief and a tremendous satisfaction when it succeeded. It was the best measurement of atomic parity violation at a time when this subject was being pursued by a number of notable atomic physics groups. Our result established both to myself and the rest of the world that I would have a career as research physicist; something that had sunk into considerable doubt during my seven years of meager accomplishments at Michigan. Shortly after the success of the PV experiment I was given tenure and promoted to Full Professor at Colorado. During the subsequent 15 years, my group has carried out two further generations of this long and difficulty experiment with ever improving accuracy.

I would be remiss if I failed to mention the tremendous benefits that I have gained in my career by having a wife who is a very talented and intelligent physicist (as well as being a wonderful person of course). Shortly after we arrived in Boulder, Sarah took a job at the NIST Boulder labs where she has worked ever since. We worked together on the PV experiments, and still collaborate on an occasional small project. Talking with Sarah about physics has always provided me with countless inspirations for new ideas, and has revealed critical flaws that I had overlooked in twice as many bad ideas. She also can understand and share in my obsession with the research and its occasional extraordinary demands. Finally, her ruthless editing has greatly improved the writing of nearly all of my papers. When we are not working, Sarah and I can usually be found running or hiking on the trails of the Boulder Mountain Parks. We can also occasionally (but not frequently enough) be found at our house on the central Oregon coast.

The development of the diode laser technology that was needed for the third generation of the parity violation experiment led to my involvement with laser cooling and trapping and ultimately BEC. Originally in about 1984 Rich Watts and I were simply looking for something fun and easy to do with the diode laser technology we had developed for the PV experiment, as a respite from the very long hard grind of that project. This resulted in our slowing atoms using lasers that were about 1% of the cost of what was used for previous work by Hall and Phillips. I then became increasingly interested in laser cooling and trapping. Initially my work on it focused largely on developing it as a useful technology for doing other atomic physics, but then I became more involved in studying the novel behavior of atoms at the unprecedented temperatures we could achieve. In the process of those studies I worked with an undergraduate Bill Swann to invent the vapor-cell MOT to replace the traditional atomic beam loading of optical traps. This provided a means to trap atoms using only inexpensive diode lasers and a small glass cell, which was a dramatic advance towards making laser trapping a simple and widely useable technology.

To me personally, this reduction in the cost and complication offered the opportunity to explore a variety of speculative directions involving laser cooled atoms with relatively little risk, since the cost and effort was now quite modest. One such quick experiment was to switch the laser cooled and trapped atoms to a magnetic trap in order to avoid the limits we had discovered were imposed by the photons in the optical trap. This worked so easily and so well – we obtained trapped atoms about 100 times colder than had been achieved previously, with a corresponding enhancement in phase space density – that it inspired me to pursue goals grander than just better trapping and cooling technology; namely, the attainment of BEC by further cooling in the magnetic trap. Fortunately Eric Cornell joined me at just that time (1990) to pursue the goal of BEC. Ours turned out to be an extraordinarily friendly and effective partnership that has continued up to the present. Our pursuit of BEC is now well-documented history.

Over the past several years I have become increasingly involved with trying to improve undergraduate physics education and have been balancing my time between that and my research. I have been examining alternative curricula and learning about the research in physics education as to how students do and do not learn. A particular concern has been improving how physics is taught to students who are not planning to become physicists, in the hope of one day making physics understandable, useful, and interesting to a large fraction of the population. My efforts have ranged from working with national organizations pursuing widespread change in undergraduate physics education to developing useful innovations in the individual courses that I teach. Because of my particular concerns, these courses have lately been large introductory courses primarily for nonscience students.

BOSE-EINSTEIN CONDENSATION IN A DILUTE GAS; THE FIRST 70 YEARS AND SOME RECENT EXPERIMENTS

Nobel Lecture, December 8, 2001

by

ERIC A. CORNELL AND CARL E. WIEMAN

JILA, University of Colorado and National Institute of Standards and Technology, and Department of Physics, University of Colorado, Boulder, CO 80309-0440, USA.

Bose-Einstein condensation, or BEC, has a long and rich history dating from the early 1920s. In this article we will trace briefly over this history and some of the developments in physics that made possible our successful pursuit of BEC in a gas. We will then discuss what was involved in this quest. In this discussion we will go beyond the usual technical description to try and address certain questions that we now hear frequently, but are not covered in our past research papers. These are questions along the lines of "How did you get the idea and decide to pursue it? Did you know it was going to work? How long did it take you and why?" We will review some our favorites from among the experiments we have carried out with BEC. There will then be a brief encore on why we are optimistic that BEC can be created with nearly any species of magnetically trappable atom. Throughout this article we will try to explain what makes BEC in a dilute gas so interesting, unique, and experimentally challenging.[1]

The notion of Bose statistics dates back to a 1924 paper in which Satyendranath Bose used a statistical argument to derive the black-body photon spectrum [1]. Unable to publish his work, he sent it to Albert Einstein who translated it into German and got it published. Einstein then extended the idea of Bose's counting statistics to the case of noninteracting atoms [2, 3]. The result was Bose-Einstein statistics. Einstein immediately noticed a peculiar feature of the distribution of the atoms over the quantized energy levels predicted by these statistics. At very low but finite temperature a large fraction

[1] This article is our "Nobel Lecture" and as such takes a relatively personal approach to the story of the development of experimental Bose-Einstein condensation. For a somewhat more scholarly treatment of the history, the interested reader is referred to E. A. Cornell, J. R. Ensher and C. E. Wieman, "Experiments in dilute atomic Bose-Einstein condensation" in "Bose-Einstein Condensation in Atomic Gases, Proceedings of the International School of Physics "Enrico Fermi" Course CXL" (M. Inguscio, S. Stringari and C. E. Wieman, Eds., Italian Physical Society, 1999), pp. 15-66, which is also available as cond-mat/9903109. For a reasonably complete technical review of the three years of explosive progress that immediately followed the first observation of BEC, we recommend reading the above article in combination with the corresponding review from Ketterle, cond-mat/9904034.

of the atoms would go into the lowest energy quantum state. In his words, "A separation is effected; one part condenses, the rest remains a 'saturated ideal gas' "[3, 4]. This phenomenon we now know as Bose-Einstein condensation. The condition for this to happen is that the phase space density must be greater than approximately unity, in natural units. Another way to express this is that the de Broglie wavelength, λ_{dB}, of each atoms must be large enough to overlap with its neighbor, or more precisely, $n\lambda_{dB}^3 > 2.61$.

This prediction was not taken terribly seriously, even by Einstein himself, until Fritz London and Laszlo Tisza [5, 6] resurrected the idea in the mid-1930s as a possible mechanism underlying superfluidity in liquid helium 4. Their work was the first to bring out the idea of BEC displaying quantum behavior on a macroscopic size scale, the primary reason for much of its current attraction. Although it was a source of debate for decades, it is now recognized that the remarkable properties of superconductivity and superfluidity in both helium 3 and helium 4 are related to BEC, even though these systems are very different from the ideal gas considered by Einstein.

The appeal of the exotic behavior of superconductivity and of superfluidity, along with that of laser light, the third common system in which macroscopic quantum behavior is evident, provided much of our motivation in 1990 when we decided to pursue BEC in a gas. These three topics all have fascinating counter-intuitive behavior arising from macroscopic occupation of a single quantum state. Any physicist would consider these phenomena among the most remarkable topics in physics. In 1990 we were confident that the addition of a new member to the family would constitute a major contribution to physics. (Only after we succeeded did we realize that the discovery of each of the original "Macroscopic Three" had been recognized with a Nobel Prize, and we are grateful that this trend has continued!) Although BEC shares the same underlying mechanism with these other systems, it seemed to us that the properties of BEC in a gas would be quite distinct. It is far more dilute and weakly interacting than liquid-helium superfluids for example, but far more strongly interacting than the noninteracting light in a laser beam. Perhaps BEC's most distinctive feature (and this was not something we sufficiently appreciated, in 1990) is the ease with which its quantum wave function may be directly observed and manipulated. While neither of us was to read C.E. Hecht's prescient 1959 paper [7] until well after we had observed BEC, we surely would have taken his concluding paragraph as our marching orders:

> The suppositions of this note rest on the possibility of securing, say by atomic beam techniques, substantial quantities of electron spin oriented H, T and D atoms. Although the experimental difficulties would be great and the relaxation behavior of such spin oriented atoms essentially unknown, *the possibility of opening a rich new field for the study of superfluid properties in both liquid and gaseous states would seem to demand the expenditure of maximum experimental effort.*[3]

[2] English translation of Einstein's quotes and the historical interpretation are from Ref. [4].
[3] Italics ours.

In any case, by 1990 we were awash in motivation. But this motivation would not have carried us far, had we not been able to take advantage of some key recent advances in science and technology, in particular the progress in laser cooling and trapping and the extensive achievements of the spin-polarized hydrogen community.

However, before launching into that story, it is perhaps worthwhile to reflect on just how exotic a system of indistinguishable particles truly is, and why BEC in a gas is such a daunting experimental challenge. It is easy at first to accept that two atoms can be so similar one to the other as to allow no possibility of telling them apart. However, confronting the physical implications of the concept of indistinguishable bosons can be troubling. For example, if there are ten bosonic particles to be arranged in two microstates of a system, the statistical weight of the configuration "ten particles in one state and zero in the other" is exactly the same as the weight of the configuration "five particles in one state, five in the other." This 1:1 ratio of statistical weights is very counterintuitive and rather disquieting. The corresponding ratio for distinguishable objects, such as the socks in drawers, that we observe every day is 1:252, profoundly different from 1:1. In the second [3] of Einstein's two papers [2–4] on Bose-Einstein statistics, Einstein comments that "The... molecules are not treated as statistically independent...," and the differences between distinguishable and indistinguishable state counting "... express indirectly a certain hypothesis on a mutual influence of the molecules which for the time being is of a quite mysterious nature." This mutual influence is no less mysterious today, even though we can readily observe the variety of exotic behavior it causes such as the well-known enhanced probability for scattering into occupied states and, of course, Bose-Einstein condensation.

Not only does the Bose-Einstein phase transition offend our sensibilities as to how particles ought best to distribute themselves, it also runs counter to an unspoken assumption that a phase transition somehow involves thermodynamic stability. In fact, the regions immediately above and immediately below the transition in dilute-gas experiments are both deep in the thermodynamically forbidden regime. This point is best made by considering the qualitative phase-diagram (Fig. 1), which shows the general features common to any atomic system. At low density and high temperature, there is a vapor phase. At high density there are various condensed phases. But the intermediate densities are thermodynamically forbidden, except at very high temperatures. The Bose-condensed region of the n-T plane is utterly forbidden, except at such high densities that (with one exception) all known atoms or molecules would form a crystalline lattice, which would rule out Bose condensation. The single exception, helium, remains a liquid below the BEC transition. However, reaching BEC under dilute conditions (say, at densities 10 or 100 times lower than conventional liquid helium) is as thermodynamically forbidden to helium as it is to any other atom.

Of course, "forbidden" is not the same as "impossible"; indeed, to paraphrase an old Joseph Heller joke, if it were really impossible, they wouldn't have bothered to forbid it. It comes down in the end to differing time scales

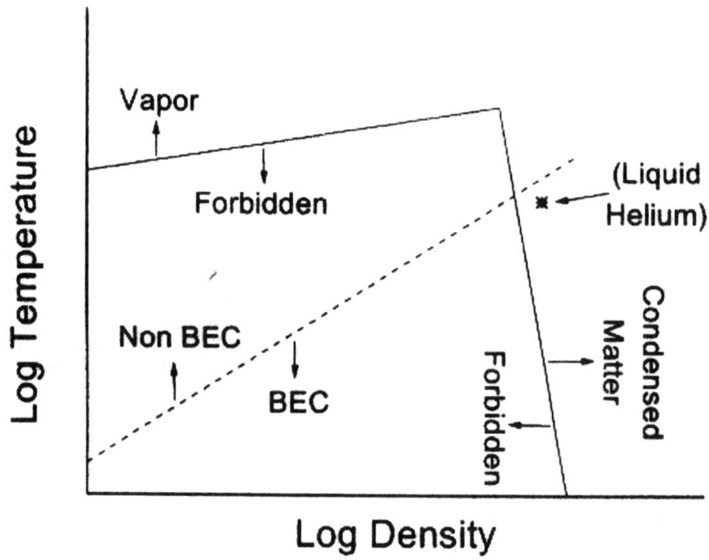

Figure 1. Generic phase-diagram common to all atoms. The dotted line shows the boundary between non-BEC and BEC. The solid line shows the boundary between allowed and forbidden regions of the temperature-density space. Note that at low and intermediate densities, BEC exists only in the thermodynamically forbidden regime.

for different sorts of equilibrium. A gas of atoms can come into kinetic equilibrium via two-body collisions, whereas it requires three-body collisions to achieve chemical equilibrium (*i.e.*, to form molecules and thence solids.) At sufficiently low densities, the two-body rate will dominate the three-body rate, and a gas will reach kinetic equilibrium, perhaps in a metastable Bose-Einstein condensate, long before the gas finds its way to the ultimately stable solid-state condition. The need to maintain metastability usually dictates a more stringent upper limit on density than does the desire to create a dilute system. Densities around 10^{20} cm^{-3}, for instance, would be a hundred times more dilute than a condensed-matter helium superfluid. But creating such a gas is quite impractical – even at an additional factor of a thousand lower density, say 10^{17} cm^{-3}, metastability times would be on the order of a few microseconds; more realistic are densities on the order of 10^{14} cm^{-3}. The low densities mandated by the need to maintain long-lived metastability in turn make necessary the achievement of still lower temperatures if one is to reach BEC.

Thus, the great experimental hurdle that must be overcome to create BEC in a dilute gas is to form and keep a sample that is so deeply forbidden. Since our subsequent discussion will focus only on BEC in dilute gases, we will refer to this simply as "BEC" in the sections below and avoid endlessly repeating "in a dilute gas."

Efforts to make a dilute BEC in an atomic gas were sparked by a 1976 paper by Stwalley and Nosanow [8]. They argued that spin-polarized hydrogen had no bound states and hence would remain a gas down to zero temperature, and so it would be a good candidate for BEC. This stimulated a number of experimental groups [9–12] in the late 70s and early 80s to begin pursuing this idea using traditional cryogenics to cool a sample of polarized hydrogen. Spin-polarized hydrogen was first stabilized by Silvera and Walraven in 1980 [9], and by the mid-80s spin-polarized hydrogen had been brought within a factor of 50 of condensing [11]. These experiments were performed in a dilution refrigerator, in a cell in which the walls were coated with superfluid liquid helium as a "nonstick" coating for the hydrogen. The hydrogen gas was compressed using a piston-in-cylinder arrangement [13] or inside a helium bubble [14]. These attempts failed however, because when the cell was made very cold the hydrogen stuck to the helium surface and recombined. When one tried to avoid that problem by warming the cell sufficiently to prevent sticking, the density required to reach BEC was correspondingly increased which led to another problem. The requisite densities could not be reached because the rate of three-body recombination of atoms into hydrogen molecules goes up rapidly with density and the resulting loss of atoms limited the density [15].

Stymied by these problems, Harold Hess from the MIT hydrogen group realized that magnetic trapping of atoms [15] would be an improvement over a cell [16, 17]. Atoms in a magnetic trap have no contact with a physical surface and thus the surface-recombination problem could be circumvented. Moreover, thermally isolated atoms in a magnetic trap would allow cooling by evaporation to far lower temperatures than previously obtained. In a remarkable 1986 paper, Hess laid out most of the important concepts of evaporative cooling of trapped atoms [15] for the attainment of BEC. Let the highest energy atoms escape from the trap, and the mean energy, and thus the temperature, of the remaining atoms will decrease. For a dilute gas in an inhomogeneous potential, decreasing the temperature will decrease the occupied volume. One can thus actually increase the density of the remaining atoms by removing atoms from the sample. The all important (for BEC) phase space density is dramatically increased as this happens because density is rising while temperature is decreasing. The Cornell University hydrogen group also considered evaporative cooling [18]. By 1988 the MIT group had demon-strated these virtues of evaporative cooling of magnetically trapped spin polarized hydrogen. By 1991 they obtained, at a temperature of 100 μK, a density that was only a factor of 5 below BEC [19]. Further progress was limited by dipolar relaxation, but perhaps more fundamentally by loss of signal-to-noise, and the difficulty of measuring the characteristics of the coldest and smallest clouds [20]. Evaporative work was also performed by the Amsterdam group [21].

At roughly the same time, but independent from the hydrogen work, an entirely different type of cold-atom physics and technology was being developed. Laser cooling and trapping has been reviewed elsewhere [22-25], but here we mention some of the highlights most relevant to our work. The idea that laser light could be used to cool atoms was suggested in early papers from Wineland and Dehmelt [26], from Hänsch and Schawlow [27], and from Letokhov's group [28]. Early optical force experiments were performed by Ashkin [29]. Trapped ions were laser-cooled at the University of

Washington [30] and at the National Bureau of Standards (now NIST) in Boulder [31]. Atomic beams were deflected and slowed in the early 80s [32-34]. Optical molasses, where the atoms are cooled to very low temperatures by six perpendicular intersecting laser beams, was first studied at Bell Labs [35]. Measured temperatures in the early molasses experiments were consistent with the so-called Doppler limit, which amounts to a few hundred microkelvin in most alkalis. Light was first used to hold ("trap") atoms using the dipole force exerted by a strongly focused laser beam [36]. In 1987 and 1988 there were two major advances that became central features of the method of creating BEC. First, a practical spontaneous-force trap, the Magneto-Optical Trap (MOT) was demonstrated [37]; and second, it was observed that under certain conditions, the temperatures in optical molasses are in fact much colder than the Doppler limit [38-40]. The MOT had the essential elements needed for a widely useful optical trap: it required relatively modest amounts of laser power, it was much deeper than dipole traps, and it could capture and hold relatively large numbers of atoms. These were heady times in the laser-cooling business. With experiment yielding temperatures mysteriously far below what theory would predict, it was clear that we all lived under the authority of a munificent God.

During the mid 80s one of us (Carl) began investigating how useful the technology of laser trapping and cooling could become for general use in atomic physics. Originally this took the form of just making it cheaper and simpler by replacing the expensive dye lasers with vastly cheaper semiconductor lasers, and then searching for ways to allow atom trapping with these low cost but also low power lasers [41, 42]. With the demonstration of the MOT and sub-Doppler molasses Carl's group began eagerly studying what physics was limiting the coldness and denseness of these trapped atoms, with the hope of extending the limits further. They discovered that several atomic processes were responsible for these limits. Light assisted collisions were found to be the major loss process from the MOT as the density increased [43]. However, even before that became a serious problem, the light pressure from reradiated photons limited the density [44, 45]. At about the same time, the sub-Doppler temperatures of molasses found by Phillips, Chu, and Cohen-Tannoudji were shown to be due to a combination of light-shifts and optical pumping that became known as Sisyphus cooling [46]. Random momentum fluctuations from the scattered photons limit the ultimate temperature to about a factor of 10 above the recoil limit. In larger samples, the minimum temperature was higher yet, because of the multiple scattering of the photons. While carrying out studies on the density limits of MOTs Carl's group also continued the effort in technology development. This resulted in the creation of a useful MOT in a simple glass vapor cell [47], thereby eliminating the substantial vacuum chamber required for the slowed atomic beam loading that had previously been used.

Seeking to take advantage of the large gains in phase space density provided by the MOT while avoiding the limitations imposed by the undesirable effects of photons, Carl and his student Chris Monroe (Fig. 2) decided to try

Figure 2. Chris Monroe examines an early hybrid MOT-magnetic trap apparatus.

loading the cold MOT atoms into a magnetic trap [47]. This worked remarkably well. Because further cooling could be carried out as the atoms were transferred between optical and magnetic trap it was possible to get very cold samples, the coldest that had been produced at that time. More importantly, these were not optical molasses samples that were quickly disappearing but rather magnetically trapped samples that could be held and studied for extended periods. These samples were about a hundred times colder than any previous trapped atom samples, with a correspondingly increased phase space density. This was a satisfying achievement, but as much as the result itself, it was the relative simplicity of the apparatus required that inspired us (including now Eric, who joined the project as a postdoc in 1990) to see just how far we could push this marriage of laser cooling and trapping and magnetic trapping.

Previous laser traps involved expensive massive laser systems and large vacuum chambers for atomic beam precooling. Previous magnetic traps for atoms were usually [17, 19] extremely complex and bulky (often with superconducting coils) because of the need to have sufficiently large depths and strong confinement. Laser traps and magnetic traps were both somewhat heroic experiments individually, to be undertaken only by a select handful of well-equipped AMO laboratories. The prospect of trying to get both traps working, and working well, in the same room and on the same day, was daunting. However, in the first JILA magnetic trap experiment our laser sources were simple diode lasers, the vacuum system was a small glass vapor cell, and the magnetic trap was just a few turns of wire wrapped around it. This magnetic field was adequate because of the low temperatures of the laser cooled and trapped samples. Being able to produce such cold and trapped samples in this manner encouraged one to fantasize wildly about possible things to do with such an atom sample. Inspired by the spin polarized hydrogen work our fantasizing quickly turned to the idea of evaporative cooling further to reach BEC. It would require us to increase the phase space density by 5 orders of magnitude, but since we had just gained about 15 orders of magnitude almost for free with the vapor cell MOT, this did not seem so daunting.

The JILA vapor-cell MOT (Fig. 3), with its superimposed ion pumped trap, introduced a number of ideas that are now in common use in the hybrid trapping business: [47, 48]: (i) Vapor-cell (rather than beam) loading, (ii) fused-glass rather than welded-steel architecture, (iii) extensive use of diode lasers, (iv) magnetic coils located outside the chamber, (v) over-all chamber volume measured in cubic centimeters rather than liters, (vi) temperatures measured by imaging an expanded cloud, (vii) magnetic-field curvatures calibrated *in situ* by observing the frequency of dipole and quadrupole (sloshing and pulsing) cloud motion, (viii) the basic approach of a MOT and a magnetic trap which are spatially superimposed (indeed, which often share some magnetic coils) but temporally sequential, and (ix) optional use of additional molasses and optical pumping sequences inserted in time between the MOT and magnetic trapping stages. It is instructive to note how a modern, IP-based BEC device (Fig. 4) resembles its ancestor (Fig. 3).

As we began to think about applying the technique of evaporative cooling

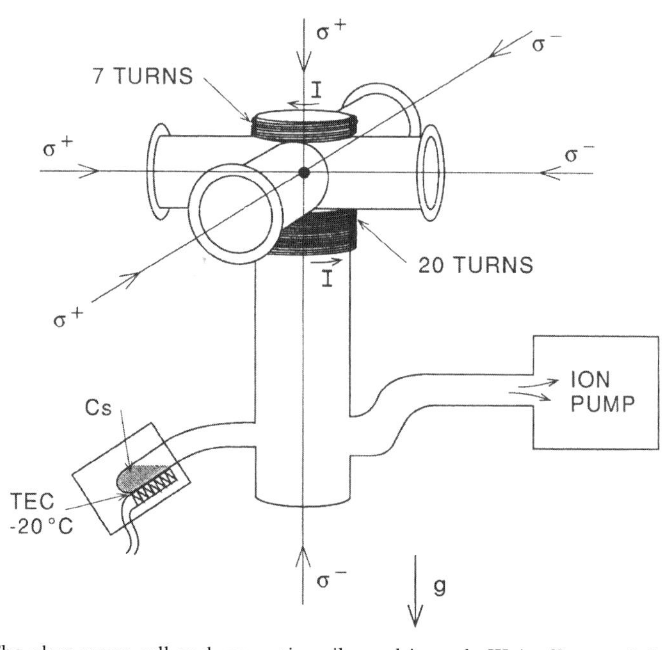

Figure 3. The glass vapor cell and magnetic coils used in early JILA efforts to hybridize laser cooling and magnetic trapping (see Ref. [47]). The glass tubing is 2.5 cm in diameter. The Ioffe current bars have been omitted for clarity.

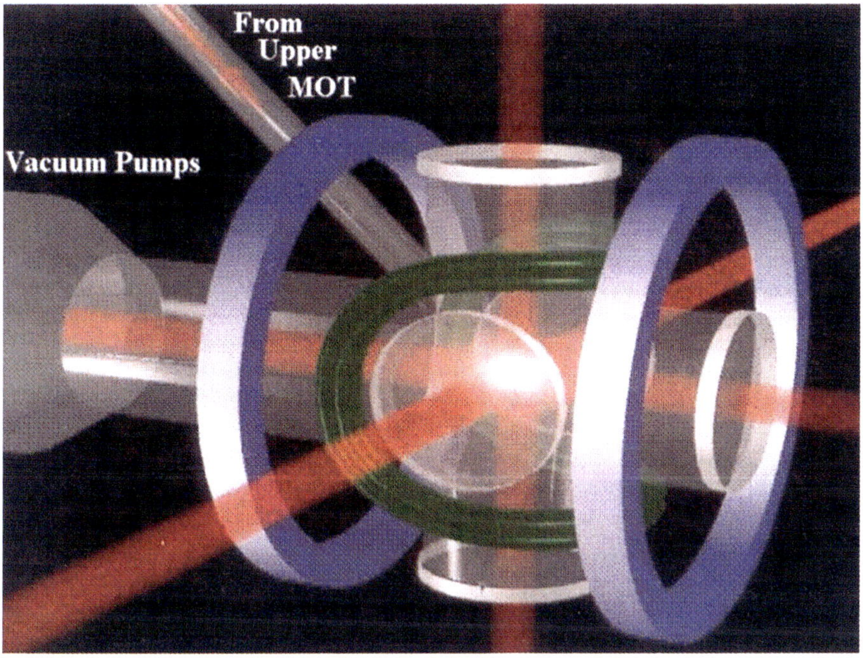

Figure 4. Modern MOT and magnetic trap apparatus, used in Ref. [77, 78]

with hydrogen to our very cold alkali atoms we looked carefully at the hydrogen work and its lessons. When viewed from our 1990 perspective the previous decade of work on polarized hydrogen provided a number of important insights. It was clear that the unique absence of any bound states for spin-polarized hydrogen was actually not an important issue (other than it being the catalyst for starting the entire field, of course!). Bound states or not, a very cold sample of spin polarized hydrogen like every other gas has a lower energy state to which it can go, and its survival depends on the preservation of metastability. For hydrogen the lower-energy state is a solid, although from an experimental point of view the rate-limiting process is the formation of diatomic molecules (with appropriately reoriented spins). Given that all atomic gases are only metastable at the BEC transition point, the real experimental issue becomes: How well can one preserve the requisite metastability while still cooling sufficiently far to reach BEC?

The realization that metastability was the key experimental challenge one should focus on was probably at least as important to the attainment of BEC as any of the experimental techniques we subsequently developed to actually achieve it. The work on hydrogen provided an essential guide for evaluating and tackling this challenge. It provided us with a potential cooling technique (evaporative cooling of magnetically trapped atoms), and mapped out many of the processes by which a magnetically trapped atom can be lost from its metastable state. The hydrogen work made it clear that it was all an issue of good versus bad collisions. The good collisions are elastic collisions that rethermalize the atoms during evaporation. The more collisions there are the more quickly and efficiently one can cool. The bad collisions are the inelastic collisions that quench the metastability. Hydrogen had already shown that three-body recombination collisions and dipole spin flip collisions were the major inelastic culprits. The fact that hydrogen researchers were fairly close to reaching BEC was also a strong encouragement – it meant that the goal was not ridiculously distant and that one only had to do a little better in the good to bad collisions to succeed.

The more we thought about this, the more we began to suspect that our heavy alkali atoms would likely have more favorable collision properties than hydrogen atoms and thus have a good chance of success. Although knowledge of the relevant collision cross sections was totally nonexistent at that time, we were able to come up with arguments for how the cross sections might scale relative to hydrogen. These are discussed in more detail below in the section discussing why collisional concerns make it likely that BEC can be created in a large number of different species. Here we will just give a brief summary consistent with our views circa 1990. The dipole spin-flip collisions that limited hydrogen involve spin-spin interactions and thus could be expected to be similar for the alkalis and for hydrogen because the magnetic moments are all about the same. The good collisions needed for evaporative cooling, however, should be much larger for heavy alkalis with their fat fluffy electron clouds than for hydrogen. The other villain of the hydrogen effort, three-body recombination, was a total mystery, but because it goes as density cubed while the good elastic collisions go as density squared, it seemed like we should always be able to find a sufficiently low density and temperature regime to avoid it (see Ref. [48]).

As a minor historical note, we might point out that during these considerations we happily ignored the fact that the temperatures required to achieve BEC in a heavy alkali gas are far lower than those needed for the same density of hydrogen. The critical temperature for ideal-gas BEC is inversely proportional to the mass. It was clear that we would need to cool to well under a microkelvin, and a large three-body recombination rate would have required us to go to possibly far lower temperatures. To someone coming from a traditional cryogenics background this would (and probably did) seem like sheer folly. The hydrogen work had been pushing hard for some years at the state of the art in cryogenic technology, and here we proposed to happily jump far beyond that. Fortunately we were coming to this from an AMO background in a time when temperatures achieved by laser cooling were dropping through the floor. Optimism was in the air. In fact, we later discovered optimism can take one only so far: There were actually considerable experimental difficulties, and further cooling came at some considerable effort and a five-year delay. Nevertheless, it is remarkable that with evaporative cooling a magnetically trapped sample of atoms, surrounded on all sides by a 300 K glass cell, can be cooled to reach temperatures of only a few nanokelvin, and moreover it looks quite feasible to reach even lower temperatures.

General collisional considerations gave us some hope that the evaporative cooling hybrid trap approach with alkali atoms would get us to BEC, or, if not, at least reveal some interesting new physics that would prevent it. Nonetheless, there were powerful arguments against pursuing this. First, our 1990-era arguments in favor of it were based on some very fuzzy intuition; there were no collision data or theories to back it up and there were strong voices in disagreement. Second, the hydrogen experiments seemed to be on the verge of reaching BEC, and in fact we thought it was likely that if BEC could be achieved they would succeed first. However, our belief in the virtues of our technology really carried the day in convincing us to proceed. With convenient lasers in the near-IR, and with the good optical access of a room-temperature glass cell, detection sensitivity could approach single-atom capability. We could take pictures of only a few thousand trapped atoms, and immediately know the energy and density distribution. If we wanted to modify our magnetic trap it only required a few hours winding and installing a new coil of wires. This was a dramatic contrast with the hydrogen experiments that, like all state of the art cryogenics experiments, required an apparatus that was the better part of two stories, and the time to modify it was measured in (large) fractions of a year. Also, atomic hydrogen was much more difficult to detect and so the diagnostics were far more limited. This convinced us that although hydrogen would likely succeed first, our hybrid trap approach with easily observed and manipulated alkali samples would be able to carry out important science and so was well worth pursuing in its own right.

From the very beginning in 1990, our work on BEC was heavily involved with cold atomic collisions. This was somewhat ironic since previously both of us had actively avoided the large fraction of AMO work on the subject of atomic collisions. Atomic collisions at very low temperatures is now a major branch of the discipline of AMO physics, but at the end of the 1980s there were almost no experimental data, and what there was came in fact from the spin-polarized hydrogen experiments [49]. There was theoretical work on hydrogen from Shlyapnikov and Kagan [50, 51], and from Silvera and Verhaar [52]. An early paper by Pritchard [53] includes estimates on low-temperature collisional properties for alkalis. His estimates were extrapolations from room-temperature results, but in retrospect, several were surprisingly accurate. As we began to work on evaporative cooling, much of our effort was devoted to determining the sizes of all the relevant good and bad collision cross sections. Our efforts were helped by the theoretical efforts of Boudewijn Verhaar who was among the first to take our efforts seriously and attempt to calculate the rates in question. Chris Greene also provided us with some useful theoretical estimates.

Starting in 1990 we carried out a series of experiments exploring various magnetic traps and measuring the relevant collision cross sections. As this work proceeded we developed a far better understanding of the conditions necessary for evaporative cooling and a much clearer understanding of the relevant collisional issues [54, 55]. Our experimental concerns evolved accordingly. In the early experiments [47, 48, 54, 56] a number of issues came up that continue to confront all BEC experiments: the importance of aligning the centers of the MOT and the magnetic trap, the density-reducing effects of mode-mismatch, the need to account carefully for the (previously ignored) force of gravity, heating (and not merely loss) from background gas collisions, the usefulness of being able to turn off the magnetic fields rapidly, the need to synchronize many changes in laser status and magnetic fields together with image acquisition, an appreciation for the many issues that can interfere with accurate determinations of density and temperature by optical methods, either fluorescence or absorption imaging, and careful stabilization of magnetic fields. The mastery of these issues in these early days made it possible for us to proceed relatively quickly to quantitative measurements with the BEC once we had it.

In 1992, we came to realize that dipolar relaxation in alkalis should in principle not be a limiting factor. As explained in the final section of this article, collisional scaling with temperature and magnetic field is such that, except in pathological situations, the problem of good and bad collisions in the evaporative cooling of alkalis is reduced to the ratio of the elastic collision rate to the rate of loss due to imperfect vacuum; dipolar relaxation and three-body recombination can be finessed, particularly since our preliminary data showed they were not enormous. It was reassuring to move ahead on efforts to evaporate with the knowledge that, while we were essentially proceeding in the dark, there were not as many monsters in the dark as we had originally imagined.

It rapidly became clear that the primary concerns would be having sufficient elastic collision rate in the magnetic trap and sufficiently low background pressure to have few background collisions that removed atoms from the trap. To accomplish this it was clear that we needed higher densities in the magnetic trap then we were getting from the MOT. Our first effort to increase the density two years earlier was based on a multiple-loading scheme [56]. Multiple MOT-loads of atoms were launched in moving molasses, optically pumped into an untrapped Zeeman level, focused into a magnetic trap, then optically repumped into a trapped level. The repumping represented the necessary dissipation, so that multiple loads of atoms could be inserted in a continuously operating magnetic trap. In practice, each step of the process involved some losses, and the final result was disappointing. Later however, as discussed below, we resurrected the idea of multiple loading from one MOT to another to good advantage [57, 58]. This is now a technique currently in widespread practice.

In addition to building up the initial density we realized that the collision rate could be dramatically increased by, after loading into a magnetic trap, compressing the atoms by further increasing the curvature of the confining magnetic fields. In a harmonic trap, the collision rate after adiabatic compression scales as the final confining frequency squared [48]. This method is discussed in Ref. [48] and was implemented first in early ground-state collisional work [54].

In fall of 1992, Eric's postdoctoral appointment concluded, and, after a tour through the job market, he decided to take the equivalent of an assistant professor position at JILA/NIST. He decided to use his startup money to build a new experimental apparatus that would be designed to put these ideas together to make sure evaporation worked as we expected. Meanwhile, we continued to pursue the possibility of enhanced collision cross sections in cesium using a Feshbach resonance. At that point our Monte Carlo simulations said that a ratio of about 150 elastic collisions per trap lifetime was required to achieve "run away" evaporation. This is the condition where the elastic collision rate would continue to increase as the temperature decreased, and hence evaporation would continue to improve as the temperature was reduced. We also had reasonable determinations of the elastic collision cross sections.

So the plan was to build a simple quadrupole trap that would allow very strong squeezing to greatly enhance the collision rate, combined with a good vacuum system in order to make sure evaporative cooling worked as expected. Clearly, there was much to be gained by building a more tightly confining magnetic trap, but the requirement of adequate optical access for the MOT, along with engineering constraints on power dissipation, made the design problem complicated.

When constructing a trap for weak-field seeking atoms, with the aim of confining the atoms to a spatial size much smaller than the size of the magnets, one would like to use linear gradients. In that case, however, one is confronted with the problem of the minimum in the magnitude of the magnetic fields (and thus of the confining potential) occurring at a local zero in the magnetic field. This zero represents a "hole" in the trap, a site at which atoms can undergo Majorana transitions [59] and thus escape from the trap. If one uses the second-order gradients from the magnets to provide the confinement, there is a marked loss of confinement strength. This scaling is discussed in Ref. [60]. We knew that once the atoms became cold enough they would leak out the "hole" in the bottom of the trap, but the plan was to go ahead and get evaporation and worry about the hole later. We also recognized that even with successful evaporative cooling, and presuming we could solve the hole in the quadrupole trap issue, there was still the question of the sign of scattering length, which must be positive to ensure the stability of a large condensate.

In setting up the new apparatus Eric chose to use rubidium. Given the "modulo" arithmetic that goes into determining a scattering length, it seemed fair to treat the scattering lengths of different isotopes as statistically independent events, and rubidium with its two stable isotopes offered two rolls of the dice for the same laser system. Eric then purchased a set of diode lasers for the rubidium wavelength, but of course we kept the original cesium-tuned diode lasers. The wavelengths of cesium and of the two rubidium isotopes are sufficiently similar that in most cases one can use the same optics. Thus we preserved the option of converting from one species to another in a matter of weeks. The chances then of Nature conspiring to make the scattering length negative, for both hyperfine levels, for all three atoms, seemed very small.

Progress in cold collisions, particularly the experiment and theory of photoassociative collisions, had moved forward so rapidly that by the time we had evaporatively cooled rubidium to close to BEC temperatures a couple of years later there existed, at the 20% level, values for several of the elastic scattering lengths. In particular, we knew that it was positive for the 2,2 state of Rb-87 [61–66].

Our original idea for the quadrupole trap experiment was to pulse a burst of rubidium into our cell, where we would catch a large sample in the MOT and then hold it as the residual rubidium was quickly pumped away leaving a long trap lifetime. We, particularly Eric's postdoc, Mike Anderson, spent many frustrating months discovering how difficult this seemingly simple idea was to actually implement in practice. The manner in which rubidium interacted with glass and stainless steel surfaces conspired to make this so difficult we finally gave up. We ended up going with a far-from-optimum situation of working with extremely low rubidium pressure and doing our best at maximizing the number of atoms captured in the MOT from this feeble vapor and enhancing the collision rate for those relatively few atoms as much as possible. We recognized that this was a major compromise, but we had been trying to evaporate for some time, and we were getting impatient! We had no stomach for building another apparatus just to see evaporation. Fortunately we were able to find two key elements to enhance the MOT loading and density. First was the use of a dark-spot MOT in which there is a hole in the center of the MOT beams so the atoms are not excited. This technique had been demonstrated by Ketterle [67] as a way to greatly enhance the density of atoms in a MOT under conditions of a very high loading rate. The number of atoms we could load in our vapor cell MOT with very low rubidium vapor was determined by the loading rate over the loss rate. In this case the loss rate was the photoassociative collisions we had long before found to be important for losses from MOTs. The dark-spot geometry reduced this two-body photoassociative loss in part because in our conditions it *reduced* the density of atoms in the MOT [68].

Figure 5. Wolfgang Petrich working on CMOT.

Using this approach we were able to obtain 10^8 atoms in the MOT collected out of a very low vapor background (so that magnetic trap lifetime was greater than 100 s). The second key element was the invention of the compressed MOT (CMOT), a technique for substantially enhancing the density of atoms in the MOT on a transient basis. For the CMOT, the MOT was filled and then the field gradient and laser detuning were suddenly changed to greatly suppress the multiple photon scattering. This produced much higher densities and clouds whose shape was a much better match to the desired shape of the cloud in the magnetic trap. This was a very transient effect because the losses from the MOT were much larger under these conditions, but that was not important; the atoms needed only to be held for the milliseconds required before they were transferred to the magnetic trap [69]. With these improvements and a quadrupole trap that provided substantial squeezing, we were able to finally demonstrate evaporative cooling in rubidium.

Cooling by evaporation is a process found throughout nature. Whether the material being cooled is an atomic nucleus or the Atlantic ocean, the rate of natural evaporation and the minimum temperature achievable are limited by the particular fixed value of the work function of the evaporating substance. In magnetically confined atoms, no such limit exists, because the "work function" is simply the height of the lowest point in the rim of the confining potential. Hess pointed out [15] that by perturbing the confining magnetic fields, the work function of a trap can be made arbitrarily low; as long as favorable collisional conditions persist, there is no lower limit to the temperatures attainable in this forced evaporation.

Pritchard [70] pointed out that evaporation could be performed more conveniently if the rim of the trap were defined by an rf-resonance condition, rather than simply by the topography of the magnetic field; experimentally, his group made first use of position-dependent rf transitions to selectively transfer magnetically trapped sodium atoms between Zeeman levels and thus characterized their temperature [71]. In our experiment we used Pritchard's technique of an rf field to selectively evaporate.

It was a great relief to see evaporative cooling of laser precooled, magnetically trapped atoms finally work, as we had been anticipating it would for so many years. Unfortunately, it worked exactly as well, but no better, than we had anticipated. The atoms were cooled to about 40 mikroK and then disappeared, at just the temperature we had estimated they would be lost, through the hole in the bottom of the quadrupole trap. Eric came up with an idea that solved this problem. It was a design for a new type of trap that required relatively little modification to the apparatus and so was quickly implemented. This was the Time Orbiting Potential (TOP) trap in which a small rotating magnetic field was added to the quadrupole field [60]. This moved the field zero in an orbit faster than the atoms could follow. It was the perfect solution to our problem.

Mike Anderson, another postdoc, Wolfgang Petrich (Fig. 5), and graduate student Jason Ensher quickly implemented this design. Their efforts were spurred on by the realization that there were several other groups who had now demonstrated or were known to be on the verge of demonstrating evaporative cooling in alkalis in the pursuit of BEC. The TOP design worked well, and the samples were cooled far colder, in fact too cold for us to reliably measure. We had been measuring temperature simply by looking at the spatial size of the cloud in the magnetic trap. As the temperature was reduced the size decreased, but we were now reaching temperatures so low that the size had reached the resolution limit of the optical system. We saw dramatic changes in the shapes of the images as the clouds became very small, but we knew that a variety of diffraction and aberration effects could greatly distort images when the sample size became only a few wavelengths in size, so our reaction to these shapes was muted, and we knew we had to have better diagnostics before we could have meaningful results. Here we were helped by our long experience in studying various trapped clouds over the years. We already knew the value of turning the magnetic trap off to let the cloud expand and then imaging the expanded cloud to get a measure of the momentum distribution in the trap. Since the trap was harmonic, the momentum distribution and the original density distribution were nearly interchangeable. Unfortunately, once the magnetic field is off, the atoms not only expand but also simply fall under the influence of gravity. We found that the atoms tended to fall out of the field of view of our microscope before they had sufficiently expanded. The final addition to the apparatus was a supplementary magnetic coil, which provided sufficient field gradient to cancel the effects of gravity while minimizing any perturbation to the relative ballistic trajectories of the expanding atoms.

Anderson, Ensher, and a new graduate student, Mike Matthews (Fig. 6), worked through a weekend to install the "anti-gravity coil" and, after an addi-

Figure 6. From left, Mike Anderson, Debbie Jin, Mike Matthews and Jason Ensher savor results of early BEC experiment.

tional day or two of trial and error, got the new field configuration "shimmed up." By June 5, 1995 the new technology was working well and we began to look at the now greatly expanded clouds. To our delight, the long-awaited two-component distribution was almost immediately apparent (Fig. 7) when the samples were cooled to the regime where BEC was expected. The excitement was tempered by the concern that after so many years of anticipating two component clouds as a signature of BEC, we might be fooling ourselves.

Almost from the beginning of the search for BEC, it was recognized [72] that as the sample starts to condense, there would be a spike in the density and momentum distributions corresponding to the macroscopic population of the ground state. This would show up as a second component on top of the much broader normal thermal distribution of uncondensed atoms. This was the signature we had been hoping to see from our first days of contemplating BEC. The size of the BEC component in these first observations also seemed almost too good to be true. In those days it was known that in the much higher density of the condensate, three-body recombination would be a more dominant effect than in the lower-density uncondensed gas. For hydrogen it was calculated that the condensed component could never be more than a few percent of the sample. The three-body rate constants were totally unknown for alkali atoms at that time, but because of the H results it still seemed reasonable to expect the condensate component might only be a mo-

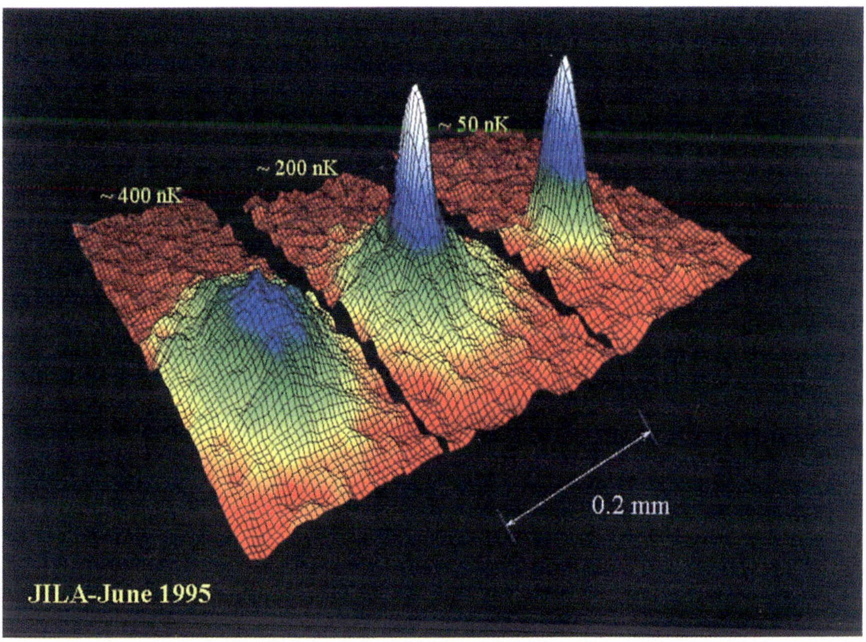

Figure 7. Three density distributions of the expanded clouds of rubidium atoms at three different temperatures..The appearance of the condensate is apparent as the narrow feature in the middle image..On the far right, nearly all the atoms in the sample are in the condensate. The original experimental data is a two-dimensional black and white shadow image, but these images have been converted to three dimensions and given false color density contours.

dest fraction of the total sample. But in our first samples we saw it could be nearly 100%! In the light of the prevailing myth of unattainability that had grown up around BEC over the years, our observations seemed too good to be true. We were experienced enough to know that when results in experimental physics seem too good to be true, they almost always are! We worried that in our enthusiasm we might confuse the long-desired BEC with some spurious artifact of our imaging system.

However, our worries about the possibility of deluding ourselves were quickly and almost entirely alleviated by the anisotropy of the BEC cloud. This was a very distinctive signature of BEC, the credibility of which was greatly enhanced to us by the fact that it first revealed itself in the experiment, and then we recognized its significance, rather than vice-versa. It was a somewhat fortuitous accident that the TOP trap provided a distinctly anisotropic trapping potential since we did not appreciate its benefits until we saw the BEC data. A normal thermal gas (in the collisionally thin limit) released from an anisotropic potential will spread out isotropically. This is required by the equipartition theorem. However, a BEC is a quantum wave and so its expansion is governed by a wave equation. The more tightly confined direction will expand the most rapidly, a manifestation of the uncertainty principle. Seeing the BEC component of our two-component distribution display just this anisotropy, while the broader "uncondensed" portion of the sample observed at the same time, with the same imaging system remained perfectly isotropic (as shown in Fig. 8), provided the crucial piece of corroborating evidence that this was the long awaited BEC.

By coincidence we were scheduled to present progress reports on our efforts to achieve BEC at three international conferences in the few weeks following these observations [74]. Nearly all the experts in the field were represented at one or more of these conferences, and the data were sufficient to convince the most skeptical of them that we had truly observed BEC. This consensus probably facilitated the rapid refereeing and publication of our results.

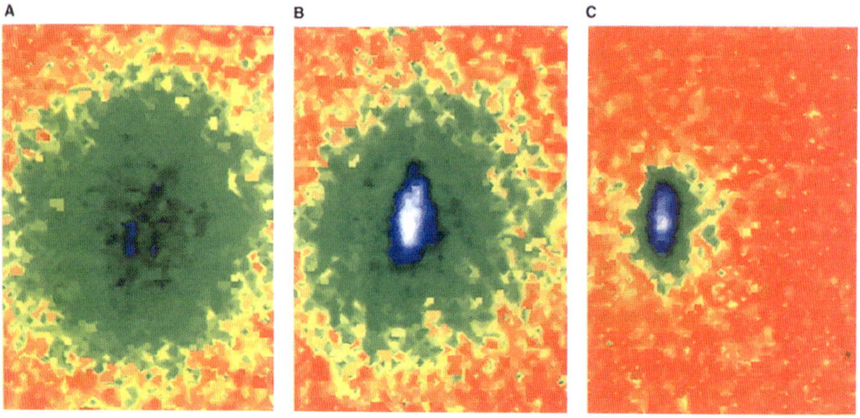

Figure 8. Looking down on the three images of Figure 7 [73]. The condensate in B and C is clearly elliptical in shape.

In the original TOP-trap apparatus we were able to obtain so-called "pure condensates" of a few thousand atoms. By "pure condensates" we meant that nearly all the atoms were in the condensed fraction of the sample. Samples of this size were easily large enough to image. Over the few months immediately following the original observation, we undertook the process of a technological shoring up of the machine, until the machine reached the level of reliability necessary to crank out condensate after reproducible condensate. This set the stage for the first generation of experiments characterizing the properties of the condensate, most notably the condensate excitation studies discussed below.

Although by 1995 and 1996 we were able to carry out a number of significant BEC experiments with the original TOP-trap machine, even by 1994, well before the original condensates were observed, we had come to realize the limitations of the single-cell design. Our efforts to modulate the vapor pressure were not very successful, which forced us to operate at a steady-state rubidium vapor pressure. Choosing the value of vapor pressure at which to operate represented a compromise between our need to fill the vapor-cell MOT with as many atoms as possible, and our need to have the lifetime in the magnetic trap as long as possible. The single-cell design also compelled us to make a second compromise, this time over the size of the glass cell. The laser beams of the MOT enter the cell through the smooth, flat region of the cell; the larger the glass cell, the larger the MOT beams, and the more atoms we could herd into the MOT from the room-temperature background vapor. On the other hand, the smaller the glass cell, the smaller the radii of the magnetic coils wound round the outside of the cell, and the stronger the confinement provided by the magnetic trap. Hans Rohner in the JILA speciality shop had learned how [75] to create glass cells with the minimum possible wasted area. But even with the "dead space" between the inner diameter of the magnetic coils and the outer diameter of the clear glass windows made as small as it could be, we were confronted with an unwelcome tradeoff.

Thus, in 1994, in parallel with our efforts to push as hard as we could toward BEC in our original, single-cell TOP trap, we began working on a new technology that would avoid this painful tradeoff. This approach was a modified version of our old multiple loading scheme in which many loads from a MOT were transferred to a magnetic trap in a differentially pumped vacuum chamber. That approach had been defeated by the difficulty in transferring atoms from MOT to magnetic trap without losing phase space density. There was no dissipation in the magnetic trap to compensate for a slightly too hard or too soft push from one trap to the other. This made us recognize the importance of having dissipation in the second trap, and so we went to a system in which atoms were captured in a large-cell MOT in a region of high rubidium pressure, and then transferred through a small tube into a second, small-cell MOT in a low-pressure region. This eliminated the previous disadvantages while preserving the advantages of multiple loading to get much larger numbers of trapped atoms in a low vacuum region. The approach worked well, particularly when we found that simple strips of plastic "refrigerator magnet" material around the outside of the transfer tube between the two traps provided an excellent guide to confine the atoms as they were pushed from one trap to the other [58].

With this scheme we were able to still use inexpensive low-power diode lasers to obtain about one hundred times more atoms in the magnetic trap than in our single MOT-loaded TOP magnetic trap and with a far longer lifetime; we saw trap lifetimes up to 1000 seconds in the double MOT magnetic trap. This system started working in 1996 and it marked a profound difference in the ease with which we could make BEC [76]. In the original BEC experiment everything had to be very well optimized to achieve the conditions necessary for run-away evaporative cooling and thereby BEC. In the double MOT system there were orders of magnitude to spare. Not only did this allow us to routinely obtain million atom pure condensates, but it also meant that we could dispense with the dark-spot optical configuration with its troublesome alignment. We could be much less precise with many other aspects of the experiment as well.

The first magnetic trap we used with the double-MOT BEC machine was not a TOP trap, but instead was our old "Baseball" style Ioffe-Pritchard trap. The Baseball coil trap is rather complementary to the TOP trap in that each has unique capabilities. For example, the geometry of the TOP trap potential can be changed over a wide range, although the range of DC fields is quite limited. In contrast, the geometry of the Baseball coil trap potential can be varied only by small amounts, but the DC bias field can be easily varied over hundreds of gauss. Thus in 1996, when we upgraded the original BEC machine to incorporate the double-MOT technology, we preserved the TOP trap coil design. Both are well suited to certain types of experiments, as will be evident in the discussions below.

With the double-MOT setups we were able to routinely make million-atom condensates in a highly reliable manner in both TOP and Baseball-type magnetic traps. These were used to carry out a large number of experiments with condensates over the period from 1996 to the present. Some of our favorite experiments are briefly discussed below.

FAVORITE EXPERIMENTS

Collective Excitations

In this section, by "excitations" we mean "coherent fluctuations in the density distribution." Excitation experiments in dilute-gas BEC have been motivated by two main considerations. First, BEC is expected to be a superfluid, and a superfluid is defined by its dynamical behavior. Studying excitations is an obvious initial step toward understanding dynamical behavior. Second, in experimental physics a precision measurement is almost always a frequency measurement, and the easiest way to study an effect with precision is to find an observable frequency that is sensitive to that effect. In the case of dilute-gas BEC, the observed frequency of standing-wave excitations in a condensate is a precise test of our understanding of the effect of interactions.

BEC excitations were first observed by Jason Ensher, Mike Matthews, and then-postdoc Debbie Jin, using destructive imaging of expanded clouds [77]. The nearly zero temperature clouds were coherently excited (see below), then allowed to evolve in the trap for some particular dwell time, and then rapidly expanded and imaged via absorption imaging. By repeating the procedure many times with varying dwell times, the time-evolution of the condensate density profile can be mapped out. From these data, frequencies and damping rates can be extracted. In axially symmetric traps, excitations can be characterized by their projection of angular momentum on the axis. The perturbation on the density distribution caused by the excitation of lowest-lying $m = 0$ and $m = 2$ modes can be characterized as simple oscillations in the condensate's linear dimensions. Fig. 9 shows the widths of an oscillating condensate as a function of dwell time.

A frequency-selective method for driving the excitations is to modulate the trapping potential at the frequency of the excitation to be excited [77]. Experimentally this is accomplished by summing a small ac component onto the current in the trapping magnets. In a TOP trap, it is convenient enough to independently modulate the three second-order terms in the transverse potential. By controlling the relative phase of these modulations, one can impose $m = 0$, $m = 2$, or $m = -2$ symmetry on the excitation drive.

There have been a very large number of theory papers published on exci-

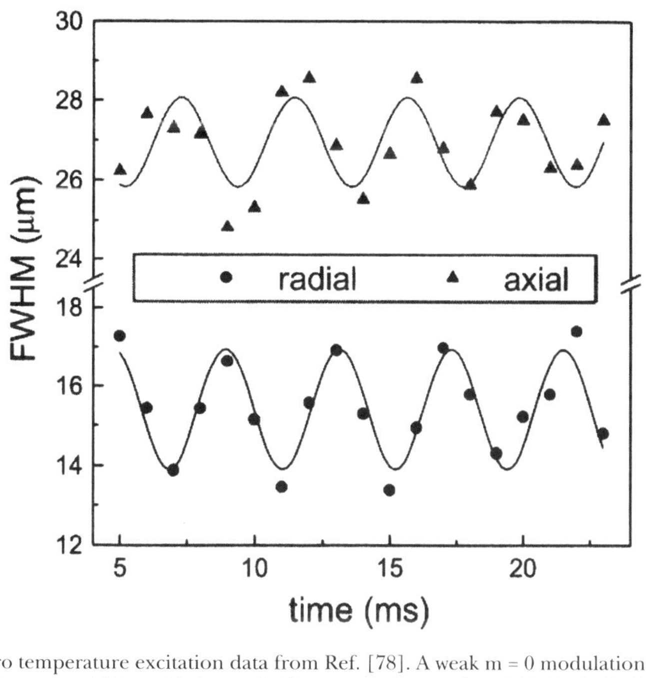

Figure 9. Zero temperature excitation data from Ref. [78]. A weak m = 0 modulation of the magnetic trapping potential is applied to a 4500-atom condensate in a 132 Hz (radial) trap. Afterward, the freely evolving response of the condensate shows radial oscillations. Also observed is a sympathetic response of the axial width, approximately 180° out of phase. The frequency of the excitation is determined from a sine wave fit to the freely oscillating cloud widths.

tations; much of this work is reviewed in Ref. [78]. All the zero-temperature, small amplitude excitation experiments published to date have been very successfully modeled theoretically. Quantitative agreement has been by and large very good; small discrepancies can be accounted for by assuming reasonable experimental imperfections with respect to the $T = 0$ and small-amplitude requirements of theory.

The excitation measurements discussed above were then revisited at non-zero temperature [79]. The frequency of the condensate excitations was clearly observed to depend on the temperature, and the damping rates showed a strong temperature dependence. This work is important because it bears on the little-studied finite-temperature physics of interacting condensates. Connection with theory [80-82] remains somewhat tentative. The damping rates, which are observed to be roughly linear in temperature, have been explained in the context of Landau damping [83, 84]. The frequency shifts are difficult to understand, in large part because the data so far have been collected in a theoretically awkward, intermediate regime: the cloud of noncondensate atoms is neither so thin as to have completely negligible effect on the condensate, nor so thick as to be deeply in the "hydrodynamic (HD) regime." In this context, "hydrodynamic regime" means that the classical mean free path in the thermal cloud is much shorter than any of its physical dimensions. In the opposite limit, the "collisionless regime," there are conceptual difficulties with describing the observed density fluctuations as "collective modes." Recent theoretical work suggests that good agreement with experiment may hinge on correctly including the role of the excitation drive [85, 86].

Two-component condensates

As mentioned above, the double-MOT system made it possible to produce condensates even if one was quite sloppy with many of the experimental parameters. One such parameter was the spin state in which the atoms are optically pumped before being loaded into the magnetic trap. As our student Chris Myatt was tinkering around setting up the evaporation one day, he noticed, to his surprise, that there seemed to be two different clouds of condensate in the trap. They were roughly at the locations expected for the 2, 2 and 1,-1 spin states to sit, but that seemed impossible to us because these two states could undergo spin-exchange collisions that would cause them to be lost from the trap, and the spin exchange collision cross sections were thought to be enormous. After extensive further studies to try and identify what strange spurious effect must be responsible for the images of two condensate clouds we came to realize that they had to be those two spin states. By a remarkable coincidence, the triplet and singlet phase shifts are identical and so at ultralow temperatures the spin exchange collisions are suppressed in ^{87}Rb by 3 to 4 orders of magnitude! This suppression meant that the different spin species could coexist and their mixtures could be studied. In early work we showed that one could carry out sympathetic cooling to make BEC by evaporating only one species and using it as a cooling fluid to chill the second spin state [76]. We also were able to see how the two condensates interacted and pushed each other apart, excluding all but a small overlap in spite of the fact that they were highly dilute gases.

These early observations stimulated an extensive program of research on two-component condensates. After Myatt's original measurements [76], our work in this field, led by postdoc David Hall, concentrated on the 1,-1 and 2, +1 states (see Fig. 10) because they could be coherently interconverted using two-photon (microwave plus RF) transitions and they had nearly identical magnetic moments and so saw nearly the same trapping potentials [87].

When the two-photon radiation field is turned off, the rate of spontaneous interconversion between the two spin species essentially vanishes, and moreover the optical imaging process readily distinguishes one species from the other, as their difference in energy (6.8 GHz) is very large compared to the excited state linewidth. In this situation, one may model the condensate dynamics as though there were two distinct quantum fluids in the trap. Small differences in scattering length make the two fluids have a marginal tendency to spatially separate, at least in an inhomogeneous potential, but the interspecies healing length is long so that in the equilibrium configuration there is considerable overlap between the two species [88, 89]. On the other hand, the presence of a near-resonant two-photon coupling drive effectively brings the two energy levels quite close to one another: on resonance, the corresponding dressed energy levels are separated only by the effective Rabi frequency for the two-photon drive. In this limit, one may in a certain sense think of the condensate as being described by a two-level, spinor field [90, 91].

We got a lot of mileage out of this system, and continue to explore its properties today. One of the more dramatic experiments we did in the two-level condensate was the creation, via a sort of wave function engineering, of a

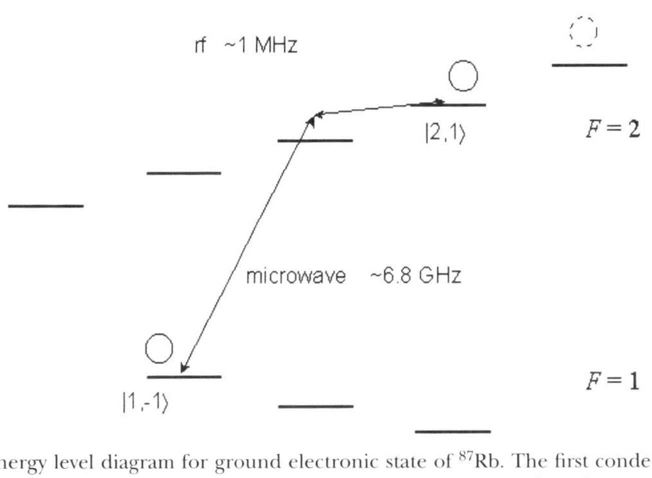

Figure 10. Energy level diagram for ground electronic state of ^{87}Rb. The first condensates were created in the 2,2 state. Mixtures containing the 2,2 and 1,-1 state were found to coexist. In later studies we created condensates in the 1,-1 state and then excited it to the 2,1 state using a microwave plus rf two photon transition.

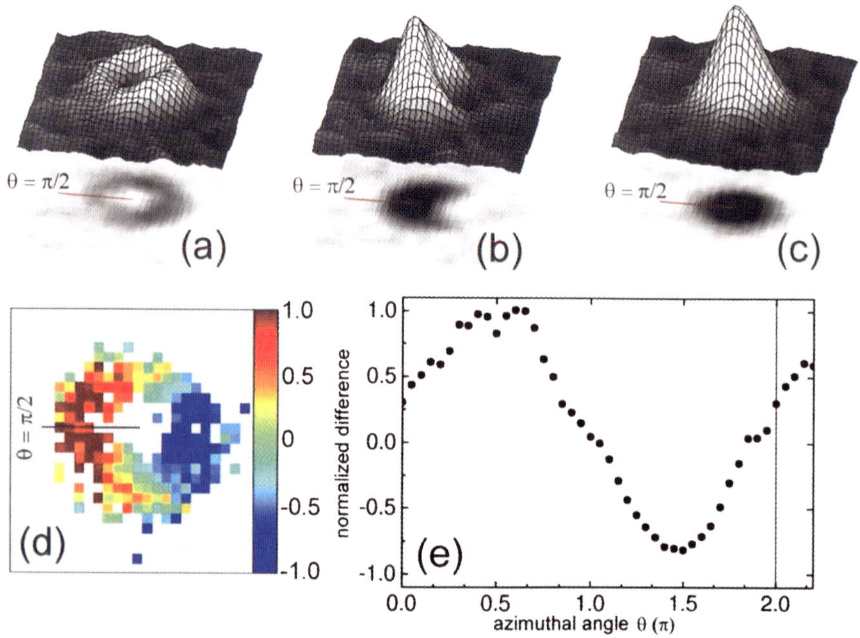

Figure 11. (from Ref. [92]). Condensate images showing the first BEC vortex and the measurement of its phase as a function of azimuthal angle. (a) Shows the density distribution of atoms in the upper hyperfine state after atoms have been put in that state in a way that forms a vortex. (b) Shows the same state after a $\pi/2$ pulse has been applied that mixes upper and lower hyperfine states to give an interferogram reflecting the phase distribution of the upper state. (c) Residual condensate in the lower hyperfine state from which the vortex was formed that interferes with (a) to give the image shown in (b). (d) A color map of the phase difference reflected in (b). (e) The radial average at each angle around the ring in (d). The data are repeated after the azimuthal angle 2π to better show the continuity around the ring. This shows that the cloud shown in (a) has the 2π phase winding expected for a quantum vortex with one unit of angular momentum.

quantized vortex. In this experiment we made use of both aspects of the two-level system – the "distinguishable fluids" and the "spinor gas." Starting with a near-spherical ball of atoms, all in the lower spin state, we applied the two-photon drive for about 100 ms. At the same time, we illuminated the atoms with an off-resonant laser beam whose intensity varied both in time and in space. The laser beam was sufficiently far from resonance that by itself it did not cause the condensate to transition from state to state, but the associated ac Stark shift was large enough to affect the resonant properties of the two-photon drive. The overall scheme is described in Refs. [92, 93]. The net effect was to leave the atoms near the center of the ball of atoms essentially unperturbed, while converting the population in an equatorial belt around the ball into the upper spin state. This conversion process also imposed a winding in the quantum phase, from 0 around to 2π, in such a way that by the time the drive was turned off, the upper-spin state atoms were in a vortex state, with a single quantum of circulation. The central atoms were nonrotating, and, like the pimento in a stuffed olive, served only to mark the location of the vortex core. The core atoms could in turn be selectively blasted away, leaving the upper-state atoms in a bare vortex configuration, whose dynamic properties were shown by postdoc Brian Anderson and grad student Paul Haljan to be essentially the same as those of the filled vortex [94].

Coherence and condensate decay

One of our favorite BEC experiments was to simply look at how a condensate goes away [95]. The attraction of this experiment is its inherent simplicity combined with the far-reaching implications of the results. Although it was well established that condensates lived for a finite period, fractions of a second to many seconds depending on conditions, no one had identified the actual process by which atoms were being lost from the condensate. To do this our co-workers Chris Myatt, Rich Ghrist, and Eric Burt simply made condensates and carefully watched the number of atoms and shape of the condensate as a function of time. From these data we determined that the loss process varied with the cube of the density, and hence must be three-body recombination. This was rather what we had expected, but it was nice to have it confirmed. In the process of this measurement we also determined the three-body rate constant, and this was more interesting. Although three-body rate constants still cannot be accurately calculated, it was predicted long ago [96] that they should depend on the coherence properties of the wave function. In a normal thermal sample there are fluctuations and the three-body recombination predominantly takes place at high-density fluctuations. If there is higher order coherence, however, as one has in macroscopically occupied quantum states such as a single mode laser, or was predicted to exist in a dilute gas BEC, there should be no such density fluctuations. On this basis it was predicted that the three-body rate constant in a BEC would be 3 factorial or 6 times lower than what it would be for the same atoms in a thermal sample. It is amusing that such a relatively mundane collision process can be used to probe the quantum correlations and coherence in this fashion. After measuring the three-body rate constant in the condensate we then repeated the measurement in a very cold but uncondensed sample. The predicted factor of 6 (actually 7.4 +/- 2.6) was observed, thereby confirming the higher order coherence of BEC [95].

Feshbach resonance physics

In 1992 Eric Cornell and Chris Monroe realized that dipole collisions at ultralow temperatures might have interesting dependencies on magnetic field as discussed in the Appendix. With this in mind we approached Boudwijn Verhaar about calculating the magnetic field dependencies of collisions between atoms in the lower F spin states. When he did this calculation he discovered [97] that there were dramatic resonances in all the cross sections as a function of magnetic field that are now known as Feshbach resonances because of their similarity to scattering resonances described by Herman Feshbach in nuclear collisions. From the beginning Verhaar appreciated that these resonances would allow one to tune the s-wave scattering length of the atoms and thereby change both the elastic collision cross sections and the self-interaction in a condensate, although this was several years before condensates had been created.

In 1992 we hoped that these Feshbach resonances would give us a way to create enormous elastic collision cross sections that would facilitate evaporative cooling. With this in mind we attempted to find Feshbach resonances in the elastic scattering of first cesium and then, with postdoc Nate Newbury, rubidium. These experiments did provide us with elastic scattering cross sections [98, 99], but were unable to locate the few-gauss-wide Feshbach resonances in the thousand-gauss range spanned by then theoretical uncertainty.

By 1997 the situation had dramatically changed, however. A large amount of work on cold collisions, BEC properties, and theoretical advances provided accurate values for the interaction potentials, and so we were fairly confident that there was likely to be a reasonably wide Feshbach resonance in rubidium 85 that was within 20 or 30 gauss of 150 G. This was a quite convenient bias field at which to operate our Baseball magnetic trap so we returned to the Feshbach resonance in the hope that we could now use it to make a BEC with adjustable interactions.

The time was clearly ripe for Feshbach resonance physics. Within a year Ketterle [100] saw a resonance in sodium through enhanced loss of BEC, Dan Heinzen [101] detected a Feshbach resonance in photoassociation in ^{85}Rb, we [102] (notably students Jake Roberts and Neil Claussen) detected the same resonance in the elastic scattering cross section, and Chu [103] detected Feshbach resonances in cesium. Our expectations that it would be as easy or easier to form BEC in ^{85}Rb as it was in ^{87}Rb and then use this resonance to manipulate the condensate were sadly naive, however. Due to enhancement of bad collisions by the Feshbach resonance, it was far more difficult and could only be accomplished by following a complicated and precarious evaporation path. However, by finding the correct path and cooling to 3 nK we were able to obtain pure ^{85}Rb condensates of 16000 atoms [104].

The scattering length of these condensates could then be readily adjusted by varying the magnetic field over a few gauss in the vicinity of the Feshbach resonance [105]. This has opened up a wide range of possible experiments ranging from studying the instability of condensates when the self-interaction is sufficiently attractive (negative a) to exploring the development of correlations in the wave function as the interactions are made large and repulsive. This regime provides one with a new way to probe such disparate subjects as molecular BECs and the quantum behavior of liquids, where there is a high degree of correlation. This work represents some of the most recent BEC experiments, but almost everything we have explored with this system has shown dramatic and unexpected results. Thus it is clear that we are far from exhausting the full range of interesting experiments that are yet to be carried out with BECs.

In the first of these Feshbach resonance experiments our students Jake Roberts, Neil Claussen, and postdoc Simon Cornish suddenly changed the magnetic field to make a negative. We observed that, as expected, the condensate became unstable and "collapsed," losing a large number of atoms [104]. The dynamics of the collapse process were quite remarkable. The condensate was observed to shrink slightly and then undergo an explosion in which a substantial fraction of the atoms were blown off [106]. A large fraction of the atoms also simply vanished, presumably turning into undetectable molecules or very energetic atoms, and finally a small cold stable remnant was left behind after the completion of the collapse. This process is illustrated in Fig. 12. Because of its resemblance (on a vastly lower energy scale) to a core collapse supernova, we have named this the "Bosenova." There is now considerable theoretical effort to model this process and progress is being made. However, as yet there is no clear explanation of the energy and anisotropy of the atoms in the explosion, the fraction of vanished atoms, and the size of the cold remnant. One of the more puzzling aspects is that the cold remnant can be far larger than the condensate stability condition that determines the collapse point would seem to allow [106].

Another very intriguing result of Feshbach resonance studies in ^{85}Rb was observed when our students Neil Claussen and Sarah Thompson and postdoc Elizabeth Donley quickly jumped the magnetic field close to the resonance while keeping the scattering length positive. They found that they could observe the sample oscillate back and forth between being an atomic and a molecular condensate as a function of time after the sudden perturbation. This

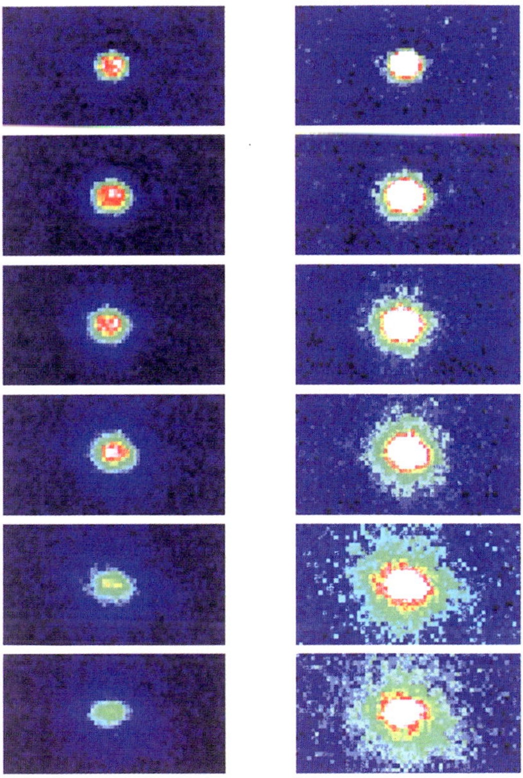

Figure 12. Bosenova explosion from Ref. [104]. From top to bottom these images show the evolution of the cloud from 0.2 to 4.8 ms after the interaction was made negative, triggering a collapse. On the left the explosion products are visible as a blue glow expanding out of the center, leaving a small condensate remnant that is unchanged at subsequent times. On the right is the same image amplified by a factor of 3 to better show the 200 nK explosion products.

curious system of a quantum superposition of two chemically distinct species will no doubt be a subject of considerable future study.

An Optimistic Appendix

Until a new technology comes along to replace evaporative cooling, the crucial issue in creating BEC with a new atom is collisions. In practice, this means that planning a BEC experiment with a new atom requires learning to cope with ignorance. It is easy to forget that essentially nothing is known about the ultralow-temperature collisional properties of any atomic or molecular species that is not an atom in the first row of the periodic table. One can not expect theorists to relieve one's ignorance – inter-atomic potentials derived from room-temperature spectroscopy are generally not adequate to allow theoretical calculations of cold elastic and inelastic collision rates, even at the order-of-magnitude level. Although the cold collisional properties of a new atom can be determined, this is a major endeavor, and in most cases it is easier to discover if evaporation will work by simply trying it.

Launching into such a major new project without any assurances of success is a daunting prospect, but we believe that, if one works hard enough, the probability that any given species can be evaporatively cooled to the point of BEC is actually quite high. The scaling arguments presented below in support of this assertion are largely the same as those that originally encouraged us to pursue BEC in alkalis, although with a bit more refinement provided by age and experience.

Although there is an extensive literature now on evaporative cooling, the basic requirement is simply that there must be on the order of 100 elastic collisions per atom per lifetime of the atoms in the trap. Since the lifetime of the atoms in the trap is usually limited by collisions, the requirement can be restated: the rate of elastic collisions must be about 2 orders of magnitude higher than the rate of bad collisions. As mentioned above there are three bad collisional processes and these each have different dependencies on atomic density in the trap, n: background collisions (independent of n), two-body dipolar relaxation ($\alpha\ n$), and three-body recombination ($\alpha\ n^2$). The rate for elastic collisions is $n\sigma v$, where n is the mean density, σ is the zero-energy s-wave cross section, and v is the mean relative velocity. The requirement of 100 elastic-to-inelastic collisions must not only be satisfied immediately after the atoms are loaded into the trap, but also as evaporation proceeds toward larger n and smaller v. With respect to evaporating rubidium 87 or the lower hyperfine level of sodium 23, nature has been kind. One need "only" arrange for the initial trapped cloud to have sufficiently large n, and design a sufficiently low-pressure vacuum chamber, and evaporation works. The main point of this section, however, is that evaporation is likely to be possible even with less favorable collision properties.

Considering the trap loss processes in order, first examine background loss. Trap lifetimes well in excess of what are needed for ^{87}Rb and Na have been achieved with standard vacuum technology. For example, we now have magnetic trap lifetimes of nearly 1000 s. (This was a requirement to achieve BEC in ^{85}Rb with its less favorable collisions.) If one is willing to accept the added complications of a cryogenic vacuum system, essentially infinite lifetimes are possible. If the background trap loss is low enough to allow evaporative cooling to begin, it will never be a problem at later stages of evaporation because nv increases.

If dipolar relaxation is to be a problem, it will likely be late in the evaporative process when the density is high and velocity low. There is no easy solution to a large dipolar relaxation rate in terms of changing the spring constant of the trap or the pressure of the vacuum chamber. Fortunately, one is not required to accept the value of dipolar collisions that nature provides. In fact, all one really has to do is operate the trap with a very low magnetic bias field in a magnetic trap, or if one uses a very far off-resonance optical trap (such as CO_2 laser), trap the atoms in the lowest spin state, for which there are no dipole collisions. The bias field dependence comes about because below a field of roughly 5 G, the dipolar rate in the lower hyperfine level drops rapidly to zero. This behavior is simple to understand. At low temperature, the incoming collisional channel must be purely s-wave. Dipolar relaxation changes the projection of spin angular momentum, so to conserve angular momentum the outgoing collisional channel must be d-wave or higher. The nonzero outgoing angular momentum means that there is an angular momentum barrier in the effective molecular potential, a barrier of a few hundred microkelvin. If the atoms are trapped in the lower hyperfine state ($F = 1$, $m_F =- 1$, in rubidium 87) the outgoing energy from a dipolar collision is only the Zeeman energy in the trapping fields and for B less than about 5 G this energy is insufficient to get the atoms back out over the angular momentum barrier. If relaxation is to occur, it can happen only at inter-atomic radii larger than the outer turning-point of the angular momentum barrier. For smaller and smaller fields, the barrier gets pushed further out, with correspondingly lower transition rates.

It is unlikely that the three-body recombination rate constant could ever be so large that three-body recombination would be a problem when the atoms are first loaded from a MOT into the evaporation trap. As evaporation proceeds, however, just as for the dipolar collisions, it becomes an increasingly serious concern. Because of its density dependence, however, it can always be avoided by manipulating the trapping potential. Adiabatically reducing the trap confinement has no effect on the phase-space density but it reduces both the density and the atom velocity. The ratio of three-body to elastic collisions scales as $1/nv$. Therefore, as long as one can continue to turn down the confining strength of one's trap, one can ensure that three-body recombination will not prevent evaporative cooling all the way down to the BEC transition.

To summarize, given (i) a modestly flexible magnetic trap, (ii) an arbitrarily good vacuum, (iii) a true ground state with $F \neq 0$, and (iv) non-pathological collisional properties, almost any magnetically trappable species can be successfully evaporated to BEC. If one is using a very far off-resonance optical trap (such as a CO_2 dipole trap) one can extend these arguments to atoms that cannot be magnetically trapped. In that case, however, current technology makes it more difficult to optimize the evaporation conditions than in magnetic traps, and the requirement to turn the trap down sufficiently to avoid a large three-body recombination rate can be more difficult. Nevertheless, one can plausibly look forward to BEC in a wide variety of atoms and molecules in the future.

We acknowledge support from the National Science Foundation, the Office of Naval Research, and the National Institute of Standards and Technology. We have benefited enormously from the hard work and intellectual stimulation of our many students and postdocs. They include Mike Anderson, Jason Ensher, Mike Matthews, Mike Renn, Debbie Jin, David Hall, Cindy Regal, Brian Anderson, Ian Coddington, Heather Lewandowski, Dave Harber, Jeff McGuirk, Yingju Wang, Peter Engels, Peter Schwindt, Tetsuo Kishimoto, Chris Monroe, Chris Myatt, Nate Newbury, Eric Burt, Neil Claussen, Jake Roberts, Liz Donley, Sarah Thompson, Scott Papp, David Sesko, Thad Walker, Michelle Stephens, Kristan Corwin, William Swann, Chris Wood, Steve Bennett, Richard Watts and Josh Zirbel. We also have had help from many other JILA faculty, including Chris Greene, John Bohn, and Murray Holland.

REFERENCES

1. Bose, S., Z. Phys., 1924. **26**: p. 178.
2. Einstein, A., Sitzungsber. Kgl. Preuss. Akad. Wiss., 1924. **261**.
3. Einstein, A., Sitzungsber. Kgl. Preuss. Akad. Wiss., 1925. **3**.
4. Pais, A., "Subtle is the Lord...". 1982, Oxford University Press: Oxford.
5. London, F., Nature, 1938. **141**: p. 643.
6. Tisza, L., Nature, 1938. **141**: p. 913.
7. Hecht, C.E., Physica, 1959. **25**: p. 1159.
8. Stwalley, W.C. and L.H. Nosanow, Phys. Rev. Lett., 1976. **36**: p. 910.
9. Silvera, I.F. and J.T.M. Walraven, Phys. Rev. Lett., 1980. **44**: p. 164.
10. Hardy, W.N., *et al.*, Physica, 1982. **109**: p. 1964.
11. Hess, H.J., *et al.*, Phys. Rev. Lett., 1983. **51**: p. 483.
12. Johnson, B.R., *et al.*, Phys. Rev. Lett., 1984. **52**: p. 1508.
13. Bell, D.A., *et al.*, Phys. Rev. B, 1986. **34**: p. 7670.
14. Sprik, R., J.T.M. Walraven, and I.F. Silvera, Phys. Rev. B, 1985. **32**: p. 5668.
15. Hess, H.F., Phys. Rev. B, 1986. **34**: p. 3476.
16. Migdall, A.L., *et al.*, Phys. Rev. Lett., 1985. **54**: p. 2596.
17. Bagnato, V.S., *et al.*, Phys. Rev. Lett., 1987. **58**: p. 2194.
18. Lovelace, R.V.E., *et al.*, Nature, 1985. **318**: p. 30.
19. Doyle, J.M., Phys. Rev. Lett., 1991. **67**: p. 603.
20. Doyle, J.M., 1991, Ph.D. Thesis, Massachusetts Institute of Technology.
21. Luitin, O.J., *et al.*, Phys. Rev. Lett., 1993. **70**: p. 544.
22. ***Proceedings of the International School of Physics 'Enrico Fermi', Course CXVIII, Laser Manipulation of Atoms and Ions.***
23. Chu, S., Rev. Mod. Phys., 1998. **70**: p. 685.
24. Cohen-Tannoudji, C.N., Rev. Mod. Phys., 1998. **70**: p. 707.
25. Phillips, W.D., Rev. Mod. Phys., 1998. **70**: p. 721.
26. Wineland, D.J. and H. Dehmelt, Bull. Am. Phys. Soc., 1975. **20**: p. 637.
27. Hansch, T.W. and A.L. Schawlow, Opt. Commun., 1975. **13**: p. 68.
28. Letokhov, V., Pis'ma Zh. Eksp. Teor. Fiz., 1968. **7**: p. 348.
29. Bjorkholm, J.E., *et al.*, Phys. Rev. Lett., 1978. **41**: p. 1361.

30. Neuhauser, W., *et al.*, Phys. Rev. Lett., 1978. **41**: p. 1978.
31. Wineland, D.J., R.E. Drullinger, and F.L. Walls, Phys. Rev. Lett., 1978. **40**: p. 1639.
32. Andreev, S.V., *et al.*, Eksp. Teor. Fiz., 1981. **34**: p. 463.
33. Prodan, J., *et al.*, Phys. Rev. Lett., 1985. **54**: p. 992.
34. Ertmer, W., *et al.*, Phys. Rev. Lett., 1985. **54**: p. 996.
35. Chu, S., *et al.*, Phys. Rev. Lett., 1985. **55**: p. 48.
36. Chu, S., *et al.*, Phys. Rev. Lett., 1986. **57**: p. 314.
37. Raab, E.L., *et al.*, Phys. Rev. Lett., 1987. **59**: p. 2631.
38. Lett, P.D., *et al.*, Phys. Rev. Lett., 1988. **61**: p. 169.
39. Dalibard, J., *et al.* in *Proceedings, 11th International Conference on Atomic Physics:* World Scientific, Singapore.
40. Chu, S., *et al.* in *Proceedings 11th International Conference on Atomic Physics:* World Scientific, Singapore.
41. Watts, R.N. and C.E. Wieman, Opt. Lett., 1986. **11**: p. 291.
42. Pritchard, D.E., *et al.*, Phys. Rev. Lett., 1986. **57**: p. 310.
43. Sesko, D., *et al.*, Phys. Rev. Lett., 1989. **63**: p. 961.
44. Walker, T., D. Sesko, and C. Wieman, Phys. Rev. Lett., 1990. **64**: p. 408.
45. Sesko, D.W., T.G. Walker, and C.E. Wieman, J. Opt. Soc. Am. B, 1991. **8**: p. 946.
46. Dalibard, J. and C. Cohen-Tannoudji, J. Opt. Soc. Am. B, 1989. **6**: p. 2023.
47. Monroe, C., *et al.*, Phys. Rev. Lett., 1990. **65**: p. 1571.
48. Monroe, C.R., Ph.D. Thesis. 1992, University of Colorado, Boulder.
49. Gillaspy, J.D., I.F. Silvera, and J.S. Brooks, Phys. Rev. B, 1989. **40**: p. 210.
50. Kagan, Y., I.A. Vartanyants, and G.V. Shlyapnikov, Sov. Phys. JETP, 1981. **54**: p. 590.
51. Kagan, Y., G.V. Shlyapnikov, and I.A. Vartanyants, Phys. Lett., 1984. **101A**: p. 27.
52. Lagendijk, A., I.F. Silvera, and B.J. Verhaar, Phys. Rev. B, 1986. **33**: p. 626.
53. Pritchard, D.E., in *Electronic and Atomic Collisions*, eds. D.C. Lorents, W.E. Meyerhof, and J.R. Peterson. 1986, North Holland, Amsterdam. 593.
54. Monroe, C.R., *et al.*, Phys. Rev. Lett., 1993. **70**: p. 414.
55. Newbury, N.R., C.J. Myatt, and C.E. Wieman, Phys. Rev. A, 1995. **51**: p. R2680.
56. Cornell, E.A., C. Monroe, and C.E. Wieman, Phys. Rev. Lett., 1991. **67**: p. 2439.
57. Gibble, K., S. Chang, and R. Legere, Phys. Rev. Lett., 1995. **75**: p. 2666.
58. Myatt, C.J., *et al.*, Opt. Lett., 1996. **21**: p. 290.
59. Majorana, E., Nuovo Cimento, 1931. **8**: p. 107.
60. Petrich, W., *et al.*, Phys. Rev. Lett., 1995. **74**: p. 3352.
61. Thorsheim, H.R., J. Weiner, and P.S. Julienne, Phys. Rev. Lett., 1987. **58**: p. 2420.
62. Miller, J.D., R.A. Cline, and D.J. Heinzen, Phys. Rev. Lett., 1993. **71**: p. 2204.
63. Lett, P.D., *et al.*, Phys. Rev. Lett., 1993. **71**: p. 2200.
64. Gardner, J.R., *et al.*, Phys. Rev. Lett., 1995. **74**: p. 3764.
65. McAlexander, W.I., *et al.*, Phys. Rev. A, 1995. **51**: p. R871.
66. Abraham, E.R.I., *et al.*, Phys. Rev. Lett., 1995. **74**: p. 1315.
67. Ketterle, W., *et al.*, Phys. Rev. Lett., 1993. **70**: p. 2253.
68. Anderson, M.H., *et al.*, Phys. Rev. A, 1994. **50**: p. R3597.
69. Petrich, W., *et al.*, J. Opt. Soc. Am. B, 1994. **11**: p. 1332.
70. Pritchard, D., *et al.* in *Proceedings, 11th International Conference on Atomic Physics:* World Scientific, Singapore.
71. Martin, A.G., *et al.*, Phys. Rev. Lett., 1988. **61**: p. 2431.
72. Lovelace, R.V.E. and T.J. Tommila, Phys. Rev. A, 1987. **35**: p. 3597.
73. Anderson, M.H., *et al.*, Science, 1995. **269**: p. 198.
74. Anderson, M.H., *et al.*, Laser Spectroscopy, XII International Conference, 1996: p. 3.
75. Rohner, H. in *Proceedings of the 39th Symposium on the Art of Glassblowing.* 1994: American Scientific Glassblowing Society, Wilmington, Delaware, p. 57.
76. Myatt, C.J, *et al.*, Phys. Rev. Lett., 1997. **78**: p. 586.
77. Jin, D.S., *et al.*, Phys. Rev. Lett., 1996. **77**: p. 420.
78. Dalfovo, F., *et al.*, Rev. Mod. Phys., 1999. **71**: p. 463.
79. Jin, D.S., *et al.*, Phys. Rev. Lett., 1997. **78**: p. 764.

80. Fedichev, P.O. and G.V. Shlyapnikov, Phys. Rev. A, 1998. **58**: p. 3146.
81. Dodd, R.J., *et al.*, Phys. Rev. A, 1998. **57**: p. R32.
82. Hutchinson, D.A.W., E. Zaremba, and A. Griffin, Phys. Rev. Lett., 1997. **78**: p. 1842.
83. Liu, W.V., Phys. Rev. Lett., 1997. **79**: p. 4056.
84. Fedichev, P.O., G.V. Shlyapnikov, and J.T.M. Walraven, Phys. Rev. Lett., 1998. **80**: p. 2269.
85. Stoof, H.T.C., personal communication.
86. Jackson, B. and E. Zaremba. **cond-mat/0201115**.
87. Matthews, M.R., *et al.*, Phys. Rev. Lett., 1998. **81**: p. 243.
88. Hall, D.S., *et al.*, Phys. Rev. Lett., 1998. **81**: p. 1539.
89. Hall, D.S., *et al.*, Phys. Rev. Lett., 1998. **81**: p. 1543.
90. Cornell, E.A., *et al.*, J. Low Temp. Phys., 1998. **113**: p. 151.
91. Matthews, M.R., *et al.*, Phys. Rev. Lett., 1999. **83**: p. 3358.
92. Matthews, M.R., *et al.*, Phys. Rev. Lett., 1999. **83**: p. 2498.
93. Williams, J.E. and M.J. Holland, Nature, 1999. **401**: p. 568.
94. Anderson, B.P., *et al.*, Phys. Rev. Lett., 2000. **85**: p. 2857.
95. Burt, E.A., *et al.*, Phys. Rev. Lett., 1997. **79**: p. 337.
96. Kagan, Y., B.V. Svistunov, and G.V. Shlyapnikov, Sov. Phys. JETP Lett., 1985. **42**: p. 209.
97. Tiesinga, E., B.J. Verhaar, and H.T.C. Stoof, Phys. Rev. A, 1993. **47**: p. 4114.
98. Newbury, N.R., C.J. Myatt, and C.E. Wieman, Phys. Rev. A, 1995. **51**: p. R2680.
99. Monroe, C.R., *et al.*, Phys. Rev. Lett., 1993. **70**: p. 414.
100. Inouye, S., *et al.*, Nature, 1998. **392**: p. 151.
101. Courteille, P., *et al.*, Phys. Rev. Lett., 1998. **81**: p. 69.
102. Roberts, J.L., *et al.*, Phys. Rev. Lett., 1998. **81**: p. 5109.
103. Vuletic, V., *et al.*, Phys. Rev. Lett., 1999. **82**: p. 1406.
104. Roberts, J.L., *et al.*, Phys. Rev. Lett., 2001. **86**: p. 4211.
105. Cornish, S.L., *et al.*, Phys. Rev. Lett., 2000. **85**: p. 1795.
106. Donley, E.A., Nature, 2001. **412**: p. 295.

WOLFGANG KETTERLE

I was born on October 21, 1957, in Heidelberg, a small town in Germany with a charming old city and a famous castle. My parents had come to Heidelberg after the second world war, when many people relocated within Germany searching for better economic opportunities. My mother's parents were farmers in Silesia, which has now become part of Poland. My father grew up in Memmingen, a small city in the southern part of Germany, where his parents had a canteen.

I enjoyed a childhood of stability and peace, in contrast to my parents who had grown up amidst the conflicts of war. When I was three, we moved from Heidelberg to the village (now city) of Eppelheim, three miles away, where my parents still live. I grew up with an older brother (Günter, 15 months older) and a younger sister (Monika, three and a half years younger).

My parents worked hard to provide security and prosperity for our family. My father first joined an oil and coal distribution company as an apprentice and retired as a director. My mother ran the household and cared for the children; later, she managed a small business distributing first-aid products. In our family, work was not regarded as sheer necessity, but as a defining feature and rewarding aspect of life.

My parents supported all our interests in music, sports and sciences. As they hadn't been exposed to many of these activities themselves, they did not steer us in certain directions, but rather observed our interests and then reinforced and supported them. That may be one of the reasons why my brother and sister are successful in quite different areas: finance and education.

My explorations of the technical world started with Legos, with which I was quite creative in constructing moving objects with the basic building blocks that were then available. (Legos have become much more fancy since then!) I remember playing with electricity kits, doing repairs of household appliances, and using my father's power tools for woodworking projects. Explorations into chemistry were done in our basement, sometimes with friends, and my parents must have had quite a bit of confidence in my abilities when they allowed me to experiment with explosive mixtures. (I was quite impressed when such a mixture was able to melt metal.) Other projects included taking old radios and a TV set apart and combining a portable radio and a vacuum tube audio amplifier to create stereo sound. I was interested in learning more about electronics, but I was disappointed that the electronic kits explained only how to put the parts together, not how they really worked. So although I explored technology and science as a child, I didn't penetrate very deeply, partially because nobody guided me, and partially because I spent a lot of time on school and sports.

W.K. (second from right) with his parents and siblings in 1971.

W.K. in December 1976.

I attended elementary school in Eppelheim and Heidelberg, and then grammar school at the Bunsengymnasium in Heidelberg. My science classes didn't involve laboratories and the variety of projects and science fairs which my children now enjoy at their schools, but they were instructive and kindled my interest. There was one mathematics teacher, Albrecht Strobel, who was inspirational. He challenged me with special problems, and tried to teach the class to approach mathematical problems in a playful rather than formal spirit. Science and mathematics did not require much of an effort for me, but I worked hard to get the highest grades in languages and other subjects. As a result, I was the best student in my class.

As a student, I liked to play soccer and basketball, and I also enjoyed trying out the various disciplines within track and field. My focus became long-distance running, but I competed occasionally in pole-vaulting. There was a year when I ran five times a week, but my talent was limited; I was occasionally well placed at town-wide events, but never in regional competitions. Still, I have maintained a passion for endurance sports until the present day. When I was around thirty, I met my own personal challenge and finished a few marathons under three hours, and I have completed many long bicycle tours.

It was clear to me early on that after high school I would go to university to study either physics, mathematics, or computer sciences. I decided on physics, as I thought it would combine the real world and mathematics. At that point, I lacked a clear idea of what modern physics was about, but my initial enthusiasm survived when I learnt more about my chosen area.

Before starting university, I received a fellowship from Studienstiftung des Deutschen Volkes (German National Merit Foundation). Part of the fellowship was the opportunity to attend special summer schools. The summer schools took place in the Italian Alps and introduced me to the beauty of mountains and to hiking. It has been one of my favorite activities ever since.

W.K. (second from left) with his parents and siblings in 1980.

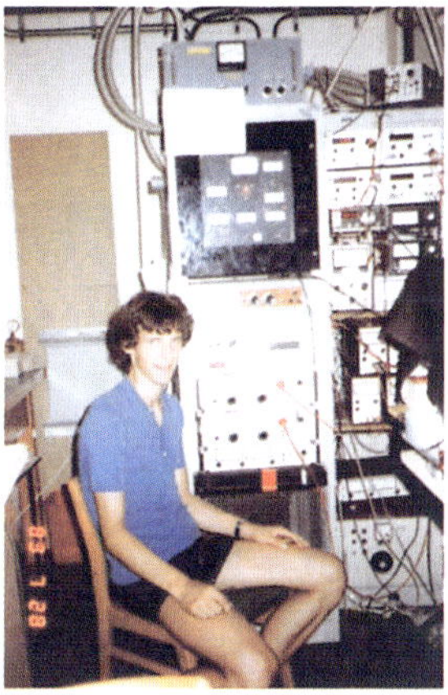

W.K. in July 1983, during the first year of his PhD research. I am sitting in front of the racks controlling an ion beam of triatomic hydrogen.

Most importantly, those schools brought together a select group of highly talented students. The interactions with them reinforced my motivation to set high goals in life for myself.

In 1976, I entered the University of Heidelberg, my home town. (This traditional university was founded in 1386 as the second German university.) It was clear to me from the outset, however, that after passing the pre-diploma (intermediate exam) in two years time I would transfer to another university and leave my parents' house. My choice was the Technical University in Munich because Munich is one of the most attractive towns in Germany and because the Technical University is excellent in many different subfields. At this stage, I wasn't certain what I would specialize in, and had only a diffuse notion of my current field, atomic physics.

Starting an independent life in a new town was a formative experience. The proximity to the Alps was an invitation to go hiking in the summer and skiing in the winter, and I loved both the small and big theaters in Munich and its arts museums. I also became involved with the peace movement and a group working on third world issues.

At the end of my diploma studies, I was very interested in theoretical physics and did my diploma thesis on spin relaxation in disordered materials under the supervision of Prof. Wolfgang Götze. I learnt a great deal from his lucidity in analyzing problems and how he obtained physical insight from mathematical solutions. The thesis project took one year and at the end I found myself at a crossroads. Up to this point I had been focusing on purely academic problems, and now wanted to gain experience with applied physics and how it connected with problems of the real world. Therefore, for my PhD. I chose an experimental project, trace analysis of semiconductors using laser spectroscopy. This project was supervised by Prof. Herbert Walther and Dr. Hartmut Figger at the Max-Planck Institute for Quantum Optics in

W.K. shortly after defending his PhD thesis in May 1986. The mortar board displays a model of the helium hydride molecule with red protons, blue neutrons and yellow electrons. One of the electrons had a blinking light built into it. Such humorous mortar boards are a German academic tradition.

W.K. around 1988 as a postdoc at the Max-Planck Institute for Quantum Optics.

Garching. After one year, it became clear that this project was too difficult to be carried out within the existing infrastructure. Since I didn't want to start over from scratch, I continued my Ph.D. in the same laboratory, and focused on the basic spectroscopy of small molecules. We generated excited neutral molecules by charge transfer to a mass-selected ion beam. This gave us much higher selectivity in observing certain molecules than the usual discharges, and we obtained almost pure fluorescence spectra of triatomic hydrogen.

Towards the end of my PhD, I applied the same method to helium hydride and observed the first discrete spectra of this molecule confirming its existence. Helium hydride is the simplest heteronuclear molecule (besides HD), yet its spectrum had not been observed. I remember my excitement when I produced helium hydride for the first time and rotated the grating of the monochromator used to record its spectrum, and there was light almost everywhere in the visible and near infrared spectral regions. In the next few months, I would decipher some of those spectra and obtain first values for the bond lengths and vibrational frequencies of this molecule. It was exciting to determine the basic properties of a new molecule, like in the old days when molecular spectroscopy was established. After earning my PhD, I stayed at the Max-Planck Institute as a postdoc, working on laser excitation of Rydberg states of triatomic hydrogen and helium hydride. I also succeeded in analyzing all the emission spectra of helium hydride, which I had discovered during my Ph.D. The analysis of the spectra was complicated because the rotation of the molecule leads to a break-down of the Born-Oppenheimer approxima-

W.K. with his PhD and postdoctoral advisors.

tion known as L-uncoupling. As a result, different electronic states are mixed. In addition, the spectrum was perturbed by interactions between s, p, and d states. Several electronic states and their perturbations had to be simulated together, before the calculated spectra started to show some similarity with the data. I regard the solution of this puzzle as one of the most challenging pieces of work I have done.

Even before finishing my PhD I already knew that I would not stay in molecular spectroscopy. I either wanted to work in a more fundamental area of physics, or focus on an area which was directly related to the needs of society. Another option was going into industry, and I had several interviews and job offers. In the end, I decided to pursue applied research in a university setting, maintaining at least some of the freedom of academic research. I joined the group of Prof. Jürgen Wolfrum at the University of Heidelberg. There, I worked in physical chemistry, focusing on combustion diagnostics with lasers. Molecular spectroscopy, in which I was an expert by this time, was used to measure temperature and molecular concentrations in a flame. One of my main projects was carried out in collaboration with the Volkswagen company. We had to transport a truckload of lasers and equipment to an engine test stand at the plant and encountered problems such as soot on the windows of the transparent engine and optics dripping with oil coming from a Diesel engine.

Changing fields was a crucial experience for me. Amazed to see how much of what I had learnt before could be applied within the new field, I realized that general skills are much more important than specific knowledge. I

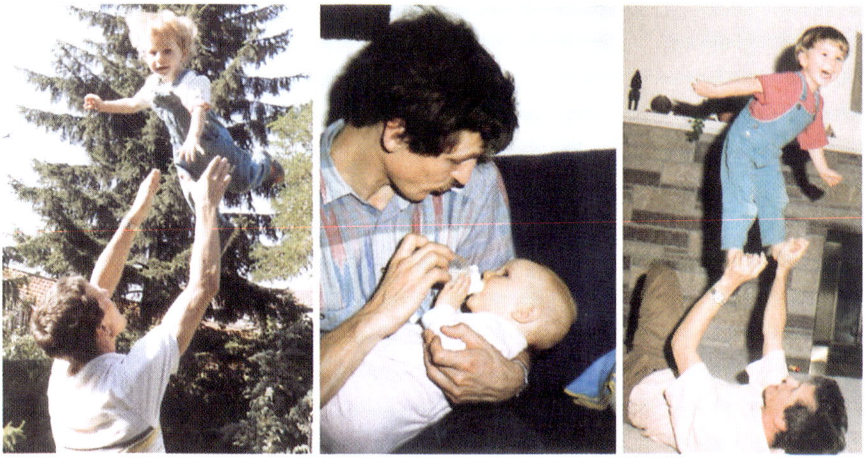

Jonas * 1986 | Johanna * 1988 | Holger * 1992

W.K. with his children.

thought it would take a long time before I became productive in my new environment, but within months, graduate students who had been working in this area for much longer came and sought my advice and leadership. This experience gave me the self-confidence to venture into new areas, and provided the impetus for my later decision to come to the United States and start once again in a new field.

When your work is directly related to cleaner and more efficient combustion, you can easily convince non-scientists of the relevance of your research. I enjoyed this, as well as the interactions with industry and engineers. However, I began to miss something, the quest for pure knowledge and the pursuit of goals which are only vaguely defined and change as the research progresses. I thus realized that my place would be in basic research.

At 32 years, I decided to change fields once more. I wanted to switch to an area of fundamental physics where I could apply some of my knowledge in optics and spectroscopy and thus identified the field of cold atoms as the most promising area. At this time, my assessment was that the field of laser cooling and trapping had reached its peak, but there was still enough to be done. I didn't anticipate that the best was still to come.

Leaving a long-term position in Germany and taking a short-term postdoc position in the U.S. was a risk for myself and my family. However, the time in Heidelberg in combustion research had helped me to discover what I really wanted to do, and also strengthened my self-confidence. I was willing to take risks which I wouldn't have taken a few years earlier. By talking to people and browsing through conference proceedings, I identified the leading groups in the field and sent out applications. I was pleased that I received two offers, despite my lack of experience with cold atoms. In the spring of 1990, I joined Dave Pritchard's group at MIT.

During the first year at MIT, I was supported by a fellowship of DAAD (Deutscher Akademischer Auslandsdienst). It is a great tradition in Germany to support scientific study abroad, but unfortunately such a tradition does not exist in the U.S. Going abroad means more than just immersing yourself in a new culture. It also means that you free yourself from your previous environment and have the opportunity to change and redefine yourself. As a foreigner in a new area of research, I didn't feel bound to a certain tradition and could develop my own personal style – in lab work, giving talks, and discussions within the group. At MIT, where half of the graduate students are foreigners, there is no prejudice, but rather a tolerance and appreciation for unconventional ideas and styles of work.

I also found a unique atmosphere in Dave Pritchard's group. Until then, I had worked in two rather large German groups. Dave's group was smaller, the interactions with him and within the group were very informal; and exciting science was pursued in an atmosphere of comradeship. Dave's knowledge of the field was enormous. During discussions, he could answer almost any question that came up, or immediately make an estimate whether a phenomenon was observable or not. Initially, I felt both intimidated and challenged by his scientific prowess, but soon we became more equal partners. This was the beginning of a wonderful collaboration that continues until the present day. Some account of it is given in the written version of my Nobel lecture.

Towards the end of my PhD studies, in 1986, I married Gabriele Sauer, whom I had known since my high school years. We had three wonderful children, Jonas, born in 1986; Johanna in 1988; and Holger in 1992, who continue to surprise me with their developing talents and personalities; they enrich my life every day. My wife and I were very different and grew apart over the years. In 2001, we separated, two months before it was announced that I was awarded the Nobel prize. Despite some difficulties, the last year has been full of joy, and the Nobel ceremonies have added extra glamour, bringing together my family, friends and colleagues. It is those people to whom I am most grateful, and they have instilled in me a passion for life and a critical, but always optimistic perspective for the future.

WHEN ATOMS BEHAVE AS WAVES: BOSE-EINSTEIN CONDENSATION AND THE ATOM LASER

Nobel Lecture, December 8, 2001

by

WOLFGANG KETTERLE*

Department of Physics, MIT-Harvard Center for Ultracold Atoms, and Research Laboratory of Electronics, Massachusetts Institute of Technology, Cambridge, Massachusetts, 02139, USA.

INTRODUCTION

The lure of lower temperatures has attracted physicists for the past century, and with each advance towards absolute zero, new and rich physics has emerged. Laypeople may wonder why "freezing cold" is not cold enough. But imagine how many aspects of nature we would miss if we lived on the surface of the sun. Without inventing refrigerators, we would only know gaseous matter and never observe liquids or solids, and miss the beauty of snowflakes. Cooling to normal earthly temperatures reveals these dramatically different states of matter, but this is only the beginning: many more states appear with further cooling. The approach into the kelvin range was rewarded with the discovery of superconductivity in 1911 and of superfluidity in helium-4 in 1938. Cooling into the millikelvin regime revealed the superfluidity of helium-3 in 1972. The advent of laser cooling in the 1980s opened up a new approach to ultralow temperature physics. Microkelvin samples of dilute atom clouds were generated and used for precision measurements and studies of ultracold collisions. Nanokelvin temperatures were necessary to explore quantum-degenerate gases, such as Bose-Einstein condensates first realized in 1995. Each of these achievements in cooling has been a major advance, and recognized with a Nobel prize.

This paper describes the discovery and study of Bose-Einstein condensates (BEC) in atomic gases from my personal perspective. Since 1995, this field has grown explosively, drawing researchers from the communities of atomic physics, quantum optics, and condensed matter physics. The trapped ultracold vapor has emerged as a new quantum system that is unique in the precision and flexibility with which it can be controlled and manipulated. At least thirty groups have now created condensates, and the publication rate on Bose-Einstein condensation has soared following the discovery of the gaseous condensates in 1995 (see Fig. 1).

* URL: http://cua.mit.edu/ketterle_group/

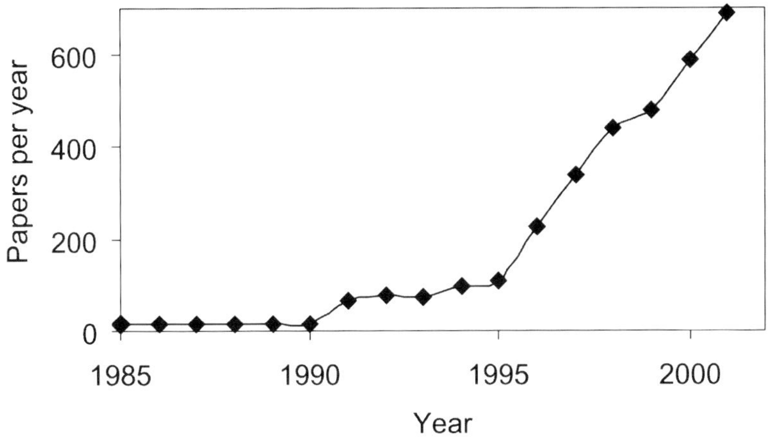

Figure 1. Annual number of published papers, which have the words "Bose" and "Einstein" in their title, abstracts or keywords. The data were obtained by searching the ISI (Institute for Scientific Information) database.

The phenomenon of Bose-Einstein condensation was predicted long ago, in a 1925 paper by Albert Einstein [1] using a method introduced by Satyendra Nath Bose to derive the black-body spectrum [2]. When a gas of bosonic atoms is cooled below a critical temperature T_c, a large fraction of the atoms condenses in the lowest quantum state. Atoms at temperature T and with mass m can be regarded as quantum-mechanical wavepackets that have a spatial extent on the order of a thermal de Broglie wavelength $\lambda_{dB} = (2\pi\hbar^2/mk_BT)^{1/2}$. The value of λ_{dB} is the position uncertainty associated with the thermal momentum distribution and increases with decreasing temperature. When atoms are cooled to the point where λ_{dB} is comparable to the interatomic separation, the atomic wavepackets "overlap" and the gas starts to become a "quantum soup" of indistinguishable particles. Bosonic atoms undergo a quantum-mechanical phase transition and form a Bose-Einstein condensate (Fig. 2), a cloud of atoms all occupying the same quantum mechanical state at a precise temperature (which, for an ideal gas, is related to the peak atomic density n by $n\lambda_{dB}^3 = 2.612$). If the atoms are fermions, cooling gradually brings the gas closer to being a "Fermi sea" in which exactly one atom occupies each low-energy state.

Creating a BEC is thus simple in principle: make a gas extremely cold until the atomic wave packets start to overlap! However, in most cases quantum degeneracy would simply be pre-empted by the more familiar transitions to a liquid or solid. This more conventional condensation into a liquid and solid can only be avoided at extremely low densities, about a hundred thousandth the density of normal air. Under those conditions, the formation time of molecules or clusters by three-body collisions (which is proportional to the inverse density squared) is stretched to seconds or minutes. Since the rate of binary elastic collisions drops only proportional to the density, these collisions are much more frequent. Therefore, thermal equilibrium of the translational degree of freedom of the atomic gas is reached much faster than chemical equilibrium, and quantum degeneracy can be achieved in an effectively metastable gas phase. However, such ultralow density lowers the temperature requirement for quantum degeneracy into the nano- to microkelvin range.

The achievement of Bose-Einstein condensation required first the identification of an atomic system which would stay gaseous all the way to the BEC transition, and second, the development of cooling and trapping techniques to reach the required regime of temperature and density. Even around 1990, it was not certain that nature would provide us with such a system. Indeed, many people doubted that BEC could ever be achieved, and it was regarded as an elusive goal. Many believed that pursuing BEC would result in new and interesting physics, but whenever one would come close, some new phenomenon or technical limitation would show up. A news article in 1994 quoted Steve Chu: "I am betting on nature to hide Bose condensation from us. The last 15 years she's been doing a great job" [3].

In brief, the conditions for BEC in alkali gases are reached by combining two cooling methods. Laser cooling is used to precool the gas. The principle of laser cooling is that scattered photons are on average blue-shifted with respect to the incident laser beam. As a result, the scattered light carries away more energy than has been absorbed by the atoms, resulting in net cooling. Blue-shifts are caused by Doppler shifts or ac Stark shifts. The different laser cooling schemes are described in the 1997 Nobel lectures in physics [4–6]. After the precooling, the atoms are cold enough to be confined in a magnetic trap. Wall-free confinement is necessary, otherwise the atoms would stick to the surface of the container. It is noteworthy that similar magnetic confinement is also used for plasmas which are too hot for any material container. After magnetically trapping the atoms, forced evaporative cooling is applied as the second cooling stage [7–9]. In this scheme, the trap depth is reduced, allowing the most energetic atoms to escape while the remainder rethermalize at steadily lower temperatures. Most BEC experiments reach quantum degeneracy between 500 nK and 2 μK, at densities between 10^{14} and 10^{15} cm^{-3}. The largest condensates are of 100 million atoms for sodium, and a billion for hydrogen; the smallest are just a few hundred atoms. Depending on the magnetic trap, the shape of the condensate is either approximately round, with a diameter of 10 to 50 μm, or cigar-shaped with about 15 μm in diameter and 300 μm in length. The full cooling cycle that produces a condensate may take from a few seconds to as long as several minutes.

After this short overview, I want to provide the historical context for the search for BEC and then describe the developments which led to the observation of BEC in sodium at MIT. Finally, some examples will illustrate the novel physics which has been explored using Bose-Einstein condensates. A more detailed account of the work of my group has been presented in four comprehensive review papers [8, 10–12].

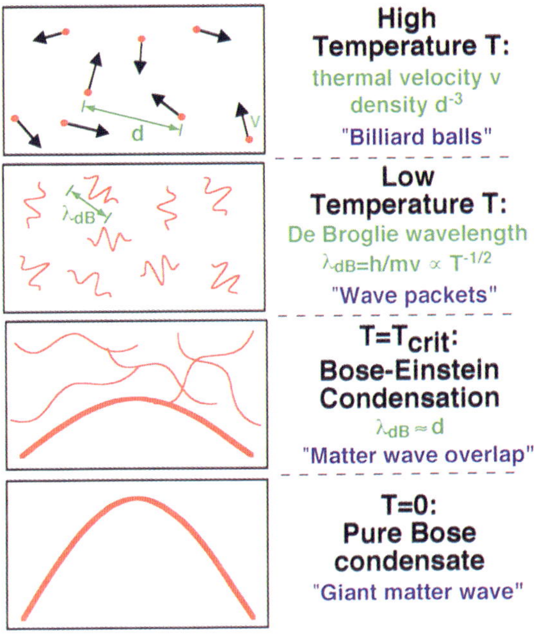

Figure 2. Criterion for Bose-Einstein condensation. At high temperatures, a weakly interacting gas can be treated as a system of "billiard balls." In a simplified quantum description, the atoms can be regarded as wavepackets with an extension of their de Broglie wavelength λ_{dB}. At the BEC transition temperature, λ_{dB} becomes comparable to the distance between atoms, and a Bose condensate forms. As the temperature approaches zero, the thermal cloud disappears, leaving a pure Bose condensate.

BEC AND CONDENSED-MATTER PHYSICS

Bose-Einstein condensation is one of the most intriguing phenomena predicted by quantum statistical mechanics. The history of the theory of BEC is very interesting, and is nicely described in the biographies of Einstein [13] and London [14] and reviewed by Griffin [15]. For instance, Einstein made his predictions before quantum theory had been fully developed, and before the differences between bosons and fermions had been revealed [16]. After Einstein, important contributions were made by, most notably, London, Landau, Tisza, Bogoliubov, Penrose, Onsager, Feynman, Lee, Yang, Huang, Beliaev and Pitaevskii. An important issue has always been the relationship between BEC and superfluidity in liquid helium, an issue that was highly controversial between London and Landau (see ref. [14]). Works by Bogoliubov, Beliaev, Griffin and others showed that Bose-Einstein condensation gives the microscopic picture behind Landau's "quantum hydrodynamics." BEC is closely related to superconductivity, which can be described as being due to Bose-Einstein condensation of Cooper pairs. Thus Bose-Einstein condensation is at the heart of several macroscopic quantum phenomena.

BEC is unique in that it is a purely quantum-statistical phase transition, i.e., it occurs even in the absence of interactions. Einstein described the transition as condensation "without attractive forces" [16]. This makes BEC an important paradigm of statistical mechanics, which has been discussed in a variety of contexts in condensed-matter, nuclear, particle and astrophysics [17]. On the other hand, real-life particles will always interact, and even the weakly-interacting Bose gas behaves qualitatively differently from the ideal Bose gas

[18]. It was believed for quite some time that interactions would always lead to "ordinary" condensation (into a solid) before Bose-Einstein condensation would happen. Liquid helium was the only counter-example, where the light mass and concomitant large zero-point kinetic energy prevents solidification even at zero kelvin. Erwin Schrödinger wrote in 1952 in a textbook on thermodynamics about BEC: "The densities are so high and the temperatures so low – those required to exhibit a noticeable departure [from classical statistics] – that the van der Waals corrections are bound to coalesce with the possible effects of degeneration, and there is little prospect of ever being able to separate the two kinds of effect" [19]. What he didn't consider were dilute systems in a metastable gaseous phase!

The quest to realize BEC in a dilute, weakly interacting gas was pursued in at least three different directions: liquid helium, excitons and atomic gases. Experimental [20, 21] and theoretical work [22] showed that the onset of superfluidity for liquid helium in Vycor has features of dilute-gas Bose-Einstein condensation. At sufficiently low coverage, the helium adsorbed on the porous sponge-like glass behaved like a dilute three-dimensional gas. However, the interpretation of these results is not unambiguous [23].

Excitons, which consist of weakly-bound electron-hole pairs, are composite bosons. The physics of excitons in semiconductors is very rich and includes the formation of an electron-hole liquid and biexcitons. As nicely discussed in refs. [24, 25], there are systems where excitons form a weakly interacting gas. However, the initial evidence for Bose-Einstein condensation in Cu_2O [26] was retracted [27]. Recent work in coupled quantum-well structures is very promising [28]. When excitons strongly interact with light in a cavity, they form polaritons. In such polariton systems, stimulated scattering and non-equilibrium condensates have been observed recently [29–31].

SPIN-POLARIZED HYDROGEN

Dilute atomic gases are distinguished from the condensed-matter systems discussed above by the absence of strong interactions. Interactions at the density of a liquid or a solid considerably modify and complicate the nature of the phase transition. Hecht[32], and Stwalley and Nosanow [33] used the quantum theory of corresponding states to conclude that spin-polarized hydrogen would remain gaseous down to zero temperature and should be a good candidate to realize Bose-Einstein condensation in a dilute atomic gas. These suggestions triggered several experimental efforts, most notably by Silvera and Walraven in Amsterdam, by Greytak and Kleppner at MIT, and by others at Moscow, Turku, British Columbia, Cornell, Harvard, and Kyoto. The stabilization of a spin-polarized hydrogen gas [34, 35] created great excitement about the prospects of exploring quantum-degenerate gases. Experiments were first done by filling cryogenic cells with the spin-polarized gas and by compressing it, and since 1985, by magnetic trapping and evaporative cooling. BEC was finally accomplished in 1998 by Kleppner, Greytak and collaborators [36]. See refs. [9, 37–39] and in particular ref. [40] for a full account of the pursuit of Bose-Einstein condensation in atomic hydrogen. Evidence for a phase transition in two dimensions was reported in 1998 [41].

The work in alkali atoms is based on the work in spin-polarized hydrogen in several respects:

- Studies of spin-polarized hydrogen showed that systems can remain in a metastable gaseous state close to BEC conditions. The challenge was then to find the window in density and temperature where this metastability is sufficient to realize BEC.
- Many aspects of BEC in an inhomogeneous potential [42–44], and the theory of cold collision processes (see e.g. [45]) developed in the 1980s for hydrogen could be applied directly to the alkali systems.
- The technique of evaporative cooling was developed first for hydrogen [7, 46] and then used for alkali atoms.

LASER COOLING

Laser cooling opened a new route to ultralow temperature physics. Laser cooling experiments, with room temperature vacuum chambers and easy optical access, look very different from cryogenic cells with multi-layer thermal shielding around them. Also, the number of atomic species that can be studied at ultralow temperatures was greatly extended from helium and hydrogen to all of the alkali atoms, metastable rare gases, several earth-alkali atoms, and others (the list of laser-cooled atomic species is still growing). A full account of the relevant laser cooling techniques and their development is given in refs. [47–49] and in the 1997 Nobel lectures of Chu, Cohen-Tannoudji and Phillips [4–6].

Some papers and proposals written in the early and mid 1980s, before and during the developments of the basic cooling and trapping techniques, listed quantum degeneracy in a gas as a visionary goal for this new emerging field [50–52]. However, major limitations of laser cooling and trapping were soon identified. Although there is no fundamental low temperature limit, the final temperature provided by polarization gradient cooling – about ten times the recoil energy – was regarded as a practical limit. Sub-recoil laser cooling techniques, especially in three dimensions, were harder to implement, and required long cooling times. The number and density of atoms were limited by inelastic, light-induced collisions (leading to trap loss [53, 54]) and by absorption of scattered laser light [55], which results in an outward radiation pressure (weakening the trapping potential and limiting the density). Furthermore, since the lowest temperatures could not be achieved at the highest densities [56–58], most trapping and cooling techniques reached a maximum phase-space density of around $n\lambda_{dB}^3 = 10^{-5}$; and a value of 2.612 is needed for BEC. This was the situation when the author joined the field of cold atoms in 1990. It was only more recently that major increases in phase-space density were achieved by laser cooling [59–61], but so far laser cooling by itself has not been able to reach BEC.

THE EFFORT AT MIT 1990–1996

Improving laser cooling

When I teamed up with Dave Pritchard at MIT in 1990 as a postdoc, the initial goal was to build an intense source of cold atoms to study cold collisions and pure long-range molecules. However, Dave and I frequently talked about the limitations in density and temperature of the current techniques and tried to develop ideas on how to get around them. One limitation of magnetic traps is that they can hold atoms only in weak-field seeking hyperfine states. Therefore, a collision between two trapped atoms can lead to a spinflip, and the Zeeman energy is converted into kinetic energy (dipolar relaxation). This process has been a major limitation to the experiments in atomic hydrogen.

First, we asked ourselves if the inclusion of electric and gravitational fields would allow the stable confinement of atoms in their lowest hyperfine states – but the answer was negative [62]. One loophole was time-dependent magnetic fields, and building on an earlier proposal [63], I designed an experiment to confine sodium atoms with ac magnetic fields which looked feasible. However, we learnt that Eric Cornell at Boulder had developed a similar idea and experimentally implemented it [64] – so we left the idea on the drawing board. It wasn't the last time that Eric and I would develop similar ideas independently and almost simultaneously!

Trapping atoms in the lowest hyperfine state was not necessary to accomplish BEC. Already in 1986, Pritchard correctly estimated the rate constants of elastic and inelastic collisions for alkali atoms [52]. From these estimates one could easily predict that for alkali atoms, in contrast to hydrogen, the so-called good collisions (elastic collisions necessary for the evaporation process) would clearly dominate over the so-called bad collisions (inelastic two- and three-body collisions); therefore, evaporative cooling in alkalis would probably not be limited by intrinsic loss and heating processes. However, there was pessimism [65] and skepticism, and the above-mentioned experimental [64] and theoretical [62] work on traps for strong-field seeking atoms has to be seen in this context.

In those years, there were some suggestions that time-dependent potentials could lead to substantial cooling, but we showed that this was not possible [66]. Real cooling needs an open system which allows entropy to be removed from the system – in laser cooling in the form of scattered photons, in evaporative cooling in the form of discarded atoms. Dave and I brainstormed about novel laser cooling schemes. In 1991, at the Varenna summer school, Dave presented a new three-level cooling scheme [67]. Inspired by these ideas, I developed a scheme using Raman transitions. Replacing the six laser beams in optical molasses by counterpropagating beams driving the Doppler-sensitive Raman transition, we hoped to realize Doppler molasses with a linewidth that was proportional to the optical pumping rate, and therefore adjustable. We had started setting up radio-frequency (rf) electronics and magnetic shields for Raman cooling when we heard that Mark Kasevich and Steve Chu were working on Raman cooling using laser pulses [68]. For this reason, and also because around the same time we had developed the idea for the Dark

SPOT (spontaneous force optical tray; see later in this section), trap, we stopped our work on Raman cooling.

Our experimental work in those years focused first on generating a large flux of slow atoms. In my first months at MIT when I overlapped with Kris Helmerson and Min Xiao, we built a sodium vapor cell magneto-optical trap (MOT). The idea was inspired by the Boulder experiment [69], and our hope was to vastly increase the loading rate by additional frequencies or frequency chirps added to the red side of the D_2 resonance line. The idea failed – we first suspected that nearby hyperfine levels of sodium may have adversely interfered, but it was later shown that it didn't work for cesium either [70] because of the unfavorable duty cycle of the chirp. Still, except for a cryogenic setup which was soon abandoned, it was the first magneto-optical trap built at MIT (Dave Pritchard's earlier work on magneto-optical trapping was carried out at Bell Labs in collaboration with Steve Chu's group). We (Michael Joffe, Alex Martin, Dave Pritchard and myself) then put our efforts on beam slowing, and got distracted from pursuing Zeeman slowing by the idea of isotropic light slowing [71]. In this scheme, atoms are sent through a cavity with diffusely reflecting walls and exposed to an isotropic light field. For red-detuned light the atoms preferentially absorb light from a forward direction and are slowed. The experiment worked very well and it was a lot of fun to do. However, the requirements for laser power and the velocity capture range of this method were inferior to Zeeman slowing, so we decided to build an optimized Zeeman slower.

We adopted the new design by Greg Lafyatis where the magnetic field increases rather than decreases as in a conventional Zeeman slower [72]. We realized that at the magnetic field maximum it would be possible to apply some additional transverse laser cooling to collimate the slow beam. Michael Joffe, a graduate student, wound a solenoid which had radial access for four extra laser beams. The collimation worked [73], but not as well as we had hoped, and we felt that the small gain was not worth the added complexity. Still, even without collimation, our Zeeman slower provided one of the largest slow-atom fluxes reported until then, and soon after we had a magneto-optical trap with a large cloud of sodium atoms. In hindsight, I am amazed at how many different schemes we considered and tried out, but this may have been necessary to distill the best approach.

The 1991 Varenna summer school on laser cooling was memorable to me for several reasons. I had joined the field of cold atoms just a year earlier, and there I met many colleagues for the first time and established long-lasting relationships. I still have vivid memories of one long afternoon where Dave Pritchard and I sat outside the meeting place, which offered a spectacular view on Lake Como, and brainstormed about the big goals of our field and how to approach them. Dave's encouragement was crucial to me and helped to increase my self-confidence in my new field of research. We considered options and strategies on how to combine laser cooling and evaporative cooling, something which had been on our mind for some time.

Following the example of the spin-polarized hydrogen experiment at MIT

[7], evaporation could be done in a magnetic trap using rf induced spin-flips, as suggested by Pritchard and collaborators in 1989 [74]. Magnetic traps and laser cooling had already been used simultaneously in the first experiments on magnetic trapping at NIST [75] and MIT [76], and on Doppler cooling of magnetically trapped atoms at MIT [74, 77]. In 1990, a magnetic trap was loaded from a magneto-optical trap and optical molasses in Boulder [69]. The laser cooling route to BEC was summarized by Monroe, Cornell and Wieman in ref. [78]. So most of the pieces to get to BEC were known in 1990, but there was doubt about whether they would fit together.

Laser cooling works best at low densities where light absorption and light-induced collisions are avoided, whereas evaporative cooling requires a high collision rate and high density. The problem is the much higher cross section for light scattering of ~ 10^{-9} cm^2, while the cross section for elastic scattering of atoms is a thousand times smaller. In hindsight, it would have been sufficient to provide tight magnetic compression after laser cooling and an extremely good vacuum to obtain a lifetime of the sample that is much longer than the time between collisions, as demonstrated at Rice University [79]. However, our assessment was that one major improvement had to be done to laser cooling to bridge the gap in density between the two cooling schemes. Dave and I discussed possibilities on how to circumvent the density-limiting processes in magneto-optical traps. We considered coherent-population trapping schemes where atoms are put into a coherent superposition state which does not absorb the light. We developed some ideas on how atoms near the center of the trap would be pumped into such a dark state, but the numbers were not too promising. A few months later, a simple idea emerged. If the so-called repumping beam of the magneto-optical trap would have a shadow in the center, atoms would stay there in the lower hyperfine state and not absorb the trapping light, which is near-resonant for atoms in the upper hyperfine state. In a MOT, the density is limited by losses due to excited-state collisions and by multiple scattering of light, which results in an effective repulsive force between atoms. When atoms are kept in the dark, the trapping force decreases by a factor which is proportional to the probability of the atoms to be in the resonant hyperfine state. However, the repulsive force requires both atoms to be resonant with the light and decreases with the square of this factor. Therefore, there is net gain in confinement by keeping atoms in the dark. Of course, there is a limit to how far you can push this concept, which is reached when the size of the cloud is no longer determined by the balance of trapping and repulsive forces, but by the finite temperature of the cloud.

The gain in density of this scheme, called Dark SPOT, over the standard MOT is bigger when the number of trapped atoms is large. So in 1992, we tweaked up the MOT to a huge size before we implemented the idea. It worked almost immediately, and we got very excited about the dark shadows cast by the trapped atoms when they were illuminated by a probe beam. We inferred that the probe light had been attenuated by a factor of more than e^{-100} [80]. This implied that we had created a cloud of cold atoms with an unprecedented combination of number and density.

Combining laser cooling and evaporative cooling

The following weeks and months were quite dramatic. What should we do next? Dave Pritchard had planned to use this trap as an excellent starting point for the study of cold collisions and photoassociation – and indeed other groups had major successes along these lines [81, 82]. But there was also the exciting prospect of combining laser cooling with evaporative cooling. We estimated the elastic collision rate in the Dark SPOT trap to be around 100 Hz [80] which appeared to be more than sufficient to start runaway evaporation in a magnetic trap. After some discussions, the whole group decided to go for the more ambitious and speculative goal of evaporative cooling. It was one of those rare moments where suddenly the whole group's effort gets refocused. Even before we wrote the paper on the Dark SPOT trap, we placed orders for essential components to upgrade our experiment to ultra-high vacuum and to magnetic trapping. All resources of the lab were now directed towards the evaporative cooling of sodium. The Dark SPOT trap was a huge improvement towards combining high atom number and high density in laser cooling. It turned out to be crucial to the BEC work both at Boulder [83] and at MIT [84] and seems to be still necessary in all current BEC experiments with sodium, but not for rubidium.

The next step was the design of a tightly confining magnetic trap. We decided to use the spherical quadrupole trap, which simply consists of two opposing coils – this design was used in the first demonstration of magnetic trapping [75]. We knew that this trap would ultimately be limited by Majorana flops in the center of the trap where the magnetic field is zero. Near zero magnetic field, the atomic spin doesn't precess fast enough to follow the changing direction of the magnetic field—the result is a transition to another Zeeman sublevel which is untrapped leading to trap loss. We estimated the Majorana flop rate, but there was some uncertainty about the numerical prefactor. Still, it seemed that Majorana flops would only become critical after the cloud had shrunk due to evaporative cooling, so they shouldn't get in the way of demonstrating the combination of laser cooling and evaporative cooling. After Michael Joffe presented our approach with the quadrupole trap at the QELS meeting in 1993, Eric Cornell informed me that he had independently arrived at the same conclusion. In 1993, my group reported at the OSA meeting in Toronto the transfer of atoms from the Dark SPOT trap into a magnetic trap, and the effects of truncation of the cloud using rf induced spinflips [85].

At about this time, I joined the MIT faculty as assistant professor. Dave Pritchard made the unprecedented offer that if I stayed at MIT he would hand over to me the running lab, including two grants. To make sure that I would receive the full credit for the work towards BEC, he decided not to stay involved in a field he had pioneered and gave me full responsibility and independence. Dave told me that he wanted to focus on his other two experiments, the single-ion mass measurement and the atom interferometry, although what he gave up was his "hottest" research activity. Even now, I am moved by his generosity and unusual mentorship. The two graduate students on the project, Ken Davis and Marc-Oliver Mewes, who had started their PhDs in 1991 and 1992, respectively, deliberated whether they should stay with Dave Pritchard and work on one of his other experiments, or to continue their work on BEC in a newly-formed group headed by a largely unknown assistant professor. They both opted for the latter and we could pursue our efforts without delay, along with Michael Andrews, who joined the group in the summer of 1993.

For a few months we got distracted from our goal of evaporative cooling. Our optical molasses temperatures were higher than those reported by the NIST group [86], and we felt that we had to learn the state-of-the-art before we could advance to even lower temperatures. We suspected that the higher density of atoms played a role, but we had to improve our technique of temperature measurements. Our goal was to characterize the interplay of parameters in "dark" molasses where most of the atoms are pumped into the dark hyperfine state. It was also a good project for the graduate students to hone their skills and develop independence. After a few months we had made some progress, but I became concerned about the delay and the competition from Boulder. We decided to drop the project and resume our work on evaporative cooling. Up to the present day, we have never implemented accurate diagnostics for the temperature obtained in laser cooling – it was just not important.

In the spring of 1994, we saw first evidence for an increase in phase-space density by evaporative cooling. We reported these results at an invited talk at International Quantum Electronics Conference (IQEC) in May 1994. At the same meeting, the Boulder group reported similar results and the limitations due to the Majorana flops as the temperature was reduced. It was clear that the next step was an improvement of the magnetic trap, to trap atoms at a finite bias field which would suppress the Majorana flops. During the meeting, I came up with the idea of plugging the hole with a focused laser beam: a blue-detuned laser beam focused onto the zero-magnetic field point would exert repulsive dipole forces onto the atoms and keep them away from this region (Fig. 3). This idea seemed so obvious to me that I expected the Boulder group to come up with something similar. It was only at the next conference (ICAP 1994) in Boulder [87], when I presented our approach, that I learnt about Eric Cornell's idea of suppressing Majorana flops with a rapidly rotating magnetic field – the so-called TOP trap [88]. However, we didn't implement the optical plug immediately. We wanted first to document our observation of evaporative cooling. We realized that our fluorescence diagnostics was inadequate and implemented absorption imaging which is now the standard technique for observing Bose-Einstein condensation. In those days, we focused on direct imaging of the trapped cloud (without ballistic expansion), and Michael Andrews and Marc-Oliver Mewes developed a sophisticated computer code to simulate absorption images in inhomogeneous magnetic fields. We thought that this would be a useful tool, but we rapidly advanced to much lower temperatures where the inhomogeneous Zeeman shifts were smaller than the linewidth, and never needed the code again after our first paper on evaporative cooling [89].

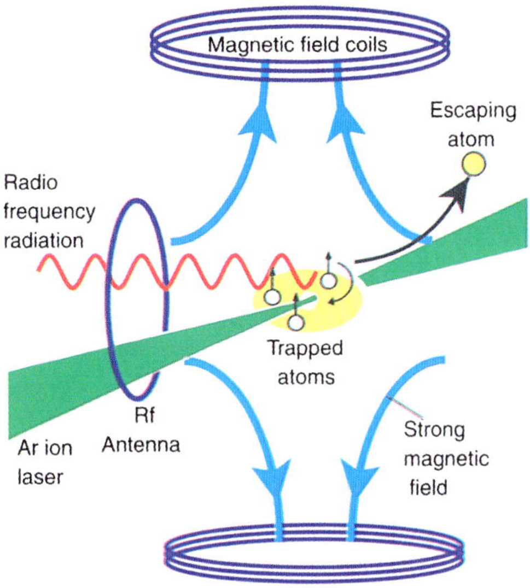

Figure 3. Experimental setup for cooling atoms to Bose-Einstein condensation. Sodium atoms are trapped by a strong magnetic field, generated by two coils. In the center, the magnetic field vanishes, which allows the atoms to spin-flip and escape. Therefore, the atoms are kept away from the center of the trap by a strong (3.5 W) argon ion laser beam ("optical plug"), which exerts a repulsive force on the atoms. Evaporative cooling is controlled by radio-frequency radiation from an antenna. The rf selectively flips the spins of the most energetic atoms. The remaining atoms rethermalize (at a lower temperature) by collisions among themselves. Evaporative cooling is forced by lowering the rf frequency.

In late 1994, we had a "core meltdown". The magnetic trap was switched on without cooling water, and the silver solder joints of the coils melted. Since in those days the magnetic coils were mounted inside the vacuum chamber, we had a catastrophic loss of vacuum and major parts of our setup had to be disassembled. I will never forget the sight of coils dripping with water behind a UHV viewport. This happened just a few hours before MIT's president, Charles Vest, visited our lab to get first-hand information on some of the research done on campus. He still remembers this event. We had lost weeks or months of work in a very competitive situation. I was despondent and suggested to the group that we go out for a beer and then figure out what to do, but the students immediately pulled out the wrenches and started the repair. I was moved to see their dedication and strength, even at this difficult time. We replaced the magnetic trap by a much sturdier one. This turned out to be crucial for the implementation of the plugged trap where the precise alignment of a laser beam relative to the magnetic-field center was important. So in hindsight the disaster may not have caused a major delay.

In early 1995, I had to tell my three graduate students that we were rapidly using up start-up money and urgently needed one of our two pending proposals approved. Otherwise we would not be able to continue spending money in the way we had done until then and would slow down. Fortunately, in April 1995, the NSF informed me that my proposal was funded. It is interesting to look at some of the reviewers comments now, seven years later: "It seems that vast improvements are required [in order to reach BEC] ...the current techniques are so far from striking range for BEC that it is not yet possible to make...an assessment..."; "The scientific payoffs, other than the importance of producing a BEC itself, are unclear". And a third reviewer: "...there have been few specific (or realistic) proposals of interesting experiments that could be done with a condensate." Despite the skepticism, all reviewers concluded that the proposed "experiments are valuable and worth pursuing". After we received the funding decision, the whole group celebrated with dinner, and a fourth graduate student (Dallin Durfee), who had expressed his interest already months earlier, could finally be supported.

In late December 1994, our paper on evaporative cooling was submitted, and we were free to focus on plugging the hole. We had to learn how to align a powerful argon ion laser beam and image it through many attenuators without major distortions. When the plug was aligned, the result was spectacular (Fig. 4). We could immediately cool down to lower temperatures and keep many more atoms. During evaporation, the cloud became so cold and small that we couldn't resolve it any more. The highest phase space density measured was a factor of thirty below BEC, but we may have been even closer. We had only a few runs of the experiment before we ran into severe vacuum problems. We focused initially on spatial imaging and became limited by resolution, whereas ballistic expansion and time-of-flight imaging would not have suffered from this limitation. We also thought that BEC would be accomplished at lower densities and in larger clouds, so we worked on adiabatic decompression and ran into problems with the zero of the magnetic field moving away from the plug.

In those months, we were plagued by vacuum problems. The coils inside the vacuum showed some strange outgassing behavior and the vacuum slowly deteriorated. We went through several bakeouts of the ultrahigh-vacuum chamber in the spring and summer of 1995. Furthermore, Ken Davis had to write his PhD thesis and stopped working in the lab. It is interesting to recall my assessment of the field in those months; I didn't realize that BEC was just

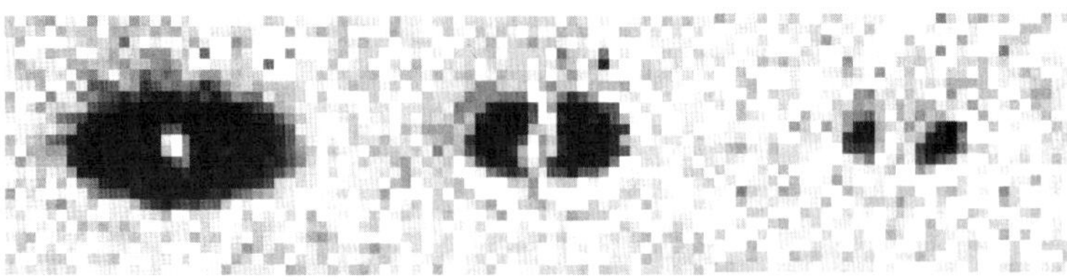

Figure 4. Absorption images of atom clouds trapped in the optically plugged trap. Cloud (a) is already colder than was attainable without the "plug" (Ar ion laser beam). Cloud (b) shows the break-up of the cloud into two "pockets" in the two minima of the potential. The size of cloud (c) reached the optical resolution of the imaging system ($\leq 10\ \mu$m) still absorbing 90 % of the probe light. This sets an upper bound on temperature ($\leq 10\ \mu$K) and a lower bound on density (5×10^{12}cm^{-3}).

around the corner. In Tom Greytak's and Dan Kleppner's group the BEC transition was approached to within a factor of 3.5 in temperature in 1991 [90], but it took several more years to advance further. So I prepared for a long haul to cover the last order of magnitude to BEC.

By this time, the group was reinforced by Dan Kurn (now Dan Stamper-Kurn), a graduate student, and Klaasjan van Druten, my first postdoc. After months of working on vacuum and other problems, we were just ready to run the machine again when we heard about the breakthrough in Boulder in June of 1995. We feverishly made several attempts with traps plugged by focused laser beams and light sheets, and tried different strategies of evaporation without success. The clouds disappeared when they were very cold. We conjectured that some jitter of the laser beam was responsible, and when accelerometers indicated vibrations of our vacuum chambers, we immediately decided to eliminate all turbo and mechanical pumps. Unfortunately, when we were exchanging the turbo pump on our oven chamber against an ion pump, we caused a leak in the ultrahigh vacuum part and had to go through another long bake-out. We also implemented a pointing stabilization for the optical plug beam. But when we finally obtained BEC, we realized that it didn't improve the cooling.

These were difficult months for me. The Rice group had cooled lithium to quantum degeneracy [79]. A new subfield of atomic physics was opening up, and I was afraid that our approach with sodium and the plugged trap would not be successful and we would miss the excitement. I considered various strategies. Several people suggested that I adopt the successful TOP trap used at Boulder. But I had already started to study several possible configurations for magnetic confinement. I realized that a highly elongated Ioffe-Pritchard trap with adjustable bias field could provide a good confinement that was equivalent or superior to the TOP trap. Around August 1995, Dan Kurn worked out an optimized configuration, which was the cloverleaf winding pattern [91]. I considered having the whole group work on this new approach, but several in my group wanted to give the plugged trap a few more attempts and at least characterize how far we could approach BEC with our original approach. Fortunately, we followed that suggestion – it is always a good idea to listen to your collaborators.

BEC in sodium

This was the situation on September 29, 1995, when we observed BEC in sodium for the first time. The goal of the run was to measure the lifetime of the trapped atoms and characterize possible heating processes. For our ultrahigh vacuum pressure, rather slow evaporation should have been most efficient, but we found out that faster evaporation worked much better. This was a clear sign for some other loss or heating process, e.g. due to fluctuations in the position of the plug. Around 11:30 p.m., an entry in the lab book states that the lifetime measurements were not reliable, but they indicated lifetimes around ten seconds, enough to continue evaporation. A few minutes later we saw some dark spots in time-of-flight absorption images, but they were quite distorted since the optical plug beam, which we couldn't switch off, pushed atoms apart during the ballistic expansion (Fig. 5). Still, the sudden appearance of dark spots meant groups of atoms with very small relative velocity. For the next few hours, we characterized the appearance of those spots, but then decided that further progress required an acousto-optical modulator to switch off the optical plug. Between 4:00 and 5:30 in the early morning, we installed optics and rf electronics and were finally able to switch off the argon ion laser beam during ballistic expansion. Fifteen minutes later, we observed the bimodal distributions that are now the hallmark of BEC. The lab book of this night captured the excitement of the moment (Fig. 6).

Those first measurements were done by imaging the atoms in the lower hyperfine ($F = 1$) state. For the next run, which took place a few days later, we prepared optical pumping and imaging on the cycling $F = 2$ transition, and obtained a much better signal-to-noise ratio in our images. The occurrence of BEC was very dramatic (Fig. 7). Our animated rendering of the data obtained in that run (done by Dallin Durfee) became well known (see [92]). We had obtained condensates with 500,000 atoms, 200 times more than in Boulder, with a cooling cycle (of only nine seconds) 40 times shorter. Our paper was quickly written and submitted only two weeks after the experiment.

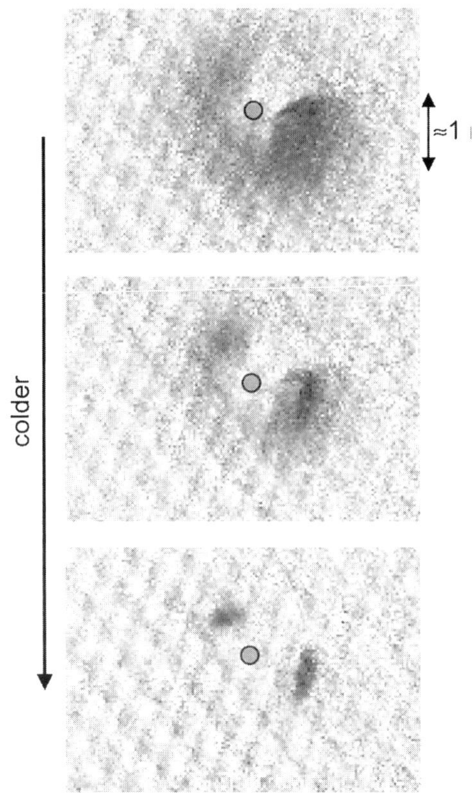

Figure 5. Time-of-flight absorption images of some of the first condensates produced at MIT in the night of September 29, 1995. After the magnetic quadrupole trap was switched off, the atom cloud expanded ballistically. However, since the optical plug (indicated by black circles) could not be turned off at the same time, it distorted the expanding cloud. Still, as the temperature was lowered from top to bottom, a distinctly sharp shadow appeared marking the presence of a condensate.

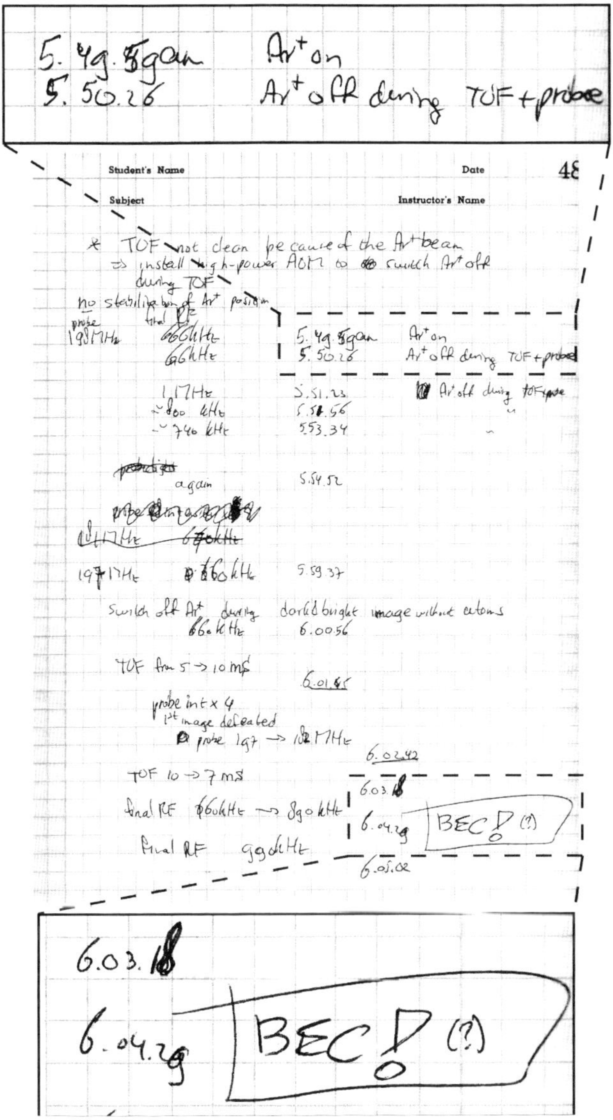

Figure 6. One page of the lab book during the night of September 29, 1995, when BEC was first observed at MIT. The handwriting is by Klaasjan van Druten. At 5:50 a.m., we had installed a new acousto-optical modulator to switch-off the optical plug (Ar ion laser beam). Fifteen minutes later, we had the first definitive evidence for BEC in sodium.

In my wildest dreams I had not assumed that the step from evaporative cooling to BEC would be so fast. Figure 8 shows how dramatic the progress was after laser and evaporative cooling were combined. Within less than two years, the number of alkali atoms in a single quantum state was increased by about twelve orders of magnitude – a true singularity demonstrating that a phase transition was achieved!

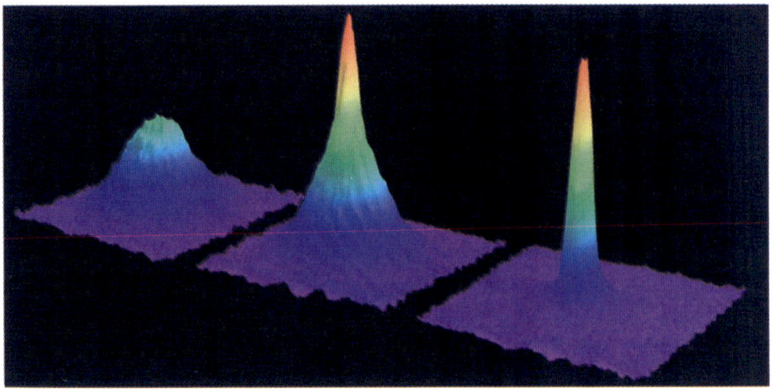

Figure 7. Observation of Bose-Einstein condensation by absorption imaging. Shown is absorption vs. two spatial dimensions. The Bose-Einstein condensate is characterized by its slow expansion observed after 6 ms time-of-flight. The left picture shows an expanding cloud cooled to just above the transition point; middle: just after the condensate appeared; right: after further evaporative cooling has left an almost pure condensate. The total number of atoms at the phase transition is about 7×10^5, the temperature at the transition point is 2 μK.

MIT with its long tradition in atomic physics was a special place to pursue the BEC work. The essential step was the combination of laser cooling and evaporative cooling. My next-door neighbors in Building 26 at MIT have been Dave Pritchard, a pioneer in laser cooling who conceived the magneto-optical trap, and Dan Kleppner, who together with Harald Hess and Tom Greytak conceived and realized evaporative cooling (see Fig. 9). I feel privileged for

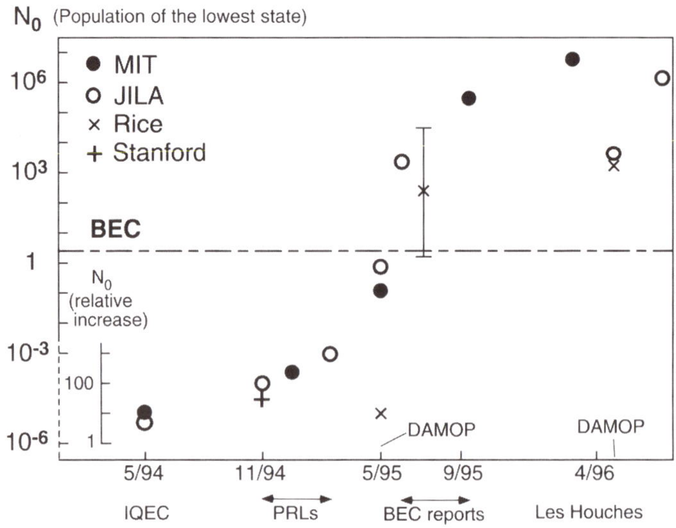

Figure 8. Progress in evaporative cooling of alkali atoms up to 1996. The number of atoms in the lowest quantum state is proportional to the phase-space density, and has to exceed a critical number of 2.612 to achieve Bose-Einstein condensation. For $N_0 < 10^{-3}$, the increase in phase-space density due to evaporation is plotted. For the Rice result of July 1995 see ref. [79] and the erratum [93].

the opportunity to combine their work and take it to the next level. It is hard to overestimate the roles which Dave Pritchard and Dan Kleppner have played for modern atomic physics. The family tree of atomic physicists (Fig. 10) shows some of the remarkable physicists that were trained and inspired by them.

Looking back, it seems that many techniques such as the Dark SPOT, compressed MOT [94], the TOP trap and the optically plugged trap were critical for first demonstrating BEC, but by no means indispensable. This is best illustrated by the experiment at Rice which used only Doppler cooling to load the magnetic trap – a technique which had been developed in the 1980s. The collision rate was slow, but an excellent vacuum made a very slow evaporation process possible [79]. So in hindsight, BEC in alkali gases did not require major innovations in cooling and trapping. It merely required enough optimism to risk a few years in the attempt to combine laser and evaporative cooling. Such an attempt needed a few years of very focused work as it involved the integration of several technologies that were not standard in the field, including ultrahigh vacuum, sensitive CCD cameras and image processing, high-current power supplies for magnetic traps, and flexible computer control of a multi-step cooling and detection process. Figure 11 compares a state-of-the-art laser cooling experiment in 1993 to a BEC experiment in 2001 using the same vacuum apparatus in the same laboratory at MIT. A lot of components have been added, and I continue to be impressed by my collaborators, who now handle experiments far more complex than I did some five years ago.

Figure 9. MIT faculty in ultralow-temperature atomic physics. Dan Kleppner, W.K., Tom Greytak and Dave Pritchard look at the latest sodium BEC apparatus.

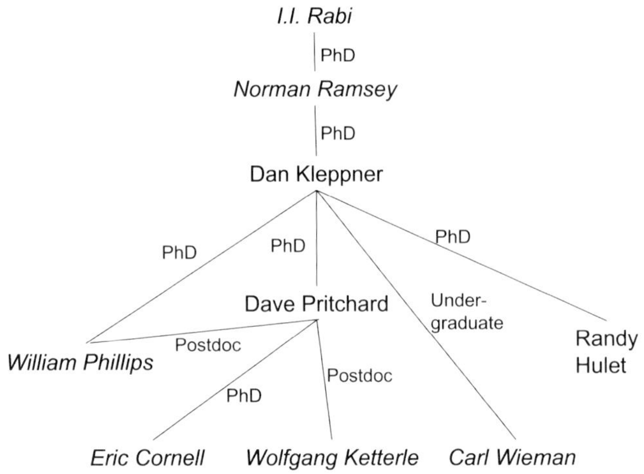

Figure 10. Family tree of atomic physicists. People with name in italics are Nobel laureates.

The cloverleaf trap

After our first observation of BEC, we made the right decision for the wrong reason. We expected many other groups to quickly upgrade their laser cooling experiments to magnetic trapping and evaporative cooling, and to join in during the next few months. Nobody expected that it would take almost two years before the next groups succeeded in reaching BEC (the groups of Dan Heinzen, Lene Hau, Mark Kasevich and Gerhard Rempe followed in 1997). I was concerned that our plugged trap would put us at a disadvantage since the trapping potential strongly depended on the shape and alignment of the laser focus. So we decided to install the cloverleaf trap instead and discontinue our plugged trap after only two experimental BEC "runs".

Since we didn't want to break the vacuum, we installed the new trap in an unfavorable geometry. The magnet coils for the plugged trap were oriented vertically in re-entrant flanges and when we replaced them with cloverleaf coils, the weakly confining axis of the Ioffe-Pritchard trap was vertical. In such a geometry, the gravitational sag would reduce the efficiency of rf induced evaporation since atoms would only evaporate at the bottom of the cloud [8, 95]. But before breaking the vacuum and reorienting the coils, we wanted to see the limitation. Around December 1995, when we were just starting to look at the efficiency of evaporation, we lost the vacuum once again due to a cracked ceramic part in an electric feedthrough and decided to reorient the whole experiment, with the weakly confining axis of the trap now aligned horizontally. Since that time, now more than six years, the machine has been under vacuum. This is in sharp contrast to the conditions in 1995, when we had to open the chamber, pump down and bake out every couple of months. Finally, we had learned from our previous mistakes and developed a very systematic procedure for pump downs and bakeouts.

I still remember the night of March 13, 1996, when the experiment was up and running, and Klaasjan van Druten and I had fine-tuned the bias field of

Figure 11. Comparison of a laser cooling and BEC experiment. The upper photograph shows the author in 1993 working on the Dark SPOT trap. In the following years, this laser cooling experiment was upgraded to a BEC experiment. The lower photograph shows the same apparatus in 2001 after many additional components have been added.

the magnetic trap, so that the switch-over to the new magnetic trap was finally completed. It was already after midnight, too late to start some serious work, when Klaasjan asked half jokingly why don't we just try to get BEC. Without knowing what our temperatures and densities were, without having ever measured the trap frequencies, we played around with the rf sweep that determines the cooling trajectory, and a condensate showed up around 2:10 a.m. We were relieved since we hadn't produced condensates for almost half a year, but also the ease at which we got the condensate in a new trap told us our setup was robust and that we were ready to switch from engineering cooling schemes and traps to the study of the condensate. The cloverleaf trap and other winding patterns for the Ioffe-Pritchard configuration are now used by almost all BEC experiments. Figure 12 shows the experimental setup during those days.

Why hadn't we considered this trap earlier and avoided the detours with the quadrupole trap, Majorana flops and plugging the hole? First, the quadrupole trap was simpler to build, and it allowed us to pursue evaporative

Figure 12. Experimental setup for cooling sodium atoms to Bose-Einstein condensation around 1996. The atoms are trapped and cooled in the center of the UHV chamber. The atomic beam oven and the Zeeman slower are to the left (outside the photo). The cloverleaf magnetic trap was mounted horizontally in re-entrant flanges. Only the leads for the current and water cooling are visible. The diagonal flange above accommodated a BNC feedthrough for radio frequency fields which were used to control the evaporative cooling. The lens and the mirror above the chamber were used to observe the condensate by dispersive or absorption imaging.

cooling faster. Second, we initially favored the quadrupole trap based on an analysis which shows that confinement by a linear potential is much stronger than by the quadratic potential of the Ioffe-Pritchard configuration [10]. However, a very elongated Ioffe-Pritchard trap provides effectively linear confinement in the two radial directions, and it was only in 1995 that I realized that it would be easy to adiabatically deform the round laser-cooled cloud to such an elongated shape.

The next weeks were exciting and dramatic: we implemented dispersive imaging and saw for the first time the condensate in the trap. We could take images non-destructively and recorded two sequential images of the same condensate. After year-long concerns of how fragile and sensitive the condensate would be once created, it was an overwhelming experience to observe the condensate without destroying it. Figure 13 shows a spatial image of a condensate; it was taken in non-destructive dispersive imaging. We first implemented dispersive imaging using the dark-ground technique [96], but soon upgraded to phase-contrast imaging, which was the technique used to record the figure.

In the first week of April 1996, there was a workshop on "Collective effects in ultracold atomic gases" in Les Houches, France, where most of the leading groups were represented. It was the first such meeting after the summer of

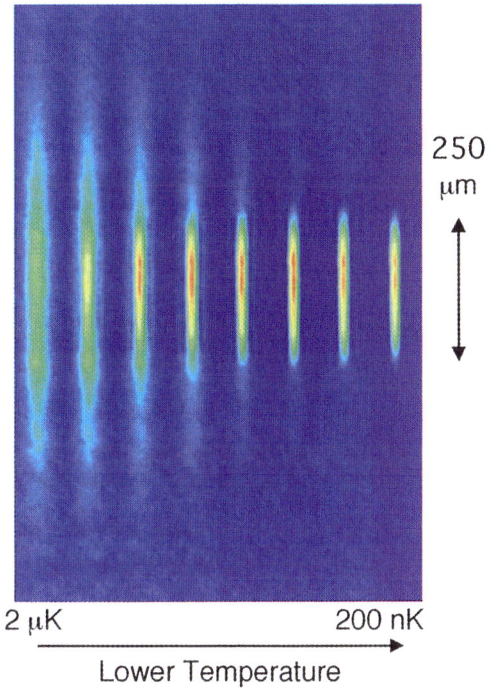

Figure 13. Phase contrast images of trapped Bose gases across the BEC phase transition. At high temperature, above the BEC transition temperature, the density profile of the gas is smooth. As the temperature drops below the BEC phase transition, a high-density core of atoms appears in the center of the distribution. This is the Bose-Einstein condensate. Lowering the temperature further, the condensate number grows and the thermal wings of the distribution become shorter. Finally, the temperature drops to the point where a pure condensate with no discernible thermal fraction remains. Each image shows an equilibrated gas obtained in one complete trapping and cooling cycle. The axial and radial frequencies are about 17 and 230 Hz, respectively.

1995, and it was not without strong emotions that I reported our results. Since no other experimental group had made major progress in BEC over the last few months, it was our work which provided optimism for further rapid developments.

Interference between two condensates

After we got BEC in the cloverleaf trap, both the machine and the group were in overdrive. After years of building and improving, frequent failures and frustration, it was like a phase transition to a situation where almost everything worked. Within three months after getting a condensate in the cloverleaf trap we had written three papers on the new trap and the phase transition [91], on non-destructive imaging [96], and on collective excitations [97]. Klaasjan van Druten left the group, shortly after Christopher Townsend had joined us as a postdoc. As the next major goal, we decided to study the coherence of the condensate. With our optical plug, we had already developed the tool to split a condensate into two halves and hoped to observe their interference, which would be a clear signature of the long-range spatial coherence.

Around the same time, the idea came up to extract atoms from the condensate using rf-induced spinflips – the rf output coupler. Some theorists regarded an output coupler as an open question in the context of the atom laser. I suggested to my group that we could simply pulse on the radio frequency source that was already used during evaporation, and couple atoms out of the condensate by flipping their spin to a non-trapped state (Fig. 14). The experiment worked the first time we tried it (but the quantitative work took awhile [98]). I have never regarded the output coupler as one of our major accomplishments because it was so simple, but it had impact on the community and nobody has ever since regarded outcoupling as a problem!

In July 1996, we had the first results on the rf output coupler, and also saw first fringes when two condensates were separated with a sheet of green light and overlapped in ballistic expansion. I was in Australia for vacation and for the IQEC conference in Sydney. By e-mail and telephone I discussed with my group the new results. The fringes were most pronounced when the condensates were accelerated into each other by removing the light sheet shortly before switching off the magnetic trap. We concluded that some of the fringes may be related to sound and other collective effects that occur when two condensates at fairly high density "touch" each other. I presented those results at the Sydney meeting only to illustrate we were able to do experiments with two condensates, but now we had to sort out what was happening.

It took us four more months until we observed clean interference between two condensates. When two condensates that were initially separated by a distance d interfere and the interference pattern is recorded after a time t of ballistic expansion, then the fringe spacing is the de Broglie wavelength h/mv as-

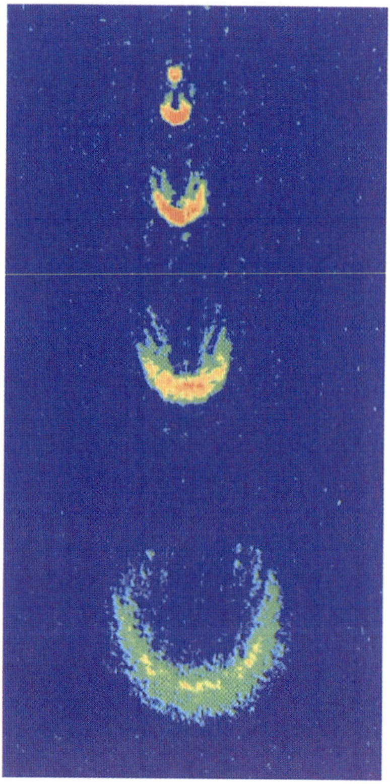

Figure 14. The MIT atom laser operating at 200 Hz. Pulses of coherent sodium atoms are coupled out from a Bose-Einstein condensate confined in a magnetic trap (field of view 2.5 × 5.0 mm^2). Every 5 ms, a short rf pulse transferred a fraction of these atoms into an unconfined quantum state. These atoms were accelerated downward by gravity and spread out due to repulsive interactions. The atom pulses were observed by absorption imaging. Each pulse contained between 10^5 and 10^6 atoms.

sociated with the relative velocity $v = d/t$. For our geometry with two condensates about 100 μm in length, we estimated that we would need at least 60 ms of time-of-flight to observe fringes with a 10 μm period, close to the resolution of our imaging system. Unfortunately, due to gravity, the atoms dropped out of the field of view of our windows after 40 ms. So we tried to gain a longer expansion time in a fountain geometry where we magnetically launched the atoms and observed them when they fell back through the observation region after more than 100 ms [99], but the clouds were distorted. We also tried to compensate gravity by a vertical magnetic field gradient. Some time later I learnt about new calculations by the theory group at the Max-Planck Institute in Garching, showing that the effective separation of two elongated condensates is smaller than their center-of-mass separation [100]. This meant that we could observe interference fringes after only 40 ms, just before the atoms fell out of the observation region. We immediately had a discussion in the group and decided to stop working on fountains and "anti-gravity" and simply let the atoms fall by 8 mm during 40 ms.

We made some ambiguous observations where we saw low-contrast fringes together with some optical interference patterns of the probe light, but the breakthrough came on November 21, 1996, when we observed striking interference patterns (Fig. 15). I still remember the situation late that night when we wondered how could we prove beyond all doubt that these were matter-wave interference patterns and not some form of self-diffraction of a condensate confined by a light sheet and then released. We came up with the idea of eliminating one of the condensates in the last moment by focusing resonant yellow light on it. Whimsically, this laser beam was dubbed the "flame thrower". If the fringes were self-diffraction due to the sharp edge in

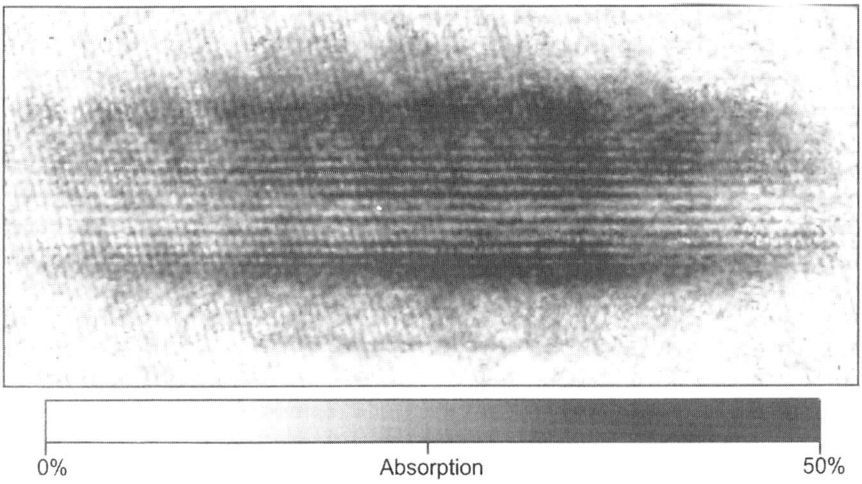

Figure 15. Interference pattern of two expanding condensates observed after 40 ms time-of-flight. The width of the absorption image is 1.1 mm. The interference fringes have a spacing of 15 μm and are strong evidence for the long-range coherence of Bose-Einstein condensates.

the confinement, they would remain; if they were true interference they would vanish. This was like a double-slit experiment in optics where you cover one of the slits. It took a few hours to align the new laser beam, and we verified in phase-contrast imaging that we were able to selectively eliminate one of the two condensates.

We had a switch in our control panel which toggled between condensate elimination on and off. Then we went back and aligned the setup for the observation of interference. When we toggled the switch we had to wait for about half a minute until a new condensate was produced. This was the moment of truth. If the fringes appeared without a second condensate, then nature would have fooled us for the whole night – but they disappeared and an enormous tension disappeared, as well. It was already early the next morning, with people arriving to work. I walked to Dan Kleppner's office and told him there was something he should see. So he shared the moment with us where we toggled the switch on alternating cooling cycles and correspondingly, the interference pattern disappeared and reappeared. Interference between two light beams is quite a sight, but with atoms it is more dramatic. Destructive interference means that atoms plus atoms add up to vacuum!

The evidence for interference was so compelling that we submitted our paper based solely on the data of one experimental run [101]. This run is memorable to me for another reason: it was to be the last time I played a major role in preparing and running an experiment. During the night, I had put in the optics for the "flame thrower". Up to then, I was familiar with every piece of equipment in the lab and never thought this could change quickly, but it was like another phase transition. Hans-Joachim Miesner had just arrived, the first postdoc who stayed for more than a year, and he soon took over much responsibility for organizing the lab. There were more demands on my time to write papers and give talks, the group grew with the addition of two more graduate students (Shin Inouye and Chris Kuklewicz), and we had intensified our efforts to build a second BEC experiment. All this coincided in a few months. After earning my PhD in 1986, I spent eleven more years in the lab during three postdoc positions and as an assistant professor, but now began to play an advisory role.

The papers on the rf output coupler [98] and the interference [101] of two condensates appeared in the same week in January 1997. Together they demonstrated the ability to create multiple pulses of coherent atoms, and have been regarded as the realization of an atom laser. The period starting with the early dreams of pursuing BEC and ending with the observation of the coherence of the condensate was remarkable. It was full of speculation, dreams, unknown physics, failures and successes, passion, excitement, and frustration. This period fused together a team of very different people who had one common denominator: the passion for experimental physics. It was a unique experience for me to work with these outstanding people (Fig. 16).

Figure 16. Team photo. This photo was taken in early 1996 in front of the MIT dome. The bottle of champagne was emptied to celebrate BEC in the cloverleaf trap. Names and photos of other collaborators during the period 1992–1996 have been added.

THE MAGIC OF MATTER WAVES

Many studies of BECs have been performed over the last several years. The progress until 1998 is nicely summarized in the Varenna summer school proceedings [102]. The studies that were most exciting for me displayed macroscopic quantum mechanics, the wavelike properties of matter on a macroscopic scale. These were also phenomena that no ordinary gas would show, and illustrated dramatically that a new form of matter had been created. The interference of two condensates presented above (Fig. 15) is one such example. In the following, I want to discuss the amplification of atoms and the observation of lattices of quantized vortices.

These two examples are representative of the two areas into which research on gaseous BEC can be divided: in the first (which could be labelled "The atomic condensate as a coherent gas" or "Atom lasers"), one would like to have as little interaction as possible – almost like the photons in a laser. The experiments are preferably done at low densities. The Bose-Einstein condensate serves as an intense source of ultracold coherent atoms for experiments in atom optics, in precision studies or for explorations of basic aspects of quantum mechanics. The second area could be labelled "BEC as a new quantum fluid" or "BEC as a many-body system". The focus here is on the interactions between the atoms that are most pronounced at high densities. The coherent amplification of atoms is an example of atom optics with condensates, and the study of vortices addresses the superfluid properties of the gas.

Amplification of atoms in a Bose-Einstein condensate

Since atoms are de Broglie waves, there are many analogies between atoms and light, which consists of electromagnetic waves. This is exploited in the field of atom optics where atoms are reflected, diffracted and interfere using various atom-optical elements [103]. One important question was whether these analogies can be extended to the optical laser, which is based on the amplification of light. When our group demonstrated a rudimentary atom laser in 1997 we had solved the problem of outcoupling (or extracting) atoms from the BEC and of verifying their coherence. The atomic amplification process happened during the formation of the Bose-Einstein condensate [104], which is quite different from the way light is amplified in passing through an active medium. It was only in 1999 that our group managed to observe the amplification of atoms passing through another cloud of atoms serving as the active medium [105] (simultaneously with the group in Tokyo [106]).

Amplifying atoms is more subtle than amplifying electromagnetic waves because atoms can only change their quantum state and cannot be created. Therefore, even if one could amplify gold atoms, one would not realize the dreams of medieval alchemy. An atom amplifier converts atoms from the active medium into an atomic wave that is exactly in the same quantum state as the input wave (Fig. 17).

The atom amplifier requires a reservoir, or an active medium, of ultracold

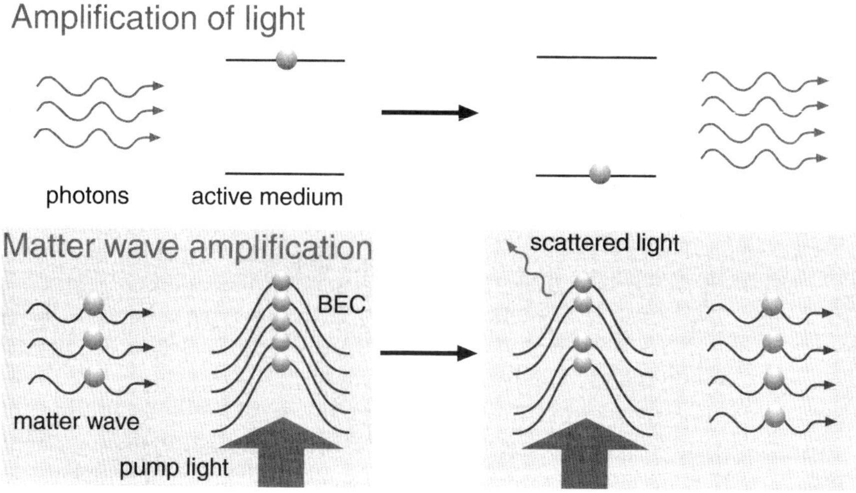

Figure 17. Amplification of light and atoms: In the optical laser, light is amplified by passing it through an excited inverted medium. In the MIT atom amplifier, an input matter wave is sent through a Bose-Einstein condensate illuminated by laser light. Bosonic stimulation by the input atoms causes light to be scattered by the condensate exactly at the angle at which a recoiling condensate atom joins the input matter wave and augments it.

atoms that have a very narrow spread of velocities and can be transferred to the atomic beam. A natural choice for the reservoir was a Bose-Einstein condensate. One also needs a coupling mechanism that transfers atoms from the reservoir at rest to an input mode while conserving energy and momentum. This transfer of atoms was accomplished by scattering laser light. The recoil of the scattering process accelerated some atoms to exactly match the velocity of the input atoms (Fig. 18). Not only were the atoms amplified, but were in exactly the same motional state as the input atoms, i.e., they had the same quantum-mechanical phase. This was verified by interfering the amplified output with a copy of the input wave and observing phase coherence.

This direct observation of atom amplification in the summer of 1999 was preceded by a surprising occurrence late one night in October 1998 when we discovered a new form of superradiance [107]. We were studying Bragg spectroscopy [108] and illuminated a BEC with two laser beams. I had no role in the running of the experiment and was working in my office, when around midnight the students came from the lab and told me that they saw atoms shooting out from the condensate with a velocity component perpendicular to the direction of the laser beams. We expected atoms to receive recoil momentum only along the laser beams, and all motion perpendicular to it to be diffuse due to the random direction of spontaneous Rayleigh scattering.

The whole lab started to discuss what was going on. With a running machine, everything could be tried out immediately. The first ideas were mundane: let's illuminate the condensate with only one laser beam and see what happen (the directional beams remained). We scrutinized the experimental setup for bouncing laser beams or beams which had not been completely

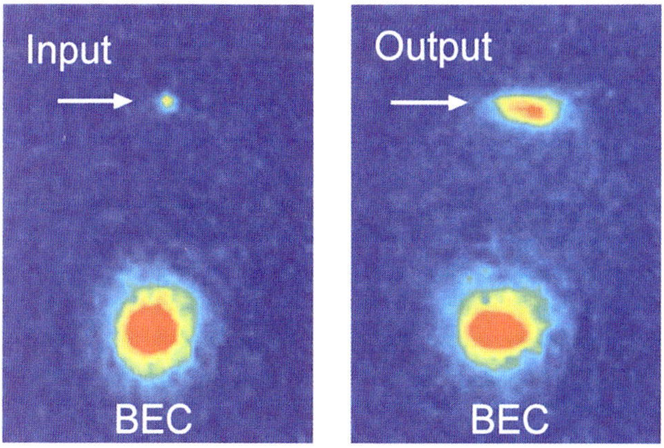

Figure 18. Observation of atom amplification. Atom amplification is probed by sending an input beam through the atom amplifier which is a Bose-Einstein condensate (BEC) illuminated with laser light. On the left side, the input beam has passed through the condensate without amplification. Some 20 ms later, a shadow picture is taken of the condensate and the input atoms. When the amplification process was activated by illuminating the condensate with laser light, the output pulse contained many more atoms than the input pulse – typical amplification factors were between 10 and 100. The field of view is 1.9×2.6 mm^2.

switched off, but we found nothing. Increasingly, we considered that the observed phenomenon was genuine and not due to some experimental artifact. Knowing that the condensate was pencil-shaped, the idea of laser emission along the long condensate axis came up, and this was already very close. We decided to stop the general discussion and continue taking data; the machine was running well and we wanted to take advantage of it. So some students, including Shin Inouye and Ananth Chikkatur, characterized the phenomenon, while Dan Stamper-Kurn and I went to a black board and tried to figure out what was going on. Within the next hour, we developed the correct semi-classical description of superradiance in a condensate. In the lab, the predicted strong dependence on laser polarization was verified. A few months later we realized how we could use the superradiant amplification mechanism to build a phase-coherent atom amplifier. However, the labs were undergoing complete renovation at this point and we had to wait until the machine was running again before the phase-coherent amplification was implemented.

The demonstration of an atom amplifier added a new element to atom optics. In addition to passive elements like beam splitters, lenses and mirrors, there is now an active atom-optical element. Coherent matter wave amplifiers may improve the performance of atom interferometers by making up for losses inside the device or by amplifying the output signal. Atom interferometers are already used as precise gravity and rotation sensors.

Observation of Vortex Lattices in Bose-Einstein Condensates

Quantum mechanics and the wave nature of matter have subtle manifestations when particles have angular momentum or, more generally, when quantum systems are rotating. When a quantum-mechanical particle moves in a circle the circumference of the orbit has to be an integer multiple of the de Broglie wavelength. This quantization rule leads to the Bohr model and the discrete energy levels of the hydrogen atom. For a rotating superfluid, it leads to quantized vortices [109]. If one spins a normal liquid in a bucket, the fluid will finally rotate as a rigid body where the velocity smoothly increases from the center to the edge (Fig. 19, left). However, this smooth variation is impossible for particles in a single quantum state. To fulfill the above-mentioned quantization rule, the flow field has to develop singular regions where the number of de Broglie wavelengths on a closed path jumps up by one. One possibility would be a radially symmetric flow field with concentric rings. Between adjacent rings, the number of de Broglie wavelengths on a circumference would change by one.

However, the energetically most favorable configuration is achieved when the singularities in the velocity field are not distributed on cylindrical shells, but on lines. This corresponds to an array of vortices. In contrast to classical vortices such as in tornados or in a flushing toilet, the vortices in a Bose-Einstein condensate are quantized: when an atom goes around the vortex core, its quantum mechanical phase changes by exactly 2π. Such quantized vortices play a key role in superfluidity and superconductivity. In superconductors, magnetic flux lines arrange themselves in regular lattices that have

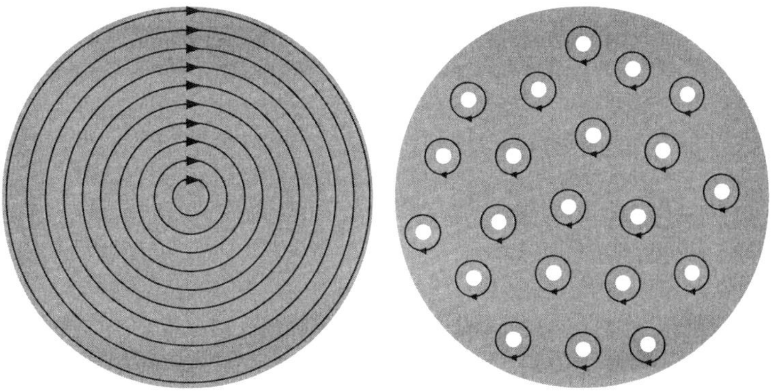

Figure 19. Comparison of the flow fields of rotating normal liquids and superfluids. A normal fluid undergoes rigid body rotation, whereas a superfluid develops an array of quantized vortices.

been directly imaged. In superfluids, previous direct observations of vortices had been limited to small arrays (up to eleven vortices), both in liquid ^{4}He [110] and in rotating gaseous Bose-Einstein condensates (BEC) by a group in Paris [111].

In 2001, our group observed the formation of highly-ordered vortex lattices in a rotating Bose-condensed gas [112]. They were produced by spinning laser beams around the condensate, thus setting it into rotation. The condensate then exhibited a remarkable manifestation of quantum mechanics at a macroscopic level. The rotating gas cloud was riddled with more than one hundred vortices. Since the vortex cores were smaller than the optical resolution, the gas was allowed to ballistically expand after the magnetic trap was switched off. This magnified the spatial structures twenty-fold. A shadow picture of these clouds showed little bright spots where the light penetrated through the empty vortex cores like trough tunnels (Fig. 20 shows a negative image).

A striking feature of the observed vortex lattices is the extreme regularity, free of any major distortions, even near the boundary. Such "Abrikosov" lattices were first predicted for quantized magnetic flux lines in type-II superconductors. However, nature is not always perfect: some of the images showed distortions or defects of the vortex lattices; two examples are shown in Fig. 21. The physics of vortices is very rich. Subsequent work by my group and others has started to address the dynamics and non-equilibrium properties of vortex structures. How are vortices formed? How do they decay? Are the vortices straight or bent? Such experiments can be directly compared with first-principle calculations, which are possible for such a dilute system. This interplay between theory and experiment may lead to a better understanding of superfluidity and macroscopic quantum phenomena.

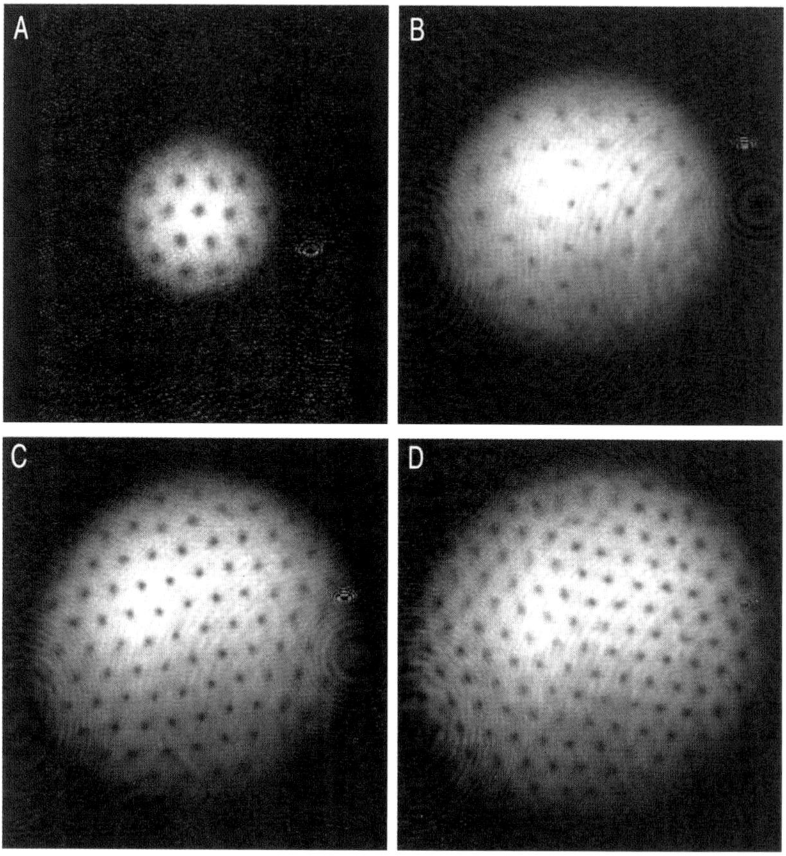

Figure 20. Observation of vortex lattices in rotating Bose-Einstein condensates. The examples shown contain (A) 16 (B) 32 (C) 80 and (D) 130 vortices as the speed of rotation was increased. The vortices have "crystallized" in a triangular pattern. The diameter of the cloud in (D) was 1 mm after ballistic expansion, which represents a magnification of twenty. (Reprinted with permission from ref. [112]. Copyright 2001 American Association for the Advancement of Science.)

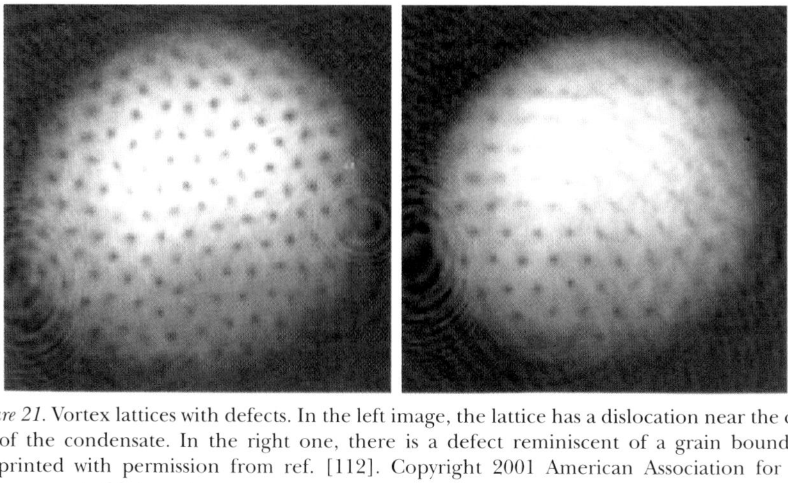

Figure 21. Vortex lattices with defects. In the left image, the lattice has a dislocation near the center of the condensate. In the right one, there is a defect reminiscent of a grain boundary. (Reprinted with permission from ref. [112]. Copyright 2001 American Association for the Advancement of Science.)

OUTLOOK

The rapid pace of developments in atomic BEC during the last few years has taken the community by surprise. After decades of an elusive search, nobody expected that condensates would be so robust and relatively easy to manipulate. Further, nobody imagined that such a simple system would pose so many challenges, not only to experimentalists, but also to our fundamental understanding of physics. The list of future challenges, both for theorists and experimentalists, is long and includes the exploration of superfluidity and second sound in Bose gases, the physics of correlations and non-classical wavefunctions (phenomena beyond the Gross-Pitaevskii equation), the study of quantum-degenerate molecules and Fermi gases, the development of practical "high-power" atom lasers, and their application in atom optics and precision measurements. These scientific goals are closely interwoven with technological advances to produce new single- or multi-species quantum-degenerate systems and novel ways of manipulation, such as using microtraps and atom chips. There is every indication for more excitement to come!

Work on BEC at MIT has been a tremendous team effort, and I am grateful to the past and present collaborators who have shared both the excitement and the hard work: J.R. Abo-Shaeer, M.R. Andrews, M. Boyd, G. Campbell, A.P. Chikkatur, J.-K. Chin, K.B. Davis, K. Dieckmann, D.S. Durfee, A. Görlitz, S. Gupta, T.L. Gustavson, Z. Hadzibabic, S. Inouye, M.A. Joffe, D. Kielpinski, M. Köhl, C.E. Kuklewicz, A.E. Leanhardt, R.F. Löw, A. Martin, M.-O. Mewes, H.-J. Miesner, R. Onofrio, T. Pfau, D.E. Pritchard, C. Raman, D. Schneble, C. Schunck, Y.-I. Shin, D.M. Stamper-Kurn, C.A. Stan, J. Stenger, E. Streed, Y. Torii, C.G. Townsend, N.J. van Druten, J.M. Vogels, K. Xu, M.W. Zwierlein, and many MIT undergraduate students. Exemplary administrative support has been provided by Carol Costa for more than twelve years. Figure 22 shows the team in November 2001. Special thanks go to Dan Kleppner and Tom Greytak for inspiration and constant encouragement. The author also acknowledges the fruitful interactions with colleagues all over the world who have contributed to this rich and exciting field. Some of these colleagues are depicted in Fig. 23, which is a group photo of the lecturers at the Varenna summer school on BEC in 1998. In particular, the yearlong competition with the group at Boulder led by Eric Cornell and Carl Wieman inspired the best from me and my team and, despite tight competition, there has been genuine collegiality and friendship. I want to thank the Office of Naval Research, the National Science Foundation, the Army Research Office, the Joint Services Electronics Program, NASA and the David and Lucile Packard Foundation for their encouragement and financial support of this work.

Figure 22. The author with his team in November 2001. Front row, from left to right: Z. Hadzibabic, K. Xu, S. Gupta, E. Tsikata, Y.-I. Shin. Middle row: A.P. Chikkatur, J.-K. Chin, D.E. Pritchard, W. K., G. Campbell, A.E. Leanhardt, M. Boyd. Back row: J.R. Abo-Shaeer, D. Schneble, J.M. Vogels, K. Dieckmann, C.A. Stan, Y. Torii, E. Streed.

Figure 23. Lecturers, seminar speakers and directors at the summer school on "Bose-Einstein Condensation in Atomic Gases" in Varenna, July 7–17, 1998. Front row: Jean Dalibard, Guglielmo Tino, Fernando Sols, Kris Helmerson. Back row: Sandro Stringari, Carl Wieman, Alexander Fetter, Tilman Esslinger, Massimo Inguscio, William Phillips, Daniel Heinzen, Peter Fedichev, Lev Pitaevskii, W. K., Allan Griffin, Keith Burnett, Daniel Kleppner, Alain Aspect, Ennio Arimondo, Theodor Hänsch, Eric Cornell.

REFERENCES

[1] A. Einstein, Sitzungsber. Preuss. Akad. Wiss., Bericht 3, p. 18 (1925).

[2] S. N. Bose, Z. Phys. **26**, 178 (1924).

[3] G. Taubes, Science **265**, 184 (1994).

[4] S. Chu, Rev. Mod. Phys. **70**, 685 (1998).

[5] C. N. Cohen-Tannoudji, Rev. Mod. Phys. **70,** 707 (1998).

[6] W. D. Phillips, Rev. Mod. Phys. **70**, 721 (1998).

[7] N. Masuhara, J. M. Doyle, J. C. Sandberg, D. Kleppner, T. J. Greytak, H. F. Hess, and G. P. Kochanski, Phys. Rev. Lett. **61**, 935 (1988).

[8] W. Ketterle and N. J. van Druten, in *Advances in Atomic, Molecular, and Optical Physics,* edited by B. Bederson and H. Walther (Academic Press, San Diego, 1996), Vol. 37, pp. 181–236.

[9] J. T. M. Walraven, in *Quantum Dynamics of Simple Systems,* edited by G.L. Oppo, S.M. Barnett, E. Riis, and M. Wilkinson (Institute of Physics Publ., London, 1996), pp. 315–352.

[10] W. Ketterle, D. S. Durfee, and D. M. Stamper-Kurn, in *Bose-Einstein condensation in atomic gases,* Proceedings of the International School of Physics Enrico Fermi, Course CXL, edited by M. Inguscio, S. Stringari, and C.E. Wieman (IOS Press, Amsterdam, 1999), pp. 67–176.

[11] D. Stamper-Kurn and W. Ketterle, in *Coherent Atomic Matter Waves,* Proceedings of the Les Houches Summer School, Course LXXII in 1999, edited by R. Kaiser, C. Westbrook, and F. David (Springer, New York, 2001), e-print cond-mat/0005001.

[12] W. Ketterle and S. Inouye, Compte Rendus de l'Académie des Sciences, Série IV - Physique Astrophysique, vol. 2, pp. 339-380 (2001); e-print cond-mat/0101424.

[13] A. Pais, *Subtle is the Lord, The Science and the Life of Albert Einstein* (Clarendon Press, Oxford, 1982).

[14] K. Gavroglu, *Fritz London: A Scientific Biography* (Cambridge University Press, Cambridge, 1995).

[15] A. Griffin, in *Bose-Einstein condensation in atomic gases,* Proceedings of the International School of Physics Enrico Fermi, Course CXL, edited by M. Inguscio, S. Stringari, and C.E. Wieman (IOS Press, Amsterdam, 1999), p. 1.

[16] A. Einstein, Sitzungsber. Preuss. Akad. Wiss., Bericht 1, p. 3 (1925).

[17] A. Griffin, D. W. Snoke, and S. Stringari (editors), *Bose-Einstein Condensation* (Cambridge University Press, Cambridge, 1995).

[18] K. Huang, *Statistical Mechanics* (Wiley, New York, 1987).

[19] E. Schrödinger, *Statistical thermodynamics* (Cambridge University Press, Cambridge, 1952), reprinted by Dover Pulblications (New York, 1989).

[20] B. C. Crooker, B. Hebral, E. N. Smith, Y. Takano, and J. D. Reppy, Phys. Rev. Lett. **51**, 666 (1983).

[21] J. D. Reppy, Physica B **126**, 335 (1984).

[22] M. Rasolt, M. H. Stephen, M. E. Fisher, and P. B. Weichman, Phys. Rev. Lett. **53**, 798 (1984).

[23] H. Cho and G. A. Williams, Phys. Rev. Lett. **75**, 1562 (1995).

[24] J. P. Wolfe, J. L. Lin, and D. W. Snoke, in *Bose-Einstein Condensation,* edited by A. Griffin, D.W. Snoke, and S. Stringari (Cambridge University Press, Cambridge, 1995), pp. 281–329.

[25] E. Fortin, E. Benson, and A. Mysyrowicz, in *Bose-Einstein Condensation,* edited by A. Griffin, D.W. Snoke, and S. Stringari (Cambridge University Press, Cambridge, 1995), pp. 519 – 523.

[26] J. L. Lin and J. P. Wolfe, Phys. Rev. Lett. **71**, 1222 (1993).

[27] K. E. O'Hara, L. Ó Súilleabháin, and J. P. Wolfe, Phys. Rev. B **60**, 10565 (1999).

[28] L. V. Butov, C. W. Lai, A. L. Ivanov, A. C. Gossard, and D. S. Chemla, Nature **417**, 47 (2002).

[29] Y. Yamamoto, Nature **405**, 629 (2000).

[30] J. J. Baumberg, Phys. World, March 2002, p. 37.
[31] M. Saba, C. Ciuti, J. Bloch, V. Thierry-Mieg, R. André, L. S. Dang, S. Kundermann, A. Mura, G. Bongiovanni, J. L. Staehli, and B. Deveaud, Nature **414,** 731 (2001).
[32] C. E. Hecht, Physica **25**, 1159 (1959).
[33] W. C. Stwalley and L. H. Nosanow, Phys. Rev. Lett. **36**, 910 (1976).
[34] I. F. Silvera and J. T. M. Walraven, Phys. Rev. Lett. **44**, 164 (1980).
[35] R. W. Cline, D. A. Smith, T. J. Greytak, and D. Kleppner, Phys. Rev. Lett. **45**, 2117 (1980).
[36] D. G. Fried, T. C. Killian, L. Willmann, D. Landhuis, S. C. Moss, D. Kleppner, and T. J. Greytak, Phys. Rev. Lett. **81**, 3811 (1998).
[37] T. J. Greytak, in *Bose-Einstein Condensation,* edited by A. Griffin, D.W. Snoke, and S. Stringari (Cambridge University Press, Cambridge, 1995), pp. 131–159.
[38] T. J. Greytak and D. Kleppner, in *New Trends in Atomic Physics,* Les Houches Summer School 1982, edited by G. Grynberg and R. Stora (North-Holland, Amsterdam, 1984), p. 1125.
[39] I. F. Silvera and J. T. M. Walraven, in *Progress in Low Temperature Physics,* edited by D.F. Brewer (Elsevier, Amsterdam, 1986), Vol. X, p. 139.
[40] D. Kleppner, T. J. Greytak, T. C. Killian, D. G. Fried, L. Willmann, D. Landhuis, and S. C. Moss, in *Bose-Einstein Condensation in Atomic Gases,* Proceedings of the International School of Physics Enrico Fermi, Course CXL, edited by M. Inguscio, S. Stringari, and C.E. Wieman (IOS Press, Amsterdam, 1999), pp. 177–199.
[41] A. I. Safonov, S. A. Vasilyev, I. S. Yasnikov, I. I. Lukashevich, and S. Jaakola, Phys. Rev. Lett. **81**, 4545 (1998).
[42] V. V. Goldman, I. F. Silvera, and A. J. Leggett, Phys. Rev. B **24**, 2870 (1981).
[43] D. A. Huse and E. Siggia, J. Low Temp. Phys. **46**, 137 (1982).
[44] J. Oliva, Phys. Rev. B **39**, 4197 (1989).
[45] H. T. C. Stoof, J. M. V. A. Koelman, and B. J. Verhaar, Phys. Rev. B **38**, 4688 (1988).
[46] H. F. Hess, Phys. Rev. B **34**, 3476 (1986).
[47] E. Arimondo, W. D. Phillips, and F. Strumia (editors), *Laser Manipulation of Atoms and Ions,* Proceedings of the International School of Physics Enrico Fermi, Course CXVIII (North-Holland, Amsterdam, 1992).
[48] H. Metcalf and P. van der Straten, Phys. Rep. **244**, 203 (1994).
[49] C. S. Adams and E. Riis, Progress in Quantum Electronics **21,** 1 (1997).
[50] V. S. Letokhov and V. G. Minogin, Optics Comm. **35**, 199 (1980).
[51] S. Chu, L. Hollberg, J. E. Bjorkholm, A. Cable, and A. Ashkin, Phys. Rev. Lett. **55**, 48 (1985).
[52] D. E. Pritchard, in *Electronic and atomic collisions,* invited papers of the XIV International Conference on the Physics of Electronic and Atomic Collisions, Palo Alto, California, 24-30 July, 1985, edited by D.C. Lorents, W.E. Meyerhof, and J.R. Peterson (Elsevier, New York, 1986), pp. 593–604.
[53] T. Walker and P. Feng, in *Advances in Atomic, Molecular, and Optical Physics,* edited by B. Bederson and H. Walther (Academic Press, San Diego, 1994), Vol. 34, pp. 125–170.
[54] J. Weiner, in *Advances in Atomic, Molecular, and Optical Physics,* edited by B. Bederson and H. Walther (Academic Press, San Diego, 1995), Vol. 35, pp. 45 – 78.
[55] T. Walker, D. Sesko, and C. Wieman, Phys. Rev. Lett. **64**, 408 (1990).
[56] M. Drewsen, P. Laurent, A. Nadir, G. Santarelli, A. Clairon, Y. Castin, D. Grison, and C. Salomon, Appl. Phys. B **59**, 283 (1994).
[57] C. G. Townsend, N. H. Edwards, C. J. Cooper, K. P. Zetie, C. J. Foot, A. M. Steane, P. Szriftgiser, H. Perrin, and J. Dalibard, Phys. Rev. A **52**, 1423 (1995).
[58] C. G. Townsend, N. H. Edwards, K. P. Zetie, C. J. Cooper, J. Rink, and C. J. Foot, Phys. Rev. A **53**, 1702 (1996).
[59] M. T. DePue, C. McCormick, S. L. Winoto, S. Oliver, and D. S. Weiss, Phys. Rev. Lett. **82**, 2262 (1999).
[60] T. Ido, Y. Isoya, and H. Katori, Phys. Rev. A **61**, 061403(R) (2000).

[61] A. J. Kerman, V. Vuletic, C. Chin, and S. Chu, Phys. Rev. Lett. **84**, 439 (2000).
[62] W. Ketterle and D. E. Pritchard, Appl. Phys. B **54**, 403 (1992).
[63] R. V. E. Lovelace, C. Mahanian, T. J. Tommila, and D. M. Lee, Nature **318**, 30 (1985).
[64] E. A. Cornell, C. Monroe, and C. E. Wieman, Phys. Rev. Lett. **67**, 2439 (1991).
[65] J. Vigué, Phys. Rev. A **34**, 4476 (1986).
[66] W. Ketterle and D. E. Pritchard, Phys. Rev. A **46**, 4051 (1992).
[67] D. E. Pritchard and W. Ketterle, in *Laser Manipulation of Atoms and Ions*, Proceedings of the International School of Physics Enrico Fermi, Course CXVIII, edited by E. Arimondo, W.D. Phillips, and F. Strumia (North-Holland, Amsterdam, 1992), pp. 473–496.
[68] M. Kasevich and S. Chu, Phys. Rev. Lett. **69**, 1741 (1992).
[69] C. Monroe, W. Swann, H. Robinson, and C. Wieman, Phys. Rev. Lett. **65**, 1571 (1990).
[70] K. Lindquist, M. Stephens, and C. Wieman, Phys. Rev. A **46**, 4082 (1992).
[71] W. Ketterle, A. Martin, M. A. Joffe, and D. E. Pritchard, Phys. Rev. Lett. **69**, 2483 (1992).
[72] T. E. Barrett, S. W. Dapore-Schwartz, M. D. Ray, and G. P. Lafyatis, Phys. Rev. Lett. **67**, 3483 (1991).
[73] M. A. Joffe, W. Ketterle, A. Martin, and D. E. Pritchard, J. Opt. Soc. Am. B **10**, 2257 (1993).
[74] D. E. Pritchard, K. Helmerson, and A. G. Martin, in *Atomic Physics 11*, edited by S. Haroche, J.C. Gay, and G. Grynberg (World Scientific, Singapore, 1989), p. 179.
[75] A. L. Migdall, J. V. Prodan, W. D. Phillips, T. H. Bergeman, and H. J. Metcalf, Phys. Rev. Lett. **54**, 2596 (1985).
[76] V. S. Bagnato, G. P. Lafyatis, A. G. Martin, E. L. Raab, R. N. Ahmad-Bitar, and D. E. Pritchard, Phys. Rev. Lett. **58**, 2194 (1987).
[77] K. Helmerson, A. Martin, and D. E. Pritchard, J. Opt. Soc. Am. B **9**, 1988 (1992).
[78] C. Monroe, E. Cornell, and C. Wieman, in *Laser Manipulation of Atoms and Ions*, Proceedings of the International School of Physics Enrico Fermi, Course CXVIII, edited by E. Arimondo, W.D. Phillips, and F. Strumia (North-Holland, Amsterdam, 1992), pp. 361–377.
[79] C. C. Bradley, C. A. Sackett, J. J. Tollet, and R. G. Hulet, Phys. Rev. Lett. **75**, 1687 (1995).
[80] W. Ketterle, K. B. Davis, M. A. Joffe, A. Martin, and D. E. Pritchard, Phys. Rev. Lett. **70**, 2253 (1993).
[81] J. Weiner, V. S. Bagnato, S. Zilio, and P. S. Julienne, Rev. Mod. Phys. **71**, 1 (1999).
[82] D. J. Heinzen, in *Bose-Einstein condensation in atomic gases*, Proceedings of the International School of Physics Enrico Fermi, Course CXL, edited by M. Inguscio, S. Stringari, and C.E. Wieman (IOS Press, Amsterdam, 1999), pp. 351–390.
[83] M. H. Anderson, J. R. Ensher, M. R. Matthews, C. E. Wieman, and E. A. Cornell, Science **269**, 198 (1995).
[84] K. B. Davis, M.-O. Mewes, M. R. Andrews, N. J. van Druten, D. S. Durfee, D. M. Kurn, and W. Ketterle, Phys. Rev. Lett. **75**, 3969 (1995).
[85] W. Ketterle, K. B. Davis, M. A. Joffe, A. Martin, and D. E. Pritchard, OSA Annual Presentation at the Meeting, Toronto, Canada, October 3–8 (1993).
[86] P. D. Lett, W. D. Phillips, S. L. Rolston, C. E. Tanner, R. N. Watts, and C. I. Westbrook, J. Opt. Soc. Am. B **6**, 2084 (1989).
[87] K. B. Davis, M. O. Mewes, M. A. Joffe, and W. Ketterle, in *Fourteenth International Conference on Atomic Physics*, Boulder, Colorado, 1994, Book of Abstracts, 1-M3 (University of Colorado, Boulder, Colorado, 1994).
[88] W. Petrich, M. H. Anderson, J. R. Ensher, and E. A. Cornell, Phys. Rev. Lett. **74**, 3352 (1995).
[89] K. B. Davis, M.-O. Mewes, M. A. Joffe, M. R. Andrews, and W. Ketterle, Phys. Rev. Lett. **74**, 5202 (1995).

Raymond Davis Jr. /AMD

RAYMOND DAVIS, JR.

I was born in Washington, D.C. on October 14, 1914. My father was a photographer at the National Bureau of Standards. A self-educated man, he never finished high school, but, in his career at the National Bureau of Standards, he made many useful inventions, and eventually became chief of the Photographic Technology Section. His early influence led me in the direction of individual experimentation and designing my own apparatus. My mother, Ida Rogers Younger, was a native of the state of Virginia. She taught me to enjoy music, although she never succeeded in making me a performer. It was to please her that I spent several years as a choirboy, in spite of my inability to carry a tune. A bit later in life, I took pleasure in attending outdoor concerts at the Watergate, in the days before air traffic grew heavy enough to drown out the music.

My constant companion in childhood was my brother Warren, only 14 months my junior. Together we played street games on summer evening, paddled a canoe on the Potomac and, after my release from the choir, spent many weekends rifle-shooting with our father. In high school and college, I gathered a number of medals for marksmanship, but I have long since abandoned this activity, having concluded that the world would be a better place with fewer sharpshooters.

As my brother and I grew older, our interests diverged. He headed for a military career, while I became more interested in science. My father encouraged my interest, bringing me chemicals for my basement experiments, and helping me become a reasonably good photographer. My favorite reading matter was Smithsonian reports on many phases of science, obtained at my local branch library. Washington offered many educational opportunities for curious young minds.

I was educated in the Washington public schools, and attended the University of Maryland as a day student, graduating in 1938 with a degree in chemistry. After working for the Dow Chemical Company in Midland, Michigan for a year, I returned to the University of Maryland to take a Master's degree, before going on to Yale to pursue a doctorate. In 1942, I received my Ph.D. in physical chemistry, and immediately entered the Army as a reserve officer. Most of my war years were spent at Dugway Proving Ground in Utah, observing chemical weapons tests and, in my spare hours, exploring and photographing the surrounding territory, which included the Great Salt Lake and geologic evidence for its much larger predecessor, Lake Bonneville.

Upon my discharge from the Army in 1945, I went to work at the Monsanto Chemical Company's Mound Laboratory, in Miamisburg, Ohio, doing applied radiochemistry of interest to the Atomic Energy Commission. In the spring of 1948, I was able to join the newly created Brookhaven National Laboratory, which was dedicated to finding peaceful uses for atomic energy.

In my first few months at Brookhaven, I lived at the Lindenmere, a summer hotel which had been leased by the laboratory to provide housing for new arrivals. It was there that I met my future wife, Anna Torrey, who was also employed at Brookhaven, in the Biology Department. Since this was a seaside community, I decided to build my own sailboat. This notion was viewed with scorn by most of my acquaintances but, with Anna's help, I built a 21-foot sloop, the Halcyon, which gave our family many years of pleasure. Now in the hands of her third owner, the Halcyon still sails the Great South Bay. Anna and I were married in late 1948 and, over the next fifteen years, five children were born to us: Andrew, a senior scientist at the University of Chicago who studies meteorites to learn about stars and the early history of the solar system, lives in River Forest, Illinois; Martha Kumler, a private tutor of high school students, lives in Honeoye Falls, New York; Nancy Klemm, a homemaker and restorer of windows in old houses, lives in Webster Groves, Missouri; Roger, a mechanical technician working on the Relativistic Heavy Ion Collider at Brookhaven National Laboratory, lives in Center Moriches, New York; and Alan, an engineer with Boeing, lives in Seattle, Washington. Among them, they have given us eleven grandchildren. We have lived in the same house, in Blue Point, New York, for over fifty years.

My first act, on arriving at Brookhaven, was to report to the chairman of the Chemistry Department, Richard Dodson, and ask him what I was expected to do. To my surprise and delight, I was advised to go to the library, do some reading and choose a project of my own, whatever appealed to me. Thus began a long career of doing just what I wanted to do and getting paid for it. In the library, I read a 1948 review paper by H. R. Crane in *Reviews of Modern Physics* which led me to decide on an experiment in neutrino physics, a field in which little was known at the time, and which seemed well-suited to my background in physical chemistry.

In early experiments, I attempted to detect neutrinos from a reactor, using the chlorine-argon detection method suggested by Bruno Pontecorvo (in 1946). In this method, a ^{37}Cl atom reacts with a neutrino to make an ^{37}Ar atom. Argon is a noble gas and is easy to separate chemically from a large amount of chlorine-rich solvent. It is radioactive with a half-life of 35 days and can be counted with a gas-filled proportional counter. A first attempt, exposing a 1000-gallon tank of carbon tetrachloride at the Brookhaven Graphite Research Reactor, failed to detect any signal, as the neutrino flux at this reactor was too small to affect a target of this size. Furthermore, a reactor emits antineutrinos, and the Pontecorvo method only detects neutrinos. It was not known at that time that the two particles were not identical. Later, I built larger experiments, using one of the Savannah River reactors as the neutrino source. I eventually set a limit on the neutrino flux that was a factor of 20 below the antineutrino flux measured by Reines and Cowan in their elegant experiment that won Fred Reines his Nobel Prize.

Other early interests included the development, with Oliver Schaeffer, of a method of geological dating using ^{36}Cl in surface rocks. With the later advent of accelerator mass spectrometry, this has become a useful tool in geochemistry, but our counting techniques were not sensitive enough to make the method work. We turned to measuring ^{36}Cl in meteorites. Measuring the ^{36}Cl radioactivity and the total accumulated decay product, ^{36}Ar, in a meteorite allowed us to determine how long the meteorite had been exposed in space. Our interest in meteorite exposure ages continued for many years. We also worked on measuring cosmic-ray production of ^{37}Ar and ^{39}Ar in a variety of freshly fallen meteorites. Our greatest success in this work was with the Lost City meteorite. The track of this meteorite was photographed as it fell, allowing its orbit to be determined. Our measurement of radioactive argon isotopes allowed us to deduce the cosmic ray intensity gradient in the inner solar system. During the era of the moon landings, I was involved in measuring ^{37}Ar, ^{39}Ar, tritium and ^{222}Rn in lunar rocks and in the lunar atmosphere (trapped in the rock boxes brought back by the astronauts). During processing of the Apollo 12 samples, one of the glove boxes in Houston leaked and I had the interesting experience of being quarantined with the astronauts and a few other unlucky scientists for two weeks until it was clear that we were not infected with any lunar diseases.

Following the Savannah River experiments, I began thinking about detecting neutrinos from the Sun. The first step was a pilot experiment located 2,300 feet underground in the Barberton Limestone Mine, near Akron, Ohio. Observing neutrinos from the Sun had the potential of testing the theory that the hydrogen-helium fusion reactions are the source of the Sun's energy. In the 1950s, however, the proton-proton chain of reactions was believed to be the principal neutrino source, but this chain only emitted low-energy neutrinos, below the threshold of the chlorine-argon reaction.

A new measurement of the nuclear reaction $^{3}He + {}^{4}He \rightarrow {}^{7}Be + \gamma$ by Holmgren and Johnston in 1958, suggested that one of the terminal reactions in the proton-proton chain would produce energetic neutrinos which could be measured by the chlorine-argon radiochemical method. Encouraged by these developments, and with the support of the Brookhaven National Laboratory and the U.S. Atomic Energy Commission, I built a much larger experiment in the Homestake Gold Mine in Lead, South Dakota. The detector itself consisted of a 100,000-gallon tank filled with perchloroethylene, a solvent most commonly used for dry cleaning of clothing. The experiment was located nearly a mile underground, at the 4850 foot level of the mine. Initially, we observed no solar neutrino signal and expressed our results only as upper limits. Subsequent refinements in technique and, particularly, in counting methods, continued over the years, producing a solar neutrino signal approximately one-third of the expected flux from the standard solar model calculated by John Bahcall. This was the genesis of the so-called "solar neutrino problem".

The solar neutrino problem caused great consternation among physicists and astrophysicists. My opinion in the early years was that something was wrong with the standard solar model; many physicists thought there was something wrong with my experiment. Years of measurements produced consistent answers and many tests showed that there were no problems with experimental procedures. Many distinguished physicists suggested explanations for the low solar neutrino flux that now seem fanciful. Trevor Pinch, a sociologist, made a study of how scientists responded to the solar neutrino problem. The disagreement between the measured solar neutrino flux and that predicted by the standard solar model was confirmed for energetic ^{8}B neutrinos by the Kamiokande II experiment in the late 1980s and for the lower energy pp neutrinos by the gallium experiments GALLEX and SAGE in the middle 1990s. Only recently, observations at the Sudbury Neutrino Observatory (SNO) in the Inco Nickel Mine in Sudbury, Ontario, Canada, have indicated that, indeed, the total number of solar neutrinos emitted agrees with the standard solar model prediction, but that two-thirds of the neutrinos change in the course of their journey to the Earth into other flavors (μ and τ neutrinos), a phenomenon known as neutrino oscillation. Only electron neutrinos can be detected with the Cl-Ar radiochemical method.

I retired from Brookhaven in 1984, but wasn't ready to give up measuring solar neutrinos, because I thought it important that the Homestake experiment measure the solar neutrino flux at the same time as new solar neutrino experiments. I transferred administration of the Homestake experiment to the University of Pennsylvania and have been a Research Professor there since that time. The experiment continued to measure the solar neutrino flux until the late 1990s, when the Homestake Mine ceased operating.

Meanwhile, to my surprise, a whole new field of neutrino physics has developed in directions I never imagined in the Homestake days.

Awards

Boris Pregel Prize, New York Academy of Sciences, 1957
Cyrus B. Comstock Prize, U. S. National Academy of Sciences, 1978
American Chemical Society Award for Nuclear Applications in Chemistry, 1979
Tom W. Bonner Prize, American Physical Society, 1988
Honorary Doctor of Science, University of Pennsylvania, 1990
W. K. H. Panofsky Prize, American Physical Society, 1992
Beatrice M. Tinsley Prize, American Astronomical Society, 1995
George Ellery Hale Prize, American Astronomical Society, 1996
Honorary Doctor of Science, Laurentian University, 1997
Bruno Pontecorvo Prize, Joint Institute for Nuclear Research, Dubna Russia, 1999
Wolf Prize, 2000
Honorary Doctor of Science, University of Chicago, 2000
U. S. National Medal of Science, 2001
Nobel Prize in Physics, 2002

A HALF-CENTURY WITH SOLAR NEUTRINOS

Nobel Lecture, December 8, 2002

by

RAYMOND DAVIS, JR.

Department of Physics and Astronomy, University of Pennsylvania, Philadelphia, PA 19104 and Chemistry Department, Brookhaven National Laboratory, Upton, NY 11973, USA.

Neutrinos are neutral, nearly massless particles that move at nearly the speed of light and easily pass through matter. Wolfgang Pauli (1945 Nobel Laureate in Physics) postulated the existence of the neutrino in 1930 as a way of carrying away missing energy, momentum, and spin in beta decay. In 1933, Enrico Fermi (1938 Nobel Laureate in Physics) named the neutrino ("little neutral one" in Italian) and incorporated it into his theory of beta decay.

The Sun derives its energy from fusion reactions in which hydrogen is transformed into helium. Every time four protons are turned into a helium nucleus, two neutrinos are produced. These neutrinos take only two seconds to reach the surface of the Sun and another eight minutes or so to reach the Earth. Thus, neutrinos tell us what happened in the center of the Sun eight minutes ago. The Sun produces a lot of neutrinos, 1.8×10^{39} per second: even at the Earth, 150 million kilometers from the Sun, about 100 billion pass through an average fingernail (1 cm^2) every second. They pass through the Earth as if it weren't there and the atoms in the human body capture a neutrino about every seventy years, or once in a lifetime. As we will see, neutrinos captured me early in my career.

I received my Ph.D. from Yale in 1942 in physical chemistry (Davis, 1942) and went directly into the Army as a reserve officer. After the war, I decided to search for a position in research with the view of applying chemistry to studies in nuclear physics. After two years with the Monsanto Chemical Company in applied radiochemistry of interest to the Atomic Energy Commission, I was very fortunate in being able to join the newly created Brookhaven National Laboratory. Brookhaven was created to find peaceful uses for the atom in all fields of basic science: chemistry, physics, biology, medicine, and engineering.

When I joined the Chemistry Department at Brookhaven, I asked the chairman, Richard Dodson, what he wanted me to do. To my surprise and delight, he told me to go to the library and find something interesting to work on. I found a stimulating review on neutrinos (Crane, 1948). This quote from Crane shows that neutrino physics was a field that was wide open to exploration: "Not everyone would be willing to say that he believes in the existence of the neutrino, but it is safe to say that there is hardly one of us who is not

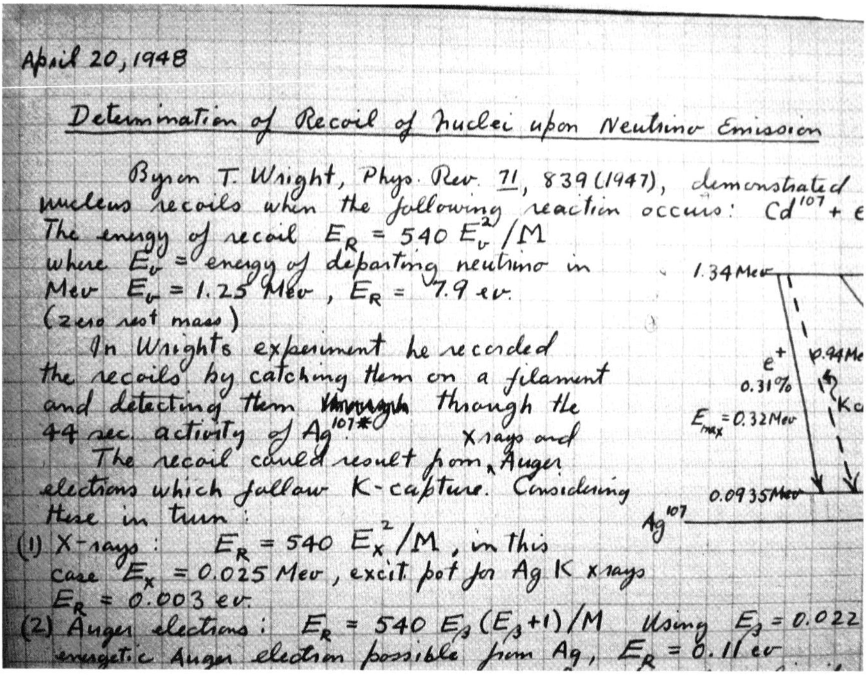

Figure 1. The first page in my first laboratory notebook at Brookhaven National Laboratory. I was hooked on neutrinos from the beginning.

served by the neutrino hypothesis as an aid in thinking about the beta-decay hypothesis". Neutrinos also turned out to be suitable for applying my background in physical chemistry. Crane had quite an extensive discussion on the use of recoil experiments to study neutrinos. I immediately became interested in such experiments (Fig. 1). I spent the first year working on the recoil of ^{107}Ag from the electron-capture decay of ^{107}Cd, but these experiments were inconclusive.

My first successful experiment was a study of the recoil energy of a ^{7}Li nucleus resulting from the electron-capture decay of ^{7}Be. In ^{7}Be decay, a single monoenergetic neutrino is emitted with an energy of 0.862 MeV, and the resulting ^{7}Li nucleus should recoil with a characteristic energy of 57 eV. A measurement of this process provides evidence for the existence of the neutrino. In my experiment, the energy spectrum of a recoiling ^{7}Li ion from a surface deposit of ^{7}Be was measured and found to agree with that expected from the emission of a single neutrino (Davis, 1952). This was a very nice result, but I was scooped by a group from the University of Illinois (Smith and Allen, 1951).

In 1951, I began working on a radiochemical experiment for detecting neutrinos using a method that was suggested by Pontecorvo (1946): capturing neutrinos with the reaction: $^{37}Cl + \nu_e \rightarrow {}^{37}Ar + e^-$. Bruno Pontecorvo's short paper was quite detailed, and the method he described, removing argon by boiling carbon tetrachloride and counting ^{37}Ar in a gas-filled counter, has many similarities to techniques I eventually used. This paper was a report from the Chalk River Laboratory in Canada and was classified by the U. S. Atomic Energy Commission, because it was feared that the method could be used to measure the power output of reactors. Luis Alvarez (1968 Nobel Laureate in Physics) proposed to use the chlorine-argon reaction to detect solar neutrinos with a large tank of concentrated sodium chloride solution (Alvarez, 1949), but did not choose to pursue the experiment. Since no one else appeared interested in attempting the chlorine-argon neutrino detection method, it seemed a natural and timely experiment for me to work on.

In the Pontecorvo method, neutrino capture on ^{37}Cl makes ^{37}Ar, a radioactive isotope that decays back to ^{37}Cl by the inverse of the capture process with a half-life of 35 days. The threshold for the capture reaction is 0.814 MeV, meaning that neutrinos with energies of less than 0.814 MeV will not be captured. There are two potential sources of neutrinos, fission reactors and the Sun, both of which were suggested as possible sources by Pontecorvo (1946).

In my first experiment using the ^{37}Cl-^{37}Ar reaction, I tried to detect neutrinos from a fission reactor, using carbon tetrachloride (CCl_4) as the target material (Davis, 1955). I exposed a 3800-liter tank of carbon tetrachloride at the Brookhaven Graphite Research Reactor for a month or two, removed the argon and counted it in a small Geiger counter. That reactor did not have a high enough neutrino flux to detect with this target size, so neutrinos were not observed. Furthermore, a reactor emits antineutrinos, and the ^{37}Cl-^{37}Ar reaction requires neutrinos. It was not clear at that time, however, whether neutrinos and antineutrinos were different particles, nor was it clear how they could differ. After all, there are other examples in nature where the particle is its own antiparticle, such as the photon and the pi-zero. The background due to cosmic rays was significant, so as part of the experiment, I buried a 3800-liter tank of carbon tetrachloride 5.8 meters underground. From the background-corrected ^{37}Ar count rate, I was able to obtain an upper limit on the solar neutrino flux of 40,000 SNU, a factor of 15,000 above my eventual result of 2.56 SNU (the solar neutrino unit, or SNU, is defined as 10^{-36} captures per target atom per second). One reviewer of my paper was not very impressed with this upper limit and commented: "Any experiment such as this, which does not have the requisite sensitivity, really has no bearing on the question of the existence of neutrinos. To illustrate my point, one would not write a scientific paper describing an experiment in which an experimenter stood on a mountain and reached for the moon, and concluded that the moon was more than eight feet from the top of the mountain." It was clear that the Brookhaven reactor was not a powerful enough neutrino source, so in 1954, I built an experiment using 3800 liters of CCl_4 in the basement of one of the Savannah River reactors, the most intense antineutrino source in the world.

One can calculate the total capture rate from all fission product antineutrinos by ^{37}Cl, presuming neutrinos and antineutrinos are equivalent particles. The sensitivity for detecting neutrinos and the flux at Savannah River were sufficiently high to provide a critical test for the neutrino-antineutrino identity. I did not detect any reactor neutrinos and found that the neutrino capture rate was a factor of five below the antineutrino capture rate. I later did a 11,400-liter experiment at Savannah River that lowered the upper limit for neutrino capture to a factor of 20 below the antineutrino capture rate (Davis, 1958). While I was at Savannah River, Frederick Reines (Nobel Prize in Physics, 1995) and Clyde Cowan and their associates were performing a beautiful experiment, the first detection of a free antineutrino (Cowan *et al.*, 1956; Reines *et al.*, 1960). This experiment was a clear demonstration that the neutrino postulated by Pauli was indeed a real particle. My experiment showed that the neutrino was not its own antiparticle.

As I mentioned earlier, the Sun's energy comes from fusion of four hydrogen atoms to make a helium atom. I would now like to explore the reactions in a little more detail and note which ones can produce neutrinos detectable with the chlorine-argon experiment. The relative rates of nuclear reactions in the Sun are based on detailed modeling of temperatures, pressures, and elemental abundances in the Sun, into which are put laboratory-measured and theoretically calculated rates of appropriate nuclear reactions. What I show in Table 1 and Fig. 2 represent current best estimates. These estimates have changed with time, but the predicted solar neutrino flux has not changed much since the late 1960s.

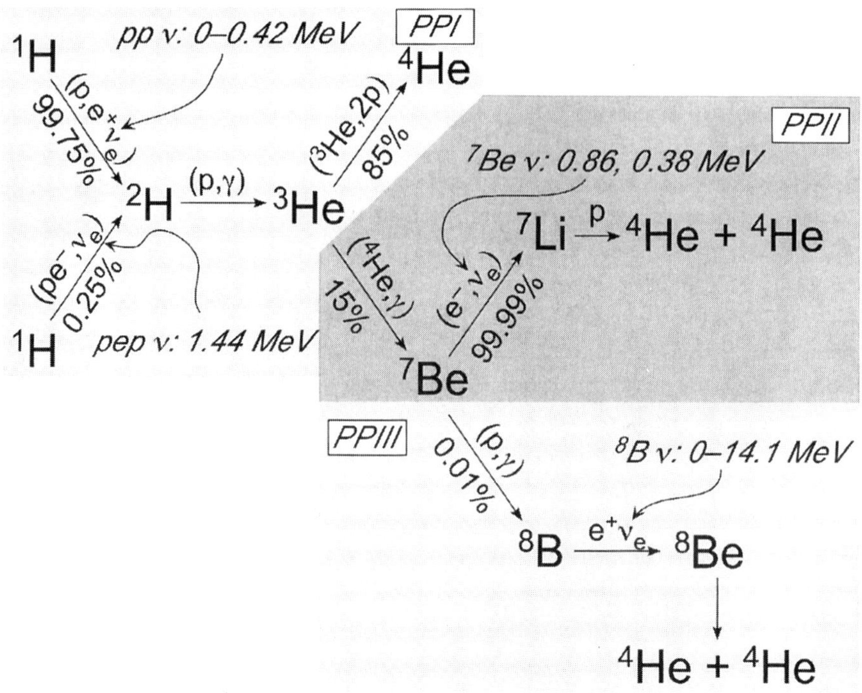

Figure 2. The main energy-producing reactions in the Sun, along with their associated neutrinos.

Table 1. Neutrino-producing reactions in the Sun.

Reaction	Frequency	Energy (MeV)	Name
PP I			
$p + p \rightarrow {}^2H + e^+ + \nu_e$	99.75%	0.0 – 0.42	pp
$p + e^- + p \rightarrow {}^2H + \nu_e$	0.25%	1.44	pep
${}^2H + p \rightarrow {}^3He + \gamma$	100%	—	
${}^3He + {}^3He \rightarrow {}^4He + 2p$	85%	—	
PP II			
${}^3He + {}^4He \rightarrow {}^7Be + \gamma$	15%	—	
$e^- + {}^7Be \rightarrow {}^7Li + \nu_e$	99.99%	0.86, 0.38	${}^7Be$
$p + {}^7Li \rightarrow {}^4He + {}^4He$	100%	—	
PP III			
$p + {}^7Be \rightarrow {}^8B + \gamma$	0.01%	—	
${}^8B \rightarrow {}^4He + {}^4He + e^+ + \nu_e$	100%	0–14.1	${}^8B$

Most of the time, two protons react to make a deuteron, a positron and an electron neutrino, but the neutrinos are below the threshold of the chlorine capture reaction (these neutrinos are known as pp neutrinos). One time out of 400, two protons react with an electron to make a deuteron and a neutrino. These so-called "pep" neutrinos are quite energetic, but there are not very many of them. All the deuterons react with a proton to make ${}^3He$ and most of the ${}^3He$, nuclei react with one another to make ${}^4He$ and two protons. The latter reaction occurs 85% of the time. This major branch is usually referred to as the PPI chain (Fig. 2, Table 1).

Fifteen percent of the time, however, ${}^3He$ and ${}^4He$ nuclei react to make ${}^7Be$. Nearly all of the ${}^7Be$ then captures an electron to make ${}^7Li$ and a neutrino. The ${}^7Li$ captures a proton to make two ${}^4He$ nuclei, completing the PPII chain, but one in 10,000 ${}^7Be$ nuclei captures a proton to make ${}^8B$ (Fig. 2, Table 1). Boron-8 decays to ${}^8Be$, which splits to make two ${}^4He$ nuclei, a positron, and an electron neutrino, completing the PPIII chain. It is these ${}^8B$ neutrinos that produce most of the solar neutrino signal I detected, but there is also some contribution from pep and ${}^7Be$ neutrinos.

In the 1950s, it was thought that the PP I chain was the only significant source of neutrinos, but the energy of the dominant pp neutrinos was too low to detect with the ${}^{37}Cl$ reaction. The branch to make ${}^7Be$ was thought not to be significant and to occur only 0.015% of the time. The CNO cycle, in which four protons make a ${}^4He$ nucleus, could produce energetic neutrinos from the ${}^{13}N(e^+,\nu_e){}^{13}C$ and ${}^{15}O(e^+,\nu_e){}^{15}N$ reactions, but it was not clear that the CNO cycle operated in the Sun. Thus, an experiment to detect solar neutrinos did not seem feasible at the time.

This situation changed dramatically when Holmgren and Johnston (1958) reported that the rate of reaction ${}^3He + {}^4He \rightarrow {}^7Be + \gamma$ is 1000 times greater than previously thought. It was immediately recognized that the path to ${}^7Be$, now about 15% of the time rather than one in 7,000, was significant. Furthermore, the ${}^7Be$ could react with a proton and become ${}^8B$. These two ra-

Figure 3. The large room where the tank will be installed in the Homestake Mine is shown about half-excavated, in the summer of 1965. From *Sharp Bits*, September 1965, published by the Homestake Mining Company.

dioactive products, ^{7}Be and ^{8}B, are sources of energetic neutrinos. The relative rates of electron and proton capture on ^{7}Be were not measured, but estimated rates allowed the possibility that virtually all ^{7}Be captured a proton to make ^{8}B. The decay of ^{8}B to ^{8}Be would be accompanied by neutrinos with typical energies of 7 MeV. W. A. Fowler and A. G. W. Cameron immediately relayed to me these developments. They pointed out that the neutrino flux from these sources could perhaps be easily observed by the chlorine-argon method (Cameron, 1958; Fowler, 1958). The initial estimate was very optimistic, several captures per day if all the ^{7}Be captured a proton to make ^{8}B, so we set up a 3800-liter tank of perchloroethylene (C_2Cl_4) in the Barberton Limestone Mine near Akron, Ohio. Our initial experimental results were disappointing: we didn't detect any clear signal of neutrino-produced ^{37}Ar.

Bahcall (1962) calculated the ^{7}Be electron capture rate after I pointed out the importance of knowing this reaction rate. At about this time, Kavanagh (1960) measured the ^{7}Be (p,γ) reaction rate, which is the one that produces ^{8}B, and found it to be disappointingly low. It is the ratio of the two reactions that consume ^{7}Be that was so discouraging. With most of the ^{7}Be capturing an electron to make ^{7}Li rather than capturing a proton to make ^{8}B (Fig. 2), it would be very hard for us to detect neutrinos. Reines (1960) wrote "The probability of a negative result even with detectors of thousands or possibly hundreds of thousands of gallons of perchloroethylene tends to dissuade experimentalists from making the attempt."

The situation brightened considerably in 1963, when John Bahcall carefully calculated the capture rate of ^{8}B neutrinos and showed that the rate was 20

Figure 4. The pieces for one end of the tank are laid out on the floor of the big room. From *Sharp Bits,* June 1966.

times higher than previously expected. This led us to propose the large chlorine experiment. The theory and experimental approach were laid out in back-to-back papers (Bahcall, 1964; Davis, 1964). The theory provided guidance as to how large a tank to build: the predicted production rate was 4 to 11 ^{37}Ar atoms per day in 100,000 gallons (378,000 liters) of perchloroethylene. Our measurement of the cosmic-ray muon background at Barberton told us how deeply it needed to be buried.

Figure 5. An X-ray source was placed at the axis of the tank and x-ray film was attached to the outside of the tank, in order to test all of the welds. From *Sharp Bits*, June 1966.

The events of the late 50s and early 60s ultimately led Brookhaven National Laboratory, with support from the chemistry office of the Atomic Energy Commission, to build a 100,000-gallon chlorine-argon neutrino detector in the Homestake Gold Mine, in Lead, South Dakota. The construction of the experiment in 1965-1966 was detailed in a series of articles in *Sharp Bits*, published by the Homestake Mining Company for its employees. A large chamber and a smaller control room were excavated at the 4850-foot (1478 m) level of the mine. In Fig. 3, the excavation of the main chamber is shown half finished. The tank was built by the Chicago Bridge and Iron Company, which was known for building municipal water tanks, reactor containment vessels, space simulation chambers, and other kinds of tanks. The tank had to be built in pieces small enough to fit in the mine hoist. Figure 4 shows one end of the tank laid out on the floor of the big tank room of the mine like flower petals. All welds were checked by X-ray. In Fig. 5 an X-ray generator is being placed at the axis of the tank. X-ray film was attached to the tank all the way around a weld and then exposed to X-rays from the source inside the tank. The next step was to vacuum leak-test the tank. The entire tank was pumped out with a 12-inch diffusion pump (Fig. 6) and then leak-tested using a helium leak detector. In order to extract argon from the tank, helium is bubbled through the perchloroethylene. This was done very efficiently using a device called an eductor, shown in Fig. 7 being tested in the Brookhaven Lab swim-

Figure 6. The entire tank was vacuum leak-tested by Chicago Bridge and Iron engineers. A small mass spectrometer capable of sensitive detection was connected to the vacuum pump and helium was introduced around the tank and all the plumbing. From *Sharp Bits*, September 1966.

ming pool. The swimming pool is filled with water, not perchloroethylene. Figure 8 shows the eductors inside the completed tank.

Figure 9 shows the layout of the experiment. During an argon extraction, which occurred every two to three months, two large pumps were used to circulate perchloroethylene through the eductors. The perchloroethylene flow rate was about 1500 liters per minute. While helium was bubbled through the perchloroethylene, the helium atmosphere in the top 5% of the tank was circulated at 17,000 liters per minute through the control room, for extraction of argon. A large condenser near the big tank froze out perchloroethylene and a charcoal trap in the control room (Fig. 10) trapped argon. All of the plumbing was controlled from the control room (Fig. 11). About 95% of the argon in the tank was removed in 20 hours of circulation. After the argon was trapped, it was purified and placed in tiny proportional counters that have internal volumes of 0.25 or 0.5 cm^3 (Fig. 12). In the early years, the samples were counted at Brookhaven, inside shields built made from battleship gun barrels made before the atomic bomb (Fig. 13). Later, the counting apparatus was moved to the Homestake Mine to take advantage of the much better shielding from cosmic rays deep underground.

In order to forecast as accurately as possible the rate of solar neutrino cap-

Figure 7. The eductors, which allowed efficient mixing of helium and perchloroethylene, are being tested here in the swimming pool at Brookhaven, using air and water. From the Brookhaven National Laboratory website, http://www.bnl.gov/bnlweb/raydavis/images/hires/11-755-04.jpg.

ture in the Homestake detector, it was necessary to measure the production cross-sections of the neutrino-producing reactions, and derive their rates in the interior of the Sun. This great effort was largely carried out at the California Institute of Technology under the leadership of William A. Fowler (Nobel Prize in Physics, 1983). Many nuclear astrophysicists and astronomers contributed to the basic physics that supported this early effort in solar neutrino astronomy. Our task at Brookhaven was far simpler and we (Don Harmer, Ken Hoffman, and myself) had the fun of building a large detector and making it work.

The Homestake experiment became operational in 1967. Even after the first run, it was clear the solar neutrino flux was lower than predicted. On August 11, 1967, I wrote to Willy Fowler, who was on a sabbatical at Cambridge University.

> *Dear Willy,*
> I do have a preliminary result from our first good run. The sample was taken June 22nd and counting has continued until today. I am now removing the sample and will rerun background. So we do have a result and during the last few weeks I have told a few people that are interested. I have, of course, had many telephone conversations with John Bahcall.

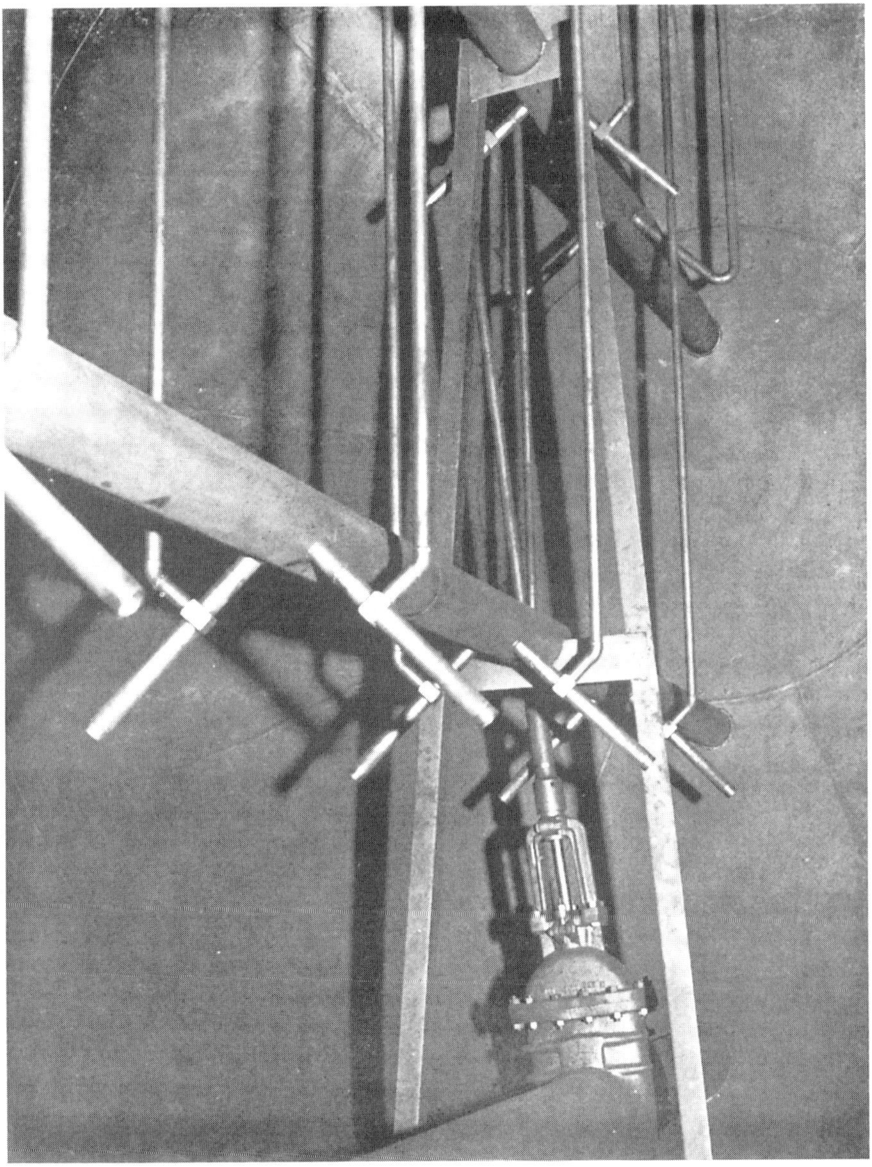

Figure 8. A double row of eductors were installed in the tank. From *Sharp Bits*, September 1966.

The tank and helium purge system seems to work quite well. The first job was to clear out the air argon and reduce its level to about 1 cc before we could make a sensitive measurement. This was accomplished finally and the first good measurement started May 5th. At this time 0.115 cm^3 of Ar^{36} carrier was introduced and the first irradiation started. It ran 48 days, and on June 22nd we recovered the Ar^{36} with a 94 percent efficiency. The sample was counted yielding the pulse height spectrum

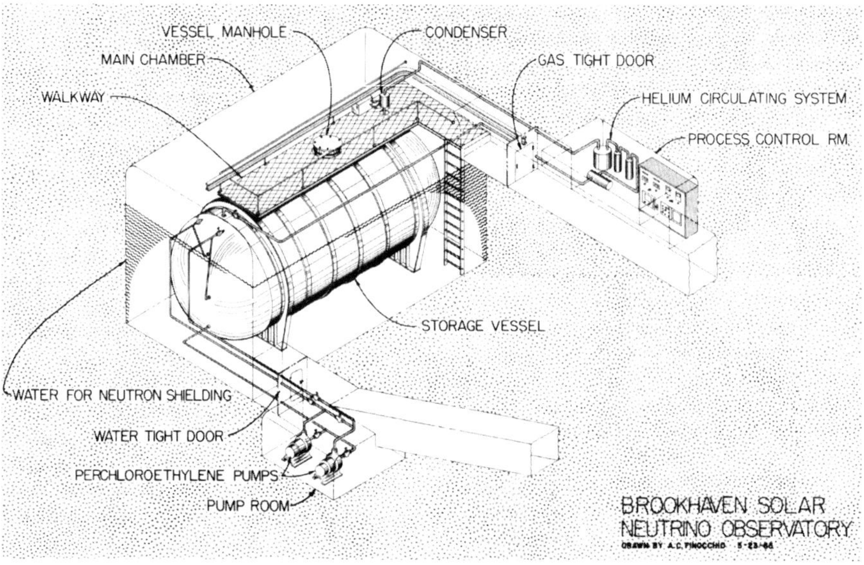

Figure 9. A drawing of the chlorine experiment. From *Sharp Bits*, Spring 1969.

on the enclosed sheet. As you see, there is no visible peak at Ar^{37} (3.0 keV). The background for this counter run just before the sample is shown also on the enclosed sheet. Comparing these we can obtain the following results:

$$\begin{aligned}
\text{Argon from } 10^5 \text{ gal tank} &= 16 \pm 4 \text{ counts (tot. 39.7 d)} \\
\text{Background} = 4 \times (39.7/11.5) &= 14 \pm 4 \text{ counts (for 39.7 d)} \\
\text{Increase} &= 2 \pm 5 \text{ counts} \\
\text{Using: } 2.1 \times 10^{30}\ Cl^{37} &\text{ atoms in tank} \\
\text{Counter efficiency} &\approx 0.50 \\
\text{Then, } \Sigma\phi\sigma &= (0.2 \pm 0.4) \times 10^{-35} \text{ sec}^{-1} \\
&\leq 0.6 \times 10^{-35} \text{ sec}^{-1} \\
\text{Using } \phi(B^8) &= 1.35 \times 10^{-42} \text{ (Bahcall)} \\
\phi B^8 &\leq 0.5 \times 10^7 \text{ cm}^{-2} \text{ sec}^{-1}
\end{aligned}$$

This limit is quite low, but according to the latest opus from Bahcall and Shaviv the B^8 flux is 1.4 (1 ± 0.6) × 10^7 cm^{-2} sec^{-1}. I hope to improve these results by improving the counter background, statistics, and longer irradiations.

Please regard these results as very preliminary. There are several points that must be checked before we are certain this is a bonafide observation. I will collect another sample in September–we are ready now, turn on the sun.

I have hopes of showing you the apparatus sometime in the future. The scenery is not be be compared with the English countryside, but it has its attractions.

Figure 10. Argon was trapped in a large liquid nitrogen-cooled charcoal trap under the table. From *Sharp Bits*, Spring 1969.

The measured rate was one-third that predicted by Bahcall and Shaviv. You can see from my plea to Willy to "turn on the Sun" that I was already concerned.

We reported our first results in September at the American Chemical Society meeting and followed this up with a paper (Davis *et al.*, 1968). The title of the paper was "Search for neutrinos from the sun". As everyone knows, when one uses "Search for" in the title of a paper, it means nothing was found. The upper limit was 3 SNU. In a companion paper, Bahcall *et al.* (1968) gave a prediction from the standard solar model of 7.5 ± 3 SNU. Collection of nearly 30 years of data and 30 years of refinement of the standard solar model have greatly improved precision: the current value from the Homestake chlorine experiment is 2.56 SNU (Cleveland *et al.*, 1998) and the current prediction of the standard solar model is 7.6 SNU (Bahcall *et al.*, 2001). The numbers haven't changed much: the Sun produces one-third as many neutrinos as expected. Thus, the "solar neutrino problem" was born in 1967 and lived until the turn of the century.

Argon-37 decays by capture of a K-shell electron and emission of an Auger electron with a characteristic energy distribution. The energy spectrum from the proportional counter from Davis *et al.* (1968) is shown in Fig. 14. Argon from two runs in the tank and the background gave about the same number of counts in the energy region corresponding to ^{37}Ar decay. It was clear that

Figure 11. John Galvin, a Brookhaven technician who worked with me for many years, and I are shown in front of the control panel. From here the entire experiment could be monitored. From *Sharp Bits*, Spring 1969.

the only way to improve the detection limit for solar neutrinos was to reduce the background.

Brookhaven electronic engineers Veljko Radeka and Lee Rogers came up with a simple, but elegant solution to this problem by devising a pulse rise-time system to discriminate ^{37}Ar decay events from background events. When a decay occurs in a proportional counter (Fig. 12), ion-electron pairs are made and the electrons move rapidly to the central anode, making an electrical pulse. The size of the pulse depends on the number of electrons generated, which in turn depends on the energy of the decay. The dominant decay mode of ^{37}Ar, K-orbital electron capture, deposits 2.82 keV of energy into the counting gas in the form of Auger electrons from atomic electron rearrangement. These Auger electrons have a range of about 100 μm and about 100 ion electron pairs are generated. All electrons have about the same drift time, resulting in a rapidly rising pulse. The principal background is from γ-ray interactions with atoms in the counter, which generates Compton electrons. These are generated along a track passing all the way across the counter, resulting in a more slowly rising pulse. The new electronics measured not only the energy of each decay, but also the rise-time of the pulse. We began using this new system in 1970, and after about one year of observations, a clear neutrino signal was observed. We did some other things to lower backgrounds. We moved the

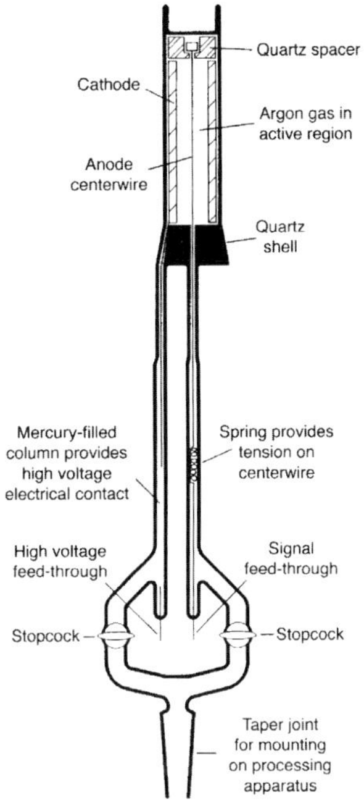

Figure 12. Small proportional counters were used to count ^{37}Ar. They typically have an overall length of 20 cm and an active region 30 mm long and 4.5 mm in diameter. Figure is from Cleveland *et al.* (1998).

counting apparatus to the mine, taking advantage of ~1500 m of rock to reduce cosmic-ray backgrounds in the counters. We filled the room with water, which eliminated a minor background.

The pulse rise-time system development gave the Homestake experiment a new life. The solar neutrino production rate was indeed lower that the solar model predictions by a factor of three. The most likely explanation, in my view at the time, was that the solar model was in error. Many physicists believed that there was something wrong with our experiment. We made many tests of the experiment to improve confidence in the results and never found any experimental reason for the low neutrino flux. We introduced ^{36}Cl-labeled perchloroethylene into a small tank of perchloroethylene. Chlorine-36 decays to ^{36}Ar and we recovered the expected amount of ^{36}Ar. There is a reentrant tube in the big tank in Homestake. We inserted a neutron source, which makes ^{37}Ar by an n,p reaction on ^{37}Cl. The amount expected was recovered. We introduced 500 atoms of ^{37}Ar into the big tank and recovered them quantitatively. I was not told how much ^{37}Ar was introduced until after this check. We introduced argon as a carrier gas at the beginning of each run. We alternately used the stable isotopes ^{36}Ar or ^{38}Ar, so we could check recovery in each run by mass spectrometry after ^{37}Ar counting was complete. Argon-40 is the dominant argon isotope in the atmosphere (99.6% of total argon), so mass spectrometric measurement of the amount of ^{40}Ar tested whether there were any leaks in the apparatus. The counting efficiency of every counter was mea-

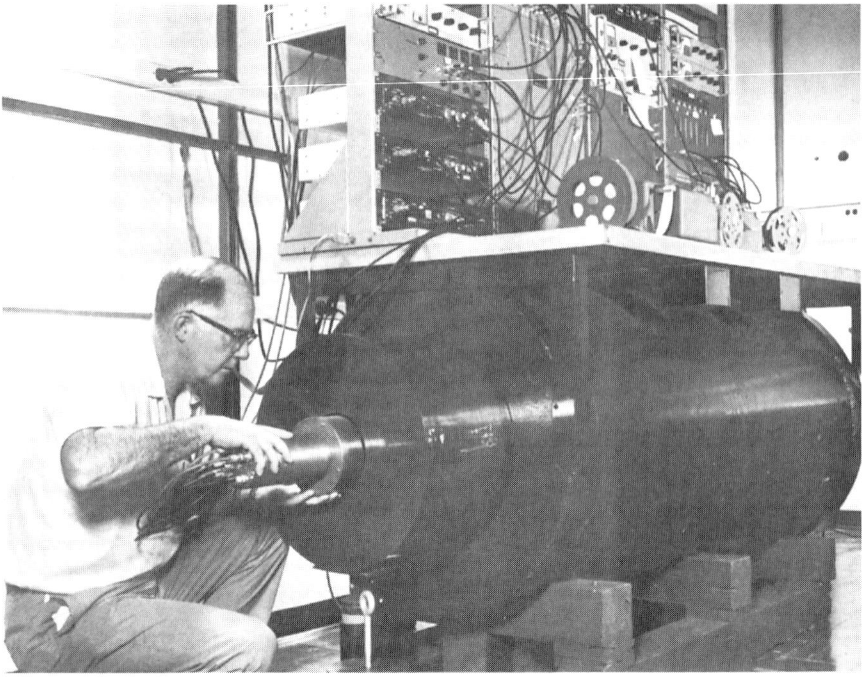

Figure 13. In the first few years of the experiment, the counters were placed in sections of pre-bomb battleship gun barrels for shielding. I am shown loading a counter into the barrel. From *Sharp Bits*, Spring 1969.

sured. These tests and more, as well as the standard operating procedure for the experiment, are described in Cleveland *et al.* (1998).

The solar neutrino problem lasted from 1967–2001. Over this period neither the measured flux nor the predicted flux changed significantly. I never found anything wrong with my experiment. John Bahcall never found anything wrong with the standard solar model, in fact, the advent of helioseismology confirmed the temperature profile in his model. The discrepancy between theory and experiment was a robust factor of three. Cleveland *et al.* (1998) summarized all of the data from the Homestake experiment. One hundred and eight runs were made after rise-time counting was implemented (Fig. 15). Over a period of 25 years, we counted a total of 2200 ^{37}Ar atoms and obtained a solar neutrino flux of 2.56 ± 0.16 (statistical error) ± 0.16 (systematic error) SNU. The current prediction from the standard solar model (Bahcall *et al.*, 2001) is $7.6^{+1.3}_{-1.1}$ SNU.

The results from the Homestake experiment provoked a great deal of activity among theorists. Here are some of the more interesting and, in retrospect amusing, alternatives to the standard solar model. Fowler (1968, 1972) and Sheldon (1969) suggested that there was a secular instability in energy production in the center of the Sun. Since light takes about 10 million years to reach the surface of the Sun, while neutrinos sample the core eigth minutes ago, the energy production could be low at the present time. Neutrino

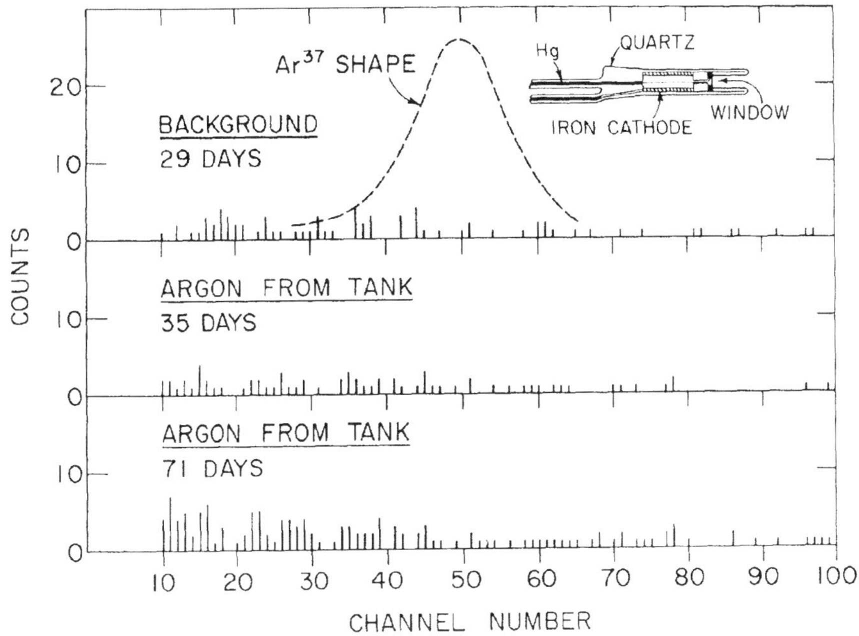

Figure 14. Pulse-height spectra from the first two runs on the chlorine experiment. No counts in excess of background were detected in either run. Figure from Davis *et al.* (1968).

oscillations were suggested by Gribov and Pontecorvo (1969) and Wolfenstein (1978) and the theory was further developed by Mikheyev and Smirnov (1985) into what is now known as the MSW effect. Although neutrino oscillations now seem to be the right answer, it should be remembered that at the time most physicists viewed this as an elegant theory, but not very likely. Some of the other possibilities are more fanciful, but all from well respected physicists and astrophysicists. Libby and Thomas (1969) and Salpeter (1970) suggested that quark catalysis could play a role. Kocharov and Starbunov (1970) suggested that there was an overabundance of ^{3}He in the present Sun. Cisneros (1969) proposed that the neutrino had a significant magnetic moment. Bahcall *et al.* (1972) suggested that neutrinos might decay. Demarque *et al.* (1973) suggested that the solar interior rotated rapidly, lowering the central pressure and temperature. Prentice (1973) proposed that the Sun was in a later stage of stellar evolution, such that hydrogen was burned out and the core was made of helium. Clayton *et al.* (1975) proposed that the Sun's energy did not come from fusion, rather from release of energy from accretion onto a black hole at the center of the Sun.

The Homestake experiment was the only measurement of the solar neutrino flux for a long time. We had to wait 23 years for the Kamiokande experiment to confirm that the solar ^{8}B neutrino flux was low (Hirata *et al.*, 1990; Fukuda *et al.*, 2001). In the 1990s, two radiochemical experiments that captured neutrinos using the inverse beta-decay of ^{71}Ga, SAGE (Abazov *et al.*, 1991; Abdurashitov *et al.*, 1999) and GALLEX (Anselmann *et al.*, 1992; Alt-

Apr. '87 to Aug. '87: Visiting Professor at DESY and University of Hamburg.

Aug. '87 to Mar. '97: Professor, Tokai University.
Sep. '87 to Aug. '88; Guest Professor, CERN, Sep. '89 to Dec. '89; Distinguished Visiting Professor, University of Chicago. Feb. '90; Regent Lecturer, University of California, Riverside, Jan. to Mar. '94; Sherman Fairchild Distinguished Scholar, California Institute of Technology. Apr. '95 to Mar. '97; Director, Washington Liaison Office, Japan Society for Promotion of Science. Sep. '96 to Mar. '97; Distinguished Visiting Scholar, George Washington University.

Jul. '98 to Aug. '99: Alexander von Humboldt Preisträger staying at DESY in Hamburg, Max-Planck-Institut für Kernphysik in Heidelberg, and Max-Planck Institut für Extraterrestrische Physik in Garching.

Present Councilor, International Center for Elementary particle Physics, The University of Tokyo.

Important publications

"Observation of a Neutrino Burst from the Supernova SN1987a" Phys. Rev. Lett., 58 (1987) 1490.

"Results from One Thousand Days of Real Time, Directional Solar Neutrino Data; Kamiokande-II Collaboration", Phys. Rev. Lett., 65 (1990) 1297.

Honors and Prizes:

Aug. '85:	Das Grosse Verdienstskreuz from the President of Federal Republic of Germany.
Dec. '87:	The Nishina Prize from the Nishina Foundation.
Jan. '88:	The ASAHI Prize from the ASAHI Press.
Nov. '88:	The Order of Culture by the Japanese Government.
Jun. '89:	The Academy Award from the Academy of Japan.
Jun. '89:	The Bruno Rossi Award from the American Physical Society.
Aug. '96:	The Special Prize from the European Physical Society.
Mar. '97:	The Alexander von Humboldt Prize from the Humboldt Foundation.
Jun. '97:	The Fujiwara Prize from the Fujiwara Science Foundation.
Nov. '97:	The Order of Cultural Merit conferred by The Emperor of Japan in person.
Jan. '99:	The second ASAHI Prize from the ASAHI Press.
Jan. '99:	The Diploma di Perfezionamento honoris causa in Fisica from The Scuola Normale Superiole, Pisa, Italy.
Jul. '99:	Doktor der Naturwissenschaften ehrenhalber from Hamburg University.

May '00:	Rochester's Distinguished Scholar Award from the University of Rochester.
May '00:	The Wolf Prize from The State President of Israel.
Jun. '00:	Citation by the Town of Kamioka.
Aug. '00:	Citation by the Governor of Gifu Prefecture.
Apr. '02:	Panofsky Prize from American Physical Society.
Dec. '02:	Nobel Prize in Physics.
Dec. '02:	Member, The Japan Academy.

BIRTH OF NEUTRINO ASTROPHYSICS

Nobel Lecture, December 8, 2002

by

MASATOSHI KOSHIBA

International Center for Elementary Particle Physics, University of Tokyo, 7-3-1-Hongo, Bunkjo-Ku, Tokyo 113-0033, Japan.

In giving this talk I am very much helped by the preceding talk because I can skip some of the topics. If you want further information, please refer to my review article, "Observational Neutrino Astrophysics," [1].

I am to talk about the birth of the neutrino astrophysics, but before the birth, there was a very important event, which was just described by Prof. Davis. [2]. It was the radiochemical work using the reaction $\nu_e + {}^{37}Cl$ going to $e + {}^{37}Ar$. He found that the observed neutrino flux was only 1/3 of the theoretically expected. This could be considered as the conception of the neutrino astrophysics and was in fact the impetus for us to begin seriously working on the solar neutrinos.

I will talk about two experiments. The first is the original KamiokaNDE, which might be called an Imaging Water Cerenkov detector with a surface coverage of 20% by photomultipliers and the total mass of the water inside this detector is 3,000 tons. It costed about 3 million U.S. dollars. This, mind you, was meant to be the feasibility experiment on the astrophysical detection of solar neutrinos. The second experiment is called Super-KamiokaNDE, the same type of detector but with a better light sensitivity, that is, 40% of the entire surface was covered by the photocathode and the total mass of the water was 50,000 tons. It costed about 100 million U.S. dollars. This was considered to be the full-scale solar neutrino observatory.

Both the experiments are situated about 1,000 meters underground in Kamioka Mine. The capital letters NDE at the end of the two experiments originally implied "Nucleon Decay Experiment." However, because of our detection of various neutrinos by these detectors, people started calling it, "Neutrino Detection Experiment".

Fig.1 shows the interior of KamiokaNDE. You can see arrays of photomultipliers on the sidewalls as well as on the top and at the bottom. When we were preparing for this KamiokaNDE experiment, we heard that a much bigger, experiment but of similar type, was being planned in the United States, [3]. We had to think very seriously about the competition with this bigger detector. Both experiments aimed at the detection of a certain type of proton decay, i.e., $e^+ + \pi^0$ mode. If we were aiming only for the detection of such particular types of proton decays, certainly the much bigger U.S. experiments would

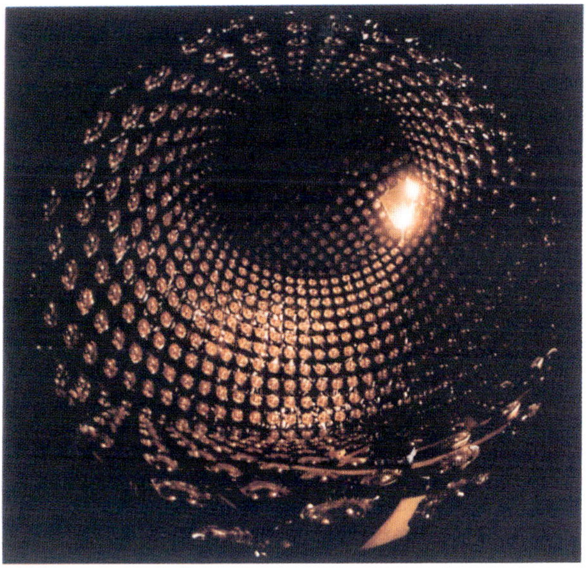

Figure 1. The interior of KamiokaNDE.

Figure 2. The newly developed large photomultiplier.

Figure 3. The interior of S-K through fish-eye lens.

find it first. Then, what could we do with a smaller detector? We thought very seriously about this competition and we came to the conclusion that the only possible way to compete with this bigger detector would be to make our detector much more sensitive than the U.S. competitors so that we could not only detect the easiest proton decay mode, but also measure other types of proton decays. Then eventually we could say that the proton decays into this mode with this branching ratio and into that mode with that branching ratio and so forth. Then our experiment would be able to point the way to the possible future, what is called the Grand Unified Theory, which is a new type of theory combining strong forces, weak forces, and electromagnetic forces.

Thanks to the cooperation of Hamamatsu Photonics Co., we jointly developed very large photomultiplier tubes [4]. I was so happy, as you can see in Fig. 2 that this tube was successfully developed.

Fig. 3 shows the fish-eye view of the Super-KamiokaNDE interior. You can see many more phototubes, a total of about 11,000 big phototubes.

Since I suppose that not many people are familiar with this type of detector, I want to show you the performance of Super-KamiokaNDE. The first example is a very slow motion picture of a cosmic ray muon passing through the detector.

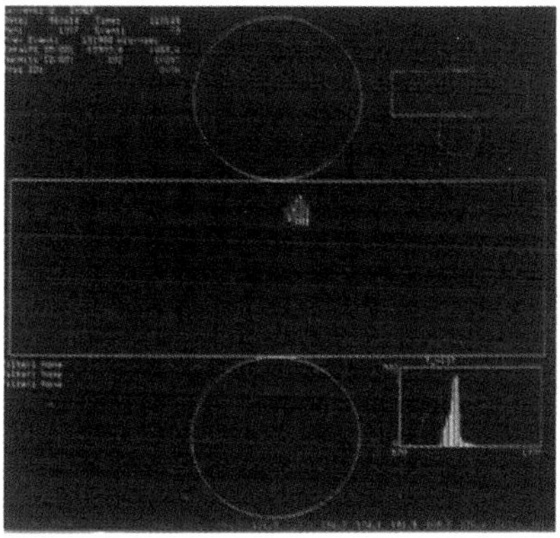

Figure 4-1. μ just entered S-K.

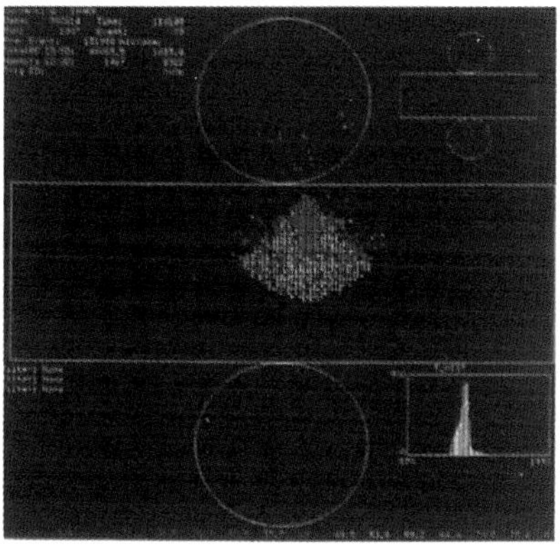

Figure 4-2. 50 nanoseconds later.

As is well known, special relativity prohibits any particle to move faster than the velocity of light in vacuum. However, in a media like water, the light velocity itself is reduced to three-quarter of its value in vacuum. Therefore, when the particle energy is very high, its velocity can exceed the velocity of light in the water. Then, what happens is that such high energy, high velocity, particles in water will generate, what might be called, a shock wave of light; the Cerenkov light. It is emitted in a cone shape with the axis on the trajectory of the moving electrically charged particle.

Fig. 4-1 shows the response of Super-KamiokaNDE when a muon just entered the detector. The Super-KamiokaNDE detector is opened up here. The sidewall is cut vertically at one point and is spread flat, the upper lid is opened up, and the bottom lid is pulled down. Each dot here represents a photomultiplier. Red light shows it received a large number of photoelectrons. The different colors indicate different numbers of received photoelectrons. At the right below is the time profile of the total number of photons received. Fig. 4-2 shows the pattern 50 nanoseconds later. You can see that the particle is moving faster than the Cerenkov light wave front. Fig. 4-3 , another 50 nanoseconds later, shows that while the Cerenkov light is still on its way the muon has already reached the bottom. You can see that the particle is traveling faster than the light velocity in water. Figs. 4-4, 4-5, and 4-6 show the subsequent development of the event. You can see that with this detector the electrically charged particle can be observed in detail. Next figure, Fig. 5, shows two events, e-event above and μ-event below. Looking at these two examples, one by an electron and the other by a muon, you can see the difference in the distribution of the detected photons, especially in the radial distribution of photons. Electrons and muons are very similar particles except that their masses are different by a factor of about 200. It means that in traversing water, the heavier μ-particle suffers much less scattering while the lighter electron gets scattered much more. Not only that, the electron emits γ-rays, which in turn get converted into electrons and positrons. Those low energy electrons, and positrons, get scattered violently. Therefore, the

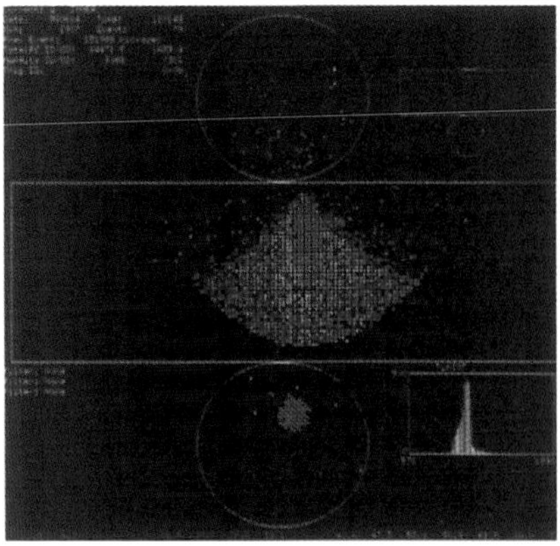

Figure 4-3. μ reached the bottom.

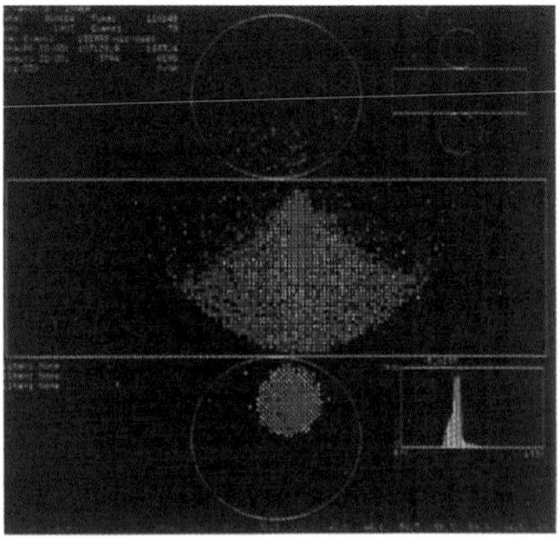

Figure 4-4.

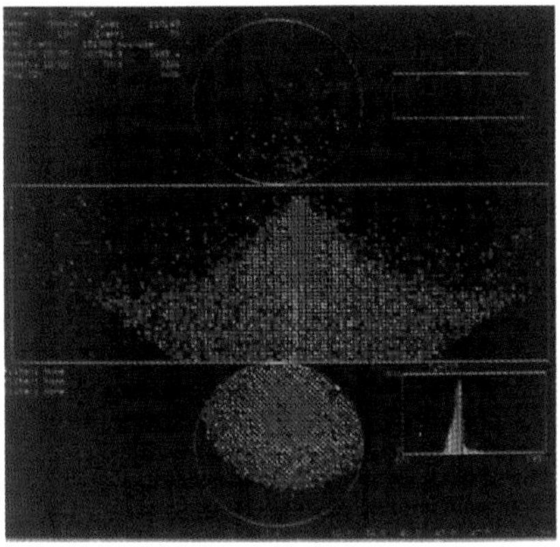

Figure 4-5.

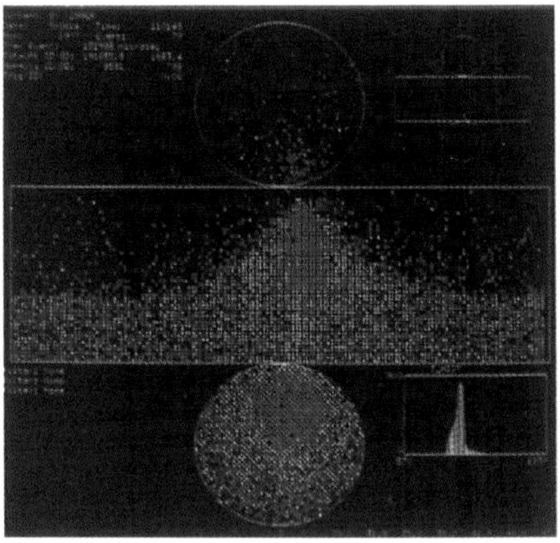

Figure 4-6.

Cerenkov light emitted by those low energy particles is widely distributed as you see in the upper event. By making a quantitative measurement of the radial distribution of those photons, you can make a very good distinction between a μ-event and an e-event with a mistaking probability of less than 1%. This is a very nice feature of this detector and led us eventually to discover what is called "the atmospheric neutrino anomaly."

The old KamiokaNDE produced four significant results.

The first is the astrophysical observation of solar neutrinos by means of ν_e-e scattering with the electron in the water, [5]. By astrophysical observation we mean that all the necessary information is available; i.e., the arrival direction, the arrival time and also the spectral information on the incoming neutrinos. In the case of ν_e-e scattering, since the electron rest mass is only 0.5 MeV, for an incoming neutrino of, say, 10 MeV, the struck electron goes almost in the dead forward direction. By observing this recoil electron, you can approximately infer the arrival direction of the neutrino. Also, the energy spectrum of the recoil electrons has a one to one relation to the original neu-

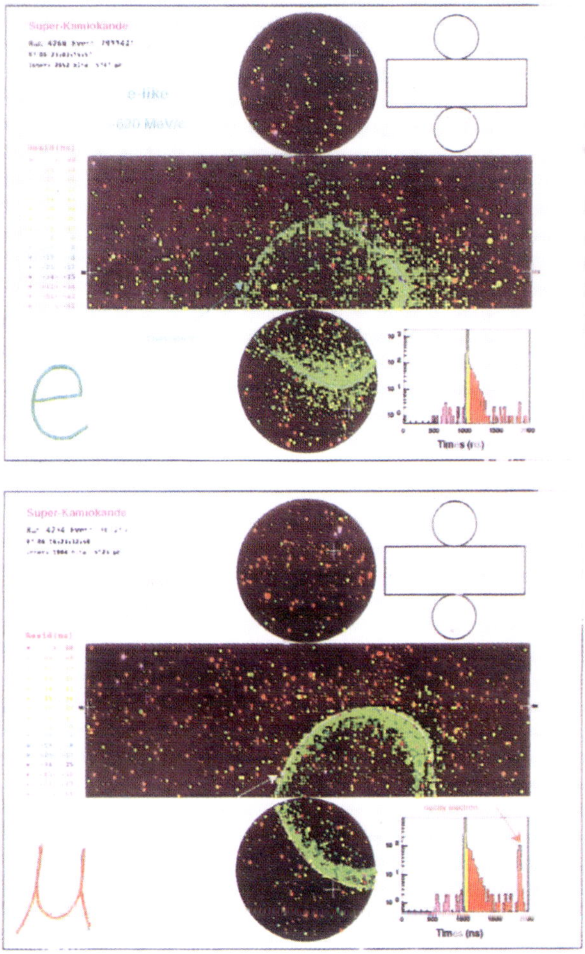

Figure 5. e-event above and μ-event below.

trino energy spectrum. The timing is accurate to better than ten nanoseconds.

The second is the observation of supernova neutrinos [6] by means of anti-ν_e on protons in water. This reaction produces an e^+ and a neutron. The e^+ is observed by the Cerenkov light it emits.

The third is the discovery of what is called the Atmospheric Neutrino Anomaly, [7]. Since we can definitely separate μ-event and e-event, as I have shown you before, we could measure the number ratio of ν_μ over ν_e very accurately by observing μ-event and e-event separately. It was the discovery of slightly more than 4 significance, but this result was later firmly confirmed at more than 9σ by the data of Super-KamiokaNDE.

Not many people are interested in proton decay any more but the non-observation of proton decay by the KamiokaNDE experiment killed the well-known Grand Unified Theory based on SU[5].

Solar Neutrinos

Standard Solar Model (SSM)

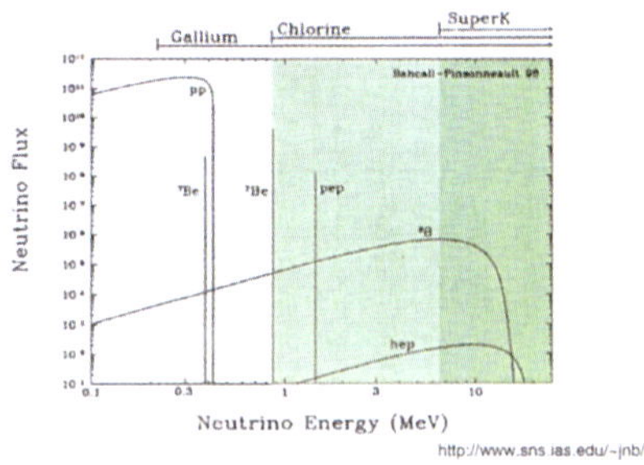

Solar Neutrino Experiments

	Target	Data / SSM (BP98)
· Homestake	^{37}Cl	0.33±0.03
· Kamiokande	e^- (water)	0.54±0.07
· SAGE	^{71}Ga	0.52±0.06
· GALLEX	^{71}Ga	0.59±0.06
· SK	e^- (water)	0.475±0.015

Figure 6.

The previous speaker showed this diagram, Fig. 6, and I am not going into the detail here but instead just ask you to notice the threshold energies of various experiments.

Fig. 7 is to shows the feasibility for KamiokaNDE of observing solar neutrinos with directional information. You can see that above the isotropic background, the accumulation of events is in the direction from the sun to the earth.

Next one, Fig. 8, shows the energy spectrum as normalized to the theoretical one. From the figure you can see that the shape is not very much differ-

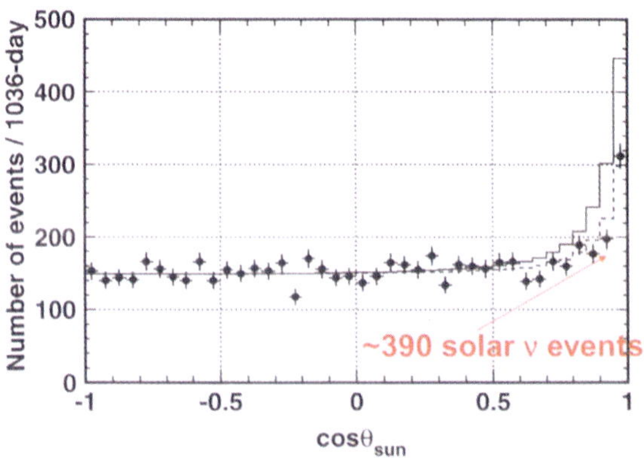

Figure 7. The directional observation of Solar neutrinos.

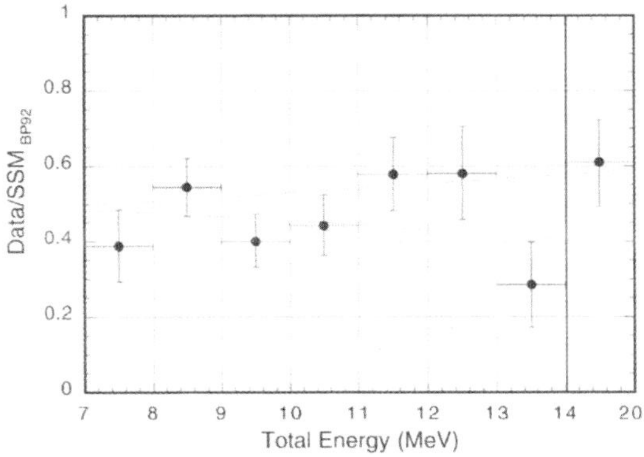

Figure 8. The normalized energy spectrum.

ent from the expected theoretical anticipation, but the intensity is almost one-half.

I now go on to the observation of supernova neutrinos. Thanks to the collaboration of Pennsylvania State University led by Prof. A. K. Mann, we could improve the performance of our detector very much by reducing the background, purifying the water, and so forth. At the very beginning of 1987; our detector was already calm enough to start taking data on the solar neutrinos. Two months later, we heard that there was a supernova explosion in the southern sky. So we immediately looked at our data and then we found the supernova neutrino signal very easily because our detector was already capable of taking solar neutrino data, which are much more difficult to observe than the supernova neutrinos; because the supernova neutrinos have considerably higher energies than the solar neutrino and furthermore those supernova neutrinos are bunched in a short period of time. It is shown in Fig. 9. You can clearly see the supernova neutrino signal of about 17 photoelectrons above the background events. This observation gave the confirmation of theoretical ideas on the supernova explosion triggered by a gravitational collapse. For instance, not only the average energy and the total number of these events agreed with the theoretical expectations, but also the time duration of about ten seconds implies that those neutrinos are emitted from a very, very dense matter, like in a nucleus.

If they were emitted from a tenuous stellar body, the time duration of the signal would have been less than one millisecond. But those neutrinos had to get diffused out of a very dense, nucleus-like, matter so that it took ten seconds to get out of this surface; probably a neutron star is responsible.

Now I come to the discussion of "the Atmospheric Neutrino Anomaly." When cosmic ray particles enter the atmosphere, they interact with the N and O nuclei to produce π-mesons and K-mesons. These mesons decay in tenuous air into μ and ν_μ. So you get one muon and one ν_μ there. If the secondary μ also decayed then you get additional ν_μ and ν_e. So if everything proceeded this way, you get two ν_μ's against one ν_e. The number ratio, N (ν_μ) / N (ν_e) is

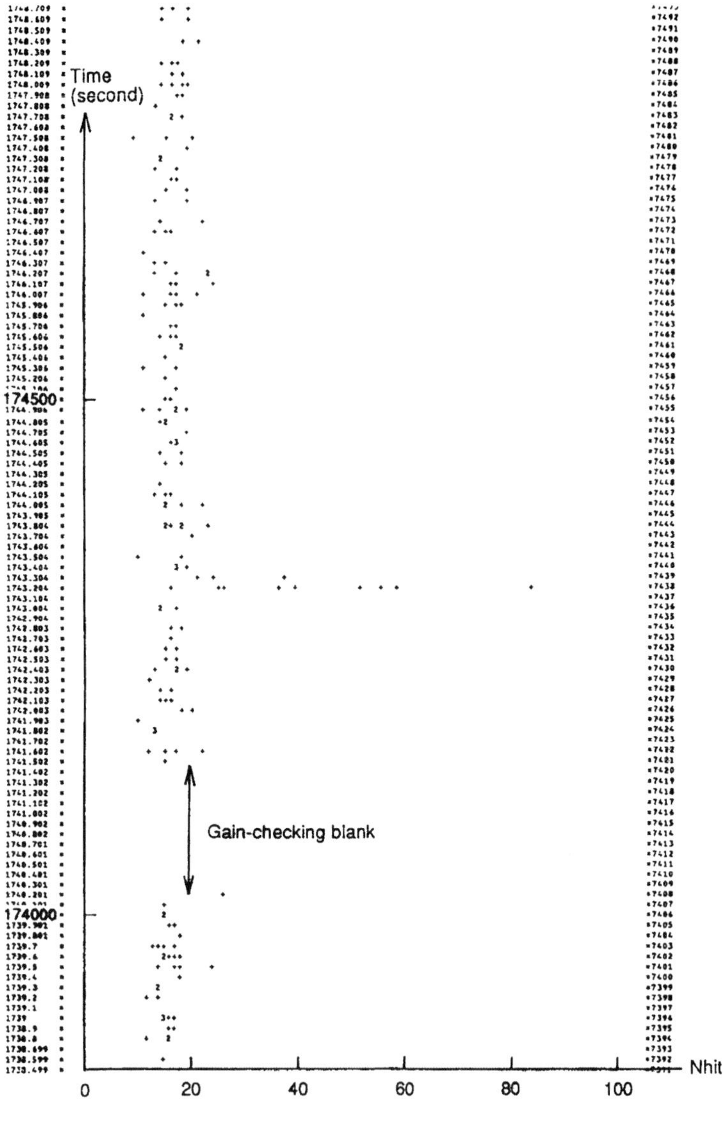

Figure 9. The SN1987A neutrino "signal" in the computer print-out.

thus two. When you go to higher energy, a μ of longer lifetime than a π-meson cannot decay. Indeed, some μ's do reach our detector, as you have seen before. In this case, you do not get additional ν_μ or ν_e. So at high energies, this ratio becomes larger than two.

In Fig. 10 are shown the above number ratio observed by KamiokaNDE together with the results of other experiments.

I now go on to the discussion of the neutrino oscillations, [9]. This may be the most difficult part of my talk. I will try to make it understandable to a first year undergraduate student.

For the sake of simplicity, we consider there are only two kinds of neutrinos

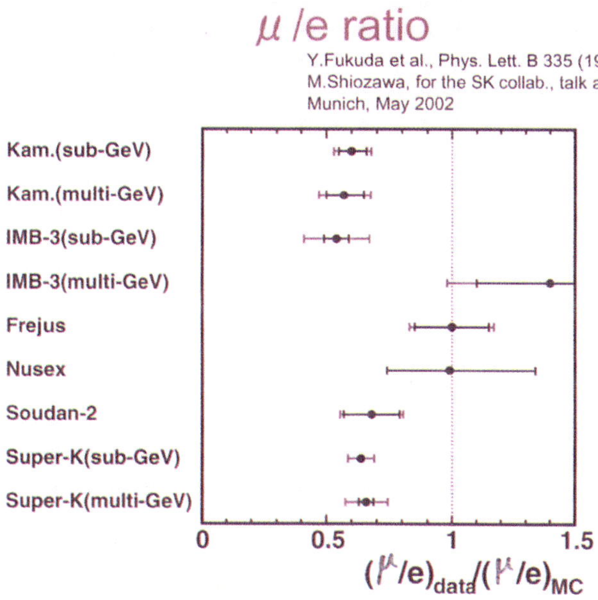

Figure 10. The number ratio $N(\nu_\mu) / N(\nu_e)$.

in nature. Then, for instance, the wave function describing the state of a neutrino can be described by a linear combination of two independent base functions. For instance, you can take the mass matrix to be diagonal and then choose the two basic vectors of mass m_1 and mass m_2, respectively. So any neutrino state can be described by a combination of ϕ_{m1} and ϕ_{m2}. $\phi_{\nu\mu} = \cos\theta\ \phi_{m1} + \sin\theta\ \phi_{m2}$. This is like two-dimensional geometry. A vector can be described by its x component and y component. So the ν_μ state is a linear combination of m_1 state and m_2 state with an angle parameter θ. The two states, ϕ_{m1} and ϕ_{m2}, oscillate with their characteristic frequencies. This frequency is proportional to the total energy of the state. If the mass m is small, then for a given momentum one can make the following approximation, $E \sim p+m^2/2p$. $E_1 - E_2$, which is proportional to the frequency difference of these two states, is then, using this approximation, proportional to $(m_1^2-m_2^2)$. This m-square difference between the two states is designated by Δm^2. When there are two oscillations of nearly equal frequencies coexist, there occurs a phenomenon known as "beat" in which the amplitudes of the two oscillations change slowly with the difference frequency. This change of the component amplitudes, ϕ_{m1} and ϕ_{m2}, induces the appearance of ν_τ-state in the original pure ν_μ state.

By using these two parameters, Δm^2 and θ, you can describe the oscillation of neutrinos from one type to the other.

In Fig. 11 is shown the result obtained by KamiokaNDE, [10] , on the atmospheric neutrino oscillations.

We now proceed to the discussion of Super-KamiokaNDE.

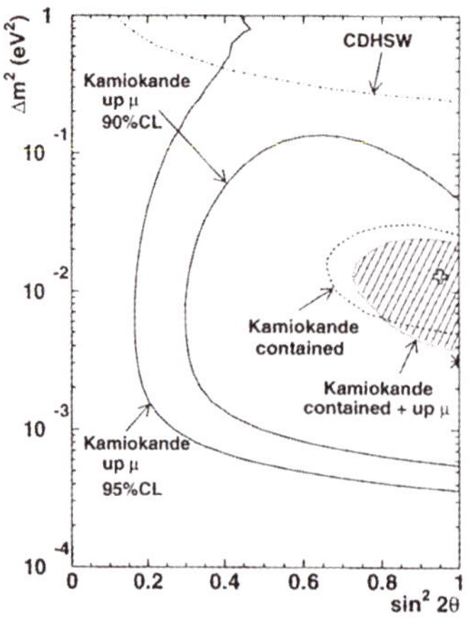

Figure 11. The allowed parameter region.

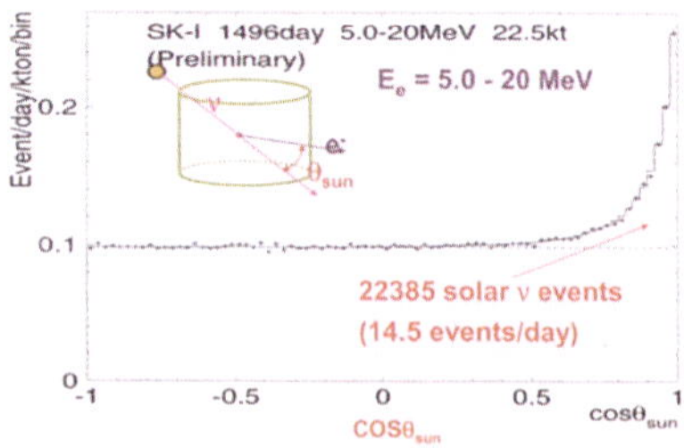

Figure 12. The directional observation.

The Super-KamiokaNDE so far produced three significant results.

The first is the astrophysical observation of the solar neutrinos with a comfortable statistics. In Fig. 12 you can see the peak of neutrinos in the direction from the sun to the earth above the isotropic background. When you break your hand you go to the doctor and get an X-ray picture taken. You then can see the inside of your hand. A bone may be broken. When you use neutrinos, with a much larger penetrability, you can see the inside of the sun. In Fig. 13

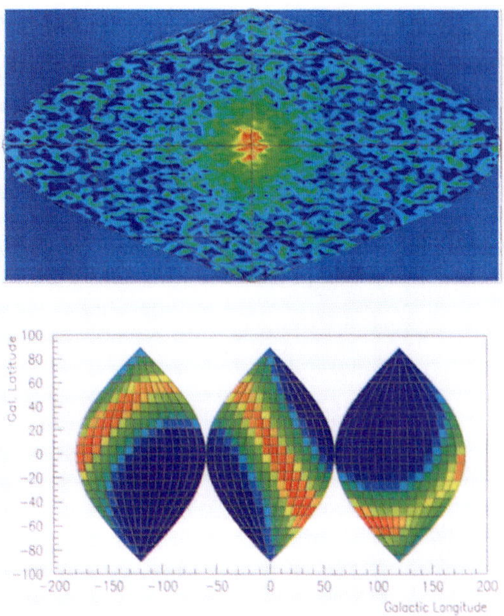

Figure 13. The neutrinograph of the sun.

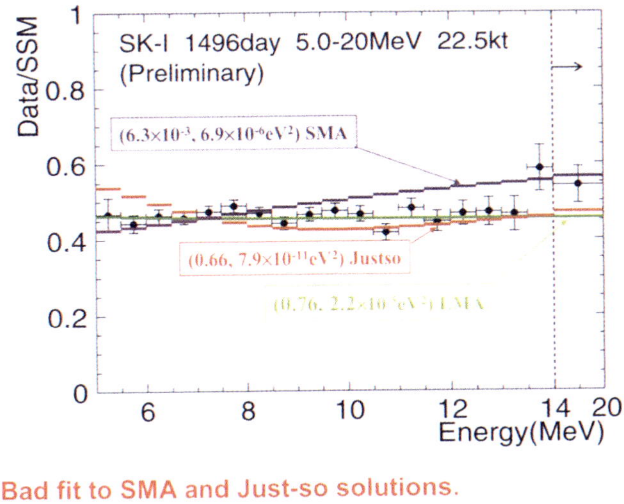

Figure 14. The energy spectrum.

is shown the first neutrinograph, rather than photograph, of the sun. Below is the orbit of the sun in the galactic coordinates as seen by the neutrinos.

This sounds very nice, but if you look at this neutrinograph carefully, you find the size of sun is much bigger than the size of sun as you see by your own eyes. The reason is, of course, that the directional accuracy of the neutrino observation is much worse than that of visible light. But you have to be patient. The neutrino astrophysics is just born. It is still in its infantile stage.

Fig. 14 shows the observation of the solar neutrino energy spectrum as compared to the theoretically expected from the Solar Standard Model.

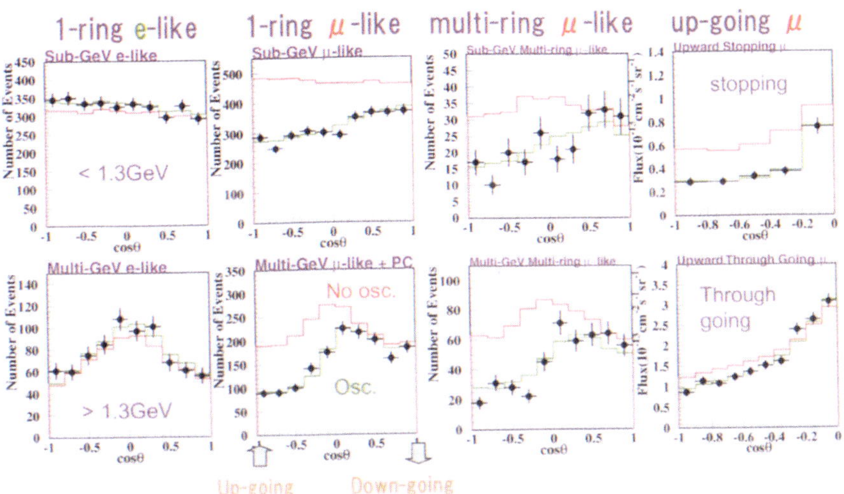

Figure 15. The change of oscillation as a function of path length.

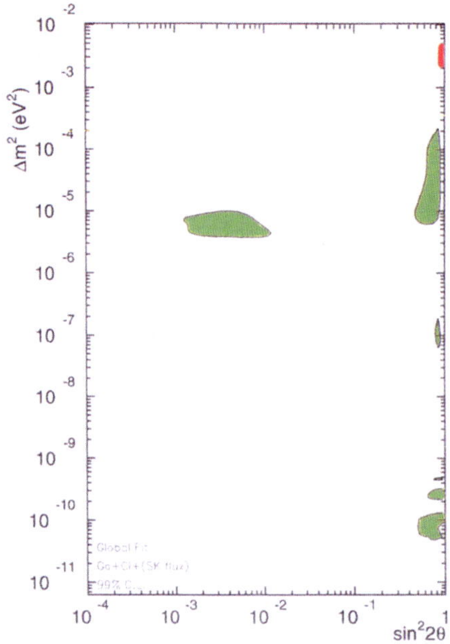

Figure 16. The allowed regions of oscillations.

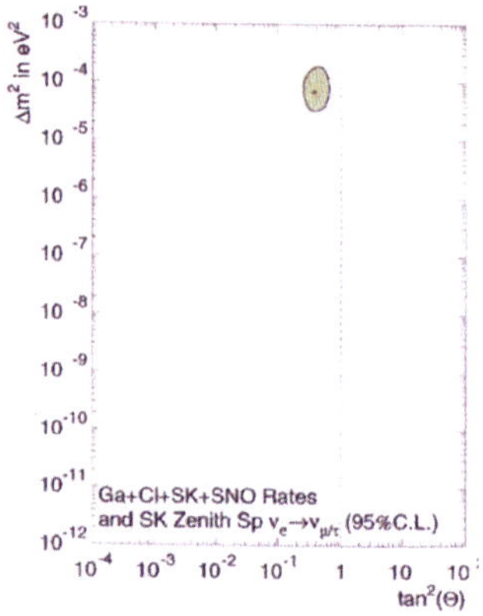

Figure 17. The allowed region for the solar neutrino oscillation.

Detailed comparison of this observed energy spectrum with the theoretical expectation gives us better information on the solar neutrino oscillation.

If the observed anomaly in the N (ν_μ) / N (ν_e) is indeed due to the neutrino oscillation, then the degree of oscillation would be different depending on the path lengths the neutrino had to traverse from its generation to our detector. When it comes from vertically above, it is only 20 kilometers. When it comes horizontally, it traveled some 1,000 kilometers. If it comes from the bottom, it was produced 13,000 kilometers away. There is a big difference in the path lengths (Fig. 15).

In the case of e-events, due to ν_e , there is no deviation from the no-oscillation expectation. Only in the case of μ-events, due to ν_μ, one sees a large reduction in the direction from the bottom. Only in the case of muon, you see this deficiency in the large distance direction. Fig. 16 shows the allowed regions for the solar neutrino oscillations, painted yellow, and that of atmospheric neutrino oscillations, painted red as determined by the data of Super-KamiokaNDE ,[11].

With the oscillation data described above of KamiokaNDE and of Super-KamiokaNDE we go on to combine them with the other available data. Next figure, Fig. 17, shows only one possible oscillation region for the solar neutrino oscillation. This was accomplished by combining all the solar neutrino experiments. Super-KamiokaNDE, SNO and other radio-chemical results [14–17].

Now that the observed Δm^2 's are definitely not zero we have to admit

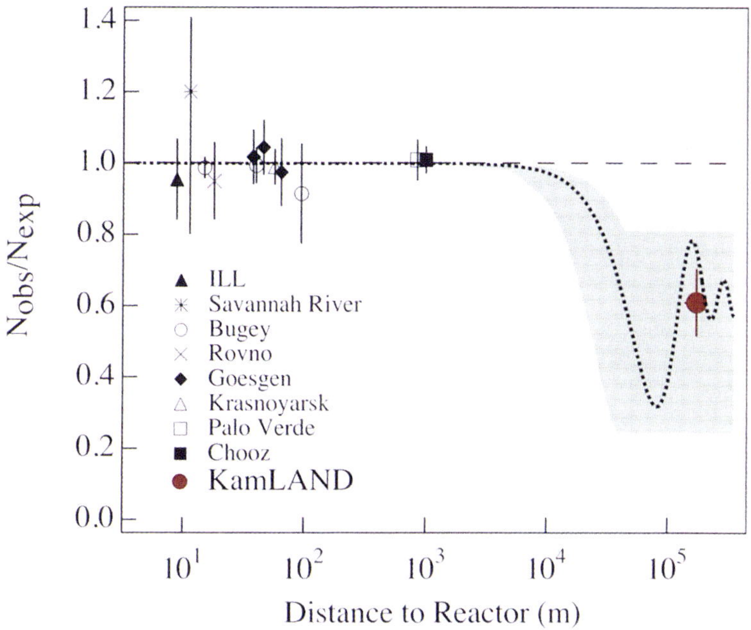

Figure 18. Results from KamLAND.

some non-zero masses for the neutrinos. This implies that the Standard Theory of elementary particles has to be modified.

Now, for the sake of giving proper credits, I give the author list of supernova neutrino detection in ref. [6] and the author list of the atmospheric neutrino paper in [12].

Lastly I show you the latest result from Kamioka. In Kamioka, there is a third generation experiment now working. This KamLAND experiment is installed in the old cave of the original KamiokaNDE and this experiment uses liquid scintillator to measure the anti-ν_e's from the reactors about 200 kilometers away. And this experiment published their first result [18] only two days ago and I got this by e-mail. The experiment is measuring the anti-neutrino flux as well as the energy spectrum. The result is shown in Fig. 18. The obtained oscillation parameters, $\sin 2\theta = 0.833$ and $\Delta m^2 = 5.5 \times 10^{-5}$ (eV)^2, are in good agreement with the solar neutrino result of Fig. 18.

Since this is a confirmation of the neutrino oscillation not for the electron neutrino but for anti-electron-neutrino, the fact that it is giving the same oscillation parameters implies that the CPT theorem is not violated. Further data accumulation may lead to some interesting insight into the CP problem within the framework of CPT invariance. Reference to this paper is given in [18]. The interesting thing is that about two-thirds of the collaborators are from the United States. Some say Kamioka is now considered as the "Mecca" for neutrino research and this pleases me very much.

Now that Neutrino Astrophysics is born, what should we do next? Of course the plan depends on whom we ask. There is a move to build a megaton Hyper-KamiokaNDE. A world network of at least three Super-KamiokaNDE's may be a good choice for supernova watching. The most challenging problem will be the observation of the Cosmic Neutrino Background of 1.9K, which would tell us the state of our universe 1 second after its birth. The non-zero masses of neutrinos imply the total reflection at low temperature of low energy neutrinos. This is a wonderful gift providing the possibility of parabolic mirror for focusing CNB. The detection, however, of such low energy neutrinos is really a formidable task.

ACKNOWLEDGEMENTS

It is my pleasure to acknowledge the technical contribution of Hamamatsu Photonics Co. for producing the 50cmϕ photomultipliers. They were the essential ingredient of the Kamioka experiments. The Ministry of Education, Culture and Science of Japan gave generous support to the Kamioka experiments, for which we are all grateful.

REFERENCES

[1] M. Koshiba, Phys. Rep. **220** (1992) 229.
[2] R. Davis, Jr., Nobel Lecture in Physics, Dec. 2002.
[3] R. M. Bionta *et al.*, Phys. Rev. Lett. **51** (1983) 27.
[4] H. Kume *et al.*, Nucl. Inst. and Meth. **205** (1983) 443.
[5] K.S. Hirata *et al.*, Phys. Rev. Lett. **63** (1989) 16.
[6] K. Hirata *et al.*, Phys. Rev. Lett. **58** (1987) 1490.
[7] K. S. Hirata *et al.*, Phys. Lett. B **205** (1988) 416.
[8] H. Georgi and S.L. Glashow, Phys. Rev. Lett. **32** (1974) 438.
[9] Z. Maki, N. Nakagawa, and S. Sakata, Prog. Theor. Phys. **28** (1962) 870.
[10] S. Hatakeyama *et al.*, Phys. Rev. Lett. **81** (1998) 2016.
[11] S. Fukuda *et al.*, Phys. Lett. B **539** (2002) 179.
[12] Y. Fukuda *et al.*, Phys. Rev. Lett. **81** (1998) 1562.
[13] Q. R. Ahmad, *et al.*, Phys. Rev. Lett. **89** (2002) 011301.
[14] B. T. Cleveland *et al.*, Astrophys. J. **496** (1998) 505.
[15] W. Hampel *et al.*, Phys. Lett. B **447** (1999) 364.
[16] J. N. Abdurashitov *et al.*, J. Exp. Theor. Phys. 95, (2002) 181.
[17] M. Altmann *et al.*, Phys. Lett. B **490** (2000) 16.
[18] K. Eguchi *et al.*, Phys. Rev. Lett. **90** (2003) 021802.

RICCARDO GIACCONI

In Memory of Marc Antonio Giacconi
and dedicated to my wife, Mirella,
and to our daughters, Anna and Guia.

I. EARLY YEARS

I was born in Genoa, Italy, on October 6, 1931, but I spent most of my life until 1956 in Milano. I was the only child. My mother, Elsa Canni Giacconi, was a teacher of Mathematics and Physics at the high school level. She was the co-author of many textbooks on geometry which were widely adopted in Italy. She held that God made geometry. My father, Antonio Giacconi, owned a small business. He had a knack of seeing historical developments clearly and to perceive when the King was naked. He was an anti-fascist and suffered for it. My mother and father were legally separated when I was eight years old.

I experienced World War II as a young teenager. I was sent to live with my aunts, Giulia and Elisa Canni, in Cremona following the 1942 bombings of Milano. The son of Giulia, Giovanni Benini, of my same age, became the only brother I ever had. I had access to a large library, including classics, and I read copiously.

I returned to Milano in 1944 and lived there through 1956. My schooling was not completely conventional. I went to a German kindergarten, then first year elementary school in a Military College, then third year in Genoa, then again third year in Milano. I cut school frequently in Genoa and I was a discipline problem in Milano. I had my best learning in Cremona. I jumped last year of high school directly to the University of Milano. I obtained the doctorate in Physics in four years. Outside of school I loved climbing, hiking, and skiing. I traveled throughout Europe and I enlisted for two weeks as a trimmer on an English trawler out of Grimsby.

II. EDUCATION

Up to and including university my interaction with the educational system was always difficult. In high school I loved to point out the mistakes by our mathematics teachers. At the university, I had difficulties following the heavy burden of lectures although I was able to pass all tests with relatively high grades. My salvation was to start doing research in the very first year of university. At the University of Milano there was at the time an active group of cosmic ray researchers interested in studying muons, lambda particles, and proton interactions. A very kind and bright young professor, Antonio Mura, put me to work doing literature searches and summaries, while Carlo Succi put me in the laboratory building power supplies with surplus U.S. parts and building and operating cloud chambers.

This training was in stark contrast to the pedantic approach of many of the older teachers. The lectures in physics, chemistry, physical chemistry, and theory were of little use to me. I still loved geometry, analysis, and physics as I saw it from a researcher's point of view.

I did my thesis work on the development of nuclear interactions by protons in the lead plates of a cloud chamber at the laboratory of Testa Grigia (3500 m). It took me two years to collect 80 proton interactions. I prepared a modest thesis confirming Fermi's model and I was finally out of formal schooling. I was hired as an assistant professor in the Physics Department and just before this time I met Giuseppe Occhialini. He was an extraordinary figure in Milano. His work was at an extremely high level and widely recognized internationally.

To my knowledge he never gave a lecture. He took a liking to me and I to him. I was able to follow his multiplexed way of thinking and talking, and I was able to moderate some of his flights of imagination by sober use of first principles. He approved of my choice of working with cloud chambers, though most of the young people worked with emulsions, with the remark that at least I would learn plumbing. He suggested I go the United States to work with R. W. Thompson for whom he and I had the most profound esteem. I obtained a Fulbright Fellowship and sailed for the U.S. in 1956 and have lived here until now, except for a seven-year period (1992–1999) in Munich.

III. FELLOWSHIPS

I spent 1956 to 1958 in Bloomington, Indiana, working on the analysis of data previously obtained by Thompson and on the construction of a new and bigger cloud chamber for cosmic ray research. Thompson was a painstaking experimentalist as well as brilliant in data analysis and theory. He was the closest in the U. S. to the ideal of the Italian school of physics as embodied by Fermi, who was both a great experimentalist and theoretical thinker. Thompson had committed a blunder in research when young and this fact haunted him for the rest of his life. He never received the recognition he deserved for his discovery of the Θ_0 mass and this troubled him greatly.

It became quickly apparent that Thompson's group was not the place for me. The search for the anti-lambda particle was unsuccessful and the new cloud chamber would take 10 years to build. I moved to Princeton University in 1958 to work in G. Reynolds' laboratory. There I conducted research in μ mesons and carried out an unsuccessful search for a new type of particle whose discovery had been claimed by Russian scientists. This search was a collaboration between Fred Hendel (a senior Austrian scientist), Herbert Gursky (a post doc), and myself. We built equipment, worked like fiends, analyzed data, and declared failure. In the meanwhile I had learned a lot about scintillation counters and image intensifiers to be used for elementary particle research at the then-envisaged Princeton-Penn accelerator.

This put me in contact with the MIT group led by Herbert Bridge. Also with American Science and Engineering (AS&E), led by Martin Annis, an ex-cosmic ray physicist. Both Bridge and Annis had been students of Bruno Rossi of MIT, who was chairman of the Board of AS&E.

My fellowship at Princeton expired in 1959. I went to visit Occhialini at CERN. He seemed to be in one of his emotional downturns seeing conspirators everywhere. CERN felt like an impersonal huge machine and offered me no prospects. I was therefore quite happy to receive an offer from AS&E to initiate for the 28-man corporation a program of space sciences. I joined the Corporation in September 1959 and I started a serious career in science.

IV. OUTLOOK AND PERSONAL CONSIDERATIONS

Cosmic ray research as a tool to study elementary particles was ending just as I was doing my thesis. The work by Thompson was destined to a dead end. The MIT cosmic ray research group led by Rossi had decided to abandon almost completely cosmic ray research and to start utilizing space to study plasmas and gamma rays from celestial objects. In particular, W. Kraushaar and George Clark carried out one of the first surveys in gamma rays.

For me the problem was much more personal. I felt I had not learned any

Mirella Giacconi, 1995

From the left: Marc, Anna, and Guia Giacconi. (To the right Ethan Schreier.) Inauguration of Space Telescope Science Institute, 1983.

useful skills while working in Milano, Indiana, and Princeton. I desperately wanted some kind of permanent position where I could learn a trade and, moreover, I felt I owed this to my family. I had married Mirella, whom I had known in Milano since the age of 16. Guia and Anna, our two daughters, needed stability and a home. The influence of Mirella on my life has been greater than that of any other person. She brought love, calm, and stability where none had existed before. She created a home for us full of beauty and tenderness. She is brighter than I am, translated many books for the MIT press, and has an uncompromising view of reality. Through thick and thin we are still together today.

V. EARLY YEARS AT AS&E

Given the task of creating a space program for the company, I started discussions with George Clark, an MIT professor and consultant and shareholder in AS&E. We discussed the possibility of searching for the ratio between alpha particles and protons in the trapped radiation belts just recently discovered by Van Allen. Another suggestion came from Bruno Rossi at a party in his house. He reported discussions at the Space Science Board of the National Academy of Sciences about the potential for x-ray astronomy and suggested AS&E might undertake it. I have always been grateful for that initial suggestion.

1959 to 1962 were among the most productive years of my life. I was involved in classified research: 19 rocket payloads, six satellite payloads, one entire satellite, and an aircraft payload, as well as four rocket payloads for geophysical research. Also in this period I produced instruments for alpha – proton ratio experiments, as well as the initial development of the x-ray telescope and the first few flights of rocket payloads for x-ray astronomy. This was a tremendous challenge. My group at AS&E went from 3 people (myself and two technicians) to 70 people in two and a half years. Frank Paolini and Herbert Gursky joined the group. Frank was a great instrumentalist. Without his contributions the successful 1962 x-ray payload could not have happened. Herbert Gursky, though quite valiant in the laboratory and in the field (he actually launched the June 12, 1962, rocket), was always more of a natural scientist. The fundamental steps in the dawn of x-ray astronomy remain in my mind: the first brief review, theoretical and experimental, of x-ray astronomy (1960), the invention of the x-ray telescope (1960), and the 1963 plan for x-ray astronomy by Gursky and me. This plan laid out a program of experiments which went from rockets to UHURU, "Einstein," and Chandra. We thought then it could be done in five years, but it was not accomplished until the year 2000.

Much of the history of the development of x-ray astronomy has been told by Richard Hirsh, by Wally Tucker and me, and by others in books and articles and I will not repeat it here. But my scientific life took an unexpected turn after 1980 and I will devote some space to my growing interest in the direction of scientific enterprises outside of x-ray astronomy.

VI. CENTER FOR ASTROPHYSICS

I had decided by 1973 that I actually loved astronomy, at least as I was practicing it. The success of UHURU gave the members of my group (Gorenstein, Gursky, Kellog, Murray, Schreier, Tananbaum, Tucker, and Van Speybroeck) a feeling of awe and gratitude at the richness and beauty of nature. We had won the NASA contract to build the Large Orbiting X-Ray Observatory (1970–1973). The contract was canceled due to cost overruns in the NASA Viking Program in 1973 and restarted the same year for what became "Einstein," at one-half the size. We felt that we wanted to operate "Einstein" as a national observatory open to astronomers of all disciplines. AS&E did not seem to be the right place to do this. Thus the move to Harvard (where I was named a full professor) with eight of my group. We were looking for a closer involvement with the rest of astronomy and we thought the Center for Astrophysics (CfA) might provide the opportunity.

As it turned out, hardly anybody at CfA cared about what we were doing and there was less support for research than we had at AS&E. (In retrospect I have often wondered if we could have been successful at all had we started our work at Harvard). When I proposed to NASA an x-ray astronomy institute to direct the construction and operation of the "Einstein" successor in 1976 (then called AXAF and now Chandra), there was little support from CfA itself, where the institute would have been located. The proposal for AXAF was written by Harvey Tananbaum and me in 1976 and started the process which led to Chandra. Harvey was the project scientist for UHURU, the scientific manager for "Einstein," and took on the leadership role in Chandra. His contributions to x-ray astronomy have never been sufficiently recognized. It took perseverance over almost 20 years to turn Chandra into reality. For me the delay between conception and execution was becoming too long. Also, after the glory of discoveries with UHURU, "Einstein" seemed relatively tame. I was ready for a change, which came with the unsought offer by Margaret Burbidge to become the first director of the Space Telescope Science Institute (STScI).

VII. HUBBLE SPACE TELESCOPE

The effort at CfA to make x-ray astronomy useful to astronomers of all disciplines spurred us to provide the user with calibrated data out of a software pipeline. The development of end-to-end data management systems stems from that beginning. As the first Director of STScI, I was able to transfer my methodology for doing science in the planning and execution of science operations of Hubble.

Every aspect of the mission was examined by myself and the distinguished staff we were able to attract to the Institute. Ethan Schreier, my comrade of UHURU and "Einstein" days, and Rodger Doxey brought the physics-oriented mentality of x-ray astronomers to the new project.

Although we were quite ignorant in optical astronomy, we quickly found that operational planning for Hubble was a disaster. Guide stars could not be found on the fly as planned, the telescope could not point to planets, simultaneous reception of data from two instruments could not be achieved, and on and on.

Even more serious was the lack of tools to schedule the complex operations of Hubble and the absence of a data reduction system capable of ingesting the very high data rates. Both were accomplished by STScI. This meant developing models for the instrument defined by parameters measured on the ground and continuously verified in orbit, determining modes of instrument operations, and defining calibration routines. We also developed a software pipeline capable of analyzing data in real time and constructed an archive of calibrated data suitable for reuse by scientists other than those who built the instruments or used the Hubble through the competitive research program. We instituted, with NASA's consent, data analysis grants and the Hubble

Fellowship program. We developed an outstanding outreach program to reach the general public as well as colleagues and students.

We took responsibility for Hubble beyond the construction of glass and metal to turn it into an outstanding scientific tool.

It would be impossible to credit all of the scientists involved in creating and running the STScI and to name only a few seems unfair to the others. Many of them have continued at the Institute until today and many have gone on to prestigious positions at outstanding research institutions.

In 1991 my son Marc died in an automobile accident. STScI, Hubble, and Baltimore were continued and painful reminders of devastating grief. When the offer came to become Director General of the European Southern Observatory (ESO), Mirella and I jumped at the opportunity.

VIII. ESO AND THE VLT

When I joined ESO in January 1993, ESO was beginning to execute the Very Large Telescope (VLT) program. Quite a bit of progress had already occurred but the program was of such a size as to equal eight times the yearly budget of ESO and was 30 times larger than the previously built New Technology Telescope.

Massimo Tarenghi, an excellent scientist and an energetic manager, needed support from the rest of ESO to carry the project through. To help him in his task we had to fully reorganize ESO and introduce modern management techniques suitable for large programs. This was reasonably straightforward. More difficult was the introduction at ESO of the concept of a single observatory in which quality rather than quantity mattered and the introduction of end-to-end software systems, calibration, and archiving, whose validity was proven on Hubble but had not yet been used in ground-based optical astronomy.

VLT became a machine to do science in which efficiency of operations and ability to use the data were as important as telescope and instrumentation performance.

Keck had the fortune of being the first to provide a 10 meter class telescope and reaped the early reward. But in a certain sense, it is operated as an old telescope for use by a restricted community. The success of VLT has placed European optical ground-based astronomy in the position to compete worldwide.

Toward the end of my stay at ESO we initiated a new cooperative program with the United States and Canada to build a large submillimeter and millimeter wave array of antennas to be placed in the Atacama Desert in Northern Chile. At the expiration of my tenure as Director General in 1999, I returned to the United States where now, as President of Associated Universities, Inc., I am working with the National Radio Astronomy Observatory as the Executive for the North American portion of the project, called the Atacama Large Millimeter Array (ALMA). The same principles and the same methods used in "Einstein," Hubble, VLT, and Chandra are being utilized.

CURRICULUM VITAE

Education:

• University of Milan Milan, Italy	Doctorate in Physics, 1954

Professional experience

• Assistant Professor of Physics University of Milan Milan, Italy	1954–1956
• Research Associate (Fulbright Fellow) Indiana University Bloomington, Indiana	1956–1958
• Research Associate, Cosmic Ray Laboratory Princeton University Princeton, New Jersey	1958–1959
• American Science & Engineering, Inc. Cambridge, Massachusetts Senior Scientist Member of the Board of Directors (1966–1973) Executive Vice President (1969-1973)	1959–1973
• Associate Director High Energy Astrophysics Division Harvard-Smithsonian Center for Astrophysics Cambridge, Massachusetts	1973–1981
• Professor of Astronomy Harvard University Cambridge, Massachusetts	1973–1982
• Principal Investigator on several NASA programs, including	
SAS-A (UHURU) SO-54 (SKYLAB) HEAO-2 ("Einstein")	1960–1981
AXAF Interdisciplinary Scientist (Chandra)	1986–
• Director Space Telescope Science Institute Baltimore, Maryland	1981–1992
• Professor of Astronomy The Johns Hopkins University Baltimore, Maryland	1981–1999
• Chairman of the Board Instituto Donegani Italy	1987–1988
• Professor of Physics and Astronomy University of Milan Milan, Italy	1991–1999

- Director General — 1993–1999
  European Southern Observatory
  Germany
- President — 1999–
  Associated Universities, Inc.
  Washington, DC
- Research Professor — 1999–
  The Johns Hopkins University
  Baltimore, Maryland

Honors:

Fulbright Fellow, 1956–1958
Helen B. Warner Award of the American Astronomical Society, 1966
Como Prize of the Italian Physical Society, 1967
Röntgen Prize in Astrophysics for 1971 of the Physikalisch-Medizinische Gesellschaft, Würzburg, Germany, 1971
NASA Medal for Exceptional Scientific Achievement, 1971
NASA Distinguished Public Service Award, 1972
NASA Exceptional Scientific Achievement Medal, 1980
Elliott Cresson Medal of the Franklin Institute, Philadelphia, 1980
Catherine Wolfe Bruce Gold Medal of the Astronomical Society of the Pacific, 1981
Dannie Heineman Prize for Astrophysics, AAS/AIP, 1981
Henry Norris Russell Lecturer of the American Astronomical Society for 1981
Gold Medal of the Royal Astronomical Society, 1982
The A. Cressy Morrison Award in Natural Sciences, The New York Academy of Sciences, 1982
Wolf Prize in Physics, 1987
Nobel Prize in Physics, 2002

Honorary degrees:

Honorary Doctor of Science, The University of Chicago, 1983
Laurea Honoris Causa in Astronomia, The University of Padua, 1984
Honorary Doctor of Science, University of Warsaw, 1996
Laurea Honoris Causa in Physics, University of Rome, 1998
Honorary Doctor of Technology and Science, University of Uppsala, Sweden, 2000

Academies:

American Academy of Arts and Sciences, Fellow
National Academy of Sciences, Member
Accademia Nazionale dei Lincei, Foreign Member
The Royal Astronomical Society, Associate

American Philosophical Society, Member
Max-Planck Society, External Member

Memberships:

American Astronomical Society
American Association for the Advancement of Science, Fellow
American Physical Society, Fellow
Astronomical Society of the Pacific
Cosmos Club

THE DAWN OF X-RAY ASTRONOMY

Nobel Lecture, December 8, 2002

by

RICCARDO GIACCONI

Associated Universities, Inc., Suite 730, 1400 16th St., NW, Washington, DC 20036, USA, and Johns Hopkins University, 3400 N. Charles Street, Baltimore, USA.

1.0 INTRODUCTION

The development of rockets and satellites capable of carrying instruments outside the absorbing layers of the Earth's atmosphere has made possible the observation of celestial objects in the x-ray range of wavelength.

X-rays of energy greater than several hundreds of electron volts can penetrate the interstellar gas over distances comparable to the size of our own galaxy, with greater or lesser absorption depending on the direction of the line of sight. At energies of a few kilovolts, x-rays can penetrate the entire column of galactic gas and in fact can reach us from distances comparable to the radius of the universe.

The possibility of studying celestial objects in x-rays has had a profound significance for all astronomy. Over the x-ray to gamma-ray range of energies, x-rays are, by number of photons, the most abundant flux of radiation that can reveal to us the existence of high energy events in the cosmos. By high energy events I mean events in which the total energy expended is extremely high (supernova explosions, emissions by active galactic nuclei, etc.) or in which the energy acquired per nucleon or the temperature of the matter involved is extremely high (infall onto collapsed objects, high temperature plasmas, interaction of relativistic electrons with magnetic or photon fields).

From its beginning in 1962 until today, the instrumentation for x-ray astronomical observations has improved in sensitivity by more than 9 orders of magnitude, comparable to the entire improvement from the capability of the naked eye to those of the current generation of 8- or 10-meter telescopes. All categories of celestial objects, from planets to normal stars, from ordinary galaxies to quasars, from small groups of galaxies to the furthest known clusters, have been observed. As a result of these studies it has become apparent that high energy phenomena play a fundamental role in the formation and in the chemical and dynamical evolution of structures on all scales. X-ray observations have proved of crucial importance in discovering important aspects of these phenomena. It was from x-ray observations that we obtained the first evidence for gravitational energy release due to infall of matter onto a collapsed object such as a neutron star or black hole. It was the x-ray emission from the high temperature plasmas in clusters of galaxies that revealed this high temperature component of the Universe, which more than doubled the amount of "visible" matter (baryons) present in clusters.

The prospects for future studies of the universe in x-rays are equally bright. The advent of new and even more powerful experimental techniques, such as non-dispersive high resolution spectroscopy and x-ray telescopes capable of focusing increasingly higher energies over wider fields, ensures a wide opportunity for new astronomical discoveries.

2.0 THE BEGINNING OF X-RAY ASTRONOMY

There had been solar x-ray observations for about 10 years by the Naval Research Laboratory (NRL) group led by Herbert Friedman and several failed attempts to find x-ray emissions from stellar objects (Hirsh, 1979)[1] when a group at AS&E (a small private research corporation in Cambridge, Massachusetts) started work in 1959 to investigate the theoretical and experimental possibilities for carrying out x-ray astronomy. Giacconi, Clark, and Rossi (Giacconi *et al.*, 1960)[2] published a document: "A Brief Review of Experimental and Theoretical Progress in X-Ray Astronomy," in which we attempted to estimate expected x-ray fluxes from several celestial sources.

The results are summarized in Table 1. The Sun produced 10^6 x-ray photons $cm^{-2}s^{-1}$ at Earth which could easily be detected with the then-available

Table 1. Estimates of fluxes from sources outside the solar system, 1960, Giacconi, R., Clark, G.W., and Rossi, B.B.

Giacconi, Clark, Rossi (1960)			
Source	*Maximum Wavelength*	*Mechanism for Emission*	*Estimated Flux*
Sun	< 20 Å	Coronal emission	~ 10^6 cm^{-2} s^{-1}
Sun at 8 light years	< 20 Å	Coronal emission	2.5 x 10^{-4} cm^{-2} s^{-1}
Sirius if $L_X \sim L_{OPT}$	< 20 Å	? No convective zone	0.25 cm^{-2} s^{-1}
Flare stars	< 20 Å	Sunlike flare?	?
Peculiar A stars	< 20 Å	B ~ 10^4 Gauss Large B Particle acceleration	?
Crab nebula	< 25 Å	Synchrotron $E_E \geq 10^{13}$ eV in B = 10^{-4} Gauss Lifetimes?	?
Moon	< 23 Å	Fluorescence	0.4 cm^{-2} s^{-1}
Moon	~ 20 Å	Impact from solar wind Electrons $\varphi_\varepsilon = 0–10^{13}$ cm^{-2} s^{-1}	0–1.6 x 10^3 cm^{-2} s^{-1}
SCO X-1	2-8 Å	?	28 ± 1.2 cm^{-2} s^{-1}

counters with sensitivities of about 10–10^2 photons $cm^{-2}s^{-1}$. But if all the stars emitted x-rays at the same rate as the Sun, we would expect fluxes at earth as small as 10^{-4} photons $cm^{-2}s^{-1}$. Other possible sources, such as supernova remnants, flare stars, peculiar A stars, etc., were considered, and great uncertainty had to be assigned to the estimates of their x-ray fluxes. It seemed that the brightest source in the night sky could be the Moon, due to fluorescent emission of lunar material under illumination of solar x-rays.

We designed an experiment capable of detecting 0.1–1 photon/$cm^{-2}s^{-1}$, 50 to 100 times more sensitive than any flown before. This increase in sensitivity was due to larger area, an anticoincidence shield to reduce particle back-

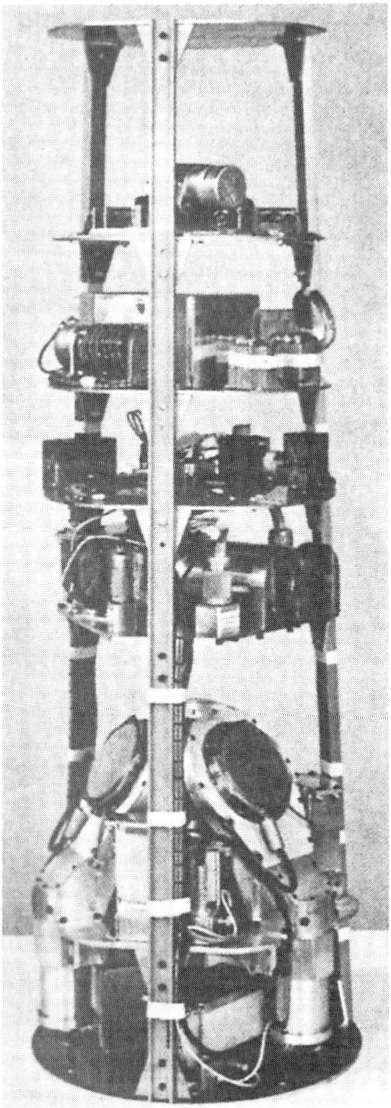

Figure 1. The payload of the June 12, 1962, AS&E rocket. From X-ray Astronomy (Eds. R. Giacconi and H. Gursky), 1974, Riedel, Dordrecht, p9.

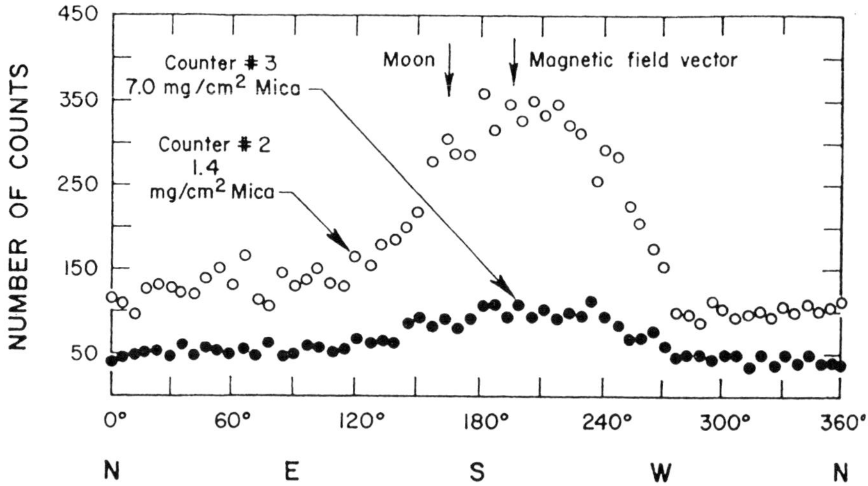

Figure 2. The first observation of Sco X–1 and of the x–ray background in the June, 12, 1962 flight. From Giacconi, *et al.*, 1962.

ground, and a wide solid angle to increase the probability of observing a source during the flight. The payload shown in Fig. 1 was successful in detecting the first stellar x-ray source in the flight of June 12, 1962 (Giacconi *et al.*, 1962).[3] An individual source (Sco X-1) dominated the night sky and was detected at 28 ± 1.2 counts $cm^{-2}s^{-1}$, just below the threshold of previous experiments (Fig. 2). No exceptionally bright or conspicuous visible light or radio object was present at that position (which, however, was very poorly known). An early confirmation of our result came from the rocket flight of April 1963 by the NRL group led by Friedman, which also discovered x–ray emission from the Crab Nebula (Bowyer *et al.*, 1964).[4]

The truly extraordinary aspect of the discovery was not that an x-ray star had been found but its extraordinary properties. The x-ray radiation intensity from the Sun is only 10^{-6} of its visible light intensity. In Sco X-1, the x-ray luminosity is 10^3 times the visible light intensity and it was later determined that the intrinsic luminosity is 10^3 the entire luminosity of the Sun! This was a truly amazing and new type of celestial object. Furthermore, the physical process by which the x-rays were emitted on Sco X-1 had to be different from any process for x-ray generation we knew in the laboratory since it has not been possible on Earth to generate x-rays with 99.9% efficiency.

Many rocket flights carried out by several groups in the 60's were able to find new stellar sources and the first extragalactical sources. The NRL group and the Lockheed group (led by Phil Fisher) continued to carry out mostly broad surveys, with the notable exception of the Crab occulation experiment by NRL in 1964 (Bowyer *et al.*, 1964).[5]

The AS&E group concentrated on the detailed study of individual x-ray sources. Most significant was the series of rocket flights which culminated in the identification of the optical counterpart of Sco X–1. First the group determined that Sco X–1 could not have the thermal spectrum that would be expected from neutron stars (Giacconi *et al.*, 1965),[6] which implied that an optical counterpart should have magnitude 13. In a first rocket flight to measure the angular size of the source it was found to be less than 7 arc sec (Oda *et al.*, 1965).[7] Thus the source had to be a visible star and not a diffused nebulosity. This led to the sophisticated measurement of the location of Sco X–1 by an AS&E-MIT group led by Herbert Gursky (Gursky *et al.*, 1966),[8] with sufficient precision to enable its identification with a 13th magnitude star (Sandage *et al.*, 1966)[9] which had spectral characteristics similar to an old nova. This renewed interest in a binary star model for Sco X–1 (Burbidge, 1967)[10] and Shklovsky proposed a binary containing a neutron star (Shklovsky, 1967).[11] However, the absence of x–ray emission from other novas, the lack of indications from either the optical spectra or the x-ray data of a binary system, and the general belief that the supernova explosion required to form the neutron star would disrupt the binary system did not lead to the general acceptance of the idea. The discovery by Hewish of pulsars in 1967 turned the attention of the theorists to pulsar models for the x–ray emitters. But such models also were not quite persuasive given the lack of observed x–ray pulsations. The solution to the riddle of Sco X–1 and similar sources was not achieved until the launch of the "UHURU" satellite, the first of a generation of x–ray observatories.

The proposal to launch a "scanning satellite," which eventually became UHURU, was contained in a document written by Herb Gursky and myself and submitted to NASA on September 25, 1963. In this document we described a complete program of x-ray research culminating in the launch of a 1.2 meter diameter x-ray telescope in 1968. This youthful dream was not fully realized until the launch of the Chandra X-Ray Observatory in 1999, which, not by chance, had a 1.2 meter diameter mirror. But while the difficult technology development that made x-ray telescopes possible was being carried out, the most fundamental advances in x-ray astronomy were made with relatively crude detectors mounted on orbiting satellites.

3.0 DISCOVERIES WITH UHURU

The total amount of time which was available for observation of the x-ray sky during the 60's was about one hour: five minutes above 100 km for each of about a dozen launches. The next step which led us from the phenomenological discoveries to those of great astrophysical relevance occurred on December 12, 1970, when UHURU, the first of the Small Astronomy Satellite series, was launched from the Italian S. Marco platform in Kenya. UHURU was a small satellite (Fig. 3) which we had labored over at AS&E for seven years between conception, development, testing, and integration.

It was the first observatory entirely dedicated to x-ray astronomy and it extended the time of observation from minutes to years or by 5 orders of magnitude (Giacconi *et al.*, 1971).[12] The field of view of the detector on board the satellite slowly rotated, examining a 5° band of the sky that shifted 1° a day. In

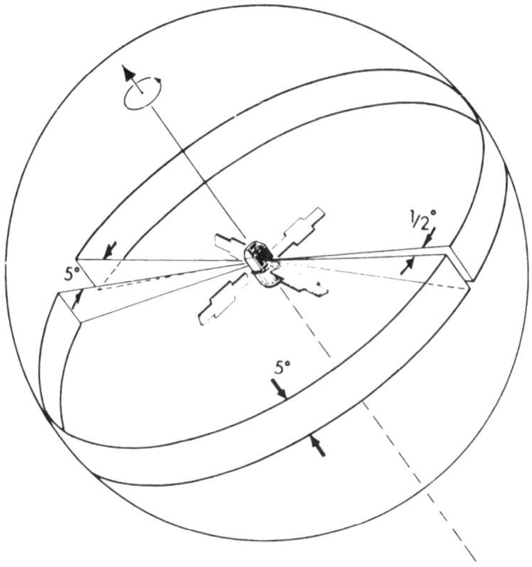

Figure 3. The fields of view of the detectors on the UHURU satellite. From H. Gursky, in A survey of Instruments and Experiments for X-ray astronomy: IAU Symposium No. 37 (Ed: L. Gratton) Riedel, Dordrecht, 1970, p19.

three months all the sky could be studied systematically and many new sources could be localized with a precision of about 1 arc minute, often permitting the identification of the x-ray sources with a visual or radio counterpart. This in turn led to an evaluation of the distance, the intrinsic luminosity, and the physical characteristics of the celestial object from which the x-rays

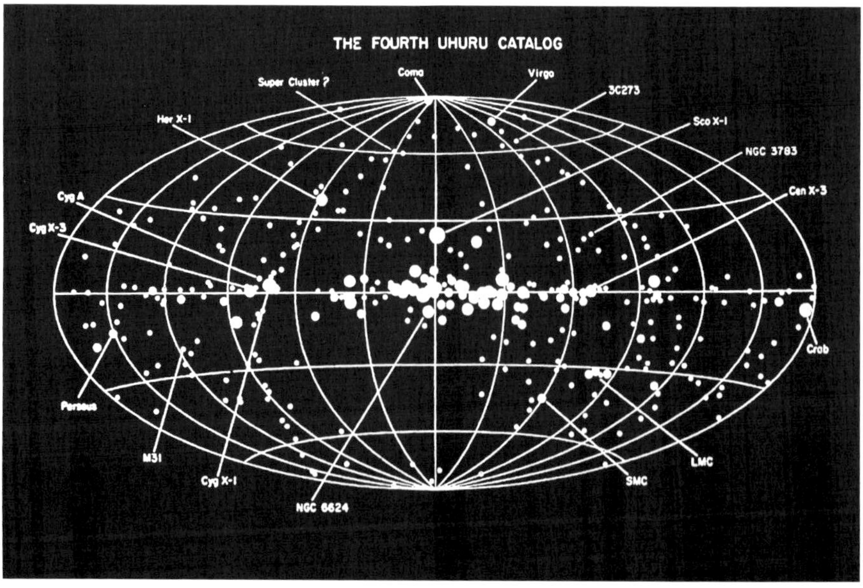

Figure 4. The x–ray sources observed by UHURU plotted in galactic coordinates. The site of the dot is proportional to intensity on a logarithmic time scale. From X-ray Astronomy (Eds. R. Giacconi, H. Gursky), 1974, Riedel, Dordrecht, p. 156.

originated. Among the 300 new sources which were discovered, we were able to identify binary x-ray sources, supernovas, galaxies, active galaxies, quasars, and clusters of galaxies (Fig. 4). But even more important from a certain point of view was the ability, which was provided by the control system, to slow down the satellite spin and spend a very long time on an individual source to study its temporal variations. It was this special ability which permitted the solution of the fundamental unresolved problem of x-ray astronomy until then, namely, the nature of the energy source capable of producing the large intrinsic luminosity of the stellar x-ray sources.

3.1 The Binary X-Ray Sources

In summary, inspection of the data revealed that some x-ray sources (Her X-1 and Cen X-3) (Fig. 5) were regularly pulsating with periods of seconds (Schreier *et al.*, 1972),[13] while others (Cyg X-1) were pulsating with an erratic behavior with characteristic times of less than a tenth of a second as first noted by Minoru Oda, who was a guest at AS&E at the time. Ethan Schreier and I noticed that the average intensity of Cen X-3 was modulated over the span of days and that the period of pulsation itself was changing as a function of the phase of the average intensity, which exhibited occulations (Fig. 6). The explanation for this behavior soon became clear: we were observing a stellar x-ray source orbiting a normal star (Fig. 7). The variation of the pulsation period was then due to the Doppler effect. In 1967 Hewish had discovered pulsars in the radio domain. Was this x-ray source a pulsar in orbit about a normal star? This seemed difficult to accept at the time. A pulsar is a neutron star whose formation is believed to be due to the collapse of a star at the end of its life. The concomitant explosion was believed to disrupt any binary system in which it took place. Joseph Taylor had not yet discovered the binary pulsar.

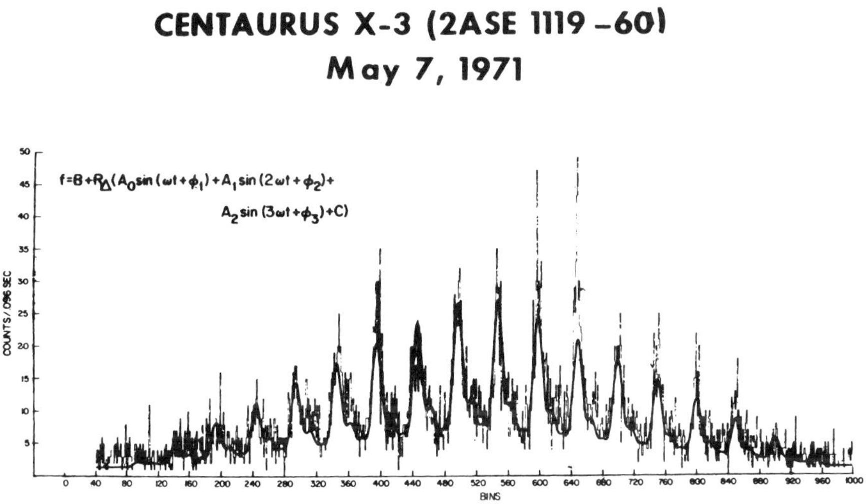

Figure 5. X–ray pulsations of Cen X–3. From Giacconi, 1974.

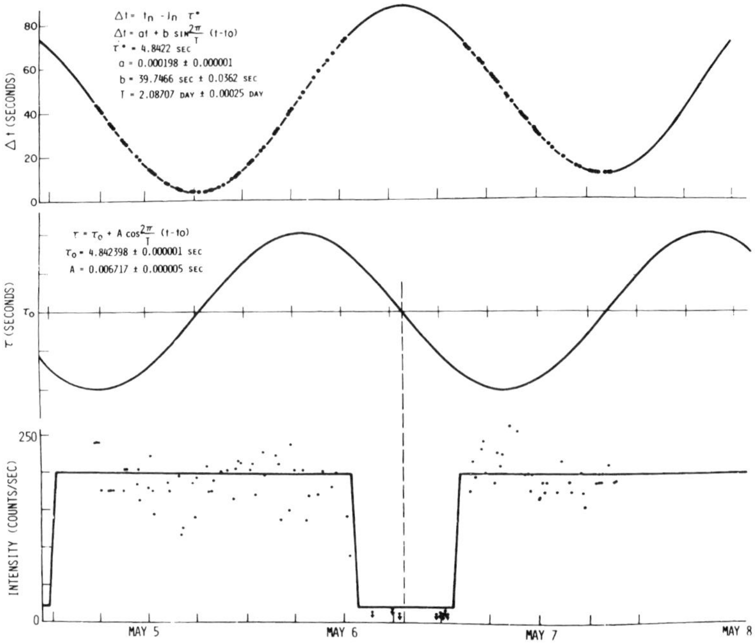

Figure 6. Period variations and occulations of Cen X–3. From Giacconi, 1974.

But a new, unexpected, and important finding came to light: the period of pulsation was decreasing rather than increasing with time (Fig. 8). This was true not only in Cen X-3, but also, as Harvey Tananbaum found, in Her X-1 (Tananbaum, 1972).[14] Now this was truly embarrassing! In a pulsar the loss of electromagnetic energy occurs at the expense of the kinetic energy of rota-

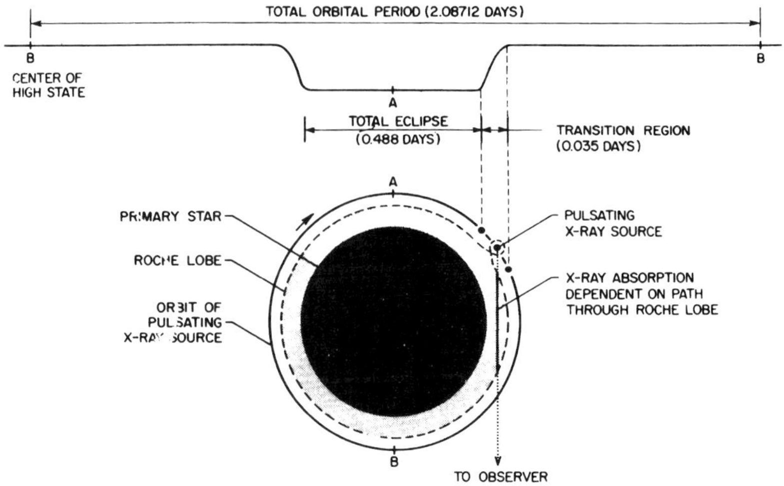

Figure 7. A model of the Cen X–3 binary system. Illustration of R. Giacconi.

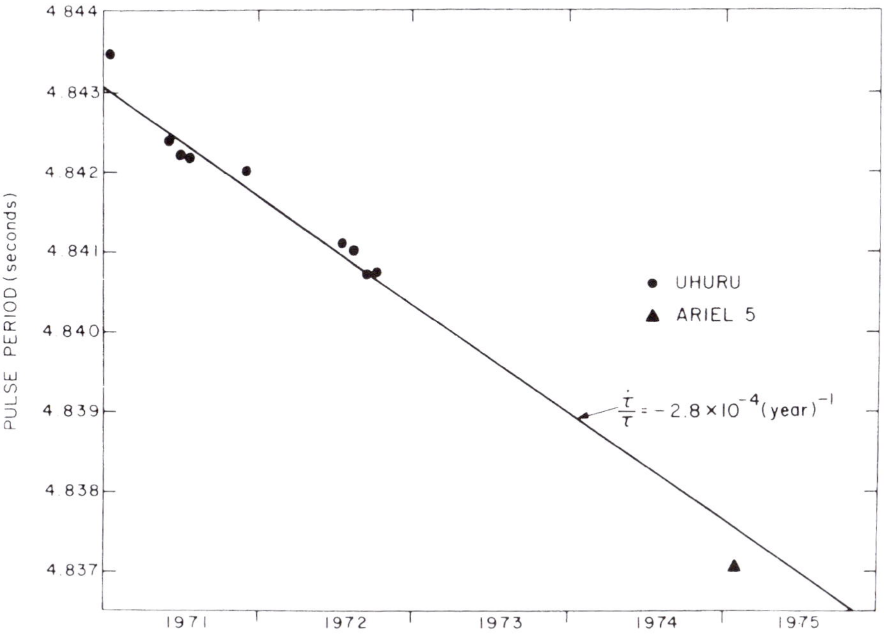

Figure 8. Annual change of Cen X–3 pulsation period. Illustration of R. Giacconi.

tion. But in the x-ray sources the neutron star was acquiring rather than losing energy! The explanation was found in the interaction of the gas in the normal star with the collapsed star. Gas from the outer layer of the atmosphere of the normal star can fall into the strong gravitational field of the collapsed star and acquire energies of order of 0.1 mc^2 per nucleon. The accel-

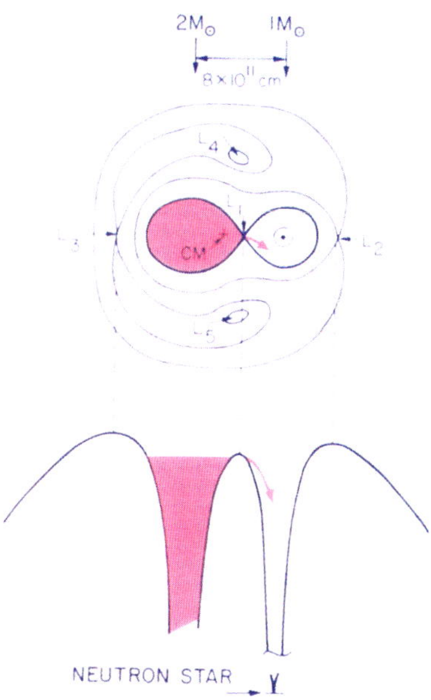

Figure 9. Representation of the equipotentials in the gravitational field of a typical binary x–ray source. Top view and cross section. Illustration of R. Giacconi.

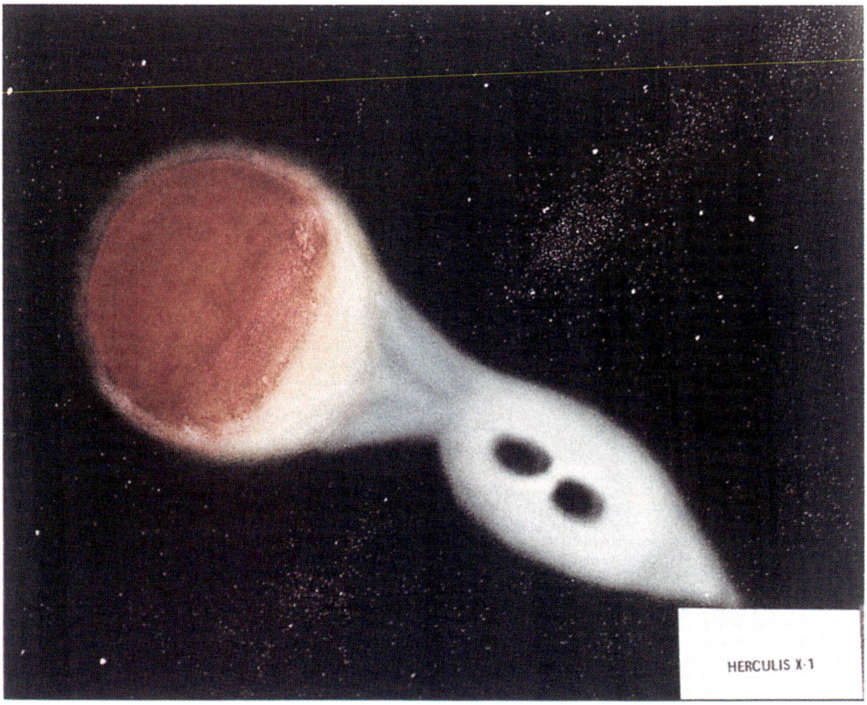

Figure 10. An artist's conception of Her X–1 with accretions occurring at the poles of the magnetic field of the neutron star. Illustration of R. Plourde.

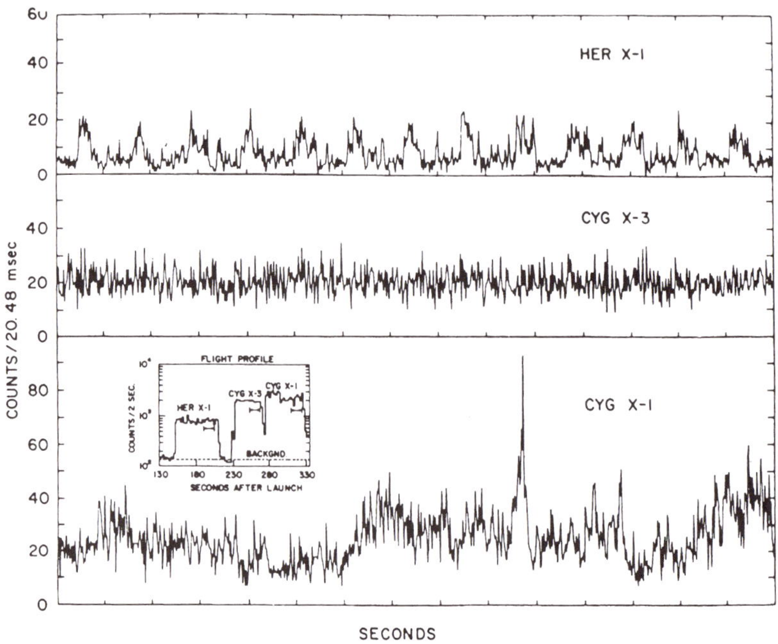

Figure 11. Comparison of the time variability of Her X–1, Cyg X–3, and Cyg X–1. Courtesy of R. Rothschild.

erated nucleons in turn heat a shock of very high temperature above the surface of the neutron star, which emits the observed x-rays (Fig. 9). It is this material infall that gives energy to the collapsed object. This is the model for Sco X–1 and most of the galactic x–ray sources. In the case of a neutron star with a strong magnetic field (10^{12} Gauss), the ionized plasma is confined to the poles of the rotating neutron star, generating the observed periodicities (Fig. 10). For a black hole, there exists no surface with particular structures and therefore the pulsation occurs chaotically (Fig. 11).

The observation by UHURU of rapid variability in Cyg X–1 reported by Oda (Oda *et al.*, 1971)[15] was soon followed by the rocket flights of the GSFC and MIT groups. These observations clarified that the observed pulsations were not periodic but chaotic (Holt *et al.*, 1971)[16] (Rappaport *et al.*, 1971).[17] By 1974, the GSFC group had achieved a temporal resolution of 1 millisecond and showed large chaotic fluctuations occurring even on this time scale (Rothschild *et al.*, 1974).[18] Such behavior could be expected to occur if the compact object in the binary system (the x–ray source) was a black hole rather than a neutron star. Stimulated by these findings, the search for optical or radio counterparts had become intense in 1971–1972. UHURU had obtained a considerably improved position for Cyg X–1 which was made available to Hjellming and Wade to aid in the search for a radio counterpart (Tananbaum *et al.*, 1971).[19] More refined positions were obtained by the Japanese group led by Oda (Miyamoto *et al.*, 1971)[20] and by the MIT group (Rappaport *et al.*, 1971)[21] by use of modulation collimators. Hjellming and Wade (Hjellming and Wade, 1971)[22] and Braes and Miley (Braes and Miley, 1971)[23] reported the discovery of a radio counterpart. The precise radio lo-

Figure 12. Artist's conception of Cyg X–1. Illustration of L. Cohen.

cation led to the optical identification by Webster and Murdin (Webster and Murdin, 1972)[24] and by Bolton (Bolton, 1972)[25] of Cyg X–1 with the 5.6 day binary system HDE 226862. The identification of the radio source with Cyg X–1 was confirmed by the observation of a correlated x–ray radio transition in Cyg X–1 (Tananbaum *et al.*, 1972).[26] Spectroscopic measurements of the velocity of HDE 226862 also permitted Webster and Murdin to establish that Cyg X–1 was indeed in a binary system. The estimated mass for the compact object was greater than 6 solar masses. Rhoades and Ruffini had shown in 1972 that black holes would have masses greater than 3.4 times the mass of the Sun (Rhoades and Ruffini, 1974).[27]

Thus we could reach conclusions regarding Cyg X–1: the Cyg X–1 x–ray emitter is a compact object of less than 30 km radius due to the rapidity of the pulsations and the fact that the pulsations are so large that they must involve the whole object (Giacconi, 1974).[28] The object has mass greater than that allowed by our current theories for neutron stars. Therefore the object is the first candidate for a black hole (Fig. 12). Currently there are at least six candidates for galactic x–ray sources containing a black hole (Tanaka, 1992).[29]

The consequences of the discovery of binary source x-rays have had far reaching consequences (Table 2). We had proven the existence of binary systems containing a neutron star and of systems containing a black hole. Black holes of solar mass size existed. The binary x-ray sources have become a sort of physical laboratory where we can study the mass, moment of inertia, and equation of state for neutron stars (density 10^{15}gr/cm^3). We had found a new source of energy for celestial objects: the infall of accreting material in a strong gravitation field. For a neutron star the energy liberated per nucleon is of order of 50 times greater than generated in fusion. The above model (of accretion of gas on a collapsed object) has become the standard explanation for the internal engines of quasars and all active nuclei. Recent data seem to confirm the model of accretion on a massive central black hole of $>10^7$ solar masses as the common denominator among all the active galaxies.

Table 2. Consequences of the discovery of binary x-ray systems.

The Discovery of X-Ray Binary Systems
• Existence of Binary Stellar Systems Containing a Neutron Star or a Black Hole
• Existence of Black Holes of Stellar Mass
• Measure of the Mass, Radius, Moment of Inertia and Equation of State for Neutron Stars (Density 10^{15} GR/CM^3)
• A New Source of Energy Due to Gravitational Infall (100 Times More Efficient per Nucleon than Fusion)
• A Model (Generally Accepted) for the Nucleus of Active Galaxies and Quasars

3.2 *The Discovery of High Temperature Intergalactic Gas*

The establishment of variability in the x–ray universe, the discovery of the existence of neutron stars and black holes in binary systems, and the discovery of accretion as a dominant energy source were only the first major accomplishments of UHURU. Among others was a second very important discovery of UHURU and x-ray astronomy, both because of its intrinsic interest and for its consequences in the field of cosmology, the detection of emission from clusters of galaxies. This emission is not simply due to the sum of the emission from individual galaxies, but originates in a thin gas which pervades the space between galaxies. This gas was heated in the past during the gravitational contraction of the cluster to a temperature of millions of degrees and contains as much mass as that in the galaxies themselves (Gursky *et al.,* 1972)[30]. In one stroke the mass of baryons contained in the clusters was more than doubled. This first finding with UHURU, which could detect only the three richest and closest galaxy clusters and with a poor angular resolution of 1/2 a degree, were followed and enormously expanded by the introduction of a new and powerful x-ray observatory, "Einstein," which first utilized a completely new technology in extrasolar x-ray astronomy: grazing incidence telescopes.

4.0 X–RAY TELESCOPES

Here I must make a short technical diversion to explain the revolution brought about in x-ray astronomy by the telescope technology. When contemplating the estimates made in 1959, I was persuaded that to ultimately succeed in x-ray astronomy, we had to develop new systems quite different from those then in use. Friedman had developed for solar studies a Geiger counter with a thin window which allowed the x-ray to penetrate the interior of the gas volume of the counter. The counter could not decide either the direction of the incoming x-ray or its energy. In order to improve the directional sensitivity, x–ray astronomers used collimators, that is, mechanical baffles which defined a field of view typically of 1°. To improve sensitivity we developed anti-coincidence shields against spurious particles and enlarged the area. This is what was done in the discovery rocket of 1962. In 1970 UHURU had a very similar detection system. Its improvement in sensitivity was due to the much larger area (800 cm^2 instead of 10) and the much longer time of observation. This led to an increase in sensitivity of about 10^4. It should be noted, however, that in the presence of a background noise, the sensitivity improvement was only proportional to the square root of the area. Thus, further improvement would have required satellites of football-stadium size. Furthermore, all attempts to gain angular resolution by clever systems of baffles (such as the modulation collimators) led to intrinsically insensitive experiments.

The solution which occurred to me as early as 1959 was to use a telescope just as it is done in visible light astronomy (Giacconi and Rossi, 1960).[31] This has the great advantage that the flux from a large area of collection is focused onto a small detector, therefore improving both the flux and the signal to noise ratio. In addition, high angular resolution can be obtained within a

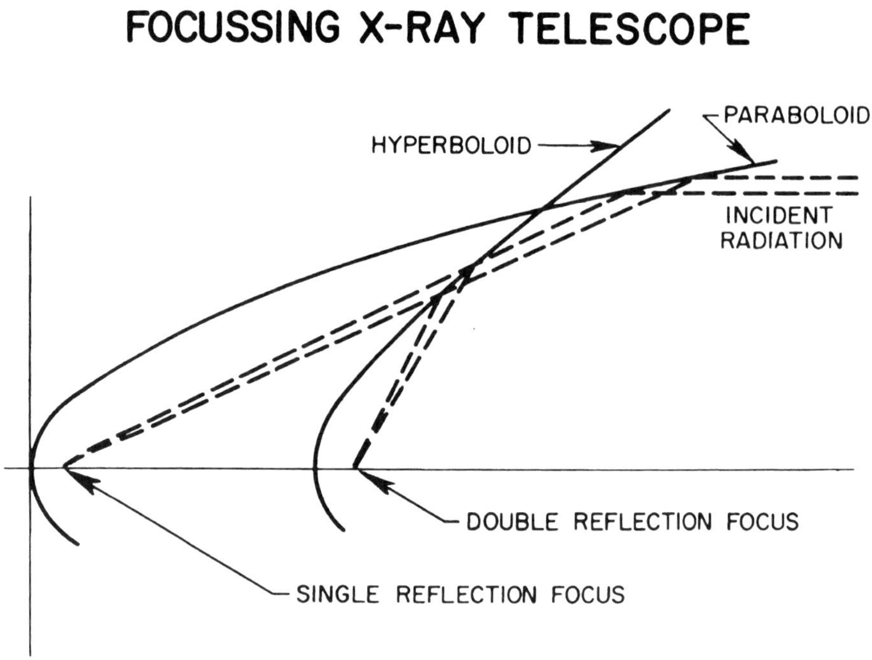

Figure 13. Principle of x–ray grazing incidence telescope. Illustration of R. Giacconi.

field which is imaged at once, without scanning or dithering motions, therefore yielding an enormous improvement in exposure time for each source in the field.

The only problem is that an x-ray telescope had to be invented and the technology necessary for its fabrication had to be developed. It ultimately took about 20 years between the conception of the x-ray telescope in 1959 and its first use for stellar x-ray astronomy in 1979. An x-ray telescope is quite different from a visible light telescope since the wavelength of x-ray photons is comparable to atomic dimensions. According to Lorentz' dispersion theory, it is clear that the index of refraction of x-rays is less than one, which makes optical systems based on refraction essentially impractical, as was realized by Röntgen himself in his classical experiment of 1895. However, x-rays can be efficiently externally reflected by mirrors, provided only that the reflection takes place at very small angles with respect to the mirror's surface. Hans Wolter had already discussed in the 40's and 50's the possibility of using images formed by reflection for microscopy. He showed that using a double reflection from a system of coaxial mirrors consisting of paraboloid and hyperboloid, one could achieve systems with a reasonably large field (1°) corrected for spherical aberrations and coma. Theoretically, therefore, the system was feasible, although the difficulties of construction given the tiny dimension of the systems for microscopy were impossible to overcome (Fig. 13). I persuaded myself, however, that in the corresponding optical de-

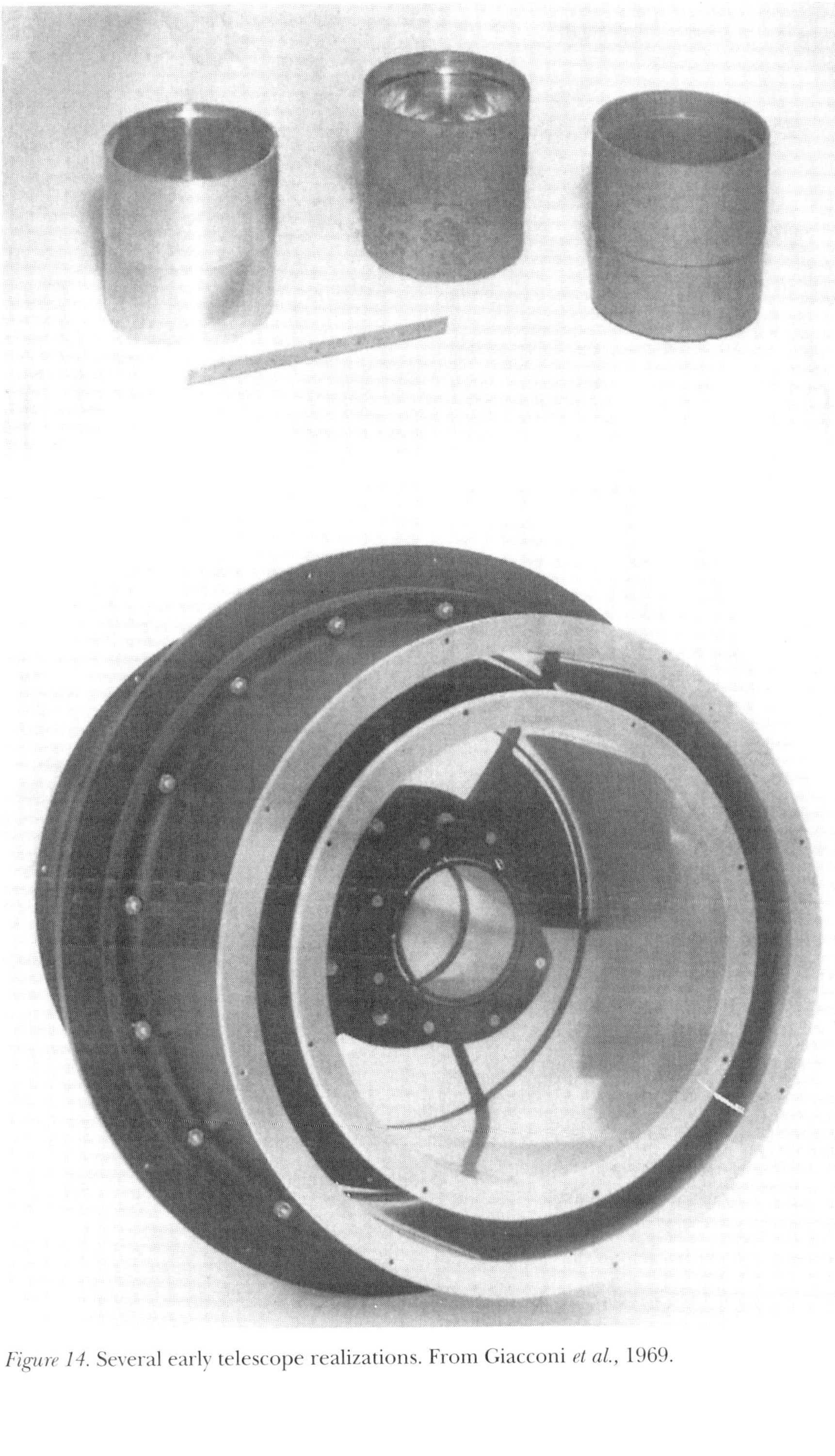

Figure 14. Several early telescope realizations. From Giacconi *et al.*, 1969.

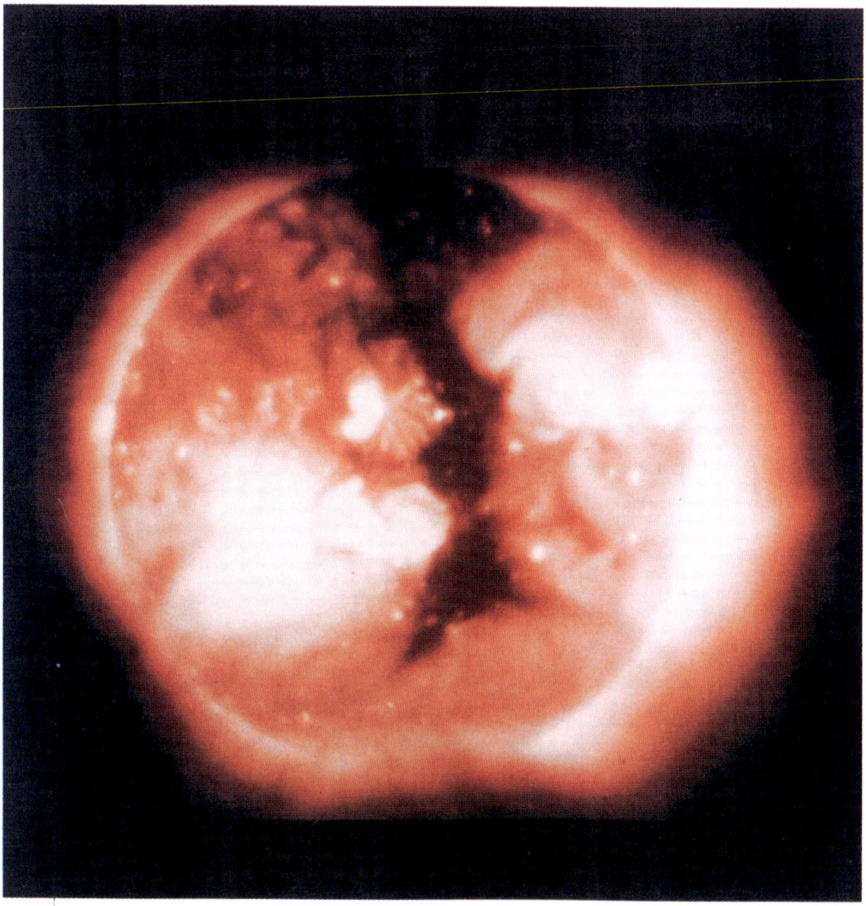

Figure 15. Picture of Sun in x–rays from Skylab. Courtesy of L. Golub, Harvard-Smithsonian Center for Astrophysics.

signs to be used for telescopes, which required much larger scales (meters rather than microns), such difficulties would not be severe.

In our 1960 paper we described a system that could achieve sensitivities of 5×10^{-14} erg $cm^{-2}s^{-1}$ and angular resolutions of 2 arc minutes. The improvement in sensitivity was 10^6–10^7 times more than for any detector then current and the improvement in angular resolution about a factor of 10^3. Unfortunately it took a long time to develop this technology (Fig. 14) (Giacconi *et al.,* 1969).[32] The first primitive pictures of the Sun with an x-ray telescope were obtained in 1965. Giuseppe Vaiana took over leadership of our solar physics program in 1967. In 1973 a high resolution x-ray telescope studied the Sun over a period of many months with a field large enough to image the disk and nearby corona and with angular resolution finer than 5 arc seconds (Fig. 15) (Vaiana and Rosner, 1978).[33]

It was not until 1979 that a fully instrumented x-ray telescope suitable for the detection and study of the much weaker stellar fluxes could be launched.

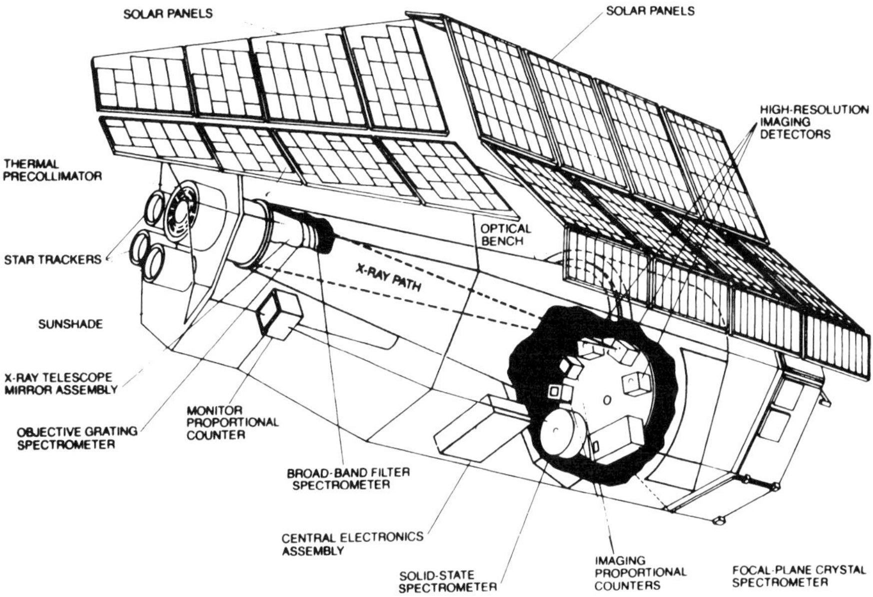

Figure 16. The Observatory Einstein schematic representation. From Giacconi *et al.*, 1979.

Table 3. Classes of objects observed with the Einstein Observatory.

Celestial Objects Observed with Einstein
• Aurora on Jupiter
• X– Ray Emission from all Types of Main Sequence Stars
• Novas and Supernovas
• Pulsars
• Binary X–ray Sources and Supernovas in Extragalactic Sources
• Normal Galaxies
• Nuclei of Active Galaxies
• Quasars
• Groups and Clusters of Galaxies
• Sources of the Extragalactic X–ray Background

The new satellite, which became known as "Einstein," was a real astronomical observatory (Fig. 16) (Giacconi *et al.*, 1979).[34] In the focal plane of the telescope one could use image detectors with angular resolutions of a few arc seconds, comparable to those used in visible light. The sensitivity with respect to point sources was increased by 10^3 with respect to UHURU and 10^6 with respect to Sco X-1. Spectroscopy could be carried out with a spectral resolving power of 500.

This substantial technical improvement made possible the detection of all types of astrophysical phenomena (Table 3). Auroras due to the Jovian Belts, main sequence stars of all types, novae and supernovae were detected. Binary x-ray sources could be studied anywhere in our own galaxy as well as in external galaxies (Fig. 17). Normal galaxies as well as galaxies with active galactic nuclei, such as Seyferts and B Lac, could be detected at very great distances. The most distant quasars ever detected in visible light or radio could be conveniently studied. The sources of the mysterious, isotropic extragalactic background could begin to be resolved.

But to come back to the study of the intergalactic plasma, it is in the study of x-ray emissions from clusters of galaxies that the Einstein observations have

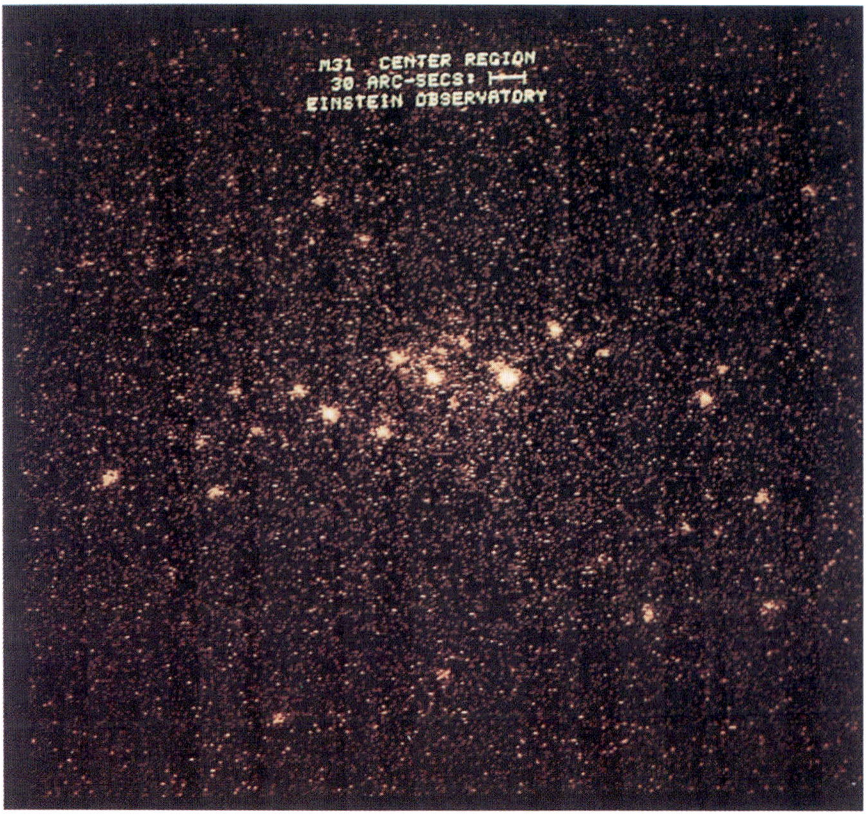

Figure 17. X–ray binaries in the galaxy M31. From the Einstein Observatory.

had some of the most profound impact. The ability to image the hot plasma has given us the means to study in detail the distribution of the gravitational potential which contains both gas and galaxies. This study could only be carried out with some difficulty by studying the individual galaxies which, even though rather numerous in rich clusters, did not yield sufficient statistical accuracy. The x-ray study has revealed a complex morphology with some clusters exhibiting symmetry and a central maximum of density, which demonstrates an advanced stage of dynamical evolution; but many others show complex structures with two or more maxima. This morphology shows that the merging of the clusters' substructures is not yet completed. The relative youth of many of these clusters had not been sufficiently appreciated previously. The discovery of x-ray emission in clusters is therefore used to study one of the most interesting open questions of modern cosmology, namely, the formation and development of structures in the early epoch of the life of the Universe. Piero Rosati (Rosati *et al.*, 2002)[35] has pushed this work to very distant clusters ($z \sim 1.2$) by use of the "ROSAT" Satellite, a splendid successor to "Einstein," built at the Max-Planck Institute for Extraterrestrial Research by Joachim Trümper and Günther Hasinger and their group.

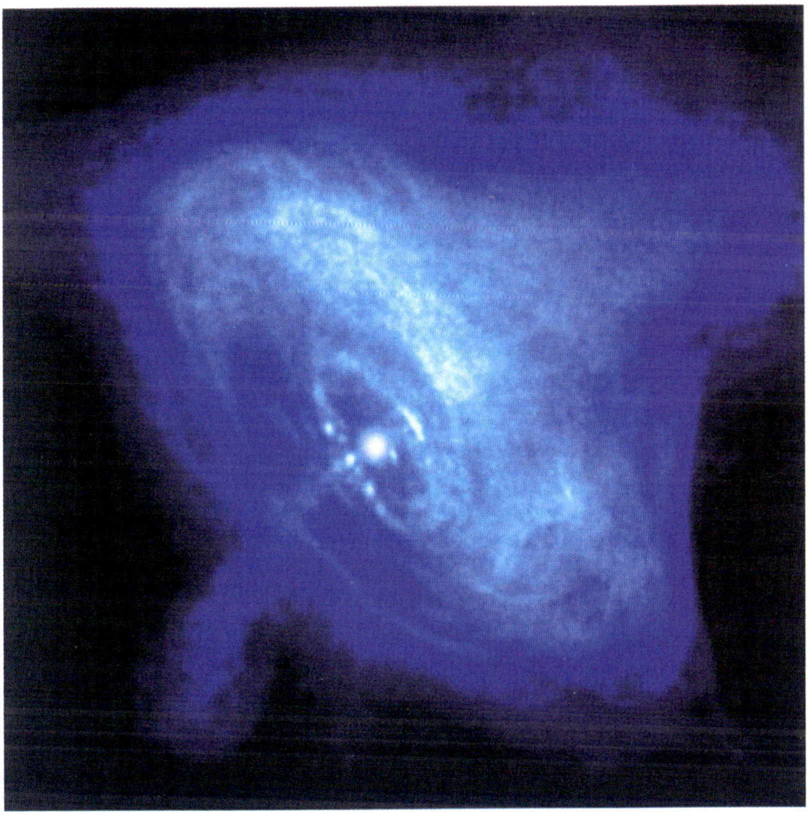

Figure 18. The x–ray image of the Crab Nebula obtained from Chandra. Courtesy of NASA.

5.0 CURRENT RESEARCH WITH CHANDRA

Harvey Tananbaum and I reproposed the concept of a 1.2 meter diameter telescope to NASA in 1976. Tananbaum, who had been project scientist on UHURU and scientific program manager on "Einstein," gave leadership to the Chandra program and brought it to successful conclusion after I left Harvard in 1981. Chandra has met or exceeded all of our expectations. Comparison of the pictures of the Crab Nebula pulsar by "Einstein" and by Chandra (Fig. 18) shows the great improvement in the sensitivity and angular resolution achieved (Tananbaum and Weisskopf, 2001).[36]

I was able to use Chandra for one million seconds to solve the problem of the sources of the x-ray background, a problem that had remained unsolved since its discovery in 1962 (Fig. 19) (Giacconi *et al.*, 2002).[37] Owing to the great sensitivity and angular resolution of Chandra, we were able to resolve the apparently diffused emission into millions of individual sources. They are

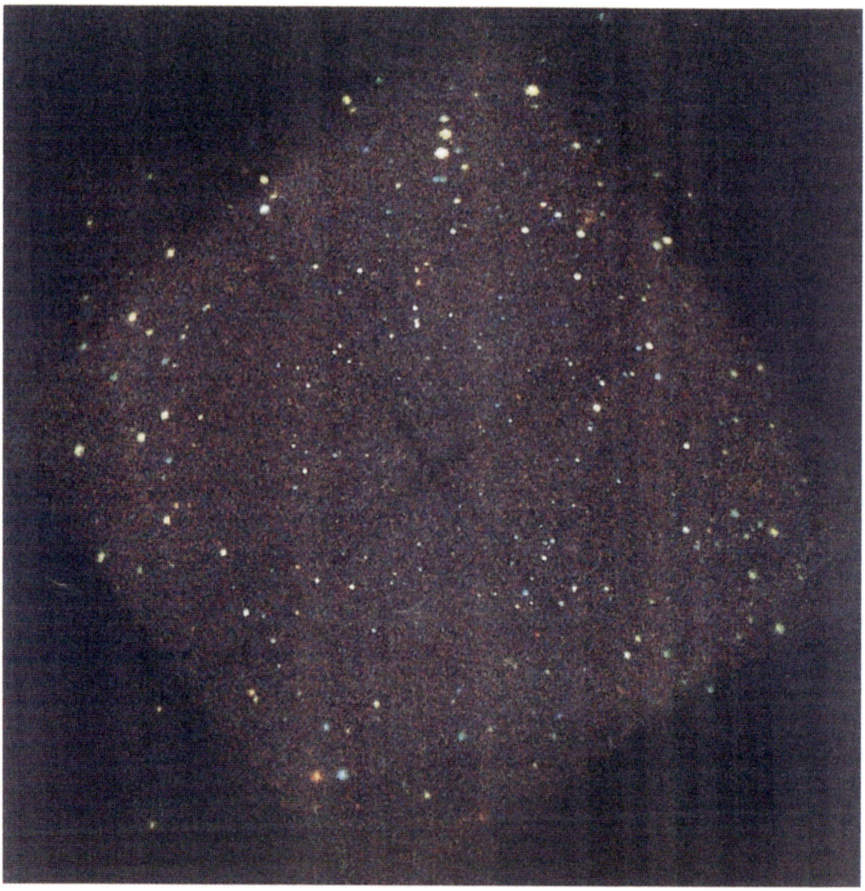

Figure 19. The one million second exposure in the deep x-ray field from Chandra (CDFS-Chandra Deep Field South).

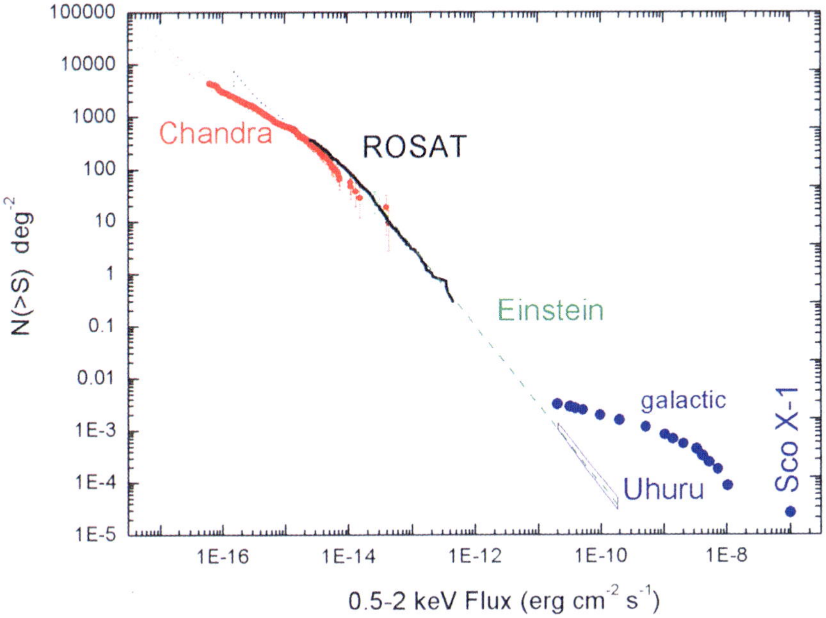

Figure 20. The sensitivity change from 1962 to 2000 (Sco X–1 to Chandra). Courtesy of G. Hasinger.

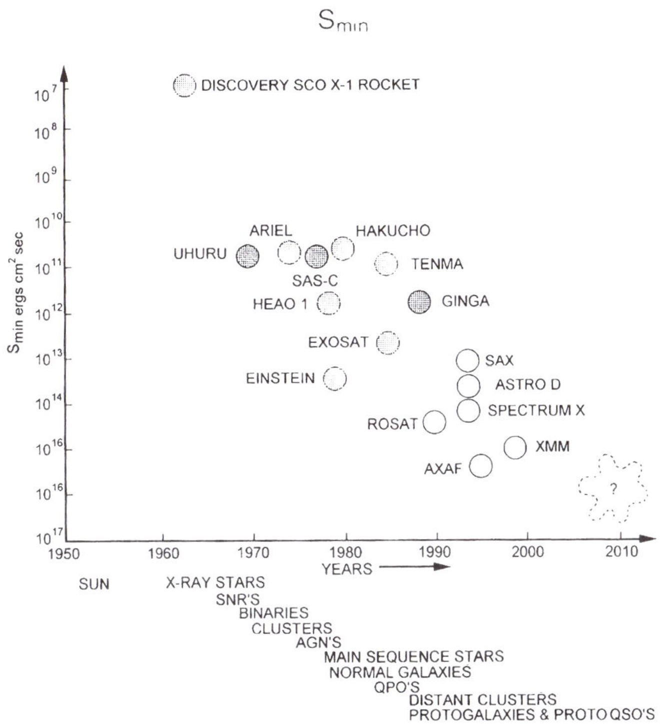

Figure 21. The sensitivity achieved by various x–ray missions (1970 to 2000). Illustration of R. Giacconi.

active galactic nuclei, quasars, and normal galaxies. The gain in sensitivity that this represents is illustrated in Fig. 20. In Fig. 21, a reminder that x-ray astronomy now involves many groups in the world with distinct and essential contributions.

6.0 THE FUTURE OF X-RAY ASTRONOMY

I would like to conclude by attempting to answer the simple question: Why is x-ray astronomy important? The reason is that this radiation reveals the existence of astrophysical processes where matter has been heated to temperatures of millions of degrees or in which particles have been accelerated to relativistic energies. The x-ray photons are particularly suited to study these processes because they are numerous, because they penetrate cosmological distances, and because they can be focused by special telescopes. This last property significantly distinguishes x-ray from γ-ray astronomy. However, in a more fundamental way, high energy astronomy has great importance in the study of the Universe because high energy phenomena play a crucial role in the dynamics of the Universe.

Gone is the classical conception of the Universe as a serene and majestic ensemble whose slow evolution is regulated by the consumption of the nuclear fuel. The Universe we know today is pervaded by the echoes of enormous explosions and rent by abrupt changes of luminosity on large energy scales. From the initial explosion to formation of galaxies and clusters, from the birth to the death of stars, high energy phenomena are the norm and not the exception in the evolution of the Universe.

REFERENCES

1. Hirsh, R.F., 1979, "Science, Technology and Public Policy: the case of X-ray astronomy, 1959–1972," Ph. D. thesis at the University of Wisconsin, Madison.
2. Giacconi, R., Clark, G.W., Rossi, B.B., 1960, "A brief review of Experimental and Theoretical Progress in X-ray Astronomy," Technical Note of American Science and Engineering, ASE-TN-49, Jan. 15 (1960).
3. Giacconi, R. *et al.*, 1962, Phys. Rev. Lett. **9**, 439.
4. Bowyer, S. *et al.*, 1964, AJ, **69**, 135.
5. Bowyer, S. *et al.*, 1964, Science **146**, 912.
6. Giacconi, R., Gursky, H., and Waters, J., 1965, Nature **207**, 572.
7. Oda, M. *et al.*, 1965, Nature **205**, 554.
8. Gursky, H. *et al.*, 1966, ApJ **146**, 310.
9. Sandage, A. *et al.*, 1966, ApJ **146**, 316.
10. Burbidge, G. R., 1967, "Theoretical Ideas concerning x–ray sources," Proceedings IAU Symposium No. 31 (Academic Press, London/New York) p. 463.
11. Shklovsky, I. S., 1967, ApJ **148**, L1.
12. Giacconi, R. *et al.*, 1971, ApJ **165**, L27.
13. Schreier, E. *et al.*, 1972, ApJ **172**, L79.
14. Tananbaum, H. *et al.*, 1972, ApJ **174**, L143.
15. Oda, M. *et al.*, 1971, ApJ **166,** L1.
16. Holt, S. *et. al.*, 1971, ApJ **166,** L65.
17. Rappaport, S. *et al.*, 1971, ApJ **168,** L43.
18. Rothschild, R. *et al.*, 1974, ApJ **189,** L13.

19. Tananbaum, H. *et al.*, 1971, ApJ **165,** L37.
20. Miyamoto, S. *et al.*, 1971, ApJ **168,** L11.
21. Rappaport, S. *et al.*, 1971, ApJ **168,** L17.
22. Hjellming, R.M. and Wade, C.M., 1971, ApJ **168,** L21.
23. Braes, L. and Miley, G., 1971, Nature **232,** 246.
24. Webster, B. and Murdin, P., 1972, Nature **235,** 37.
25. Bolton, C., 1972, Nature **235,** 271.
26. Tananbaum, H. *et al.*, 1972, ApJ **177,** L5.
27. Rhoades, C. and Ruffini, R., 1974, Phys. Rev. Lett. **32,** 324.
28. Giacconi, R., 1974, "Binary X-Ray Sources," Proceedings of IAU Symposium No. 64 (Reidel, Dordrecht) p. 147.
29. Tanaka, Y., 1992, "Black Hole X–ray Binaries," Proceedings of ISAS Symposium on Astrophysics (Inst. of Space and Astronomical Science, Tokyo) p. 19.
30. Gursky, H. *et. al.*, 1972, ApJ **173,** L99.
31. Giacconi, R. and Rossi, B., 1960, J. of Geophys. Res. **65,** 773.
32. Giacconi, R. *et al.*, 1969, Space Science Reviews **9,** 3.
33. Vaiana, G.S. and Rosner, R., 1978, Annu. Rev. of Astron. and Astrophy. **16,** 393.
34. Giacconi, R. *et al.*, 1979, ApJ **230,** 540.
35. Rosati, P. *et al.*, 2002, Annu. Rev. of Astron. and Astroph. **40,** 539.
36. Tananbaum, H. and Weisskopf, M., 2001, in "New Century of X-Ray Astronomy," ASP Proceedings Vol. **251** (Astron. Soc. Pacific, San Francisco) p. 4.
37. Giacconi, R. *et al.*, 2002, ApJS **139,** Issue 2, 369.

Physics 2003

ALEXEI A. ABRIKOSOV, VITALY L. GINZBURG and ANTHONY J. LEGGETT

"for pioneering contributions to the theory of superconductors and superfluids"

THE NOBEL PRIZE IN PHYSICS

Speech by Professor Mats Jonson of the Royal Swedish Academy of Sciences. Translation of the Swedish text.

Your Majesties, Your Royal Highnesses, Ladies and Gentlemen,

This year's Nobel Prize in Physics is all about order! Not in the everyday world we see around us but in the microworld of electrons and atoms. This world, ruled by the laws of quantum physics, displays many spectacular phenomena not normally observable in the macroworld. In certain cases, however, a special type of order among electrons and atoms can amplify quantum phenomena of the microworld and make them visible, even to the naked eye. This year's Nobel Prize in Physics goes to three scientists who have contributed to our understanding of the relation between order and those macroscopic quantum phenomena we call superconductivity and superfluidity.

As early as 1911, the Dutch physicist Heike Kamerlingh Onnes discovered that the electrical resistance of mercury disappears at a few degrees above absolute zero temperature. He called it superconductivity. The Russian physicist Pyotr Kapitsa later coined the analogous expression superfluidity to describe how liquid helium, at even lower temperatures, flows without internal friction or resistance.

From the very beginning it was felt that superconductivity may become of widespread importance in modern industrial society. By winding a superconducting wire into a coil, for example, it was thought possible to make powerful electromagnets that would not suffer energy losses. But unfortunately the first superconductors turned into ordinary metals even in weak magnetic fields. Later on, however, superconductors of a different type were discovered. By allowing superconductivity and magnetism to coexist they remained superconducting even in strong magnetic fields. The superconducting magnets which could therefore be made have become of great importance in modern society and are used, for example, in high resolution magnetic imaging devices for medical diagnostics and at large accelerator facilities used by particle physicists.

Two of this year's Laureates, Vitaly Ginzburg and Alexei Abrikosov, have made decisive contributions to our understanding of how superconductivity and magnetism can coexist. Ginzburg formulated together with Lev Landau a theory that in more detail than before could describe how superconductivity disappears at certain "critical" values of electrical current and magnetic fields. They introduced a measure for the order among the electrons, which they called the superconducting order parameter. Guided by a deep physical intuition they went on to formulate mathematical equations whose solution determines the order in a superconductor. They found good agreement with what had been measured for superconductors known at the time. It is worth pointing out that the reasoning behind this Ginzburg-Landau theory was of such general validity that it is used today to gain new knowledge in many of the subfields of physics.

Soon, however, experiments were performed on new superconducting materials with results that were not expected. It was Abrikosov who discovered the more complicated type of order that explains these deviations. By an insightful analysis of the Ginzburg-Landau equations he was able to show how vortices may form in the spatial distribution of the order parameter and how a magnetic field through these can penetrate the superconductor. The vortices are essentially of the same type as those we can see form in the water when we empty a bath tub. Abrikosov was able to give a complete explanation of how superconductivity and magnetism coexist in a new class of what he called type II superconductors. This amounted to a breakthrough in the study of novel superconducting materials.

An order of an even more complicated kind can be found in the superfluid liquid of helium-3, which was discovered in the early 1970s. This was long after Kapitsa's discovery of superfluidity in a liquid of naturally occurring helium, which is completely dominated by the heavier isotope helium-4.

There is an essential difference between the two isotopes. Helium-4 belongs to a class of particles called bosons that can be directly ordered into a superfluid by what we call Bose-Einstein condensation. Helium-3, on the other hand, is a fermion just like the electron. Such particles must first form pairs before they can reach the superfluid state.

It turns out that the atoms in liquid helium-3 form pairs with internal degrees of freedom. This is partly because the atoms rotate around each other, partly because of their magnetic properties – or spin, as the physicists say. As a consequence the Ginzburg-Landau order parameter has here not only two – as in the superconductors – but eighteen components and the superfluid is anisotropic; it can have different properties in different directions.

It was Anthony Leggett who succeeded in explaining the relationship between the properties of the new superfluid and the many different types of order that this complicated order parameter allows. His theory helped the experimentalists to interpret their measurements and provided a framework for systematic investigations. Leggett's theory has also been used in other fields, as diverse as liquid crystal physics and cosmology.

Professor Abrikosov, Professor Ginzburg, Professor Leggett,

You have been awarded the 2003 Nobel Prize in Physics for your pioneering contributions to the theory of superconductors and superfluids. It is an honour for me to convey to you the warmest congratulations of the Royal Swedish Academy of Sciences. I now ask you to step forward to receive your Nobel Prizes from the hands of His Majesty the King.

ALEXEI A. ABRIKOSOV

I was born in 1928, June 25, in Moscow, USSR (now Russia). My parents were physicians. I graduated from high school in 1943 and was accepted as student of the Institute for Power Engineers. In 1945 I transferred to the Physics Department of the Moscow State University, from which I graduated in 1948 with a diploma (M. Sc. degree). After that I was accepted, as a Postgraduate (Ph. D. student) to the Institute for Physical Problems (now P. L. Kapitza Institute). My scientific adviser was L. D. Landau. After I defended in 1951 a thesis on thermal diffusion in completely and incompletely ionized plasmas, I got the Candidate of Science (Ph. D.) degree and was taken to the staff of the Institute, as Junior Scientist.

In 1951–1952 I worked with the experimentalist of the same institute, N. V. Zavaritskii on the experimental verification of the predictions of the recently published Ginzburg-Landau theory of superconductivity on the critical magnetic field of thin films. This resulted in our discovery of the "superconductors of the second group" (now *Type II superconductors*). After that I started to work on the magnetic properties of bulk Type II superconductors, and came to the conclusion that the transition from superconducting to the normal state happens gradually in increasing field with two limiting critical fields. Between these two values the field gradually penetrates the superconductor forming thin threads of magnetic flux surrounded by vortex currents. The array of these quantum vortices forms a regular structure (now referred in the literature, as *Abrikosov vortex lattice*). I compared my results with the magnetization curves obtained experimentally in the 1930s for superconducting alloys, and there was a perfect fit. The experimentalist explained their data, as due to inhomogenity of their samples. My paper was published in 1957 but the experimentalists accepted the vortex lattice only 10 years later, after it was demonstrated by decoration experiments.

In the middle of 1950s I worked also on the transition from the insulating molecular phase into the atomic metallic phase in hydrogen and on the structure of hydrogen planets. Another my topic was quantum electrodynamics at high energies. The latter works became my Doctor of Science thesis (this degree is similar to Habilitation in Germany), which I defended in 1955.

In the end of 1950s – beginning of 1960s we worked with L. Gor'kov on the microscopic theory of superconductivity. We constructed the theory of superconductors in a high-frequency field (with I. M. Khalatnikov) and the theory of superconductors with magnetic impurities, where we discovered the possibility of the so-called *gapless superconductivity*. We also solved the mystery of the finite Knight shift at zero temperature, taking into account the spin-orbit scattering. Simultaneously we worked with I. M. Khalatnikov on the theory of nonsuperfluid He^3: thermodynamics, kinetics, sound dispersion, light and γ-ray scattering, etc. These works were based on the theory of a Fermi-liquid by L. Landau. I worked also during this time on the theory of strongly compressed matter.

In 1961 we published a book with L. Gor'kov and I. Dzyaloshinskii "Quantum field theory methods in statistical physics". Originally written in Russian, it was translated into English, German, Chinese, Japanese, and became (and still is) the main textbook on the subject.

In 1962–63 with my postgraduate L. Falkovsky we constructed the theory of semimetals of the Bi type. These substances have a very small number of charge carriers (in Bi ~ 10^{-5} per atom) and a very peculiar crystalline lattice differing from a simple cubic lattice by two small deformations. The resulting lattice has two atoms per unit cell, and in principle it could be an insulator. However, a simple cubic lattice has one atom per unit cell, and should be a "good" metal with the number of carriers of the order of one per atom, and small deformations cannot transform it into an insulator. This paradox can be resolved by constructing an artificial phase, which at zero deformation has an energy higher than a conventional metal but the energy decreases with deformation, so that eventually this phase becomes energetically favorable. This gives the opportunity to approach the (almost) insulating phase continuously. In this series of papers the energy spectrum was calculated, and the metal-insulator transition with the vanishing of the energy gap was predicted. The infrared properties were analyzed and the transparency thresholds in the frequency were established. All this was confirmed by experiments.

In 1962 our dear teacher, L. D. Landau, got in a car accident and suffered heavy injuries. His life was saved but his brain was damaged, and he never returned to science after that. He died in 1968 from remote consequences of the accident. After the accident the attitude towards theorists at the Institute for Physical Problems changed, and the Landau Group started to think about leaving.

In 1964 I was elected Corresponding Member of the Academy of Sciences, USSR (now Russian Academy of Sciences). In 1966 I was awarded the Lenin Prize together with V. L. Ginzburg and L. P. Gor'kov "for the theory of superconductivity in strong magnetic fields".

In 1965 I became the head of the Condensed Matter Theory Department in the newly organized Institute for Theoretical Physics (later named L. D. Landau Institute). I was one of its organizers.

In 1965–68 I published several papers on the Kondo effect at low temperatures, where I established the appearance of a resonance in the scattering amplitude of an electron from a magnetic impurity atom (now *Abrikosov-Suhl resonance*).

In 1971 I published a book "Introduction to the Theory of Normal Metals", which was translated into English. In 1972 I was awarded the International Fritz London Award for my works on low-temperature physics.

In 1970–75 I constructed the theory of gapless semiconductors, where I showed, that in substances of the type of HgTe, a strong interaction region close to the band matching point exists where the dependencies of various quantities on temperature and magnetic field are described by nontrivial power laws. At the same time I worked on the theory of an excitonic transition in Bi in strong magnetic fields. The predictions were in complete agreement with experimental data.

In 1975 I was awarded the title Doctor of Sciences Honoris Causa by the University of Lausanne (Switzerland).

In 1977–81 with my postgraduate, I. A. Ryzhkin we constructed the theory of one-dimensional and quasi-one-dimensional metals. The main results were a) the probability distribution function of resistivity of a one-dimensional wire, where due to mesoscopic effects there was no self-averaging, and b) the conclusion that suppression of superconductivity in $(TMTSF)_2PF_6$ by non-magnetic defects was an evidence of triplet pairing. Later this was confirmed. During the same years I worked on the theory of spin-glasses with short-range interaction, including semiconductor-based spin-glasses.

In 1982 with a group of experimentalists I was awarded the State Prize of USSR, for the works on semimetals and gapless semiconductors. In 1987 I was elected Full Member of the Academy of Sciences. In 1988 I published the book "Fundamentals of the Theory of Metals" on which I worked three years. It was translated into English and Japanese. The same year I was elected Director of the High Pressure Physics Institute in Troitsk, Moscow District. In 1989 L. Gor'kov, I. Dzyaloshinskii and myself were awarded the L. D. Landau award of the Academy of Sciences, USSR, for our book "Field Theory Methods in Statistical Physics".

In 1991 I accepted the offer of the Argonne National Laboratory, USA, and became Distinguished Argonne Scientist. Since then I continue to work at the same position. In the same year, together with V. L. Ginzburg and L. P. Gor'kov I was awarded the International John Bardeen Award and was elected as Foreign Honorary Member to the American Academy of Arts and Sciences. In 1992 I was elected Fellow of the American Physical Society.

Being at Argonne I became interested by the high-T_c layered cuprates. This interest resulted in a theory which was based on the Bardeen-Cooper-Schrieffer approach but took into account the specific features of the electron spectrum, namely the quasi-two-dimensionality and the existence of the "extended saddle point singularities", or "flat regions", which made the motion of quasiparticles in some regions of the Fermi surface quasi-one-dimensional. At the same time these regions had the maximal density of states. Another idea was the resonant tunneling mechanism of the electron transport between the CuO_2 planes. On the basis of these ideas I was able to explain almost all the unusual behavior of the high-T_c layered cuprates, including the isotope effect, neutron scattering, pseudogap and the metal-insulator transition.

In 1998 in connection with the experiments performed at Argonne and the University of Chicago I introduced a new phenomenon: "*Quantum Linear Magnetoresistance*". The analysis of experimental data showed that it was first

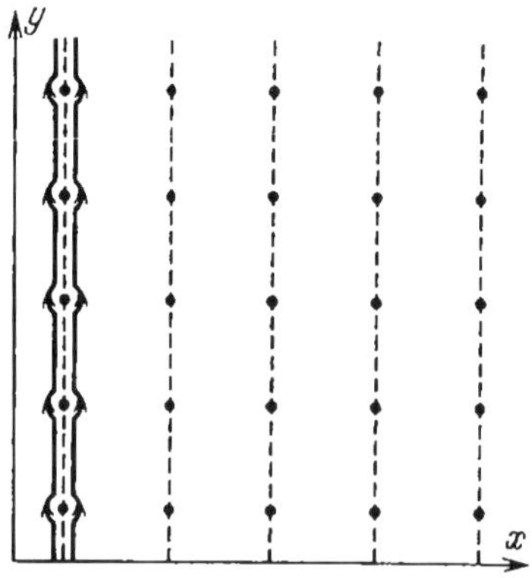

Figure 1. The dots correspond to zeros of the order parameter (square lattice). The dashed lines are the branch cuts introduced in order to make the phase single valued. The gradient of the phase has a discontinuity at every branch cut (see text).

order parameter cannot have a systematic growth, the growth of the vector potential has to be compensated. This can be done by the phase of the order parameter.

If the phase is taken into account, i.e. $\Psi = |\Psi| e^{i\chi}$, then χ enters the GL equations in the following combination with the vector potential:

$$\mathbf{A} - \frac{\hbar c}{2e} \nabla \chi \quad . \tag{7}$$

Consider the behavior of the complex order parameter in the coordinate plane (fig. 1). In order to define the phase unambiguously, branch cuts are introduced into this plane, going through the zeroes of the order parameter parallel to the y-axis.

If we move along the left shore of such a branch cut, the phase varies according to the formula

$$\chi_{left}(y) = \chi_{reg} - \pi \frac{y}{a} \quad ,$$

where the first term is the regular part, and the second is due to the rapid change of the phase in the vicinity of a zero of Ψ; a is the period. Along the right shore it changes according to the formula

$$\chi_{right}(y) = \chi_{reg} + \pi \frac{y}{a} \quad .$$

From these two expressions it can be concluded that the gradient of the phase has a discontinuity at every branch cut:

$$\Delta \left(\frac{\partial \chi}{\partial y} \right) = \frac{2\pi}{a} \quad . \tag{8}$$

If the magnetic field is directed along z, and we chose $A_y = Hx$, then the compensation of the growth of the vector potential, according to (7), can be achieved, if $Ha = \frac{\pi \hbar c}{ea}$, or

$$a = \sqrt{\frac{\pi\hbar c}{eH}} \ . \tag{9}$$

It follows

$$Ha^2 = \frac{\pi\hbar c}{e} \equiv \Phi_0 \ . \tag{10}$$

From these formulas two conclusions can be made: a) the period of the structure grows with decreasing magnetic field, and b) the flux of the magnetic field through one elementary cell is a universal constant, which is called "*magnetic flux quantum*". It is equal to 2.05×10^{-7} $Oe \times cm^2$ and was first introduced by F. London in 1950 [5].

The increase of the period with decreasing magnetic field happens not only close to H_{c2} but also at any field. Indeed, the reasoning, leading to fig. 1 and corresponding conclusions, remains justified, except that the vector potential is no more a linear function of the coordinate, and the compensation condition has to be reformulated. This leads to the substitution of the magnetic field by its average value $B = \frac{1}{a^2}\int_0^a\int_0^a Hdxdy$. Hence, we get the same result, as before, with B instead of H.

From this it can be concluded that even far from H_{c2} the period of the structure increases with decreasing magnetic field and its value, H_{c1}, at which $B = 0$, or $a = \infty$ is the boundary between a purely superconducting phase and the phase with a partial penetration of the magnetic field, which I called "*mixed state*". The boundary with the purely superconducting phase corresponds to the magnetic field

$$H_{c1} = \frac{H_{cm}}{\kappa\sqrt{2}}(\ln\kappa + 0.5) \tag{11}$$

According to (1) with increasing κ the upper critical field H_{c2} grows, and simultaneously the lower critical field H_{c1} decreases.

Since the distance between the zeros of Ψ becomes infinite at H_{c1}, in its vicinity it is large, and only one such point can be considered. According to the GL theory the current can be written in a form

$$\mathbf{j} = \frac{\hbar e}{\mathrm{m}}|\Psi|^2\left(\nabla\chi - \frac{2e}{\hbar c}\mathbf{A}\right) \ . \tag{12}$$

In the vicinity of $\Psi = 0$ $\chi = \varphi$, and $\nabla\chi$ has only the φ-component, which is equal to $\frac{1}{\rho}\frac{\partial\chi}{\partial\varphi} = \frac{1}{\rho}$. Hence it is much larger than the second term in (12), and the current forms a vortex. In the general case these vortices from a lattice. The lines of the current in the vicinity of H_{c2} are presented at fig. 2.

A very similar structure has the triangular lattice, which for an isotropic model has a slightly lower energy. Since the energy difference is very small, in real substances the crystalline symmetry can make the square lattice more favorable. Due to this structure, the mixed state is sometimes called "*the vortex lattice phase*".

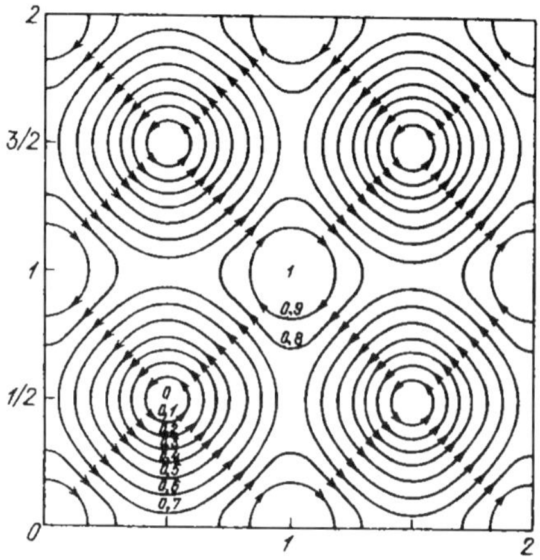

Figure. 2. The lines of current coinciding with the lines of constant $|\Psi|$ for a square lattice.

In the microscopic Bardeen-Cooper-Schrieffer (BCS) theory, as well as in the GL theory, which, as shown by Gor'kov [6], is the limiting case of the BCS theory at $T \to T_c$, there exist two characteristic lengths: the smaller "coherence length" ξ, which is the size of a Cooper pair, and the larger "penetration depth" λ. The GL parameter κ is, essentially, their ratio. For a pure superconductor at $T \to T_c$

$$\kappa = 0.96 \frac{\lambda_L}{\xi_0} \quad , \tag{13}$$

where $\lambda_L = (mc^2/4\pi ne^2)^{1/2}$ is the London penetration depth (n is the electron density), and $\xi_0 = 0.18\ (\hbar v/T_c)$ is the coherence length at $T = 0$ (v is the electron velocity). In the case $\kappa >> 1$ $\lambda >> \xi$ (extreme type II, or London type, superconductor) every vortex has a "core" of the size ξ, where the order parameter varies rapidly, and the outer region of the size λ where the magnetic field decays to zero. According to formula (6), in the vicinity of the vortex axis the order parameter grows linearly with distance. The vanishing of Ψ at the center is due to the fact that this is the only way to avoid ambiguity of Ψ. Beyond distances of the order of ξ the order parameter approaches the equilibrium value at zero field. The overall shape of the behaviour of the order parameter and magnetic field in a vortex is presented at fig. 3.

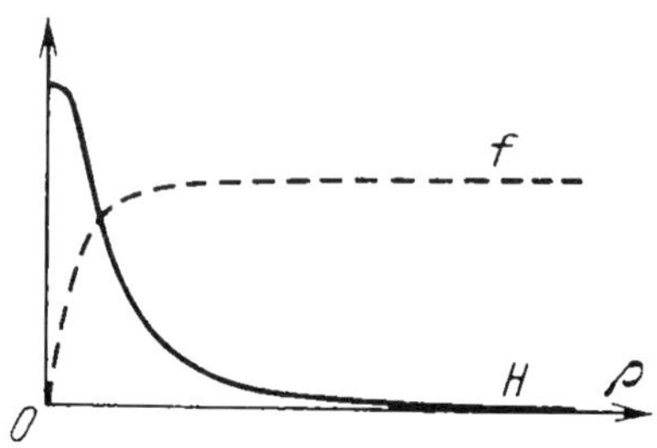

Figure 3. The plot of the magnetic field (bulk line) and $|\Psi|$ in a vortex.

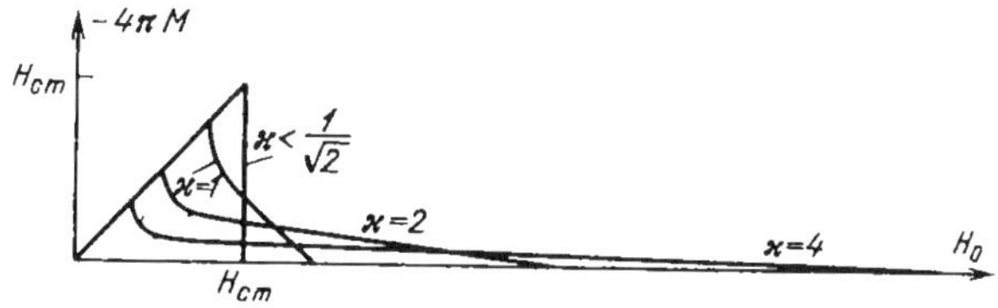

Figure 4. Dependence of magnetization on magnetic field for different values of κ.

The theory permitted also to define the macroscopic characteristics, namely, the dependence of the magnetization on external field. The latter is presented at fig. 4 for different values of κ.

For $\kappa < \frac{1}{\sqrt{2}}$ the dependence is a "triangle", reflecting an ideal diamagnetism below H_{cm} and no magnetization in the normal phase. At higher values of κ the vortex phase appears, and with increasing κ its lower boundary decreases, whereas its upper boundary increases. The limiting formula for the magnetization in the vicinity of the upper critical field is

$$-4\pi M = \frac{H_{c2} - H_0}{(2\kappa^2 - 1)\beta_A} \quad . \tag{14}$$

I compared the theoretical predictions about the magnetization curves with the experimental results obtained by Lev Shubnikov and his associates on Pb-Tl alloys in 1937 [7], and there was a very good fit.

Here I would like to describe the situation with experiments. The magnetization of superconducting alloys was first measured by De Haas und Casimir-Jonker in 1935 [8], and they got a gradual transition from the superconducting to the normal state with two critical fields. They explained it by inhomogenity of their samples. Shubnikov, who worked previously with De Haas, decided to make better samples, and his group annealed the alloys a long time at temperatures close to the melting point. After that the X-ray diffraction studies, performed at room temperature, did not show any inhomogenity. Since the authors could not imagine another explanation of the gradual transition, they wrote in their paper that the precipitation of another phase must happen at lower temperature. Unfortunately, L. V. Shubnikov accused of an attempt to organize an "anti-Soviet strike", was arrested and executed by the KGB the same year. I am sure that, given the opportunity, he would have discovered that there appeared a new phase, and that there exist a special kind of superconductors. I would like to pay here a tribute to Shubnikov, whose data gave me real inspiration. I have never met him but I heard about him from Landau, who was his close friend.

I made my derivation of the vortex lattice in 1953 but the publication was postponed since Landau first disagreed with the whole idea. Only after R. Feynman published his paper on vortices in superfluid helium [9], and Landau accepted the idea of vortices, he agreed with my derivation, and I

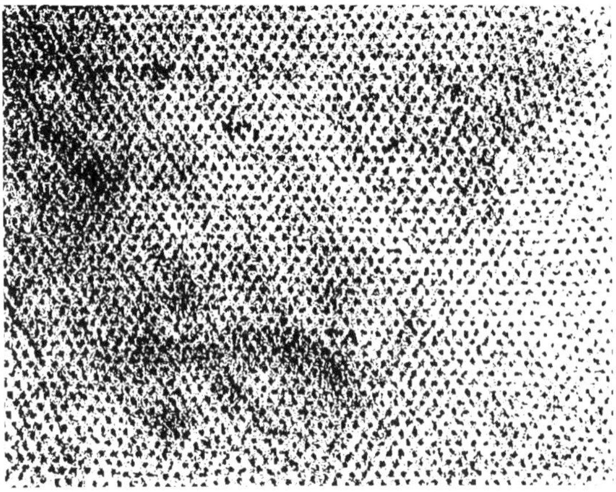

Figure 5. First decoration picture of vortices by Essmann and Traeuble (1967).

published my paper in 1957 [10]. Even then it did not attract attention, in spite of an English translation, and only after the discovery in the beginning of the 1960s of superconducting alloys and compounds with high critical magnetic fields there appeared an interest in my work. Nevertheless, even after that the experimentalists did not believe in the possibility of existence of a vortex lattice incommensurable with the crystalline lattice. Only after the vortex lattice was observed experimentally, first by neutron diffraction [11] and then by decoration [12], (fig. 5) they had no more doubts. Now there exist many different ways to get images of the vortex lattice. Apart from those, which I mentioned, there are electron holography, scanning tunneling microscopy (fig. 6) and magneto-optics.

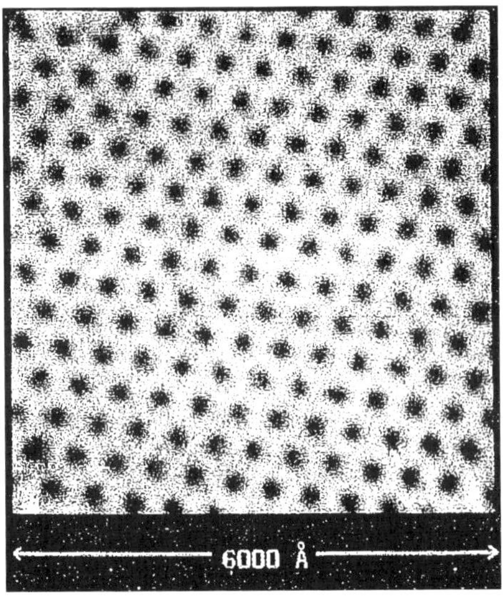

Figure 6. Vortices in $NbSe_2$ defined by scanning tunneling microscopy (STM).

After that I made only one work on vortices, namely I defined the lower critical field of thin films and the vortex lattice in its vicinity [13].

Although I worked in many different fields of theoretical physics afterwards, superconductivity was my favorite. In the beginning of the 1960s we did several works together with Lev Gor'kov. These were based on his Green functional presentation of the BCS theory, which permitted to extend the microscopic theory to spatially inhomogeneous problems. We studied the behaviour of superconductors in a high frequency field (with I. M. Khalatnikov) [14], the role of magnetic impurities [15], where we discovered the so called *"gapless"* superconductivity, and managed to solve the mystery of the finite Knight shift at low temperatures, introducing the spin-orbit scattering [16].

After the discovery of the high-temperature superconductivity in layered copper oxides by J. G. Bednorz and K. A. Mueller [17] I became interested in their properties. There existed many different approaches to these unusual substances and virtually all of them postulated some exotic mechanism of superconductivity. I based my approach on the BCS theory taking into account the specific features of the electron spectrum, mostly, the quasi-two-dimensionality and the so called "extended saddle point singularities", or "flat regions" in the electron spectrum [18]. Another idea was the resonant tunneling connection between the CuO_2, layers [19], which are responsible for conductivity and superconductivity. On this basis I was able to explain most of the experimental data about layered cuprates without dividing them into "good" ones, which should be mentioned on every possible occasion, and "bad" ones, which should be forgotten. As a result I can state that the so called "mystery" of high-T_c superconductivity does not exist.

REFERENCES

[1] V. L. Ginzburg and L. D. Landau, Zh. Eksp. Theor. Fiz **20,** 1064 (1950).

[2] L. D. Landau, Phys. Z. der Sowjet Union, **11**, 26 (1937); ibid. **11**, 129 (1937).

[3] F. London and H. London, Proc. Roy Soc. London, Ser. A **149,** 71 (1935).

[4] A. A. Abrikosov, Doklady Akademii Nauk SSSR **86**, 489 (1952).

[5] F. London, "Superfluids", V. 1, New York, 1950.

[6] L. P. Gor'kov, Soviet Phys. – JETP **9**, 1364 (1959); ibid. **10**, 998 (1960).

[7] L. V. Shubnikov *et al.,* Zh. Eksp. Teor. Fiz. **7.** 221 (1937).

[8] J. M. Kasimir-Jonker and W. J. De Haas, Physica **2**, 943 (1935).

[9] R. P. Feynman: in "Progress in Low Temperature Physics", ed. By D. F. Brewer, North-Holland, Amsterdam, 1955, V, 1, Ch. 11.

[10] A. A. Abrikosov, Sovjet Physics – JETP **5**, 1174 (1957).

[11] D. Cribier, B. Jacrot, L. M. Rao and B. Farnoux, Phys, Lett. **9**, 106 (1964).

[12] U. Essmann and H. Traeuble, Phys. Lett. A **24**, 526 (1967).

[13] A. A. Abrikosov. Sov. Phys. – JETP **19,** 988 (1964).

[14] A. A. Abrikosov, L. P. Gor'kov and I. M. Khalatnikov, Sov. Phys. – JETP **8**, 182 (1958); ibid. **10**, 132 (1959).

[15] A. A. Abrikosov and L. P. Gor'kov, Sov. Phys. – JETP **12**, 1243 (1961).

[16] A. A. Abrikosov and L. P. Gor'kov, Sov. Phys – JETP **15**, 752 (1962).

[17] J. G. Bednorz and K. A. Mueller, Zs. Physik. B **64**, 189 (1986).

[18] A. A. Abrikosov, Physica **C 341-348**, 97 (2000).

[19] A. A. Abrikosov, Physica **C 317-318**, 154 (1999).

V. Ginzburg

VITALY L. GINZBURG

Apart from several monographs on physics and astrophysics, I have published two books which are collections of various articles of scientific, semi-scientific and general, social and political character[1]. Some of these articles contain rather much biographical material. So I do not have any wish to come back to my autobiography. However, it is the wish of the Nobel Foundation that the Nobel Prize winners, together with the lecture, wrote their autobiography. I respect the wishes of the Foundation, so I am writing. Of course, I could confine myself to brief biographical information, while writing a detailed autobiography is rather dangerous – it can bring reproach for "exhibitionism" and immodesty. Nevertheless, I have decided to write in a rather detailed and frank way, as it corresponds to my habits and tastes. One more reason to justify this decision is that I am already 87 and will hardly ever have another occasion to write about myself and my views.

1. I was born in Moscow, on 4th October, 1916, that is, as far back as in the tsarist Russia (at that time even the calendar was different, so the date of my birth was 21st September). My father, Lazar' Efimovich Ginzburg, almost half of his life lived in the 19th century, he was born in 1863, that is only two years after the abolition of serfdom in Russia. Having graduated from Riga Polytechnic, he was an engineer engaged in purification of water and had a number of patents. My mother, Avgusta Veniaminovna Vil'dauer-Ginzburg, was a doctor, she was born, in 1886, in Mitava (Latvia). I was the only child in the family. Mother died of typhoid in 1920, I only remember one episode at her bedside and her funeral. Mother's younger sister, Rosa, who started to live with us, did everything she could for me. She died in 1948. Father had died still earlier (in 1942), when we were living in Kazan', where we had been evacuated. Below I will tell about this period in a more detailed way. And here I will only note that, as is well known, life in Russia during World War I, and especially during the period of military communism, and later in the Soviet Union was hard, in many cases even very hard. But in Moscow, where we lived all the time, with the exception of about two years (1941–1943) spent in

[1] I mean the book "O Fizike I Astrofizike" (On Physics and Astrophysics) (Moscow: Byuro Kvantum, 1995); it is the third edition. The English translation: "The Physics of a Lifetime" (Springer-Verlag, 2001). The second book is "O Nauke, o Sebe i o Drugikh" (About Science, Myself and Others) (Moscow: Fizmatlit, 2003): it is the third edition. The English translation "About Science, Myself and Others" will, I hope, be published at the publishing house Bristol: IOP Publ. (2004). Below these books are referred to as (I) and (II) respectively.

The parents of V.L. Ginzburg: Lazar' Efimovich Ginzburg and Avgusta Veniaminovna Vil'dauer-Ginzburg.

Kazan', material conditions were better than in the majority of other regions of the country. However, one of my childhood memories is a wagon loaded with half-covered coffins with dead bodies and pulled by a horse past our house in the center of Moscow. Another memory is of buying fresh meat "for the kid", which turned out to be the meat of a dog. Normally dogs had never been eaten in Central Russia. However, in general we did not starve, but we lived in a so-called communal apartment – into my father's four-room apartment two more families had been placed after the revolution. Still, it is not these hardships, more or less general, but my loneliness that really sticks in my mind. It was exacerbated by my being sent to school only at the age of 11 (in 1927 I entered the 4th form). My parents (or, to bemore exact, father and aunt) must have been afraid that school at the Soviet times had become quite bad, and sending children to school was not obligatory. I am not acquainted with the general state of education in the USSR in the twenties. But school No. 57 belonging to the Sokol'niki district department of education, which I entered, does not seem bad to me. It was a former French grammar school, with many good old teachers still working there. True, as far as I remember, history was practically not taught, being substituted with rendering the contents of some reports of comrade Stalin and, it seems, some other material like that. What was really bad came later. In 1931, just at the time when I finished the 7th form, yet another school reform happened, and the full high school (I do not remember whether it had 9 or 10 forms) was abolished. After the 7th form pupils were to enter vocational schools (FZU), which were supposed to train skilled factory workers. After a FZU one could, in principle, enter directly higher educational establishments (VUZes), including universities – through "rabfaks", that is worker preparatory departments. I am putting it somewhat vaguely, because, not at all willing to enter a FZU, I did not follow this way. So for a while I, then a boy of 15, remained rather lost and unhappy. My aunt, already mentioned, was working in an organization dealing with purchases of foreign scientific literature. One of their customers was Evgeni Bakhmet'ev, a professor of a technical VUZ. He was a very picturesque figure – a former submarine sailor, a Bolshevik, who became a specialist in the field of X-ray structure analysis. His fate, like the fates of many other people of such kind, was tragic – later he died, crushed by the millstones of Stalins's terror. But in 1931 he helped me to get a job of a laboratory assistant in the X-ray laboratory of the technical VUZ where he was teaching. I will not go into details. I will confine myself to saying that the key figure in the laboratory was Veniamin (Venya) Tsukerman, quite a young man (only three years older than me) with truly uncommon inventiveness and initiative. With him and his school friend and peer, Lev Al'tshuler, we communicated and worked together for about two years. Both Tsukerman and Al'tshuler became physicists, from 1946 they were very close to their superior, Y.B. Khariton, who headed the main center where the atomic and hydrogen bombs were created (now it is the town of Sarov, as this place was called as far back as in the tsarist time; after a nuclear center was created there it was called by some code name, which I have already forgotten, and later its name was Arzamas-16). Tsukerman and Al'tshuler have a lot of awards, and, the main thing, many works of theirs are already declassified and well-known. Both of them are people with interesting fates, in part already described in literature. Giving a more detailed description here is evidently impossible (I will only say that Tsukerman died in 1993, and on 9th November, 2003 we celebrated Al'tshuler's 90th birth anniversary).

What I acquired at the laboratory is not so much some concrete knowledge as the taste for work, for physics, for inventiveness. I also remember reading, with interest, the book by O.D. Khvol'son "Fizika Nashikh Dnei" (The Physics of Our Days) – a popular writing about the achievements of physics, a rare book for that time. Anyway, I decided to become a physicist, especially as I did not have any talents, but in physics I was at least interested. So I decided to enter the physical department of the Moscow State University (MGU). Just at that time, from 1933, it became possible to enter the MGU by open competition (previously for several years the entrance procedure was somehow different – for instance, through rabfaks or in some other way, but certainly not by competition). But for entering the University it was necessary to have finished the complete high school course, while I had finished only 7 forms. So I had to learn in three months (partly with a tutor) what had to be learned in the senior forms. I am going into these details because all my life I have wished I had got a normal school education. And I want to warn those who think that school wisdom is not worth spending many years on it. Indeed, though having, as I am convinced, merely average abilities, I managed to do in three months the program of three school years. But at what cost? The cost is, first of all, a total lack of automatism in elementary mathematics and in the orthography even of the Russian language. Specifically, at school I would have solved, for instance, 100 problems in algebra, trigonometry, etc., while on my own I solved only about 10. Why solve more if it is not demanded? Such things lead to lack of skills in the further learning of mathematics. The same with orthography. Illiteracy "in the people" was at that time being combated, though, and, as far as I know, there were some good results in this field. But at the higher level things were much worse. Suffice to say that already at the second year at the university we had a check (a dictation) in Russian, and half of the students of the year (!), me including, got unsatisfactory marks. After that an obligatory course of Russian was introduced for those who had failed this exam. I doubt that it gave any noticeable benefit – making up for what has been missed in childhood is very difficult. Neither do I know foreign languages, although later I had to master English, but only in the limits necessary for my doing physics. I hope that the given information is not superfluous, both in terms of understanding the life in the USSR in the years described and in terms of my biography.

I did not fail any of the entrance exams at the MGU, in 1933, but on the whole I did not make out brilliantly. As a result, I was not admitted, although admitted were some people with slightly lower marks but more advantageous personal particulars (I was neither a member of the Young Communist League nor a worker, and my parents also were not proletarians, and so on). But still this result was not discrimination or a sign of anti-Semitism, which flourished after the war. Some of my comrades who were not admitted either decided to wait for a year, but I had already left work and had somehow become keen on studying. That is why I entered the external department, which proved possible. And here again I painfully felt my sore spot. In 1934 I managed to get transferred to the internal department, to the second year, that is, I caught up with my fellow students and started to study like everybody. But I learned how much richer and brighter their life was, with all sorts of optional courses and so on. Not to mention that, strangely, I never got acquainted with the courses of astronomy and chemistry, which I did not have to do at the external department and somehow was not obliged to sit when being transferred to the internal department.

I want to finish the story of the first stage of my life by mentioning just one of many episodes which show that man's fate is merely a chain of chances and it is like a pathetic little boat in the sea waves, which may capsize at any moment. What I mean here is the following. At the second year at the MGU all students were enrolled in groups of two types, military and civil. Only men were placed in military groups, where they were trained and qualified as military officers. So, when being transferred to the internal department I was sent to a military hospital in order to decide where I should be placed. I was rather not athletic (height 180 cm, weight about 60 kilograms). A military doctor, who seemed an elderly man to me, poked his finger at my neck, uttered the diagnosis "struma" and sent me to a civil group. Struma is some swelling of the thyroid gland, which I have not noticed up to now. I am telling about that because practically all my fellow students who got into military groups were killed at the war. To finish with this unpleasant topic, I shall note that there were three more similar episodes. I cannot say that I was bursting to go to the front but I was not in the least trying to avoid it. Telling about all incidents would be too long, so I shall limit myself to only one. In 1941, after the beginning of the war (for the USSR it began on 22nd June,1941) the Lebedev Physical Institute of the USSR Academy of Sciences (FIAN), where I was working at the time (and were I have been working up till now), was evacuated to Kazan' together with the greater part of the Academy of Sciences. And, what was quite natural for the war time, at least in our country, the workers of the Institute were used for different works. So once we were sent to unload barges with logs or firewood on the Volga. By the way, everybody took part in that, for instance, I.E. Tamm, also mentioned in my Nobel Lecture. My task was bringing logs from the barge to the bank. I was wearing the so-called "koza" (which means "a goat") – something like a rucksack with a step at the rear. Such "kozas" were widely used in Russia at one time for carrying heavy weights, and they are effective indeed. Two men put a log on a "koza" and the one who was wearing it carried the log to the bank. I am still surprised what heavy logs can be carried in such a way. But still I must have overstrained myself and on the next day I had bleeding at the throat, though not profuse. Some small blood vessel must have burst. I was sent to hospital, where some petrified spots were found in my lungs. The bleeding stopped, and, like the struma, it did not show itself anymore. At the same time voluntary enlistment into some airborne landing forces was announced and I, then a YCL member, enlisted. But I was not admitted, although I had not mentioned any illnesses and, if I remember correctly, even did not know that I was on the medical books. By the way, I can say that I was not officially exempted from military service – at least, in the first years of the war. But there was some order not to mobilize untrained graduates of some professions or some specialists in general. I was never conscripted, but, expecting to be conscripted all the time, I was trying to write my thesis as soon as possible. In peace time I would not have hurried to do it, but then I wanted, though it looks a bit silly, to be able to finish before the draft what had been started. So I defended this thesis in the spring of 1942 (it was the D.Sc. thesis, as the Ph.D. one I had defended in 1940).

The life in Kazan' was hard, the four of us (my father, my aunt, my wife and I, as my daughter with her grandmother had been evacuated to another place) lived in one room. We felt rather cold and rather hungry. But I worked a lot, as did all my fellow workers. I was exploring the propagation of radio waves in the atmosphere and something else which seemed useful for the defense. I also tackled some other problems, as the scientific life was going on.

2. However, let me come back to the student days. Starting from the second year, I did well in my study and, so to say, enjoyed studying. I liked theoretical physics, but I thought, and still think, that my ability for mathematics is average, at most; besides, I had difficulties solving problems and making calculations (see above) – while it was considered, and rather justly, that a theoretical

physicist should be good at mathematics. In short, because of all that, when at the end of the third year or at the beginning of the fourth year (now I do not remember when it was) I had to choose my specialization, I did not dare to take up theoretical physics and chose optics. The choice was not accidental. At the MGU physical department at that time there were rather many really good specialists, but there were some retrogrades as well. They were struggling with each other, and this struggle had a political coloring. Just a beginner at that time, I was far from their controversy. I remember only one public debate between the partisans of the contemporary physics and the so-called "mechanicists", who criticized the theory of relativity and so on. It took place in 1935 or maybe the first half of 1936. I do not remember the date and I did not try to find it out, having a sad possibility to say something in this respect, as I know the fate of B. Gessen – a participant of this dispute, a physicist and a philosopher, who was the dean of the physical department at one time. He was also an old Bolshevik (so they called those who joined the party whether before October 1917 or shortly after) and shared the lot of the majority of his comrades – arrested on 21st August, 1936 and sentenced to be shot on 20th December, 1936, at the private meeting of the Military Board of the Supreme Court. The sentence was carried out on the same day. Of course, he was posthumously rehabilitated in view of a complete "absence of *corpus delicti*" – let it be a consolation for the representatives of "progressive, leftist intelligentsia" defending today, because of their much talked about "political correctness", dictators, terrorists and hooligans.

Returning to the subject, I can say that, despite a complete absence of any personal connections or acquaintances, I understood from the very beginning, putting it in the words from a popular Russian children's poem, "what is good and what is bad". In short, I decided to specialize in optics because the chair of optics was headed by G.S. Landsberg, one of the people grouping around the remarkable person and physicist, Leonid Mandel'shtam. The associate professor of this chair, Saul Levi, was, luckily, appointed my tutor. I remember him with a warm feeling. I write about him and my relationship with him in the article "Notes of an Amateur Astrophysicist"[2]. There is also some material repeated above and further on. Maybe it would be more convenient for the readers if everything was repeated here, but I have decided not to do it, for this autobiography is becoming very long as it is. But still I shall tell how I left optics and experiment in general and became a theoretical physicist. When I was waiting, in September 1938, to be conscripted (as a matter of fact, I had already been conscripted and was waiting to be summoned to come with my things to the district military committee), which I again avoided, one can say, as luck would have it (it was the last time when the MGU post-graduates were granted a conscription deferment), I did not want to sit in the room

[2] This article was published under the number 17 in collection II. Earlier it had been published, though without some small modifications, in *Annual Review of Astronomy and Astrophysics* **28** 1 (1990).

without windows and with the walls painted black, where I was trying to measure the spectrum of radiation of canal rays. So I made an attempt to explain the effect we were looking for, notably some asymmetry in the radiation of the mentioned canal rays. Notably, I thought that the electro-magnetic field of moving charges could cause induced radiation. Such an assumption was erroneous, as the field of a charge is not equivalent to free (light) field. But I did not understand it at once and turned with the corresponding question to I. Tamm, who at that time was heading the chair of theoretical physics at the MGU and reading lectures to us. Fortunately, I. Tamm also did not notice at once that my idea was erroneous, he listened to me in a very friendly manner, advised to look something up and in general supported me. The latter was so important for me, suffering from the inferiority complex and not at all sure of my ability to obtain any theoretical results. I write about it in more detail in the article "A Scientific Autobiography – an Attempt" published in book I. In short, it turned out that I managed, without any complicated mathematics, to sort out some problems of quantum electrodynamics. Specifically, such an undoubtedly highly knowledgeable theoretical physicist, and mathematically minded at that, as V. Fock, when doing quantum calculations had come to the conclusion that an electron uniformly moving radiates electromagnetic waves. This conclusion surprises, as we in the classical theory are used to thinking that a uniformly moving charge (electron) does not radiate. But the thing was merely in formulating the problem in a different way: in the classical case it is usually a stationary problem, while the quantum calculation had been done with the inclusion of quantum interaction at a certain instant of time $t = 0$. But the latter is equivalent to the fact that when $t < 0$ there is an electron without the field it entrains, while at $t = 0$ it acquires a velocity v and starts to interact with the field. Physically this is equivalent to the assumption that the electron at the moment $t - 0$ is instantly accelerated to a velocity v. It is clear that this process is accompanied by radiation. In general, using the so-called Hamiltonian method I managed to elucidate some questions and solve a number of electrodynamic problems both in vacuum and in the movement of charges in a medium, for example, the problem of the Vavilov- Cherenkov radiation in the passage of charges through crystals. Here I did not have to use any complicated mathematics. Of course, I found what had long been known – however close are physics and mathematics their connection can be most different, in particular, it is possible to make progress in theoretical physics using but a very modest mathematical apparatus, for instance, not exceeding the limits of what is taught at physical departments of universities. By contrast, in some cases theoretical physicists both use most complicated modern mathematics and develop it.

So I realized that I could work, achieve results, think up new possibilities. Such awareness brings a great joy, it is happiness. And I worked a great deal, I wrote a thesis (which is a candidate thesis with us and a Ph.D. according to the western standards) practically in a year and defended it in 1940. Thus I finished the post-graduate course at the MGU physical department in two years instead of the regular three; it seems that there it was the first case of

Daughter Irina Vital'evna Dorman with husband Lev Isaakovich Dorman.

such kind. After that they wanted to leave me, in some capacity, at the MGU Institute of Physics, but the atmosphere there was bad and, most fortunately, on 1st September, 1940 I was taken to the FIAN as a person working for the degree of doctor (such a person has to already have the first scientific degree, Ph.D., and is to prepare, officially in three years, the second thesis for which the D.Sc. is given). I had been doing my postgraduate studies under the guidance of G. Landsberg, as I had been supposed to take up optics. But he quite nobly had not hindered me from doing something quite different. And I.Tamm, who was considered to be my tutor in my doctor work, also did not hamper my doing what I wanted. In general such is the style characteristic of theoretical physicists of the USSR and Russia, at any rate, in many places and undoubtedly at the FIAN.

Thus I have been doing theoretical physics since 1938, and I have done many dozens of research works on different physical and astrophysical subjects. As for articles, I have already written hundreds of them, because for me writing is fairly easy and writing articles is an integral part of my work. Some colleagues condemned and may still be condemning me for writing such a great number of articles. And indeed, there are people who write articles in order to increase the number of references to their works. But I, believe me, have never been guided by such considerations, besides, I have got very many references as it is. For instance. I have recently seen in a certain reference book that after 1985 there have been 8962 references to my articles, in spite

Granddaughter Maria Dorman.

of the fact that the article which we wrote together with Landau and which was published, in 1950, only in Russian, is almost always mentioned without placing the reference in the list of literature; the same source informs that after 1945 there have been 19519 references to my articles. I began to publish them in 1939. By the way, Landau judged, though not very seriously, the age of physicists by the time when their first publication appeared. For instance, he said that I was 13 years younger than him, because his first article appeared in 1926; whereas he was born in 1908, therefore by calendar years I am only 8 years younger. I think that it is not worth attaching too much importance to the number of references, sometimes it is quite deceptive, especially if people refer to some sensational assertion which may prove to be wrong. The number of publications may be slightly more reliable in this respect, but it characterizes the style of working rather than its quality – because it is clear that one publication containing an important result is more substantial than a great number of articles with less important results. Here I will not write about the content of my works, as I can refer the readers to the article "A Scientific Autobiography – an Attempt" published in book I. Let others judge the quality of these works, although below I shall make one remark in this respect.

3. Now I must come back to my biography. In 1937 I married my fellow student, Olga Zamsha. In 1946 we divorced. We have one daughter, Irina Dorman, born in 1939; she graduated from the Physics Department of the

Granddaughter Viktoria Dorman with husband Mikhail Petrov and great-grand-children (twins) Grigori and Elizaveta.

Moscow State University, defended her Ph.D. thesis, and was occupied with the history of physics. Her husband Lev Dorman is a D.Sc., a specialist in the area of cosmic rays. They have two daughters, Maria Dorman and Viktoria Dorman. The former is not married and lives in Israel. Viktoria has got a Ph.D. in physics in Princeton, she is married to M. Petrov, also a physicist, they have twins, Grigory and Elizaveta, who are now already three years old. So they are my great-grand-son and great-grand-daughter. They live in Princeton, USA. In 1946 I married the second time, to Nina Ermakova, who became Nina Ginzburg. Thus we have been together already for 57 years. Unfortunately, we have no children. As this is not quite an ordinary marriage, at least for the Soviet Union, I have written rather much about it in a number of articles published in book II. The thing is that Nina's father, a prominent engineer, was arrested as far back as before the war and died of starvation, in 1942, in a prison of the city of Saratov. It seems that he died in the same cell and almost at the same time as the renowned biologist N.I. Vavilov. And Nina was arrested in 1944 as a member of a group of young people allegedly going to kill comrade Stalin himself. Of course, it was merely an invention of the KGB, but I am writing about it here, especially as a warning to contemporary "revolutionaries" and terrorists, so they could foresee their fate in case of their victory. Under a totalitarian regime the punitive organs deal out justice arbitrarily, and innocent people suffer. Quite often a regular theatre of the absurd adds to that. In Nina's case, she suffered, according to the KGB's sce-

Nina Ivanovna Ginzburg (Ermakova).

nario, mainly because she lived on Arbat, the street where the great leader sometimes drove by. It was from the window of her flat that he was to be shot at. But the valorous fighters against counter-revolutionaries had not taken the trouble to learn that after the head of the family had been arrested only one room in their apartment had been left to Nina and her mother, with the windows overlooking the yard. The clearing up of this circumstance, which happened already after the arrest and ruled out the charge with terrorism, and also some other circumstances resulted in the sentence surprisingly lenient for that time – "only" three years in camps. And if her room had overlooked Arbat, we would probably have never met. In 1945, Nina was released by an amnesty, but without the right to live in Moscow and in a number of other cities. Her aunt lived in the city of Nizhni Novgorod (then Gorki), so she went there, in fact, into exile, especially as in Gorki itself she had no right to live as well and was officially registered in the village of Bor, on the opposite bank of the Volga[3].

In 1945 I was invited by a group of physicists headed by Alexander Andronov and working at Gorki University to become a visiting professor at the newly organized radiophysical department of this university. I was 29 and

[3] For foreigners it should be explained that in the USSR a person could live somewhere only with the permission of the militia and had to be registered there.

full of energy, but in Moscow I did not have any place to teach (I. Tamm and many others had been actually banished from the MGU by that time). During the war I had worked, among other things, on the propagation of radio waves in the ionosphere. So the invitation to head the chair of propagation and radiation of radio waves was quite natural, and I accepted it. I first came to Gorki at the end of 1945, it was then that I met Nina, and in the summer of 1946 we got married. Of course, I was permanently putting in applications to the KGB (to be more exact, it could be done once a year) with the request to permit my wife to return to Moscow and live where she was born and where her mother lived. But all the time I was refused, despite the fact that my applications were also signed by the directors of our institute, at first by S.I. Vavilov and then by D.V. Skobel'tsyn. I have written about it in a rather detailed way in "A Story of Two Directors" published in book I. My wife managed to move to Moscow only in1953, after Stalin's death and the amnesty that followed. It goes without saying that later (in 1956) she was rehabilitated, like all the members of the imaginary counter-revolutionary group – "owing to the absence of *corpus delicti*". Unfortunately, not all of them lived to see this moment, not to mention the long years of imprisonment for some of those who did.

In 1942 I joined the communist party (CPSU). It was just when the Germans reached the Volga, and our view of the future was far from optimistic. Thus I can hardly be suspected of any careerist considerations, not to mention the fact that I certainly hated the Nazis and all the faults of the communist rule were then receding. At the same time, I cannot help saying, with great regret and bitterness, that for many years I was virtually blind in my estimation of communism and Bolshevism. In general, I believed in "a radiant communist future", not understanding that here we had, in fact, a regime of a Nazi type, headed by a criminal no less mean and bloodthirsty than Hitler. Having written the previous phrase I remembered the observation of Churchill that Stalin and Hitler differed only in the shape of their mustache. Anyway, I was sharing the lot of millions of people who did not understand the inevitable fate of the totalitarian regime sliding to lawlessness and terror. I have written a lot on this subject (see a number of articles in book II), and it is not possible for me to dwell on it here. I will come back to my own fate.

The years from 1946 to 1953 were far from easy for me. Living mainly in Moscow I went to Gorky when I could, to read lectures and train a number of postgraduates and workers in physics with an astrophysical and astronomical bias. All that would not have been too bad, but the clouds began to gather over my head. State anti-Semitism and different kinds of persecution of the so-called cosmopolitans, who supposedly worshipped the West, were blossoming out in the country. An offensive against modern science was in progress – first of all, in biology, but accusations of idealism and of other sins were raining down on us in physics as well. I proved to be a good target for all sorts of attacks. Indeed: a member of the party, married to a former prisoner, who had been accused of counter-revolutionary activities and therefore deprived of many rights. And a Jew at that. So finally they started to accuse me of idealism, cosmopolitism and so on, and so forth. Appearing as some culmination of all that was an article published in a newspaper widely distributed at that time, the *Literaturnaya Gazeta* (Literary Newspaper), on 4th October 1947, that is, on my birthday, under the heading "Against Servility". In the article inspired by the rather well-known physicist D.D. Ivanenko[4], I was accused of all sins together with the biologist A.R. Zhebrak, an opponent of academician Lysenko. On the same day (what a coincidence!), on the initiative of the same Ivanenko, the High Assessment Commission of the ministry of higher education did not confirm promoting me to the rank of professor, for which I had been nominated by Gorki University. After that my name began to be mentioned wherever possible as a negative example. Even our institute evidently had to remove me from its academic council for "strengthening" it. I can only guess what fate awaited me in this situation at that time. I think that it would have cost me dear, but I was saved by the hydrogen bomb.

4. The history of creating atomic and hydrogen bombs and my role in it have been extensively described. So I have the more reasons to be brief. The Soviet atomic bomb was first blown up on 29th August, 1949, and the hydrogen bomb, on 12th August, 1953. The hydrogen bomb had been arousing interest even before the tests of the atomic bomb, but then it was completely unclear how to make it, nor was it topical, as far as I understand. It was merely something not to be missed. So in 1948 I. Tamm was attached to this work, though the authorities did not have any special trust in him (once he belonged to the Men'sheviks, who were opponents of the Bolsheviks, and his only brother had been arrested and shot). Only recently have I learned from some jubilee article (the jubilee took place on 12th August, 2003) that at first I had evidently not been admitted to the work in Tamm's group, it only happened some time later. It is a miracle that it happened at all – that someone having the wife in exile was recruited for a top secret job ("top secret, special file"). Of course, I do not know the reasons. I think that both the task was not considered very important (see above) and I had a high "rating". But recruiting my closest friend and a superb physicist, E.L. Feinberg, turned out to be impossible for Tamm, as Feinberg's wife had once lived in the USA. By the way, I cannot help telling why the services of A. Sakharov were enlisted, though earlier I already wrote about it somewhere, and Sakhrov himself mentions it in his "Memories"[5]. From 1945 Sakharov was a postgraduate in the FIAN theoretical department headed by I.Tamm, in 1947 or 1948 he defended his Ph.D. thesis and was going to carry on with peaceful science at the FIAN. He had a small daughter and did not have an apartment of his own, renting a room somewhere, and in general his life was rather hard. So the director of our institute, S.I. Vavilov, asked Tamm to include Sakharov in his special group in order to

[4] As I am mentioning his name I must say that I am writing about the fact established beyond all doubt (see book II, article 18, commentary 10 at the end of the article).

[5] A.D. Sakharov. "Vospominaniya" (Memories). Moscow: "Prava Cheloveka" (1996). There is also an English edition.

get a room for him, which they managed to do (Sakharov got a small room with the area of 14 square meters in a communal apartment). If he had had a room of his own before that, many fates, including both his fate and mine, would have probably turned out quite differently. But it happened as it happened. From the middle of 1948 we started working, at first reading the materials which we had and which, as far as I remember, did not raise any hopes that this problem would be solved. Then Sakharov put forward "the first idea" and I, "the second idea", which together made it possible to build the first Soviet hydrogen bomb. It is ridiculous, but until the very death of Sakharov in 1989, that is 40 years later, all this material had been classified. That is why Sakharov in his memories writes about the 1st and the 2nd ideas without revealing their content. Now it has long been known that the 1st idea is "the layer structure" (I will not explain it here), and my 2nd idea is using as fuel ^{6}Li (or, if you wish, ^{6}LiD) for obtaining tritium $^3\mathrm{H} \equiv \mathrm{t}$ as a result of the reaction $^6\mathrm{Li} + \mathrm{n} \rightarrow \mathrm{t} + {}^4\mathrm{He} + 4.6$ MeV. These two "ideas" were recognized as opening the possibility of creating the hydrogen bomb. With the aim of carrying out this task Tamm and Sakharov were sent, in 1950, to "the site" (Arzamas-16), and I was not admitted there, obviously for the reasons mentioned above (it was a higher level of secrecy), and I remained in Moscow as the head of a small "group of support", but with a sentinel sitting at out door. It goes without saying that I regard this turn of the fate as a greatest luck. I was sure of my safety, so I could visit my wife in Gorki, and could do science. It was not that I was neglecting my work, but our task was mainly performing different calculations, which is not my strong suit. Taking an interest in other people's secrets was not customary with us in general, and I was especially indifferent to them. So I remember my surprise, probably, already in the seventies or the eighties, when by chance I asked Sakharov how "the layer structure" had "worked" or developed. He said that it had "not worked", which I did not understand. Now it is known that the first two Soviet hydrogen bombs were made as "layer structures." But such a construction proved to be able to increase the power of the bombs, I do not remember exactly, whether 20 or some other, comparatively small number of times more than that of the bomb dropped on Hiroshima. Somebody thought for some reason that it was not enough, and the 3rd idea (by Sakharov's terminology) consisting in radiative compression was used. While "my" 2nd idea is probably still working. Or rather, not working, to our great luck, but merely figuring.

Trying to be worth my classified salt, as it has already been said, I was glad when Tamm or Sakharov (I do not remember exactly who of them), having once come to Moscow in 1950, told me of the problem of controlled thermonuclear synthesis and the "Tokamak" system offered by them. I took up this problem, managed to write several reports and, by the way, in 1962, when all that was declassified, even published some material from these reports. But I did it as a sort of some rather silly "revenge" for having been dismissed from this work in 1951, which must have been considered too secret for me (for more detail see book II, article 18).

Then especially terrible time came. Stalin went totally insane, repressions were going on, culminating in "the case of doctors" and a bestial anti-Semitism related to it. It seems that the corresponding documents have not been found yet. They may not have existed, as it was clear even to bandits that it was better not to leave trace. According to the rumors, "the doctor killers" were whether to be hanged on the Red Square or exterminated in some other way, and all Jews were to be exiled into some camps already built. Some "necessary" people, possibly I among them, would have been left in "sharashkas" later described by Solzhenitsyn and others (a "sharashka" is in fact a prison were scientific and technical research was conducted). It was a tremendous luck that the Great Leader did not have enough time to carry out what he had planned to do and died, or was killed, on 5th March, 1953. In the former USSR many people (at any rate, my wife and I) have up till now been celebrating this day as a great festival.

Everything in the country started to change very quickly, suffice to mention the rehabilitation of "the doctor killers" and the shooting up of Beriya, who was the head of the Soviet "atomic project" (by the way, he was a good organizer and probably not more of a bandit than all the rest). At the USSR Academy of Sciences, contrary to its Charter, there had been no election since 1946, evidently, because it had not been allowed by Stalin. Already in 1953 such an election took place, and I was elected corresponding member of the Academy. As they say, "from rags to riches" – although, in my case this saying looks somewhat exaggerated. But still, both in the USSR and now in Russia being a member of the Russian Academy of Sciences means a rather privileged position. In addition, I got some rather high governmental awards[6] and in general turned into a VIP, although second rate. No less important was that my wife managed to return to Moscow, after 8 years of half-exile, not to mention approximately a year and a half of prison and camp.

Since that time life has been going on more or less normally, but still it made you think of a popular joke: "Question: What is permanent under the Soviet power? Answer: temporary difficulties are." The Sakharov story became a temporary difficulty for me. In 1969, dismissed from classified work, he returned to our theoretical department of the FIAN. As for me, in 1971, after Tamm's death, I had to become head of this department. "Had to become" indeed, because I did not want it and in general I do not at all like work like that. But the workers of the department had asked me, and it was really necessary, because due to numerous "temporary difficulties" it was important that the de-

[6] I must say I do not feel like enumerating all these awards, including prizes and being elected members of some academies. Enumerating all of them would be too long, and mentioning only the main ones would mean lack of respect for others. The corresponding information, although incomplete, can be found in books I and II. Let me note that I have never strived for awards, never aspired for them, though I was glad to get them. Besides, the Nobel Prize has exceeded everything. I know that they had started to nominate me for this prize as far back as about thirty years ago. However, I had long ago come to the conclusion that it had been decided not to give it to me. By the way, I was able to understand it – both because of the certain circumstances of my life and because I knew that the work of the Nobel committee is extremely difficult. That is why the news of the prize having been awarded to me, in 2003, was quite unexpected.

partment should be headed by a person of some rank. And in our department only Sakharov and I were academicians (I had been elected full member of the Academy in 1966). But Sakharov had already become a dissident, had taken up politics, and was totally unsuitable for the post. So I had to take over. And it was the same Sakharov who first of all became a problem, one can say, especially after his being exiled to Gorki in 1980. I have written about it in the article "The Sakharov Phenomenon", which can be found both in book I and in the book of memories about Sakharov. I will not repeat myself. I will only say that I do not see anything for which I could reproach myself in this matter.

However, I must note that "the case" of Sakharov aggravated my already not temporary but permanent problem which I faced when under the Soviet regime I either would not be allowed to go abroad at all or would be allowed with great difficulty. The explanations always referred to the reasons of our famous "secrecy", but it was an obvious pretext. For instance, Tamm, who really did know some secrets (while I actually knew none), went abroad many times after 1953. Of course, I was very glad for him. But I was refused permission to go because of the "sins" of my wife and my own sins. My greatest "achievement" in this area was made in 1959. That year a big international physical conference took place in the USSR, in Kiev, it seems that it was called the Rochester conference. So, some people, me including, were not allowed to go to this conference (!) under the pretext of secrecy. Landau was also not allowed at first, but he announced that he would go all the same and would make a scandal. Well done! They retreated and he went there. And I cannot make scandals and, even if I could, I would perhaps not have obtained anything by it. So I did not go to this conference, did not see many good physicists, did not learn many interesting things. I was very bitter about it at that time, and I am bitter now when I remember this insult. And the greatest pity for me is that I could not meet Einstein and speak with him – for he died in 1955 when I was already 39, understood something in physics and astronomy already and could well have had a professional talk with him if I had lived in a free world. But I was not fated to. I only remember asking L. Infeld, who came to the USSR, to give my regards to Einstein. I do not know if Infeld had an opportunity to do it.

Going on with the subject, I shall describe two latest experiences characterizing the conditions in which we lived before the collapse of the USSR. In 1984 the Danish Academy of Sciences, whose foreign member I have been since 1977, invited me to come to Copenhagen for a week, I do not remember on what occasion. I started the painful procedure of "registration": for going abroad one had to fill out many papers and pass some commissions. Then all the documents went to the "instance", and sometimes you were not informed until the very last moment whether or not your trip had been permitted. This time I was also told shortly before the departure that I could go but without my wife (!). I refused to go. Such a natural reaction was a rare thing at that time, so I even had a telephone call from the president of the USSR Academy of Sciences, who expressed his disapproval by saying that he, for example, did go without his wife – so how can I be so shrewish instead of being grateful for the great honor done to me and the trust given (to be honest, all the last words were not said but, in my opinion, this call could be interpreted only in such a way). A year later, in 1985, again in Copenhagen, Niels Bohr's birth centenary was to be celebrated. And again I was invited, this time, if my memory does not fail me, being the only person from the USSR who was to make a report at a plenary meeting of the corresponding conference. And again I started to go through the "registration" with my wife. And again at the last moment I was told that I was allowed to go but without my wife. Probably, someone was afraid that if we went together, we might not return. By the way, we had never thought of anything of the kind. However, this time I went, though I was indignant. The thing was that I had already prepared my report, which meant quite a lot of work done, taking into account that I had to prepare it both in Russian and in English and then to pass it through censorship.[7] And, what was even more important, I certainly had a great respect for Bohr (I had heard his reports and was acquainted with him, because he had been to the USSR) and I wanted to take part in the jubilee conference. In my report at the conference and especially in private talks, I somehow expressed my indignation at our lack of freedom. I do not remember the details, but I do remember the bitter after-taste left by the indifference of western colleagues. But I was probably wrong. In the West they knew about the Soviet arbitrariness and were used to it, they protested and defended some people. And in general, why should they have taken my case to heart. Besides, I spoke in a reserved manner, without any attempts to arrange a rally, so it was easier to turn a deaf ear.

By a chance coincidence, this year, 1985, was a turning point, as M.S. Gorbachev came to power, and soon the "perestroika" began. I took part in it – although not very active, but I did. I was a deputy of the USSR Supreme Soviet (from 1989 to 1991, the year when it was dissolved), elected from the Academy of Sciences. I resigned from the party in 1991, since then I have never again been a member of any party and am not going to be one. Of course, I have always been a partisan of the democratic forces but I have not always identified myself with concrete actions of some of their representatives, Sakharov in particular. However, I do not remember anything important and interesting from my "political" activities, which would be worth describing here.

In 1988 I finally managed to free myself from heading the I.E. Tamm Department of Theoretical Physics, as it began to be called. The thing is that according to the rule which had just been introduced and which had long been necessary, people over 70 were not allowed to hold some positions. Having remained at the FIAN in the position of advisor of the Academy of Sciences, I have a small group of subordinates. Besides, I have been heading, since 1968, the chair of problems of physics and astrophysics, created that

[7] Fortunately, now things are quite different!

fy everything. Later, when I got to know him, he himself looked at these classifications, for instance, of women, with irony. I shall only dwell on his classification of theoretical physicists who worked in the 20th century. It was done according to a logarithmic scale, with the base 10, so a 2nd class physicist was 10 times inferior to a 1st class physicist. It was a question of accomplishments, and not of the level of knowledge, of pedagogical ability or oratorical talent. Class 0.5 was given only to Einstein. Put into class 1 were Bohr, Dirac, Heisenberg, Schrödinger, de Broglie, Feynman. I have undoubtedly forgotten several names, but, of course, I will not supplement this list in the way I see it. I remember Pauli being put to class 1.5. Landau himself was placed in his own classification, as far as I remember, at first to class 2.5 and then to class 2. Once E. Lifshits told me that Landau had upgraded himself to class 1.5. Why does this classification seem interesting to me? First, it refutes the legends about Landau's immodesty and conceit. Secondly, it is important that he put an emphasis on the record of accomplishments, achievements. For instance, I remember arguing with Landau about de Broglie, who did not seem to me a particularly powerful figure. But Landau stood firm, and I think he was right. The wave properties of the electron and other particles is a guess of genius, although this idea occurred to de Broglie under the influence of Einstein's notion of photons and the equations $\mathbf{p} = -\hbar\mathbf{k}$ and $k = \omega/c$. By the way, I am surprised that Landau put Feynman higher than himself and in general put him into the 1st class. There is no doubt that Feynman was a brilliant physicist and lecturer but it seems to me that his accomplishments cannot be compared with those of other "first-class" physicists. Probably, Landau especially valued the diagrammatic technique, thinking that he himself would not have been able to hit upon it. What looks interesting and important, from my point of view, is an enormous gap which may exist and often exists between accomplishments and knowledge, command of the apparatus and so on. I think, though I cannot assert it, of course, that "by knowledge and command of the apparatus" Landau stood even higher than Einstein, but "by accomplishments" he sober-mindedly estimated himself as much inferior to him. I am certainly not going to give my estimation of myself here, but it is undoubtedly much lower "by command of the apparatus" than "by accomplishments".

It seems to me that working in science, in physics – at any rate, in a position of at least some authority – is absolutely impossible without having a stand on important issues of our existence and thinking about philosophical questions. Unfortunately, I have not had the chance to really acquaint myself with the philosophy and methodology of science. To a certain extent it can be explained by the fact that I have always worked at concrete tasks and not at, for instance, problems of interpretation of quantum mechanics and so on. That is why the atmosphere of dogmatism and censorship in philosophical issues, which prevailed in our country, has not had an especially pernicious effect on me, as well as on the majority of my colleagues in the USSR. I do not even want to name illiterate rogues who dictated to us under the sign of dialectical materialism how the laws of physics or, for instance, of genetics should be understood. But it does not in the least prove that dialectical materialism in itself is unacceptable (among its advocates there were quite decent and reasonable people, as for instance, the already mentioned B. Gessen). I have always understood the core of this philosophical trend as, first of all, materialism, that is the conviction that nature and matter exist beyond our consciousness and independently of it. Secondly, studying nature, which is what natural sciences do, requires a flexible approach based on facts and not on dogmas. It is such an approach which is dialectical. I do not mean anything else, although philosophers may consider my words naive, insufficient or maybe even worse.[8] A really important point is understanding that materialism is "an intuitive judgment" (using the term especially widely used by E.L. Feinberg) which can be neither proved nor disproved. Atheism and belief in the existence of God are also intuitive judgments. I am an atheist, that is, I think nothing exists except and beyond nature. Within the limits of my, undoubtedly insufficient knowledge of the history of philosophy, I do not see in fact any difference between atheism and the pantheism of Spinoza. That is why I think that Einstein was also an atheist, because in 1929, when asked what he believed in, he answered: "I believe in Spinoza's God, who shows himself in the harmony of all that exists, but not in a God who takes care of the fate and actions of people." Einstein, however, used the term "cosmic religion" and reckoning him among atheists may be not quite right. At any rate, I, like many others, do not have any "cosmic feeling", and I do not see any place for God, that is, for something which is beyond nature and has created this nature (such is the opinion of deists). But evidently it is impossible to prove that God does not exist. This itself leads to the conclusion that the principle of the freedom of conscience is fair, that is, people should have the unimpeded right to believe in God and, if they so desire, to practice some religion (of course, I am not speaking about wildly fanatical sects and the beliefs justifying banditry and terrorism). The Bolshevik – communists were not merely atheists but, according to Lenin's terminology, militant atheists. This term, being not quite clear for broad masses, was changed to the name "the militant godless". These pursued believers, especially priests, destroyed temples (churches, mosques, synagogues) turning some of them into warehouses, stables and so on. Identifying atheists with "the militant godless", which is done in Russia by unscrupulous people or simply by illiterate demagogues, is absolutely groundless. It is the same as identifying a respectable catholic with a partisan of the inquisition or considering all members of the Orthodox Church to be partisans of the brutal persecution of Old Believers or other "heretics". Unfortunately, in the post-Soviet time in Russia a clerical offensive has been going on, while the voice of atheists is completely stifled. That is why

[8] As it was explained to me, a more exact interpretation of dialectical materialism is seeing it as a combination of materialism with the recognition of the laws of dialectics formulated by Hegel. Whereas I regard these laws as scholastic to a considerable degree. Perhaps the most correct name for the views that I hold would be scientific materialism.

since 1998 I have been defending atheism in the press, and after being awarded the Nobel Prize I managed to say about that on television as well. Here it is not a question of combating religion, as this would be at variance with the principle of the freedom of conscience, and, besides, religion sometimes does good, calling for good and for worthy behaviour. It is merely a question of atheistic enlightenment, for instance, of scientific elucidation of a complete falsehood of creationism. In general, an abstract belief in the existence of God (concretely, deism) should not be confused with theism (Christianity, Islam, Judaism) involving belief in miracles and the sanctity of the Bible, Koran and so on. By the way, no cultured person will deny the great artistic and historical value of the Bible. But a belief in Biblical miracles is another thing altogether, being incompatible with a scientific outlook, because a miracle, by definition, is something at variance with scientific data and results of scientific research (for some more details see a number of articles in book II).

7. It remains for me to write about politics, or rather about my opinion on certain questions in this respect. I have always remembered the line from a song by the popular poet and singer Alexander Galich (by the way, his real surname was Ginzburg but we are not relatives): "Be only afraid of the one who will say: I know what should be done". Lenin, Hitler and Stalin knew "what should be done" and this knowledge cost the lives of millions of people. Even if I knew "what should be done", I would not be able to influence the course of history. But the main thing is that I do not know this. I have some opinions, but there is a world of difference between being sure that you are right and acting correspondingly, in spite of everything – and having an opinion and understanding that it may prove to be wrong. So, my opinion is well-known enough: I am convinced that only a democratic form of rule is acceptable. Churchill was certainly right when he said, with the clarity of thought inherent in him, that (unfortunately, I do not remember his exact words) democracy is a very bad form of rule but we do not know any better. Indeed, the hereditary monarchy is presently something which remains merely by tradition. Even so, the monarchic regimes remaining in some countries of Europe do not hamper the democratic order, as far as I know. Where the totalitarian regime is concerned, Lenin, Stalin, Hitler, Mao Tse-tung and many leaders of a lower rank proved convincingly enough that this form of rule is inhuman and is inevitably leading to tragic circumstances. Millions of people, and I am not an exception, believed, putting it in a more or less modern language, in the possibility of "a communism with a human face". To tell the truth, I still do not understand why it is impossible. But there is "a criterion of practice", in exact sciences absolutely unshakeable and enabling us to have indisputable knowledge, for instance, that astrological forecasts are false and the law of conservation of energy is correct (though for the latter this is with a reservation about the whole known range of phenomena). In social life there is certainly no similar indisputability, but the whole experience of mankind shows that totalitarianism and dictatorship inevitably slide to the rails of arbitrary rule, crimes and atrocities. However, for democracy a high price is to be payed too, and arguments emerge about this price – how far can we go, where

is the border? The whole history of the rule of the Bolsheviks and the Nazis is full of examples how democrats, progressive intellectuals, liberals, pacifists and all public like that often did not understand elementary things, and step by step they in actual fact encouraged bandits and consolidated their power. Much has been written on this theme, and I also could write a whole book. I will confine myself to just one example of "deep understanding" of Stalinism, which I have recently come across. Herbert Wells wrote, characterizing Stalin, that he had never met a person more sincere, decent and honest; that in Stalin there was nothing dark and sinister, and it was these qualities that should account for his tremendous power in Russia(!) According to Wells, Stalin was a Georgian totally devoid of cunning and insidiousness, and his sincere orthodoxy guaranteed the safety of his comrades-in-arms. And this was written in the thirties, already after the horrors of Stalin's collectivization, known all over the world. Several years later Stalin's comrades-in-arms who Wells had probably met were also shot. In all, the "Georgian devoid of cunning and insidiousness" signed by his own hand death sentences (shooting lists) for 44000 people![9]

There are really lots of examples of "understanding" reality in the way demonstrated above. But I have an impression that the "leftist progressive" public "has forgotten nothing and has learned nothing". Under the slogans of "political correctness", following democratic principles and so on, they follow the same road of protection and justification of hooliganism, banditry and terror. Even the tragic events of 11th September, 2001 were far from having a sobering effect on each and everyone. A spectacular example is the attitude to the events in Iraq. The Iraqi regime of Saddam Hussein used chemical weapons in the war with Iran, it occupied and plundered Kuwait, it committed innumerable atrocities inside the country. But they say that it has not been proved whether Saddam did or did not manage to build up a stock of chemical, bacteriological and maybe also of nuclear weapons. It has been proved, however, that he aspired for that. All vileness of this regime and its crimes have also been proved. However, the action of the USA and Britain taken in order to overthrow this regime is condemned by many people. In their opinion, they should have waited until Saddam became stronger and was able to drop something on their heads. I understand that the Anglo-American action was clumsy and this issue is complicated. But, by the way, I do not understand why the elimination of the regime of Talibs in Afghanistan had been considered more legitimate from formal point of view. It is just that 11th September was closer, and bin Laden and Afghanistan, farther. I have already said that I do not know "what should be done". But I know, as it seems to me, that if civilized countries follow the principles of "political correctness"

[9] See A. Yakovlev. "Sumerki" (Twilight). Moscow: "Materik", 2003. The German translation has already been published, the English one is being prepared. The citation from H. Wells is made by re-translating from the Russian text in the Russian edition of book II in comment 12* to article 25.

and observing laws literally, they will again incur innumerable calamities, like those from Hitler and Stalin. The wild fanatics killing absolutely innocent people, women and children, cannot be stopped and rendered harmless without soiling one's hands.

As a Jew I cannot go without dwelling on the "Jewish question" here, although it is still not entirely clear to me why this quite small, suffering people turned out to be in the focus of the world politics. For understanding of what follows I must make several observations. The Jews have remained as a nation, they have not assimilated, as it is customary to think, due to their attachment to Judaism and to considering themselves a chosen nation. I am an atheist and internationalist, that is, I do not regard any people as "chosen", in particular, I do not think that Jews are better than Arabs. I do not know any Jewish language (Hebrew or Yiddish); by the way, I wish I knew, but I do not have ability for languages and my native tongue is Russian. One would think that I must have assimilated. But it is absolutely wrong, I have never been able to even think of giving up my native people. What are the reasons? I do not know and understand them quite well myself. Of course, family roots are essential, there were some Jewish traditions in my family. No less essential is anti-Semitism. Although I have not suffered from it directly, I also used to be called "a Yid". Not to mention the time when after the war anti-Semitism flourished with us. Anyway, because of all these and maybe some other reasons which I do not understand (what if it is merely a question of genes, their role is not quite clear yet), I am a bearer, so to say, of the Jewish national feeling. This is by no means nationalism, for I see nationalism as the opinion of the superiority of "one's own" nation or, at least, striving to justify and defend "one's own" people. I categorically deny having such feelings. On the contrary, with me the most spectacular display of the Jewish national feeling are shame and indignation when I face a Jew who is a scoundrel and on the whole a bad person. At the same time I am glad if a worthy person turns out to be a Jew. For instance, I am happy that Einstein was a Jew, as well as many other outstanding people. In these feelings I see nothing to be ashamed of. It is shameful to promote somebody "one's own" at someone else's expense, it is shameful to pardon "one's own" scoundrel. But taking pride in "one's own" good person is not shameful to me. However, I cannot clearly explain and understand these feelings of mine. But this is another question. By the way, to avoid suspicion of being insincere, let me say that my first and my second wives are Russian. The daughter's husband is a Jew. The granddaughter's husband is Russian. I do not see any problem here, life is like that.

I have burst out telling the above partly in order to explain why I was interested in Israel and its fate. I am very glad that there is such a state where at last the Jews are not a minority, often persecuted and humiliated. At the same time, many things there I do not like. Not in the order of importance, but I cannot help noting clericalism. In the history of the Jews synagogues played not only the role of prayer houses but also of the centers of a community. So some state support of religion in Israel can be understood. But it is important to know the measure. I cannot see any reasons justifying the absence (at least, in a number of places) of public transport on Saturdays and a number of other, sometimes more important, restrictions or consequences of religious character. In such a way atheists are discriminated. Still more important are a complete absence of unity in the country and the presence of a great number of cases of abusing the advantages of democracy which I mentioned above. That is why, I think, the very existence of the state of Israel is under threat, while the elimination of this state would be a new catastrophe of the Holocaust sort. I am writing about it, although I am aware that it might be irrelevant here, as I am indignant at the support of Arafat and his gangs by anti-Semitic and at the same-time progressive "left" forces of the West. And it is not because of my supposedly anti-Arab stand. Stalin and Hitler would have solved the Palestinian problem in 48 hours. They would eliminate either Israel or, which is much less likely, the Palestinian autonomy, deporting the disagreeable population to some distant place. Of course, in the civilized world such a decision is inadmissible. I am of the opinion that it is necessary and possible to have two completely isolated states. All those who know the lessons of history (for instance, the history of "friendship" of Catholics and Protestants in Northern Ireland) or who have seen on television dancing crowds of Palestinians of all ages while the television was showing the terrorist acts of 11th September, which took away the lives of thousands of innocent people, cannot have any illusions about the friendship and love between Palestinians and Israelis. To be more exact, only absolutely brainless bankrupts who had once obtained the celebrated agreements in Oslo can assume (or pretend to assume) that this might be possible in the foreseeable future. That is why I am convinced that a friendly coexistence of Israel and the state of Palestine is nowadays impossible. The only way out, in my view, is a complete isolation of these states from each other. It cannot be done with the majority of Israeli "settlements" remaining on the Palestinian territory. They should be removed, and it is not a question of law. One Israeli man told me that the land for these "settlements" had been bought and not annexed. So much the better, but it merely means that they can try to "exchange" these settlements for the land of the Arabs living in Israel. These Israeli Arabs are the only true obstacle to a complete division of the two states[10]. But this obstacle is not decisive either. Of course, an impenetrable "wall" dividing the two states is also necessary. Using Palestinian workers on the territory of Israel is absolutely inadmissible. This is motivated, as I heard, by humanist considerations, allegedly by the care about poor Arabs. While in actual fact, I think, many Israelis do not want to do manual or menial dirty work, and their own workers are, in spite of the unemployment, either more expensive or un-

[10] There also is a certain problem because Jerusalem is a very important center from the religious point of view not only for the Jews but also for the Christians and the Moslems. However, I do not see why a free religious life cannot be provided under a correspondent agreement (and, perhaps, also under international control) and in the conditions when Jerusalem is the capital of only Israel.

available. But the Palestinians working in Israel can't help hating their rich masters, which is a source of additional antagonism. The history of importation of black slaves to America could be a good lesson. And taking care of Palestinians should be carried out by their brothers in rich Arab countries.

As for the question of the Golan Heights, it seems to me completely made up. I have been there myself and seen the ruins of a big ancient synagogue. So why is it Syrian land "from time immemorial"? Syria attacked Israel, suffered a crushing defeat and lost the Golan Heights. Now there is no Syrian population on this territory, its loss is Syria's pay for aggression. Why those who are indignant at this situation do not demand that Konigsberg and its environs, renamed into Kaliningrad and the Kaliningrad region, should be returned to Germany? Such is a result of Germany's attack on the USSR, and nobody, including the Germans, is not going to revise it. By the way, I am indignant that a part of Eastern Prussia and Konigsberg were named after Kalinin, this nonentity, who licked Stalin's boots, while Stalin sent his completely innocent wife to a concentration camp.

Being a realist, I am sure that such "Ginzburg plan" will not be realized and will only arouse spite and mockery of pseudo-democrats and "peace-makers". Why, that is their business, while I wanted to express my opinion, taking advantage of the freedom of speech.

In conclusion, one observation of general character. A tremendous progress of science has led to its deep internationalization. There is no such thing as American, Russian, Jewish or whatever else national physics. There is only one physics in the world, and when we are speaking, for instance, about British or Russian physics, we only mean the organization or the state of physics in Britain or in Russia. The Aryan science of the Nazis and the Marxist and Leninist science of the communists have long been forgotten. Whereas in the field of the social sciences and the sciences related to them, like sociology, psychology, economics and so on, the true depth and internationalization are still far from being achieved. But, I hope, here also the time of great success is near. Such is one of the factors which enable us, as it seems to me, to look into the future with hope. Another factor is the slowdown of the growth of the population of the Earth. Also positive, though this assertion is disputable, is the increase of the average life expectancy and thus the increase of the average age of the population. I remember the saying: if a man is not a communist at the age of 20, it means that he has no heart, but if he is a communist at 50, it means that he has no head. The bitter statement "what history teaches is merely that it does not teach anything" is not without reason. But still to completely agree with this statement would mean to totally lose the faith in mankind.

What has been said explains why I am still inclined to believe in the radiant future of mankind. Today on the road to it there are many obstacles, first of all, the Islamic (terrorist) threat, poverty and the lack of education of great masses of population, AIDS and other diseases. But let us remember the situation, for example, in 1943, sixty years ago. Europe was under Hitler's heel, the USSR, though heroically resisting, was living under the Stalinist yoke.

America was not so strong, and the world war was raging. Was it easier and better than now? The forces of democracy have coped with it, saved the civilized society and nowadays both the Nazism and the communism have almost sunk into oblivion. That is why we can hope for the ultimate triumph of the democratic system and the secular humanism all over the world. The necessary conditions for that are the presence of historical memory and the development of science.

23 November 2003

ON SUPERCONDUCTIVITY AND SUPERFLUIDITY

Nobel Lecture, December 8, 2003

by

VITALY L. GINZBURG

P. N. Lebedev Physics Institute, Russian Academy of Sciences, Moscow, Russia.

INTRODUCTION

First of all I would like to express my heartfelt gratitude to the Royal Swedish Academy of Sciences and its Nobel Committee for physics for awarding me the 2003 Nobel Prize in physics. I am well aware of how difficult it is to select no more than three Laureates out of the far greater number of nominees. So all the more valuable is this award. Personally, I have two additional motives for appreciating the award of the Prize. First, I am already 87, the Nobel Prize is not awarded posthumously, and posthumous recognition is not all that significant to me since I am an atheist. Second, the 1958 and 1962 Nobel Prizes were awarded respectively to Igor' Evgen'evich Tamm and Lev Davidovich Landau. Outside of high school, the notion of a teacher is very relative and is quite often applied by formal criteria: for instance, it is applied to the supervisor in the preparation of a thesis. But I believe that the title real teacher can appropriately be given only to those who have made the greatest impact on your work and whose example you have followed. Tamm and Landau were precisely these kind of people for me. I feel particularly pleased, because in a sense I have justified their good attitude toward me. Of course, the reason lies not with the Prize itself, but with the fact that my receiving the award after them signifies following their path.

Now about the Nobel Lecture. It is the custom, I do not know whether by rule or natural tradition, that the Nobel Lecture is concerned with the work for which the Prize was awarded. But I am aware of at least one exception. P.L. Kapitza was awarded the 1978 Prize for "his basic inventions and discoveries in the area of low-temperature physics". But Kapitza's Lecture was entitled "Plasma and the Controlled Thermonuclear Reactions". He justified his choice of the topic as follows: he had worked in the field of low-temperature physics many years before he had been awarded the Prize and he believed it would be more interesting to speak of what he was currently engaged in. That is why P.L. Kapitza spoke of his efforts to develop a fusion reactor employing high-frequency electromagnetic fields. By the way, this path has not led to success, which is insignificant in the present context.

I have not forgotten my "pioneering contributions to the theory of superconductors and superfluids" for which I have received the Prize, but I would

I.E. Tamm.

like not to dwell on them. The point is that in 1997 I decided to sum up my activities in the corresponding field, and I wrote a paper entitled "Superconductivity and superfluidity (what was and what was not done)" [1, 2]. In particular, this article set out in detail the story of quasi-phenomenological superconductivity theory constructed jointly with Landau [3]. Under the circumstances, it would be unnecessary, and above all tedious to repeat all that. Furthermore, the Ginzburg–Landau theory of superconductivity, which I call the Ψ-theory of superconductivity, is employed in the work of A.A. Abrikosov [4], and he will supposedly dwell on it in his Nobel Lecture. This is to say nothing of the fact that the Ψ-theory of superconductivity has been covered in many books (see, for instance, Refs [5, 6]). At the same time, there are several problems bearing on the field of superconductivity and superfluidity which I have taken up and which have not been adequately investigated. This is why I decided to dwell on these two most important problems in my lecture.

The case in point is thermoelectric effects in the superconducting state and the Ψ-theory of superfluidity. However, before I turn to these issues, I will nevertheless cover briefly the entire story of my activities in the field of superconductivity. At the end of the lecture I will allow myself to touch on some educational program for physicists (the issue of a 'physical minimum'), which has been of interest to me for more than thirty years.

L.D. Landau.

BRIEF ACCOUNT OF MY ACTIVITY IN THE FIELD OF SUPERCONDUCTIVITY PRIOR TO THE ADVENT OF HIGH-TEMPERATURE SUPERCONDUCTORS.

Lev Landau was in prison for exactly one year and was released on April 28, 1939 primarily due to the efforts of Kapitza, who became his 'personal guarantee'.[1] Landau resided in this state until his premature death in 1968. The Landau 'case' was officially discharged by virtue of *"corpus delicti"* ("absence of a basis of a crime") only in 1990 (!). The imprisonment had a strong effect on Landau, but fortunately it did not bereave him of his outstanding capabilities as a physicist. That is why he 'justified the confidence', as they said at that time, of those who released him on bail instead of shooting him or leaving him to rot in jail (Landau personally told me that he had not been far from death) by constructing his superfluidity theory [7]. I was present at his report on this topic in 1940 or maybe in 1941 (the paper was submitted for publication on May 15, 1941). Also considered at the end of this paper was super-

[1] For more details, see for example article 10 in the book [2].

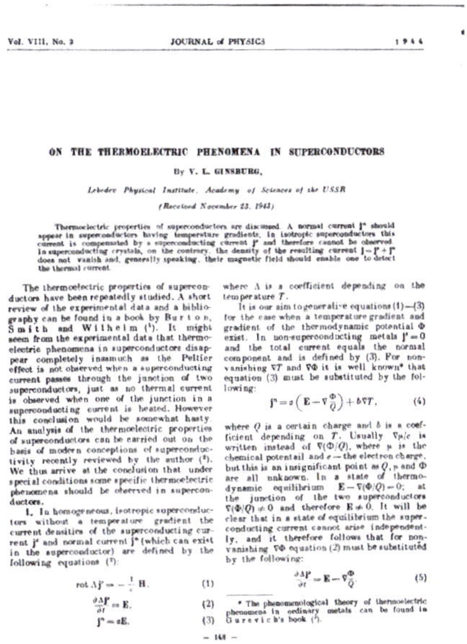

Vol. VIII, No. 3 JOURNAL of PHYSICS 1944

ON THE THERMOELECTRIC PHENOMENA IN SUPERCONDUCTORS

By V. L. GINSBURG,

Lebedev Physical Institute, Academy of Sciences of the USSR

(Received November 23, 1943)

Thermoelectric properties of superconductors are discussed. A normal current j^n should appear in superconductors having temperature gradients; in isotropic superconductors this current is compensated by a superconducting current j^s and therefore cannot be observed. In superconducting crystals, on the contrary, the density of the resulting current $j = j^s + j^n$ does not vanish and, generally speaking, their magnetic field should enable one to detect the thermal current.

The thermoelectric properties of superconductors have been repeatedly studied. A short review of the experimental data and a bibliography can be found in a book by Burton, Smith and Wilhelm (1). It might seem from the experimental data that thermoelectric phenomena in superconductors disappear completely inasmuch as the Peltier effect is not observed when a superconducting current passes through the junction of two superconductors, just as no thermal current is observed when one of the junction in a superconducting current is heated. However this conclusion would be somewhat hasty. An analysis of the thermoelectric properties of superconductors can be carried out on the basis of modern conceptions of superconductivity recently reviewed by the author (2). We thus arrive at the conclusion that under special conditions some specific thermoelectric phenomena should be observed in superconductors.

1. In homogeneous, isotropic superconductors without a temperature gradient the current densities of the superconducting current j^s and normal current j^n (which can exist in the superconductor) are defined by the following equations (2):

$$\mathrm{rot}\,\Lambda j^s = -\frac{1}{c}\mathbf{H}, \qquad (1)$$

$$\frac{\partial \Lambda j^s}{\partial t} = \mathbf{E}, \qquad (2)$$

$$j^n = \sigma \mathbf{E}, \qquad (3)$$

where Λ is a coefficient depending on the temperature T.

It is our aim to generalise equations (1)—(3) for the case when a temperature gradient and gradient of the thermodynamic potential Φ exist. In non-superconducting metals $j^s = 0$ and the total current equals the normal component and is defined by (3). For non-vanishing ∇T and $\nabla\Phi$ it is well known* that equation (3) must be substituted by the following:

$$j^n = \sigma\left(\mathbf{E} - \nabla\frac{\Phi}{Q}\right) + b\nabla T, \qquad (4)$$

where Q is a certain charge and b is a coefficient depending on T. Usually $\nabla\mu/e$ is written instead of $\nabla(\Phi/Q)$, where μ is the chemical potentail and e — the electron charge, but this is an insignificant point as Q, μ and Φ are all unknown. In a state of thermodynamic equilibrium $\mathbf{E} - \nabla(\Phi/Q) = 0$; at the junction of the two superconductors $\nabla(\Phi/Q) \neq 0$ and therefore $\mathbf{E} \neq 0$. It will be clear that in a state of equilibrium the superconducting current cannot arise independently, and it therefore follows that for non-vanishing $\nabla\Phi$ equation (2) must be substituted by the following:

$$\frac{\partial \Lambda j^s}{\partial t} = \mathbf{E} - \nabla\frac{\Phi}{Q}. \qquad (5)$$

* The phenomenological theory of thermoelectric phenomena in ordinary metals can be found in Gurevich's book (3).

— 148 —

conductivity, which was treated as the superfluidity of electron liquid in metals.

That work impressed me, of course, but at that time I was enthusiastic about quite a different range of questions, namely, the theory of higher spin particles. That is why I did not take up the low temperature subject right away, and shortly after our lives radically changed when the war broke out (as is well known, for the USSR it began on June 22, 1941). The Physical Institute of the USSR Academy of Sciences, where I was working and still work, was evacuated from Moscow to the town of Kazan, where many difficulties were encountered, which I describe in my autobiography. In any case, it was not until 1943 that I made an attempt to do, in the spirit of the Landau theory of superfluidity [7], something of the kind as applied to superconductivity.[2] That work [9] is of no great value today, but I believe there were some interesting points in it, for Bardeen considered it at length in his famous review [10]. Even at that time I was aware the work was poor and therefore did not submit it to a journal in English, which we would normally have done at that time (the journal – *Journal of Physics USSR* – was terminated in 1947 during the cold war). My next paper was concerned with thermoelectric effects in the superconducting state [11], and its destiny seems to be unusual and strange. The point is that 60 years have passed, but some predictions made in that work have never been verified and thermoelectric effects in the superconducting state have not been adequately investigated. I myself returned to these problems more than once, but made no significant progress. Appeals

[2] True, somewhat earlier I had considered the problem of light scattering in Helium II [8] on the basis of the Landau theory [7].

addressed to other physicists have only a minor effect, for the problem is out of fashion. Here I would like to take advantage of my last opportunity to bring it to the attention of physicists. Section 4 below is concerned with this question.

However, the question of thermoelectric effects in superconductors, while interesting, is still a particular problem, which evidently emerges only in the presence of a temperature gradient. Furthermore, at that time there existed no thorough theory of superconductivity even under thermodynamic equilibrium. The fact is that the well-known Londons theory advanced in 1935 [12] (it will also be discussed in Section 4 of this lecture) yielded much, and is widely employed under certain conditions even nowadays [5, 6, 13], but it is absolutely insufficient. The last-mentioned circumstance was largely elucidated in my next work performed as far back as 1944 [14]. Specifically, the Londons theory is inapplicable in a strong magnetic field (in the theory of superconductivity, the field is termed strong when it is on the order of the critical magnetic field H_c; we are dealing with type-I superconductors). From the Londons theory it follows also that the surface energy at the interface between the normal and superconducting phases is negative, and to attain positiveness one is forced to baselessly introduce some additional and, moreover, high surface energy of non-electromagnetic origin. Therefore, it became evident that the Londons theory had to be generalized. This problem was solved in 1950 in the Ψ-theory of superconductivity [3].[3] This brings up the question, which has been repeatedly addressed to me: why did it take five years after the work in Ref. [14], in which the necessity of generalizing the Londons theory was recognized, to construct the Ψ- theory? Of course, I cannot answer this question as regards other physicists. As to myself, to some extent I was nearing my objective, as described in the article [1]. But, I believe, the main reason for the slowness of this process lay with the fact that I did not focus my attention on the theory of superconductivity. Theoretical physicists have the good fortune to be able to work almost simultaneously in different directions and in general to move from one subject to another. Specifically, in the period from 1944 to 1950, apart from superconductivity and superfluidity, I was engaged in radio wave propagation in the ionosphere (plasma), radiofrequency solar radiation, light scattering in liquids, the theory of transition radiation (I.M. Frank and I jointly called attention to the existence of this effect), the relativistic theory of higher-spin particles (in part jointly with Tamm), undulator radiation, the theory of ferroelectrics, and other things. Of special note is the fact that ferroelectric effects (as applied primarily to $BaTiO_3$) were considered [15] on the basis of the Landau theory of phase

[3] As already mentioned, this theory is commonly referred to as the Ginzburg–Landau theory. But I resort to the term ψ-theory of superconductivity, because it seems to me that using one's own name rings, at least in Russian, somewhat pretentiously. Furthermore, a similar theory, as applied to superfluidity, was jointly elaborated in my work not with Landau, but with L.P. Pitaevskii and A.A. Sobyanin.

transitions, and this direction subsequently progressed (see article 5 in the collection [2]).[4]

The Ψ-theory of superconductivity [3] is, if you like, an application of the Landau theory of phase transitions to superconductivity. In this case, some scalar complex Ψ function fulfils the role of the order parameter. By virtue of the foregoing I restrict myself to giving the equations employed for Ψ and the vector electromagnetic field potential **A** (as is generally known, rot **A** = **H**, where **H** is the magnetic field strength, which does not differ from the magnetic induction B in this case; furthermore, advantage is taken of the gauge div **A** = 0):

$$\frac{1}{2m^*}\left(-i\hbar\nabla - \frac{e^*}{c}\mathbf{A}\right)^2 \Psi + \alpha\Psi + \beta|\Psi|^2\Psi = 0\,, \tag{1}$$

$$\Delta\mathbf{A} = -\frac{4\pi}{c}\mathbf{j}_s\,,$$

$$\mathbf{j}_s = -\frac{ie^*\hbar}{2m^*}(\Psi^*\nabla\Psi - \Psi\nabla\Psi^*) - \frac{(e^*)^2}{m^*c}|\Psi|^2\mathbf{A}\,. \tag{2}$$

We consider an equilibrium or, in any case, a stationary state, and we assume that the normal current density in the superconductor is $\mathbf{j}_n = 0$ (the total current density is $\mathbf{j} = \mathbf{j}_s + \mathbf{j}_n$, where $\mathbf{j}_s$ is the superconducting current density). Furthermore, at the superconductor-vacuum interface we impose the boundary condition

$$\mathbf{n}\left(-i\hbar\nabla - \frac{e^*}{c}\mathbf{A}\right)\Psi = 0\,, \tag{3}$$

where **n** is the normal to the interface.

In the vicinity of the critical temperature T_c, at which there occurs the normal-to-superconducting phase transition in the equilibriumcase, in the Ψ-theory it can (and even must) be assumed that

$$\alpha = \alpha'_c(T - T_c)\,, \quad \beta = \beta(T_c) \equiv \beta_c > 0\,, \quad \alpha'_c > 0 \tag{4}$$

and the superconductor behavior is determined by the parameters

$$\delta_0 = \sqrt{\frac{m^* c^2 \beta_c}{4\pi(e^*)^2|\alpha|}}\,, \quad \varkappa = \frac{m^* c}{e^*\hbar}\sqrt{\frac{\beta_c}{2\pi}} = \frac{\sqrt{2}\,e^*}{\hbar c} H_{cm}\delta_0^2\,. \tag{5}$$

Here, δ_0 is the depth of penetration of the weak magnetic field $H \ll H_{cm}$ and H_{cm} is the critical magnetic field for massive samples (earlier, mention was made of the critical field H_c, which, say, for films is stronger than H_{cm}).

Since the Ψ-theory is phenomenological, the values of mass m^* and charge

[4] For more details on the above-mentioned and other works of mine, see the article "A Scientific Autobiography – an Attempt" in the book [16].

e^* are beforehand unknown. In this case, since Ψ is not an observable quantity (among the observable quantities are, in particular, the δ_0 and H_{cm} quantities), the mass can be arbitrarily selected: it is not among the measurable (observable) quantities. The question of choice of the e^* value is very interesting and intriguing. It seemed to me from the outset that e^* is some effective charge, which may be different from the electron charge or, as is said on occasion, the free-electron charge e. However, Landau did not see why e^* should be different from e, and in our paper [3] it is written as some compromise that "there are no grounds to believe that the charge e^* is different from the electron charge". I remained of my opinion and saw the way to solve this question was to compare the theory with experiment. Specifically, the charge e^* enters in expression (5) for $\varkappa$, where δ_0 and H_{cm} are measured by experiment; at the same time, $\varkappa$ enters into the expression for the surface energy σ_{ns}, for the depth of penetration in the strong field (the field $H \gtrsim H_{cm}$), and for the limiting fields of the overcooling and overheating of superconducting samples. Following the path of comparing the theory with experiment, I arrived at the conclusion [17] that $e^* = (2–3)\ e$. When discussed this result with Landau, he raised an objection, which he had evidently been guided by before, though had not advanced it. Specifically, with the charge e^* assumed to be an effective quantity like, say, the effective mass m_{eff} in the theory of metals and semiconductors, the effective charge may and, generally speaking, will depend on the coordinates, because the parameters that characterize the semiconductor are functions of the temperature, the pressure, and the composition, which in turn may depend on the coordinates $\mathbf{r}$. If $e^*(\mathbf{r})$, the gauge (gradient) invariance of equations (1)–(2) of the Ψ-theory is lost. I did not find objections to this remark, and in article [17] outlined the situation (reporting Landau's opinion, naturally, with his permission). The solution, however, was quite simple. After the advent of the Bardeen–Cooper–Schrieffer (BCS) theory in 1957 [18], it became clear that in superconductors there occurs 'pairing' of electrons with opposite momenta and spins (I imply the simplest case). The resultant 'pairs', which are sometimes referred to as the Cooper pairs, possess zero spin and are Bose particles or, to be more precise, quasi-particles. The Bose-Einstein condensation of these pairs is responsible for the origin of superconductivity. By the way, as early as 1952 I noted [19] that the charged Bose gas would behave like a superconductor, but did not arrive at the idea of pairing. Interestingly, it had been advanced [20, 21] even before Cooper [22]. It is immediately apparent from the BCS theory that the role of charge in the theory of superconductivity should supposedly be played by the pair charge, i.e., $2e$. This fact was proved by Gor'kov [23], who derived the Ψ-theory equations from the BCS theory. Therefore, Landau was right in the sense that the charge e^* should be universal and I was right in that it is not equal to e. However, the seemingly simple idea that both requirements are compatible and $e^* = 2e$ occurred to none of us. After the event one may be ashamed of this blindness, but this is by no means a rare occasion in science, and it is not that I am ashamed of this blindness, but I am rather disappointed that it did take place.

Many results were obtained in our work [3]. For small values of the parameter $\varkappa$ we calculated the surface energy σ_{ns} and pointed out that it lowers with increasing $\varkappa$ and vanishes when $\varkappa = \varkappa_c = 1/\sqrt{2}$. Relying on the available experimental data we believed that for pure superconductors $\varkappa < \varkappa_c$, and this is generally correct. In any case, we considered in detail only the superconductors with $\varkappa < \varkappa_c$, which now are termed type-I superconductors. Subsequently I would also restrict myself to the investigation of type-I superconductors (a certain exception is Ref. [24]). In 1950, as well as previously, the superconducting alloys were known to usually behave in a significantly different manner than pure superconductors. Particularly clear data concerning alloys were obtained by L.V. Shubnikov[5] and his collaborators in Kharkov in the mid-30s (see references and the results in [25]; this material was also touched upon in [26]; for more details see [27]). In [27], use is made of the term 'Shubnikov phase' for the alloys investigated by Shubnikov. However, an understanding of the situation was lacking, and Landau and I, like many others, believed that alloys are an 'unsavory business', and did not take an interest in them, restricting ourselves to the materials with $\varkappa < \varkappa_c$ for which $\sigma_{ns} > 0$, i.e., type-I superconductors. True, as noted in A. Abrikosov's paper [4] and in [5], Landau hypothesized that alloys are the ones where $\varkappa > 1/\sqrt{2}$, i.e., they are type-II superconductors according to present-day concepts.

The solution of different problems on the basis of Ψ-theory equations was our concern in the bulk of our work [3]. Apart from the above-mentioned question of the energy σ_{ns}, we considered primarily the behavior of superconducting plates and films in the external magnetic field and in some cases in the presence of current, and in doing this compared the theory with experiment. Subsequently, Landau took no interest in such calculations and in general in the development of the Ψ-theory. My own effort made in this direction is described in [1]. Here, I restrict myself to the mention of a fairly evident yet important generalization of the Ψ-theory [3], in which superconductors were assumed to be isotropic, to the anisotropic case [28]. Furthermore, investigations were made of the overheating and overcooling of superconductors in the magnetic field [29] and of the quantization of magnetic flux in the case of a superconducting cylinder with an arbitrary wall thickness [30], and the Ψ-theory was compared with experiment after the construction of the BCS theory [31]. Of special note is Ref. [32], which was developed in [33], had little bearing on the Ψ-theory, and applied to ferromagnetic superconductors. Such superconductors had not been discovered by that time, and Ref. [32] put forward the explanation for this fact related to the inclusion of magnetic energy. Subsequently (after the construction of the BCS theory), it became clear that the emergence of superconductivity in ferromagnetics is also hampered due to spin interaction. I was not engaged in that problem, but would like to call attention to the following. Certain considerations were given in [32], which allowed changing the role of the magnetic factor (the use

[5] In 1937, when Stalin's terror was in full swing, L.V. Shubnikov was arrested and shot.

of thin films and materials with a relatively strong coercive force). I do not think that anyone has given attention to these possibilities, for old papers are seldom read. Of course, I do not feel sure that at the present stage one can find something of interest in [32, 33] – I would just like these papers to be looked at.

In long ado 1943, I engaged in the study of superconductivity because at that time this phenomenon appeared to be the most mysterious one in the physics of the condensed state. But after the construction of the Ψ-theory, and especially of the BCS theory, the picture generally became clear as regards the materials known at that time. That is why I lost particular interest in superconductivity, though I worked in this area episodically (see, for instance, [30, 34]). My interest was rekindled in 1964 in connection with the formulation of the problem of the feasibility of high temperature superconductors (HTSCs). Mercury – the first superconductor discovered in 1911 – possesses $T_c = 4.15$ K, while the boiling temperature of ^{4}He at atmospheric pressure is $T_{b,\,^4\mathrm{He}} = 4.2$ K. By the way, from 1908 to 1923, for fifteen long years, liquid helium was obtained only in Leiden, and low-temperature physics research was pursued on a very small scale, judged by present-day standards. For the example it would suffice to note that the bibliography given at the end of monograph [26] contains about 450 references to the papers on superconductivity (or, sometimes, related problems) over the period from 1911 to 1944; among them, only 35 references fall within the 1911–1925 period. Meanwhile, after 1986–1987, when high-temperature superconductivity was discovered, during the 10 subsequent years approximately 50,000 papers were published, i.e., about 15 papers per day (!).

There can be no doubt that immediately after the discovery and first investigations of superconductivity the question arose of why this phenomenon is observed only at low temperatures or, in other words, helium temperatures. Naturally, there was no way to provide the answer until the nature of superconductivity was understood, i.e., till the construction of the BCS theory in 1957 [18]. The following expression was derived for the critical temperature in this theory:

$$T_c = \theta \exp\left(-\frac{1}{\lambda_{\mathrm{eff}}}\right), \tag{6}$$

where $k_B\theta$ is the energy range near the Fermi energy $E_F = k_B\theta_F$, in which the conduction electrons (more precisely, the corresponding quasi-particles) are attracted together, which is responsible for pair production and the instability of the normal state; furthermore, in the simplest case, $\lambda_{\mathrm{eff}} = \lambda = N(0)\,V$, where $N(0)$ is the electronic level density near the Fermi surface in the normal state and V is some average matrix element of electron interaction which corresponds to the attraction. In the BCS theory, in its initial form, the 'coupling constant' λ_{eff} and, specifically, λ is assumed to be small ('weak coupling'), i.e.,

$$\lambda \ll 1\,. \tag{7}$$

As regards the temperature θ, in the BCS theory it was assumed that

$$\theta \sim \theta_D , \tag{8}$$

where θ_D is the Debye temperature of the metal, for the interelectron attraction was thought to be due to electronphonon interaction (as is generally known, the highest phonon energy in a solid is of the order of $k_B\theta_D$). Typically, $\theta_D \lesssim 500$ K and $\lambda \lesssim 1/3$; whence it follows, according to (6), that $T_c \lesssim 500 \exp(-3) = 25$ K or more generally

$$T_c \lesssim 30 - 40 \text{ K} . \tag{9}$$

Defining all this more precisely would be out of place here. But it seems to me that the aforesaid will suffice to understand why the condition (9) is fulfilled for typical metals, and even safely fulfilled. In particular, prior to the discovery of high temperature superconductivity in 1986–1987, all attempts to discover or produce a superconductor with the highest possible critical temperature had led in 1973 to the production of only the Nb_3Ge compound with $T_c = 23$–24 K (of course, I do not endeavor to find the exact values of various parameters; they depend on the purity and processing of samples, etc.).

ON HIGH-TEMPERATURE AND ROOM-TEMPERATURE SUPERCONDUCTORS (HTSC AND RTSC)

The advent of the BCS theory made it possible to envisage the feasibility of a radical elevation of the critical temperature. It may be that I am not familiar with some facts, but to my knowledge this question was clearly and constructively posed for the first time by Little in 1964 [35] Being forced to outline the following part of this section quite schematically owing to the lack of space, I can mention that Little proposed considering the possibility of replacing the phonon mechanism of attraction between conduction electrons with the same attraction arising from the interaction with bound electrons present in the same system. I call this mechanism excitonic or electron-excitonic; to state it in plain terms, we are dealing with the replacement of phonons with excitons – excitations in the system of bound electrons. True, this term is not universally used in the literature. In his case, Little employed a quasi-one-dimensional model, in which some conducting 'spine' was surrounded by side 'polarizers', say, organic molecules. For electronic excitons or, in other words, for the excited states of bound electrons, the characteristic temperatures $\theta_{ex} = E_{ex}/k_B \lesssim \theta_F \sim 10^4$–$10^5$ K and, in any case, the values $\theta_{ex} \sim 10^4$ K are quite realistic. It is therefore evident that replacing $\theta \sim \theta_D$ in (6) with $\theta \sim \theta_{ex}$ gives us the values $T_c \lesssim 10^3$ K (when, say, $\lambda \sim 1/3$). Of course, these are no more than words, for it is still unclear how to realize the Little model, and this has never been accomplished. Furthermore, it became clear that the fluctuations in quasi-one-dimensional systems are so strong that the transition to the superconducting state is unlikely to occur. However, having familiarized myself with the paper [35], I put forward straight away [36] a quasi-two-dimensional model, wherein a plane conductor is in contact with a dielectric, say, a dielectric film. We termed the development of this version – the alternation of thin conducting layers with dielectric layers – a 'sandwich'. Going over from the quasi-one-dimensional model to the quasi-two-dimensional model was not accidental, for immediately before the work [36] D.A. Kirzhnits[6] and I had considered [37], not in connection with the high-temperature superconductivity problem, the problem of two-dimensional (surface) superconductivity. By the way, this problem is still of interest in itself, but I cannot enlarge on it for the lack of space and I will restrict myself to giving references [37, 38].

Compared to quasi-one-dimensional systems, quasi-twodimensional systems have the advantage that they exhibit significantly weaker fluctuations that destroy superconductivity. We took up the quasi-two-dimensional version [36, 39]. More precisely, at FIAN (the P.N. Lebedev Physical Institute of the USSR Academy of Sciences) a group of theorists turned to the high-temperature superconductivity problem in the broad sense, considering all issues and possibilities known to us. The fruits of this labor were represented in the monograph [40]; even its English version (1982) appeared 4–5 years before the experimental realization of high-temperature superconductors [41, 42] in 1986–1987. If the consideration of different models and possibilities is omitted, the most significant quantitative finding of our work, which is primarily due to Kirzhnits, is the crystal stability condition. The point is that the main objection against the possibility of developing a high-temperature superconductor was the anxiety that the crystal lattice will be unstable for the metal parameter values required to obtain a high-temperature superconductor, i.e., for a material with $T_c > T_{b,\,N_2} = 77.4$ K[7]. When the problem is formulated in terms of the longitudinal material permittivity $\varepsilon(\omega, \mathbf{q})$, where ω is the frequency and $\mathbf{q}$ is the wave vector (we restrict our consideration to an isotropic body here), the production of electron pairs necessitates, roughly speaking, that the interelectron interaction $V = \dfrac{e^2}{\varepsilon(0, q)\, r}$ should be negative, i.e., should correspond to attraction. But this corresponds to the requirement that $\varepsilon(0, q) < 0$. Meanwhile, on the ground of some considerations it was believed that the lattice would be stable when

$$\varepsilon(0, q) > 0\,. \qquad (10)$$

True, on closer examination (see [1, 40]) it was found that superconductivity is also possible under the condition (10), but the T_c values would turn out

[6] Unfortunately, the outstanding theoretical physicist D.A. Kirzhnits deceased untimely in 1998.
[7] I do not know whether there exists the commonly accepted definition of what superconductor can be regarded as a high-temperature one. In my opinion, HTSC takes place when $T_c > 77.4$ K, i.e. is higher than the boiling temperature of nitrogen at atmospheric pressure.

to be moderate, even below the estimate (9). In [40] and references therein it was found that the correct stability condition for $q \neq 0$ is of the form

$$\frac{1}{\varepsilon(0, q)} \leqslant 1 , \tag{11}$$

i.e. is fulfilled when either of two inequalities

$$\varepsilon(0, q) > 1 , \qquad \varepsilon(0, q) < 0 . \tag{12}$$

takes place. In other words, any negative values of $\varepsilon(0, q)$ are admissible from the standpoint of stability and there are no limitations on T_c. To be more precise, up to now we do not know of such limitations. The following conclusion was drawn from our work, which is contained in Chapter 1 in book [40] written by me:

"On the basis of general theoretical considerations, we believe at present that the most reasonable estimate is $T_c \lesssim 300$ K; this estimate being, of course, for materials and systems under more or less normal conditions (equilibrium or quasi-equilibrium metallic systems in the absence of pressure or under relatively low pressures, etc.). In this case, if we exclude from consideration metallic hydrogen and, perhaps, organic metals, as well as semimetals in states near the region of electronic phase transitions, then it is suggested that we should use the exciton mechanism of attraction between the conduction electrons.

In this scheme, the most promising materials – from the point of view of the possibility of raising T_c – are, apparently, layered compounds and dielectric–metal– dielectric sandwiches. However, the state of the theory, let alone the experiment, is still far from being such as to allow us to regard other possible directions as being closed, in particular, the use of filamentary compounds. Furthermore, for the present state of the problem of high-temperature superconductivity, the most sound and fruitful approach will be one that is not preconceived, in which attempts are made to move forward in the most diverse directions.

The investigation of the problem of high-temperature superconductivity is entering into the second decade of its history (if we are talking about the conscious search for materials with $T_c \gtrsim 90$ K with the use of the exciton and other mechanisms). Supposably, there begins at the same time a new phase of these investigations, which is characterized not only by greater scope and diversity, but also by a significantly deeper understanding of the problems that arise. There is still no guarantee whatsoever that the efforts being made will lead to significant success, but a number of new superconducting materials have already been produced and are being investigated. Therefore, it is, in any case, difficult to doubt that further investigations of the problem of high-temperature superconductivity will yield many interesting results for physics and technology, even if materials that remain superconducting at liquid nitrogen (or even room) temperatures will not be produced. Besides, as has been emphasized, this ultimate aim does not seem to us to have been discredited in any way. As may be inferred, the next decade will be crucial for the problem of high-temperature superconductivity." This was written in 1976. Time passed, but the multiple attempts to find a reliable and reproducible way of creating a high-temperature superconductor have been unsuccessful. As a result, after the flash of activity came a slackening which gave cause for me to characterize the situation in a popular paper [43] published in 1984 as follows:

"It somehow happened that research into high-temperature superconductivity became unfashionable (there is good reason to speak of fashion in this context since fashion sometimes plays a significant part in research work and in the scientific community). It is hard to achieve anything by making admonitions. Typically it is some obvious success (or reports of success, even if erroneous) that can radically and rapidly reverse attitudes. When they smell success, the former doubters, and even dedicated critics, are capable of turning coat and becoming ardent supporters of the new work. But this subject belongs to the psychology and sociology of science and technology.

In short, the search for high-temperature superconductivity can readily lead to unexpected results and discoveries, especially since the predictions of the existing theory are rather vague."

I did not expect, of course, that this 'prediction' would come true in two years [41, 42]. It came true not only in the sense that high-temperature superconductors with $T_c > T_{b, N_2} = 77.4$ K were obtained, but also, so-to-say, in the social aspect: as I have mentioned above, a real boom began and a 'high-temperature superconductivity psychosis' started. One of the manifestations of the boom and psychosis was the almost total oblivion of everything that had been done before 1986, as if the discussion of the high-temperature superconductivity problem had not begun 22 years before [35, 36]. I have already dwelt on this subject above and in the papers [44, 45] and would not like to return to it here. I will only note that J. Bardeen, whom I have always respected, treated the high-temperature superconductivity problem with understanding both before and after 1986 (see [46]; this article was also published in book [16]).

The foregoing in no way implies that our group or I pretend to a practical contribution of great importance to the development of high-temperature superconductivity. At the same time I believe that Little's works and ours have played a significant role in the formulation of the problem and have drawn attention to it. The solution of the problem was obtained to a large measure accidentally. The proposal to employ layered compounds was reasonable and promising, but neither I nor, to my knowledge, anybody else proposed the use of precisely the cuprates. Other layered compounds investigated do not belong to high-temperature superconductors. The following fact serves to illustrate the accidental, to a certain measure, character of discovery of high-temperature superconductivity. As far back as 1979, in one of the institutes in Moscow they produced and investigated [47] a $La_{1.8}Sr_{0.2}CuO_4$ ceramic, which was close to that investigated by Bednorz and Muller, with $T_c \simeq 36$ K [48]. However, the authors of [47] measured the resistance of their samples at temperatures not lower than the liquid-nitrogen temperature and therefore did not discover their superconductivity. From the above one may draw a trivial conclusion that all newly produced materials should be 'tested' for superconductivity. Also evident is another conclusion, namely, that even nowadays it is possible to make a major discovery and next year be awarded a Nobel Prize for it without gigantic facilities and the work of a large group. This should be a source of inspiration, particularly for young people.

The present situation in solid-state theory does not allow us to calculate the value of T_c nor of other superconductor parameters, with the possible exception of a metallic hydrogen yet to be produced. Moreover, for more than 15 years the mechanism of superconductivity in cuprates has remained obscure. I should remark that, despite the fact that I counted on the excitonic mechanism in high-temperature superconductivity research, the role of this mechanism in the known high-temperature superconductors is still completely unclear. In this case, in high-temperature superconductors (in cuprates) with $T_c < 170$ K (the highest-known value $T_c \simeq 165$ K was attained back in 1994 in the $HgBa_2Ca_2Cu_3O_{8+x}$ cuprate under high pressure), as I see it, the electron–phonon mechanism of pairing may prove to be the dominant one. This possibility has previously been underestimated (in particular, by me), since the estimate (9) has served as a guide. But it is valid only for a weak coupling (7). For a strong coupling (i.e., when $\lambda_{eff} \gtrsim 1$), formula (6) is no longer applicable, but even from this formula it is clear that T_c increases with λ_{eff}. The generalization of the BCS theory [18] to the strong-coupling case [49] enables us to investigate the corresponding possibilities. Their analysis (see particularly [50] and references therein and in [1]) suggests that the electronphonon mechanism in cuprates may well ensure superconductivity with $T_c \lesssim 200$ K owing to the high θ_D and λ_{eff} values. At the same time, the electron-phonon interaction alone is supposedly insufficient in the context of so-called d pairing and maybe other special features of superconductivity in cuprates. However, the role of other possibilities (spin interactions, excitonic interaction) is unclear. Of course, it would be out of place to discuss this vital topical problem here. I only want, on the one hand, to emphasize that the longstanding disregard of electron-phonon interaction in cuprates has always seemed and now seems unjustified to me (see [51]). On the other hand, the likelihood of attaining, on the basis of the electron – phonon mechanism, the values $T_c \sim 300$ K, and this is room-temperature superconductivity (RTSC), appears to be small, as with the use of the spin mechanism. At the same time, the excitonic mechanism, as far as I know, does not provoke objections for $T_c \sim 300$ K, either. That is why I pin my hopes on precisely this mechanism for the attainment of room-temperature superconductivity. However, all this is no more than an intuitive judgment.

The creation of high-temperature superconductivity had been my dream for 22 years, even with no guarantee that the goal was at all attainable and, in particular, attainable in the foreseeable future. In my view, obtaining room-temperature superconductivity now occupies the same place.

THERMOELECTRIC PHENOMENA IN THE SUPERCONDUCTING STATE

The first attempt to observe thermoelectric phenomena and, specifically, thermoelectric current or thermal electromotive force in a nonuniformly heated circuit of two superconductors, to my knowledge, was made by Meissner [52] in 1927. He arrived at the conclusion that the thermoelectric effect is completely absent for superconductors. When I took an interest in this problem in 1943, this viewpoint was generally accepted (see, for instance, [53] and especially the first and later editions of the book [25]). However, I encountered this statement more recently as well. Meanwhile, this conclusion is erroneous, which was pointed out in my work [11] published as far back as 1944.

The point is that the superconducting state can carry, apart from a superconducting current $\mathbf{j}_s$, a normal current $\mathbf{j}_n$ as well. This normal current is carried by 'normal electrons', i.e., electron- or hole-type quasi-particles present in the metal in both the normal and superconducting states. In the superconducting state, the density of such normal quasiparticles depends strongly on the temperature and, generally, tends to zero as $T \to 0$. These notions, which are sometimes referred to as the two-liquid model, can be traced back to paper [54]. An isotropic non-superconductor or, more precisely, an isotropic metal residing in a normal state, can carry only the current with a density

$$\mathbf{j} = \sigma\left(\mathbf{E} - \frac{\nabla\mu}{e}\right) + b\nabla T, \tag{13}$$

where μ is the chemical potential of the electrons and **E** is the electric field. In the superconducting state, for a normal current we have (for more details, see [55])

$$\mathbf{j}_n = \sigma_n \left(\mathbf{E} - \frac{\nabla \mu}{e} \right) + b_n \nabla T. \tag{14}$$

At the same time, the superconducting current density $\mathbf{j}_s$ in the Londons theory [12] approximation, to which we restrict ourselves here (naturally, this is precisely the approximation used in [11]), obeys the equations

$$\text{rot}\,(\Lambda \mathbf{j}_s) = -\frac{1}{c}\mathbf{H}, \tag{15}$$

$$\frac{\partial(\Lambda \mathbf{j}_s)}{\partial t} = \mathbf{E} - \frac{\nabla \mu}{e}, \tag{16}$$

where $\Lambda = m/(e^2 n_s)$ is somewhat a constant, with n_s being the 'superconducting electron' density (so that $\mathbf{j}_s = en_s\mathbf{v}_s$, where $\mathbf{v}_s$ is the velocity); in this scheme, the field penetration depth is

$$\delta\lambda = \sqrt{\frac{\Lambda c^2}{4\pi}} = \sqrt{\frac{mc^2}{4\pi e^2 n_s}}.$$

Notice that this is some simplification, for different chemical potentials μ_n and μ_s should in fact be introduced in Eqns (14) and (16), respectively, for the normal and superconducting electrons. In addition, yet another term (generally, not large) proportional to ∇j_s^2 (see [55]) figures in Eqn (16).When the superconductor is nonuniform, the parameter Λ depends on the coordinates.

As is clear from Eqn (16), in the stationary case, in the superconductor

$$\mathbf{E} - \frac{\nabla \mu}{e} = 0 \tag{17}$$

and, in view of Eqn (14),

$$\mathbf{j}_n = b_n(T)\nabla T. \tag{18}$$

Therefore, the thermoelectric current $\mathbf{j}_n$ in no way vanishes in the superconducting state. However, this current in not directly observable in the simplest case, because it is compensated for by the superconducting current $\mathbf{j}_s$. Let us consider a uniform superconducting rod, one end of the rod residing at a temperature T_2 and the other at a temperature $T_1 < T_2$ (Fig. 1). Then, in the normal state (i.e., when $T_1 > T_c$), since there is no closed circuit, from Eqn (13) we have (see Fig. 1a)

$$\mathbf{j} = 0\,, \quad \mathbf{E} - \frac{\nabla \mu}{e} = -\frac{b}{\sigma}\nabla T\,. \tag{19}$$

In the superconducting state (for $T_2 < T_c$),

$$\mathbf{j} = \mathbf{j}_s + \mathbf{j}_n = 0\,, \quad \mathbf{j}_s = -\mathbf{j}_n = -b_n \nabla T\,,$$
$$\mathbf{H} = 0\,, \quad \mathbf{E} - \frac{\nabla \mu}{e} = 0\,. \tag{20}$$

True, near the rod ends, where $\mathbf{j}_s$ transforms to $\mathbf{j}_n$ or vice versa, uncompensated charges (charge imbalance effect) emerge, and therefore the field **E** is not equal to $\nabla\mu/e$; in what follows I ignore this feature.

An important point is that the thermoelectric current $\mathbf{j}_n$ exists in the uniform case in the superconducting state (Fig. 1b), but the field $H = 0$. When the superconductor is nonuniform or anisotropic, the currents $\mathbf{j}_s$ and $\mathbf{j}_n$ do not in general compensate each other completely, and an observable thermoelectric magnetic field emerges, which was noted in [11]. In days of old (60 years ago!), as noted above, the case of alloys was considered to be unsavory and it was even unclear whether the Londons equation could be applied to alloys. That is why I restricted myself to a brief consideration of a bimetallic plate (say, of two different superconductors fused or soldered together: this juncture is the alloy) in the presence of a temperature gradient (see also § 16 in [26] and [55]). In this case, because the parameter Λ depends on the coordinates (evidently, the Λ parameter is different for different metals), along the junction line there emerges an uncompensated current **j** and hence the magnetic field **H**, which is perpendicular to the plate and the junction line (Fig. 2). Considered in greater detail in [11] and [26] was the case of an anisotropic superconductor. To this end, the Londons equations were generalized in a rather trivial way by replacing the scalar Λ with the tensor Λ_{ik} (for isotropic and cubic metals, $\Lambda_{ik} = \Lambda\delta_{ik}$). When the temperature gradient ∇T in a plateshaped noncubic superconducting crystal is not directed along the symmetry axis, there emerges a current **j** flowing around the plate and a magnetic field $\mathbf{H}_T$ transverse to the plate and proportional to $(\nabla T)^2$. In principle, this field is not difficult to observe with modern techniques. Curiously enough, this is an interesting effect, which in addition makes it possible to measure the thermoelectric coefficient $b_n(T)$ or, more precisely, the components of its generalized tensor $b_{n,\,ik}(T)$. More than 30 years ago I managed to convince W. Fairbank to stage the corresponding experiment, and its results remain, as far as I know, the only ones on this subject [56]. Unfortunately, this work did not make things clear [55, 57]. I am amazed by the fact that nobody has taken an interest in this question even after the fabrication of strongly anisotropic high-temperature superconductors. Evidently, such is the force of fashion in science, too.

True, a certain interest was attracted precisely by the isotropic superconductors, in essence, as applied to a more or less conventional thermoelectric current (Fig. 3a). For this circuit is equivalent to the 'circuit' of Fig. 3b. For

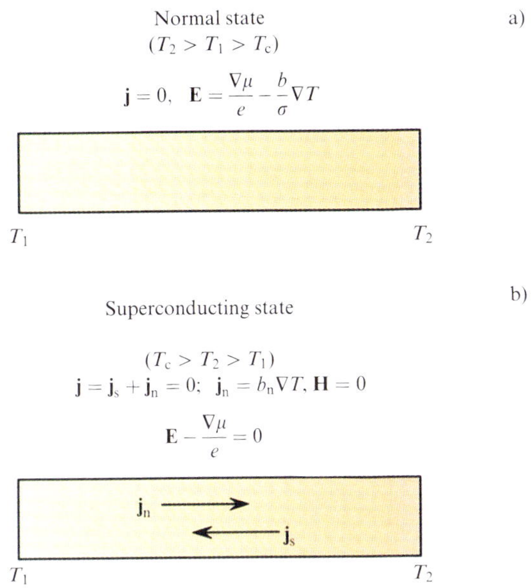

Figure 1.

this circuit it is easy to show [58, 59] (the derivation is also given in [55]) that the magnetic flux $\Phi = \int \mathbf{H} d\mathrm{S}$ through the opening is

$$\Phi = k\Phi_0 + \Phi_{\mathrm{T}}\,, \quad \Phi_{\mathrm{T}} = \frac{4\pi}{c}\int_{T_1}^{T_2} (b_{\mathrm{n,II}}\delta_{\mathrm{II}}^2 - b_{\mathrm{n,I}}\,\delta_{\mathrm{I}}^2)\,\mathrm{d}T\,,$$

$$\Phi_0 = \frac{hc}{2e} = 2 \times 10^{-7}\ \mathrm{G\ cm}^2\,, \quad k = 0, 1, 2, 3 \ldots . \tag{21}$$

Here, the indices I and II refer to the superconducting metals I and II, δ_{I} and δ_{II} are the field penetration depths for these metals, $b_{\mathrm{n,I}}$ and $b_{\mathrm{n,II}}$ are the corresponding coefficients $b_{\mathrm{n}}(T)$ in formula (18), and Φ_0 is the so-called flux quantum. The configuration in Fig. 3b is essentially equivalent to the bimetallic plate in Fig. 2 with $k = 0$, i.e., without an opening. Unfortunately, I did not recognize this at the time (i.e., in [11, 26]).

If we assume for simplicity that $(b_{\mathrm{n}}\delta^2)_{\mathrm{II}} \gg (b_{\mathrm{n}}\delta^2)_{\mathrm{I}}$ and $\delta_{\mathrm{II}}^2 = \delta_{\mathrm{II}}^2(0)\,(1 - T/T_{\mathrm{c,II}})^{-1}$, from expression (21) we obtain

$$\Phi_{\mathrm{T}} = \frac{4\pi}{c}\, b_{\mathrm{n,II}}(T_{\mathrm{c}})\,\delta_{\mathrm{II}}^2(0)\,T_{\mathrm{c}}\,\ln\left(\frac{T_{\mathrm{c}} - T_1}{T_{\mathrm{c}} - T_2}\right). \tag{22}$$

If we substitute the known values $b_{\mathrm{n}}(T_{\mathrm{c}})$ and $\delta(0)$ for $\ln(T_{\mathrm{c}} - T_1)/(T_{\mathrm{c}} - T_2) \sim 1$ in expression (22) we arrive at an estimate $\Phi_{\mathrm{T}} \sim 10^{-2}\Phi_0$. This flux is easy to measure, which was done in several papers (see [1, 55] and references therein). However, the flux Φ_{T} observed in some more complex configuration of the superconducting circuit was found to be orders of magnitude higher than the flux given by expressions (21)–(22) and to possess a different

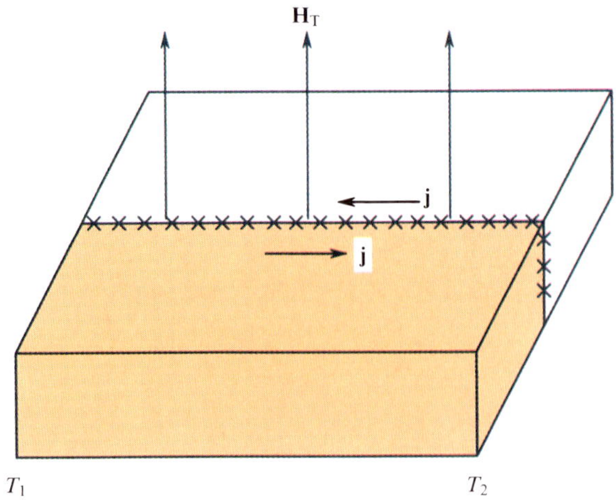

Figure 2.

temperature dependence [60]. The reason for this result has not been elucidated, and different assumptions have been made on that score [61, 62]; see also other references in [1].

It is also pertinent to note that expression (21) and the ensuing formula (22) are obtained under the assumption that the equality $\mathbf{j} = \mathbf{j}_s + \mathbf{j}_n = 0$ is fulfilled throughout the circuit depth (the current $\vec{\mathbf{j}}$ flows only near the surface). Meanwhile, as T_c is approached, the field penetration depth δ increases; as $T \to T_c$, the depth $\delta \to \infty$ and the current density $\mathbf{j}_n$ tends to the thermoelectric current density in the normal state, i.e., for $T > T_c$. In these conditions, a more detailed analysis is required to include the charge imbalance effect. This interesting question has not been investigated (for more details, see [1]).

The aforesaid is not the whole story. Even in the simplest case of a uniform superconductor, the existence of a temperature gradient (see Fig. 1b) affects the thermal conduction: since $\mathbf{j}_n \neq 0$, there is bound to be an additional (convective) heat flux $\mathbf{q}_c = -\varkappa_c \nabla T$ similar to that occurring in a superfluid liquid. This was noted even in [11] and was, in fact, the initial idea in this work.

The total heat flux in the superconducting state $\mathbf{q} = -\varkappa_c \nabla T$, $\varkappa = \varkappa_{ph} + \varkappa_e + \varkappa_c$, where $\varkappa_{ph}$ is the thermal conductivity coefficient related to the lattice (phonons), $\varkappa_e$ is the electron contribution in the absence of convection (circulation), i.e., subject to the condition $\mathbf{j}_n = 0$, and, as already noted, $\varkappa_c$ is the contribution of circulation. As is generally known, the thermal conductivity coefficient in the normal state is, by definition, measured for $\mathbf{j} = 0$, and it is valid to say that $\varkappa_c = 0$ (see [8]). When estimating the $\varkappa_c$ coefficient, I, like others, got tangled up, and now I will restrict myself to a reference to paper [1] and

[8] It is another matter that, for instance, a semiconductor subjected to the condition $\mathbf{j} = 0$ in the presence of electron and hole conduction can simultaneously carry electron $\mathbf{j}_e$ and hole $\mathbf{j}_h = -\mathbf{j}_e$ currents; we ignore these possibilities.

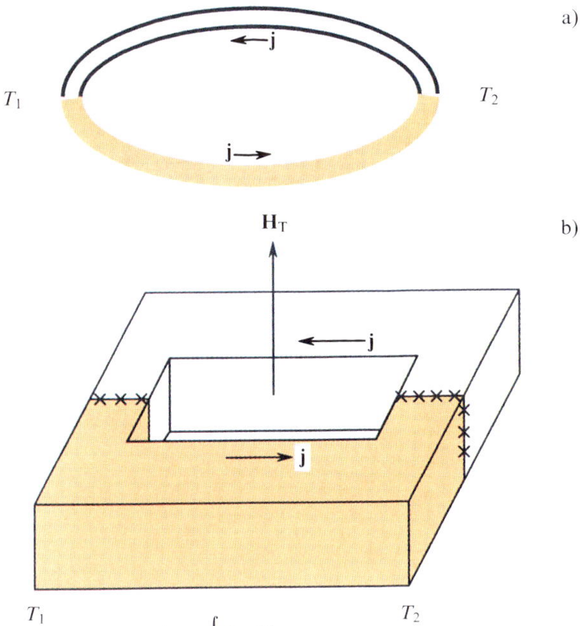

$$\Phi_{\mathrm{T}} = \int \mathbf{H}_T \, d\mathbf{S}$$

$$\Phi = k\Phi_0 + \Phi_{\mathrm{T}}, \ \Phi_0 = \frac{hc}{2e} = 2 \times 10^{-7}\ \mathrm{G\,cm^2}$$
$k = 0, 1, 2, \ldots$

Figure 3.

a remark that in ordinary (not high-temperature) superconductors supposedly $\varkappa_c \ll \varkappa_e$. The role of $\varkappa_c$ in high-temperature superconductors is unclear to me. Most important of all, it is not clear how to extract $\varkappa_c$, even if it were possible to determine separately $\varkappa_{ph}$ and $\varkappa_{e,\,tot} = \varkappa_e + \varkappa_c$ (the total thermal conductivity coefficient $\varkappa$ is measured directly; on the separation of $\varkappa_{ph}$ from $\varkappa_{e,\,tot}$, see [1]).

We have no way of dwelling on the thermoelectric effects in the superconducting state. My aim is to draw attention to this range of questions, which came under the scrutiny of science back in 1927 (see [52] as well as [25]) and under mine in 1944 [11], but which remains largely unclear to date. This is so in spite of a multitude of papers concerned with superconductivity.

SUPERFLUIDITY RESEARCH. Ψ-THEORY OF SUPERFLUIDITY

Superconductivity is, if you please, the superfluidity of a charged liquid or, equivalently, superfluidity is the superconductivity of a noncharged liquid. It is therefore natural that the investigations of both effects have been interrelated. My first work in this area [8], concerned with light scattering in Helium II, was already mentioned above. By the way, there is good reason to revert to this question in light of modern understanding of the fluctuations near the λ point. Several other papers were dealt with in [1]; here, I will consider only the Ψ-theory of superfluidity, albeit with one exception. Namely, I would like

E. F. Zharkov.

D. A. Kirzhnits.

to mention also a proposal made jointly with A.A. Sobyanin[9] and partly with G.F. Zharkov [63, 64], and then mention the study of the feasibility of observing the thermomechanical circulation effect in a superfluid liquid.

In an annular vessel filled with a superfluid liquid (specifically, the case in point was Helium II), which has two different 'bottlenecks' (for instance, narrow capillaries), under a temperature gradient there is bound to emerge a circulation – a superfluid flow engulfing the entire vessel (Fig. 4). By the way, we made the inference about the existence of this effect [63] on the basis of analogy with the thermoelectric effect in a superconducting circuit. As to the inference about the existence of thermoelectric current in a superconducting circuit, I made it [11] at the time on the strength of analogy with the behavior of Helium II under a temperature gradient. The above thermocirculation effect in Helium II has been observed [65] and discussed [64], and, in my view, interesting possibilities were pointed out for future research [64]. However, nobody, as far as I know, has taken an interest in this question during the past 20 years.

After the development of the Ψ-theory of superconductivity [3], the transfer of something similar to the superfluidity case appeared to be rather obvious. At the same time, even before (see, for instance [9]) I was concerned about the behavior of Helium II near the λ point, and the question of the boundary condition for superfluid component velocity $\mathbf{v}_s$ was obscure. By the way, Landau, the originator of the theory of phase transitions and superfluidity, for some reason was never concerned with this range of questions, as far as I know. In the Landau theory of superfluidity [7], the velocity $\mathbf{v}_s$ along the

[9] The talented theoretical physicist and public figure Aleksandr Sobyanin prematurely died at the age of 54 in 1997.

E. G. Maksimov.

wall (unlike the normal component velocity $\mathbf{v}_n$ along the wall) does not vanish at the wall: there is some kind of discontinuity. But in this case, it seemed to me, this discontinuity was bound to be related to some surface energy σ_s [66]. However, dedicated experiments [67] showed that the σ_s energy is nonexistent or, in any case, is many orders of magnitude lower than the expected energy [66]. I saw a way out in the assumption that the superfluid component density at the wall $\rho_s(0)$ is zero. Then, the superfluid component flux $\mathbf{j}_s = \rho_s \mathbf{v}_s$ at the wall vanishes despite the fact that $\mathbf{v}_s$ have a discontinuity at the wall. In the Ψ-theory of superfluidity, evidently,

$$\rho_s = m|\Psi|^2 , \tag{23}$$

where it may be assumed that $m = m_{He}$ is the mass of a helium atom (we imply the superfluidity of Helium II) and, in view of the foregoing, the boundary condition at the wall is

$$\Psi(0) = 0 , \tag{24}$$

instead of the condition (3) for superconductors. At this stage, as far as I remember, it turned out that L.P. Pitaevskii had independently taken up the Ψ-theory of superfluidity and, naturally, we combined efforts. As a result, the work [68] emerged; I speak of the Ψ-theory of superfluidity constructed in that work as 'initial' because I consider below the 'generalized' Ψ-theory of superfluidity elaborated together with Sobyanin [69, 70] (see also several other references in [1]).

The initial Ψ-theory of superfluidity [68] is quite similar to the Ψ-theory of superconductivity [3], of course, with the use of the boundary condition (24) and in the absence of the electric charge. In this case, the scalar complex function $\Psi = |\Psi| \exp(i\varphi)$ obeys the equation

$$\oint \mathbf{v}_s \, d\mathbf{l} = 2\pi \frac{\hbar}{m} k, \quad k = 0, \pm 1, \pm 2, \ldots$$

$$\frac{2\pi\hbar}{m_{^4\mathrm{He}}} \approx 10^{-3} \ \mathrm{cm}^2 \ \mathrm{s}^{-1}$$

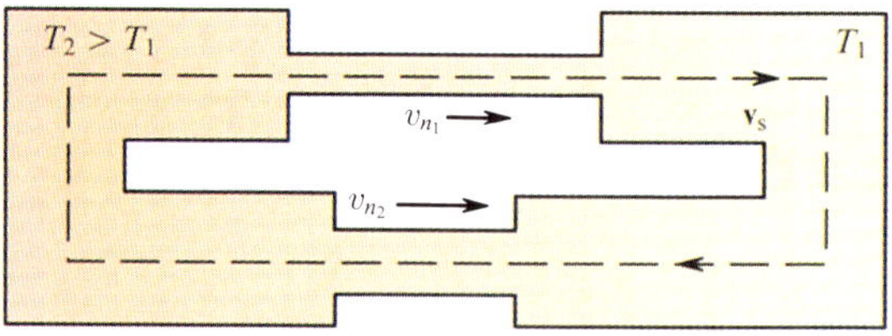

$$-\frac{\hbar^2}{2m} \Delta\Psi + \alpha(T)\,\Psi + \beta_\lambda |\Psi|^2\, \Psi = 0 \tag{25}$$

and

$$\mathbf{j}_s = \rho_s \mathbf{v}_s = -\frac{i\hbar}{2}(\Psi^* \nabla \Psi - \Psi \nabla \Psi^*) = \hbar |\Psi|^2 \, \nabla\varphi \,, \tag{26}$$

i.e., $\mathbf{v}_s = (\hbar/m)\nabla\varphi$, with $m = m_{\mathrm{He}}$ irrespective of how Ψ is normalized (see [1, 68]).

Furthermore, the correlation length ξ denoted as l in [68] is (T_λ is the temperature of the λ point)

$$\xi(T) = \frac{\hbar}{\sqrt{2m|\alpha|}} = \xi(0)\, \tau^{-1/2}, \quad \tau = \frac{T_\lambda - T}{T_\lambda} \,. \tag{27}$$

The estimate of Ref. [68], based on experimental data, for ^{4}He, i.e., for Helium II, leads to a value $\xi(0) \sim 3 \times 10^{-8}$ cm. At the same time, the Ψ-theory is applicable only when the macroscopic Ψ function varies only slightly over atomic-scale distances. Hence, there follows the condition $\xi(T) \gg a \sim 3 \times 10^{-8}$ cm (here a is the average interatomic distance in liquid helium). The Ψ-theory can therefore be adequate only near the λ point (for $\tau \ll 1$), say, for ($T_\lambda - T$) < (0.1–0.2) K. A similar condition also takes place in the case of Ψ-theory of superconductivity, which is also appropriate, generally speaking, only near T_c. It is of prime importance that the Landau theory of phase transitions, which is a mean-field theory, for superconductors (i.e., the Ψ-theory of superconductivity) is also correct in the immediate vicinity of T_c. This is due to the relatively large value of $\xi(0)$ in superconductors (the length $\xi(0)$ is on the order of the dimension of the Cooper pairs, i.e. in ordinary superconductors is on the order of, say, 10^{-5} cm). The point is that the temperature range near T_c (or T_λ), in which fluctuations are already large and the mean field approximation is inappropriate, is proportional to $[\xi(0)]^{-6}$ (see [1] and references therein, particularly [34]). In Helium II, the fluctuations near T_λ are relatively strong due to the smallness of $\xi(0)$, and the Ψ-theory [68] can be used only for $(T_\lambda - T) \gg 10^{-3}$ K [1]. Meanwhile, the temperature range significantly closer to T_λ is of special interest. That the meanfield theory is inapplicable in the region of the λ transition in ^{4}He is testified too by the very existence of the λ singularity in the temperature dependence of the heat capacity. This circumstance might not, at least on the face of it, be related too the temperature dependence of the density $\rho_s(T)$, which was proportional to $|\Psi|^2$ [see expression (23)]. That is why in 1957, when the work [68] was carried out, we did not see the drawbacks to our theory right away. However, this became clear somewhat later, when it was found out that in Helium II to a good approximation

$$\rho_s(\tau) = \rho_{s0}\tau^\zeta, \quad \zeta = \frac{2}{3}\,. \tag{28}$$

In the mean-field theory,

$$\zeta = 1\,. \tag{29}$$

In experiment, by the way, the index ξ is not exactly equal to 2/3 but is very close to it. For instance, according to [71], $\xi = 0.6705 \pm 0.0006$.

Therefore, the initial Ψ-theory of superfluidity [68] is poorly applicable to liquid ^{4}He in a quantitative sense. At the same time, several results based on it were obtained in [68], which were also of significance for Helium II in a qualitative sense. The case in point is the density distribution $\rho_s(z)$ near the solid wall and in films with a thickness d in relation to this thickness. Also solved were the problems of velocity $\mathbf{v}_s$ circulation about a vortex line at the axis of which $\Psi = 0$, of the energy of this filament, and of the surface energy at the interface between Helium II and the solid wall. No less significant is the fact that liquid ^{4}He is not the only existing superfluid liquid. Such a liquid is also encountered in the case of ^{3}He – ^{4}He solutions, liquid ^{3}He, neutron stars, and maybe in other cases. In these cases, however, the Ψ function may prove to be no longer scalar but, on the other hand, the length $\xi(0)$ is relatively large (in liquid ^{3}He, for instance, $\xi(0) \sim 10^{-5}$ cm), and the fluctuation region is rather small. Finally, the theory of Ref. [68] had played, so far as I can judge, a significant role in the construction and elaboration of the Gross–Pitaevskii theory, which is widely used in the investigation of Bose–Einstein condensation (see [72]).

Liquid ^{4}He, i.e., helium II, has always occupied and still occupies the leading position in the physics of superfluidity, both historically and regarding the scale of investigations. The Landau theory [7], which describes its behavior, is primarily macroscopic or, if you like, quasi-macroscopic. But it does not provide answers to several questions, particularly near the λ point. At the same time, a microtheory of the BCS type for superconductors does not exist for Helium II. On the other hand, Helium II near the λ point is interesting from various viewpoints, in particular, in the investigation of two-liquid hydrodynamics near the λ point, in the modeling of some cosmological situations [73], etc. It is likely that the initial Ψ-theory of superfluidity [68, 74] can be used to some extent for the solution of these problems, though with the

L. P. Pitaevskii.

A. A. Sobyanin.

above significant limitation arising from the inapplicability of the mean field approximation, i.e., from the neglect of fluctuations. The generalized Ψ-theory of superfluidity [69, 70] was intended to eliminate these drawbacks. It is based on some semiempirical generalization of the Landau theory of phase transitions (see, for instance, [75]). In the Landau theory of phase transitions and, in particular, in the Ψ-theory of superconductivity, i.e., when the Ψ function is selected as the order parameter, the free energy density of the ordered phase near the transition point T_λ is written in the form

$$F_{\rm II} = F_{\rm I} + \alpha|\Psi|^2 + \frac{\beta}{2}|\Psi|^4 + \frac{\gamma}{6}|\Psi|^6 ; \tag{30}$$

away from the tricritical point it being safe to assume that

$$\alpha = \alpha'_\lambda(T - T_\lambda) = -a_0\tau , \quad \beta = \beta_\lambda ,$$
$$\gamma = 0 , \quad \tau = \frac{T_\lambda - T}{T_\lambda} . \tag{31}$$

In the generalized theory

$$F_{\rm II} = F_{\rm I} - a_0\tau|\tau|^{1/3}|\Psi|^2 + \frac{b_0}{2}\tau^{2/3}|\Psi|^4 + \frac{g_0}{3}|\Psi|^6 . \tag{32}$$

When selecting expression (32), for small $|\Psi|^2$ in the equilibrium $|\Psi_0|^2 = -\alpha/\beta = (\alpha_0/b_0)\tau^{2/3}$, i.e., there occurs a temperature dependence which agrees with the observed one [see (28)]. Evidently, expression (32) is selected for precisely the attainment of this goal.

The generalized Ψ-theory of superfluidity [69, 70] formally differs from the initial theory [68, 74] just by the replacement of expressions (30)–(31)

with expression (32). Several expressions and inferences were derived on this basis. For instance, for a thin film of Helium II of thickness d, the λ-transition temperature is

$$T_\lambda(d) = T_\lambda - 2.53 \times 10^{-11} \left(\frac{3+M}{M} \right) d^{-3/2} K, \tag{33}$$

where $T_\lambda = T_\lambda(\infty)$ is the λ-transition temperature in massive helium (as is well known, $T_\lambda = 2.17$ K) and M is the parameter of the theory proportional to the g_0 coefficient in expression (32). When $M < 1$, the λ transition is of the second kind (by comparison with experiment, only a crude estimate was obtained for Helium II: $M = 0.5 \pm 0.3$). By the way, if we consider a cylindrical capillary of diameter d instead of a plane film, the coefficient 2.53 in formula (33) should be replaced with 4.76. Quite a number of other expressions were also derived [69, 70, 76].

Unfortunately, the generalized Ψ-theory of superfluidity has not come to the attention of either experimenters or theorists. True, some pessimistic judgments regarding it were expressed in the literature (they were mentioned in [1]). Sobyanin and I also abandoned the superfluidity research during the period of rapid changes in the USSR and Russia that set in after 1985–1988. Only in [1] did I review our work.

Undeniably the generalized Ψ-theory of superfluidity is not a lofty *ab initio* theory. At the same time, its simplicity (at least in comparison with other known methods) suggests that the Ψ-theory of superfluidity (initial as well as generalized) can still yield much in the study of superfluidity. In any case, the opposite opinion is not substantiated at all. This section of the lecture has been written precisely with the aim of attracting the attention of physicists engaged in the corresponding areas to the Ψ-theory of superfluidity. It may well be, in my view, that the lack of attention is a delusion. It is conceivable, on the contrary, that I am in error myself, though.

'PHYSICAL MINIMUM' – WHAT PROBLEMS OF PHYSICS AND ASTROPHYSICS SEEM NOW TO BE ESPECIALLY IMPORTANT AND INTERESTING IN THE BEGINNING OF THE XXI CENTURY?

I have encountered the viewpoint that my work in the area of superconductivity and superfluidity is a matter of the remote past. There is no question that the work of Ginzburg and Landau [3] performed back in 1950 stands out. But on the whole, as is clear from the foregoing and particularly from [1], I have been occupied with this field of physics since 1943 until the present time. In this case, it seems to me, several questions and problems have also been posed which have not been solved and which deserve attention. Of course, presently the most urgent problems in the area of superconductivity are the elucidation of the mechanism and several features of hightemperature superconductivity and the creation of room-temperature superconductivity. More precisely, what is wanted in the latter case is to elucidate the po-

16. Standard model. Grand unification. Superunification. Proton decay. Neutrino mass. Magnetic monopoles.
17. Fundamental length. Particle interaction at high and superhigh energies. Colliders.
18. Nonconservation of CP-invariance.
19. Nonlinear phenomena in vacuum and in superstrong magnetic fields. Phase transitions in a vacuum.
20. Strings. M-theory.
21. Experimental verification of the general theory of relativity.
22. Gravitational waves and their detection.
23. The cosmological problem. Inflation. Λ-term and 'quintessence'. Relationship between cosmology and high energy physics.
24. Neutron stars and pulsars. Supernova stars.
25. Black holes. Cosmic strings (?).
26. Quasars and galactic nuclei. Formation of galaxies.
27. The problem of dark matter (hidden mass) and its detection.
28. The origin of superhigh-energy cosmic rays.
29. Gamma-bursts. Hypernovae.
30. Neutrino physics and astronomy. Neutrino oscillations.

The singling out of 30 particular problems (more precisely, items in the 'list') is of course absolutely conditional. Moreover, some of them might be subdivided. In my first 'list' published in 1971 there were 17 problems [77]. Subsequently their number would grow (for some more details, see [2]). It would supposedly be well to add some new subjects to the 'list', given about, for instance, those concerning quantum computers and advances in optics. But I cannot do this with adequate comprehension.

Any 'list' is undoubtedly not a dogma, something can be discarded and something added, depending on the preferences of lecturers and authors of corresponding papers. More interesting is the question of the evolution of the 'list' with time, as it reflects the development of physics. In the 'list' of 1970–1971 [77] quarks were given only three lines in the enumeration of the attempts to explain the mass spectrum. This did not testify to my perspicacity. However, at that time (in 1970) quarks were only five or six years old (I mean the age of the corresponding hypothesis), and the fate of the concept of the quark was indeed vague. Now the situation is of course quite different. True, the heaviest t-quark was discovered only in 1994 (its mass, according to the data of 1999, is $m_t = 176 \pm 6$ GeV). The list [77] naturally contains no fullerenes, which were discovered in 1985, no gamma-bursts (the first report of their discovery was published in 1973). High-temperature superconductors were synthesized in 1986–1987, but in the list [77] this problem was nonetheless considered rather thoroughly for it had been discussed since 1964 (this was discussed in greater detail in the previous sections of the lecture). Generally, much has been done in physics over the past 30 or 35 years, but, I believe, not very much essentially new has appeared. In any case, the 'lists' in [77, 16], as well as that presented above, characterize to a certain extent the development and the state of physical and astronomical problems from 1970–1971 to the present day.

It should be added that three 'great problems' of modern physics are also to be included in the 'physics-minimum', included in the sense that they should be singled out in some way and specially discussed, and development of these 'great problems' should be reviewed. This is discussed at some length in [2]. The 'great problems' are, first, the increase in entropy, time irreversibility, and the 'time arrow'. Second is the problem of interpretation of nonrelativistic quantum mechanics and the possibility of learning something new even in the field of its applicability (I personally doubt this possibility but believe that one's eyes should remain open). And third is the question of live-to-liveless reduction, i.e., the feasibility of explaining the origin of life and thought on the basis of physics alone. On the face of it, how could it be otherwise? But until the questions are elucidated, one cannot be quite sure of anything. I think that the problem of the origin of life will unreservedly be solved only after 'life in a test-tube' is created. Until then, this will be an open question.

One more concluding remark. In the past century, and even nowadays, one could encounter the opinion that in physics nearly everything had been done. There allegedly are only dim 'cloudlets' in the sky or theory, which will soon be eliminated to give rise to the 'theory of everything'. I consider these views as some kind of blindness. The entire history of physics, as well as the state of present-day physics and, in particular, astrophysics, testifies to the opposite. In my view we are facing a boundless sea of unresolved problems. It only remains for me to envy the younger members of the audience, who will witness a great deal of new, important, and interesting things.

REFERENCES

1. Ginzburg V L *Usp.* Fiz. Nauk **167** 429 (1997); **168** 363 (1998) [*Phys.* Usp. **40** 407 (1997); **41** 307 (1998)].
2. Ginzburg V L O *Nauke, o Sebe i o Drugikh* (About Science, Myself, and Others) (Moscow: Fizmatlit, 2003)[10] [Translated into English (Bristol: IOP Publ., 2004) Article 7 (to be published)].
3. Ginzburg V L, Landau L D *Zh. Eksp. Teor. Fiz.* **20** 1064 (1950); This paper was published in English in the volume: Landau L D *Collected Papers* (Oxford: Pergamon Press, 1965) p. 546.
4. Abrikosov A A *Zh. Eksp. Teor. Fiz.* **32** 1442 (1957) [*Sov. Phys. JETP* **5** 1174 (1957)].
5. Lifshitz E M, Pitaevskii L P *Statisticheskaya Fizika* (Statistical Physics) Pt. 2 *Teoriya Kondensirovannogo Sostoyaniya* (Theory of Condensed State) (Moscow: Nauka, 1978, 1999) [Translated into English (Oxford: Pergamon Press, 1980)].
6. Tinkham M *Introduction to Superconductivity* 2nd ed. (New York: McGraw Hill, 1996).
7. Landau L D *Zh. Eksp. Teor. Fiz.* **11** 592 (1941); *J. Phys. USSR* **5** 71 (1941).
8. Ginzburg V L *Zh. Eksp. Teor. Fiz.* **13** 243 (1943); *J. Phys. USSR* **7** 305 (1943).
9. Ginzburg V L *Zh. Eksp. Teor. Fiz.* **14** 134 (1944).

[10] Article 7 published in this Collection is the somewhat edited article given above in Ref. [1].

10. Bardeen J, in *Kältephysik* (Handbuch der Physik, Bd. 15, Hrsg. S von Flügge) (Berlin: Springer-Verlag, 1956) p. 274 [Translated into Russian: Bardeen J, in *Fizika Nizkikh Temperatur* (Ed. A I Shal'nikov) (Moscow: IL, 1959) p. 679].
11. Ginzburg V L *Zh. Eksp. Teor. Fiz.* **14** 177 (1944); *J. Phys. USSR* **8** 148 (1944).
12. London F, London H *Proc. R. Soc. London Ser. A* **149** 71 (1935); *Physica* **2** 341 (1935).
13. Waldram J R *Superconductivity of Metals and Cuprates* (Bristol: Institute of Physics Publ., 1996).
14. Ginzburg VL *Zh. Eksp. Teor. Fiz.* **16** 87 (1946); *J. Phys. USSR* **9** 305 (1945).
15. Ginzburg V L *Zh. Eksp. Teor. Fiz.* **15** 739 (1945); *J. Phys. USSR* **10** 107 (1946).
16. Ginzburg V L *The Physics of a Lifetime. Reflections on the Problems and Personalities of 20th Century Physics* (Berlin: Springer-Verlag, 2001).[11]
17. Ginzburg V L *Zh. Eksp. Teor. Fiz.* **29** 748 (1955) [*Sov. Phys. JETP* **2** 589 (1956)].
18. Bardeen J, Cooper L N, Schrieffer J R *Phys. Rev.* **108** 1175 (1957).
19. Ginzburg V L *Usp. Fiz. Nauk* **48** 25 (1952); *Fortschr. Phys.* **1** 101 (1953).
20. Ogg R A (Jr) *Phys. Rev.* **69** 243; **70** 93 (1946).
21. Schafroth MR *Phys. Rev.* **96** 1149 (1954); **100** 463 (1955).
22. Cooper L N *Phys. Rev.* **104** 1189 (1956).
23. Gor'kov L P *Zh. Eksp. Teor. Fiz.* **36** 1918; **37** 1407 (1959) [*Sov. Phys. JETP* **9** 1364 (1959); **10** 998 (1960)].
24. Ginzburg V L *Zh. Eksp. Teor. Fiz.* **31** 541 (1956) [*Sov. Phys. JETP* **4** 594 (1957)].
25. Shoenberg D *Superconductivity* 3rd ed. (Cambridge: Cambridge Univ. Press, 1965) [Translated into Russian (Moscow: IL, 1955)].
26. Ginzburg V L *Sverkhprovodimost'* (Superconductivity) (Moscow-Leningrad: Izd. AN SSSR, 1946).
27. Buckel W *Supraleitung* (Weinheim, Bergster: Physik-Verlag, 1972) [Translated into English: Buckel W *Superconductivity: Fundamentals and Applications* (Weinheim: VCH, 1991); Translated into Russian (Moscow: Mir, 1975)].
28. Ginzburg V L *Zh. Eksp. Teor. Fiz.* **23** 236 (1952).
29. Ginzburg V L *Zh. Eksp. Teor. Fiz.* **34** 113 (1958) [*Sov. Phys. JETP* **7** 78 (1958)].
30. Ginzburg V L *Zh. Eksp. Teor. Fiz.* **42** 299 (1962) [*Sov. Phys. JETP* **15** 207 (1962)].
31. Ginzburg V L *Zh. Eksp. Teor. Fiz.* **36** 1930 (1959) [*Sov. Phys. JETP* **9** 1372 (1959)].
32. Ginzburg V L *Zh. Eksp. Teor. Fiz.* **31** 202 (1956) [*Sov. Phys. JETP* **4** 153 (1957)].
33. Zharkov G F *Zh. Eksp. Teor. Fiz.* **34** 412 (1958); **37** 1784 (1959) [*Sov. Phys. JETP* **7** 278 (1958); 10 1257 (1959)].
34. Ginzburg V L *Fiz. Tverd. Tela* **2** 2031 (1960) [*Sov. Phys. Solid State* **2** 1824 (1961)]
35. Little W A *Phys. Rev.* **134** A1416 (1964).
36. Ginzburg V L *Phys. Lett.* **13** 101 (1964); *Zh. Eksp. Teor. Fiz.* **47** 2318 (1964) [*Sov. Phys. JETP* **20** 1549 (1965)].
37. Ginzburg V L, Kirzhnits D A *Zh. Eksp. Teor. Fiz.* **46** 397 (1964) [*Sov. Phys. JETP* **19** 269 (1964)].
38. Ginzburg V L *Phys. Scripta* **T27** 76 (1989).
39. Ginzburg V L, Kirzhnits D A *Dokl. Akad. Nauk SSSR* **176** 553 (1967) [*Sov. Phys. Dokl.* **12** 880 (1968)].
40. Ginzburg V L, Kirzhnits D A (Eds) *Problema Vysokotemperaturnoi Sverkhprovodimosti* (The Problem of High-Temperature Superconductivity) (Moscow: Nauka, 1977) [Translated into English: Ginzburg V L, Kirzhnits D A (Eds) *High-Temperature Superconductivity* (New York: Consultants Bureau, 1982)].
41. Bednorz J G, Muller K A *Z. Phys. B* **64** 189 (1986).
42. Wu M K, Ashburn J R, Torng C J, Hor P H, Meng R L, Gao L, Huang Z J, Wang Y Q, Chu C W *Phys. Rev. Lett.* **58** 908 (1987).

[11] This book is, for the most part, a translation of the book: Ginzburg V L *O Fizike i Astrofizike* (Moscow: Byuro Kvantum, 1995).

43. Ginzburg V L *Energiya* (a Scientific and Popular Journal) (9) 2 (1984).
44. Ginzburg V L *Prog. Low Temp. Phys.* **12** 1 (1989).
45. Ginzburg V L, in *From High-Temperature Superconductivity to Microminiature Refrigeration* (Eds B Cabrera, H Gutfreund, V Kresin) (New York: Plenum Press, 1996).
46. Ginzburg V L *J. Supercond.* **4** 327 (1986).
47. Shaplygin I S, Kakhan B G, Lazarev V B *Zh. Neorg. Khim.* **24** 1476 (1979).
48. Cava R J *et al. Phys. Rev. Lett.* **58** 408 (1987).
49. Eliashberg G M *Zh. Eksp. Teor. Fiz.* **38** 966; 39 1437 (1960) [*Sov. Phys. JETP* **11** 696 (1960); **12** 1000 (1961)].
50. Maksimov E G *Usp. Fiz. Nauk* **170** 1033 (2000) [*Phys. Usp.* **43** 965 (2000)].
51. Ginzburg V L, Maksimov E G *Sverkhprovodimosti: Fiz., Khim., Tekh.* **5** 1543 (1992) [*Superconductivity: Phys., Chem., Technol.* **5** 1505 (1992)].
52. Meissner W Z. *Ges. Kälteindustr.* **34** 197 (1927).
53. Burton E F, SmithGH, Wilhelm J O *Phenomena at the Temperature of Liquid Helium* (American Chemical Society: Monograph Ser., No. 83) (New York: Reinhold Publ. Corp., 1940).
54. Gorter C J, Casimir H *Phys. Z.* **35** 963 (1934).
55. Ginzburg V L, Zharkov G F *Usp. Fiz. Nauk* **125** 19 (1978) [*Sov. Phys. Usp.* **21** 381 (1978)]
56. Selzer P M, Fairbank W M *Phys. Lett. A* **48** 279 (1974).
57. Ginzburg V L, Zharkov G F *Pis'ma Zh. Eksp. Teor. Fiz.* **20** 658 (1974) [*JETP Lett.* **20** 302 (1974)].
58. Gal'perin Yu M, Gurevich V L, Kozub V N *Zh. Eksp. Teor. Fiz.* **66** 1387 (1974) [*Sov. Phys. JETP* **39** 680 (1974)].
59. Garland J C, Van Harlingen D J *Phys. Lett. A* **47** 423 (1974).
60. Van Harlingen D J *Physica B + C* **109–110** 1710 (1982).
61. Arutyunyan R M, Ginzburg V L, Zharkov G F *Zh. Eksp. Teor. Fiz.* **111** 2175 (1997) [*JETP* **84** 1186 (1997)]; *Usp. Fiz. Nauk* **167** 457 (1997) [*Phys. Usp.* **40** 435 (1997)].
62. Galperin Y M *et al. Phys. Rev. B* **65** 064531 (2002).
63. Ginzburg V L, Zharkov G F, Sobyanin A A *Pis'ma Zh. Eksp. Teor. Fiz.* **20** 223 (1974) [*JETP Lett.* **20** 97 (1974)]; Ginzburg V L, Sobyanin A A, Zharkov G F *Phys. Lett. A* **87** 107 (1981).
64. Ginzburg V L, Sobyanin A A *Zh. Eksp. Teor. Fiz.* **85** 1606 (1983) [*Sov. Phys. JETP* **58** 934 (1983)].
65. Gamtsemlidze G A, Mirzoeva M I *Zh. Eksp. Teor. Fiz.* **79** 921 (1980); **84** 1725 (1983) [*Sov. Phys. JETP* 52 **468** (1980); **57** 1006 (1983)].
66. Ginzburg V L *Zh. Eksp. Teor. Fiz.* **29** 254 (1955) [*Sov. Phys. JETP* **2** 170 (1956)].
67. Gamtsemlidze G A *Zh. Eksp. Teor. Fiz.* **34** 1434 (1958) [*Sov. Phys. JETP* **7** 992 (1958)].
68. Ginzburg V L, Pitaevskii L P *Zh. Eksp. Teor. Fiz.* **34** 1240 (1958) [*Sov. Phys. JETP* **7** 858 (1958)].
69. Ginzburg V L, Sobyanin A A *Usp. Fiz. Nauk* **120** 153 (1976) [*Sov. Phys. Usp.* **19** 773 (1976)]; *J. Low Temp. Phys.* **49** 507 (1982).
70. Ginzburg V L, Sobyanin A A *Usp. Fiz. Nauk* **154** 545 (1988) [*Sov. Phys. Usp.* **31** 289 (1988)]; *Jpn. J. Appl. Phys.* **26** (Suppl. 26–3) 1785 (1987).
71. Golder L S, Mulders N, Ahlers G *J. Low. Temp. Phys.* **93** 131 (1992).
72. Pitaevskii L, Stringari S *Bose-Einstein Condensation* (Intern. Series of Monographs on Physics, Vol. 116) (Oxford: Clarendon Press, 2003).
73. Zurek W H *Nature* **382** 296 (1996).
74. Pitaevskii L P *Zh. Eksp. Teor. Fiz.* **35** 408 (1958) [*Sov. Phys. JETP* **8** 282 (1959)].
75. Mamaladze Yu G *Zh. Eksp. Teor. Fiz.* **52** 729 (1967) [*Sov. Phys. JETP* **25** 479 (1967)]; *Phys. Lett. A* **27** 322 (1968).
76. Ginzburg V L, Sobyanin A A, in *Superconductivity, Superdiamagnetism, Superfluidity* (Ed. V L Ginzburg) (Moscow: MIR Publ., 1987) p. 242.
77. Ginzburg V L *Usp. Fiz. Nauk* **103** 87 (1971) [*Sov. Phys. Usp.* **14** 21 (1971)].

Tony Leggett

A. J. LEGGETT

I was born in Camberwell, South London, on the 26th of March, 1938. I am told I only made it into the world on the date in question by seven minutes, thereby exhibiting at this early stage the tendency to procrastination which I fear has characterized much of my subsequent career. Not a great deal is known about my ancestry beyond a few generations, though one tidbit of information which has been passed down is that an ancestor on my father's side served for a time as a cook on Nelson's flagship, the Victory (it is unknown whether he was actually present at the battle of Trafalgar). As far as we can trace them, my father's forebears were village cobblers in a small village in Hampshire, though his father broke with this tradition to become a greengrocer; my father would tell me how he used to ride with him to buy vegetables at the Covent Garden market in London. My mother's parents were both of Irish stock; her father had emigrated to England and worked as a clerk in the naval dockyard in Chatham. My maternal grandmother, who survived into her eighties (and my twenties), was a remarkable person; sent out to domestic service at the age of twelve, she eventually married my grandfather and raised a large family, then in her late sixties emigrated to Australia to join her daughter and son-in-law, and finally returned to the UK for her last years. She was a very warm person, and I still have very fond memories both of my meetings with her (unfortunately few in her last years, because of the geography) and of the many letters she wrote me; since she had had no formal education, she simply wrote down on paper exactly what she would have said to you in person, and to read a letter from her was like having her stand in the room with you. It is interesting that that kind of spontaneity in the written word, which seemed to have been lost forever under the influence of universal secondary education, is now returning in the guise of e-mail.

My father and mother were each the first in their families to receive a university education; they met and became engaged while students at the Institute of Education at the University of London, but were unable to get married for some years because my father had to care for his own mother and siblings. He worked as a secondary-school (high-school) teacher of physics, chemistry and mathematics. My mother also taught secondary-school mathematics for a time, but had to give this up when I was born. I was eventually followed by two sisters, Clare and Judith, and two brothers, Terence and Paul (both now deceased). My parents were both Catholics (in my mother's case ancestrally, in my father's I believe because his own father had converted), so that we children were brought up in that faith, making us members of a small and somewhat embattled minority in the England of those days. Although I ceased to be a practicing Catholic in my early twenties, I still wonder from time to time how far the experience, in childhood and adolescence, of maintaining and defending, sometimes in public and in the face of some ridicule, beliefs and attitudes not shared by the vast majority of my compatriots may have influenced my subsequent attitude to physics and indeed to life in general (which, at least as regards the former, is I suspect sometimes regarded by my colleagues as reflecting a degree of iconoclasm verging on counter-suggestibility).

Soon after I was born, my parents bought a house in Upper Norwood, just outside the southern boundary of London proper (but well within the London conurbation). However, when I was eighteen months old, war broke out and we were "evacuated" to Englefield Green, a small village in Surrey on the edge of the great park of Windsor Palace, where we stayed for the duration of the war. The village was fortunate to escape the bombs that fell on towns all around (including Slough, perhaps in response to John Betjeman's couplet!) and my war-related memories are relatively few: barrage balloons tethered above the Thames, lying in bed at night listening to the German "doodle-bugs" droning overhead (and praying they would not cut out immediately above our house), and the day I was "officially" informed that there was a war on (and decided that at no cost should my sister Clare, a year younger than me, find out about it). When in later life I read the memoirs of people who lived through those same years in the ravaged cities of continental Europe or Asia, or even through the Blitz in London, I realize how extraordinarily lucky we were.

Although Englefield Green is only a few miles outside the western borders of the Greater London conurbation, it is (or was!) in some sense in the countryside, and many of my memories of those years are of long walks through the fields and parks surrounding the village. However, one of my main activities, and the one which gained me most notoriety among our neighbors, was digging deep holes, I think on occasion deeper than my height at the time, in our front garden (not the easiest activity in a soil which was basically London clay). Exactly what motivated me to do this, and what deep insights it reveals into my psyche, is something I have never figured out, but it was perhaps loosely connected with my then choice of future career: after a brief flirtation with the idea of becoming a railway signalman, I decided firmly that I wanted to be an explorer (you have to remember that in those days there were still odd patches of the planet which had never been trodden by human feet, nor mapped by GPS). In the end I suppose that, like many academics, I achieved this ambition though in a more abstract way.

After the end of the war, we returned to the Upper Norwood house and lived there until 1950; my father taught at a school in north-east London (unfortunately a long commute) and my mother by that time had her hands full looking after what were eventually five of us children. I attended the local Catholic elementary (grade) school, and later, following a successful performance in the "eleven-plus" examination which I took rather earlier than most, transferred to the College of the Sacred Heart in Wimbledon (a "grammar", i.e. state-financed academic-stream, school). This required a slightly complicated commute (bus, train, then another bus), which had the advantage that I could get a fair amount of reading done on the journey to and fro. (A few years ago, while on a brief return visit to the UK, I had for a reason I forget, to repeat exactly this journey and got completely lost – such have been the changes in the transportation system of southwest London in 50 years). Although I took part in the required physical sports at school, in particular in long-distance running, my main recreation in those years was chess, an interest I kept throughout my teens and into my Oxford years (I had a brief moment of glory when some years later, I was picked for the English team to compete against Scotland, Wales, and Ireland in the (under-16) Glorney Cup). Another practice I developed in those years was to take long hikes or cycle rides alone into the countryside surrounding southwest London; in later years I kept up this habit in much more untamed terrain, sometimes steering by compass, in thick mist, over many miles of trackless Welsh or Scottish mountainside. (Since I often did this alone and did not always leave details of my itinerary, and there are plenty of accidents in those areas even in good weather, I am perhaps lucky still to be here!)

In 1949 my father, who had become increasingly unhappy with the long commute to his job in north-east London, obtained a job teaching physics and chemistry at Beaumont College, a school run by the Jesuits near Windsor. This was a "public" school in the British sense of the word (i.e., "private" in the US sense) and we could certainly not have afforded the fees for even one of us three boys to attend it; however, my father negotiated as part of his conditions of service that all three of us could go to Beaumont free of charge, so we did. At the same time we moved to a large, rambling and somewhat broken-down house in Staines, on the far western edge of the London conurbation, which remained our family home until I was well into my thirties; this house was a delight to my father, who in another life would have been a handyman, but probably less so to my mother who had to endure the consequences of all the things he had not yet got around to fixing. One of the prime attractions, at least for us children, was the large garden and in particular the huge chestnut tree, in which we built many camps and "forts".

Even within the somewhat esoteric world of British public schools, Beaumont was a rather unusual place. One feature which is, to my knowledge, unique to Jesuit schools, is that while the teaching staff includes both laymen, such as my father, and priests, those who are priests typically have not been school teachers for all or even most of their careers: they may have been for example parish priests, missionaries or even university teachers, and be doing the teaching job only for a few years before going back to one of those occupations. Also, in the case of "boarding" schools such as Beaumont, it is common for priests who are retired to live on the premises, whether or not they have been teachers there. As I will relate, this circumstance had totally unforeseen consequences for my career.

Not long after my transfer to Beaumont, at the age of thirteen, I had to make what under normal circumstances would have amounted to an irrevocable decision about my career in life. I imagine that to anyone reading this 50 years later, in the context of a European (even British) or North American school system, this will seem quite incredible; but in those days the degree of "channeling" even in progressive schools in Britain (and Beaumont was far from progressive!) was extreme: you had to choose at an early age (normally 15, but in my case 13) between specialization in classics, modern languages, mathematics and science. That choice then essentially determined which kinds of degree you could take at university (assuming you went), and that in turn put severe constraints on the kind of job you could reasonably apply for. At Beaumont, the unspoken tradition was that if you showed any signs of academic ability, as I apparently did, and had no strong predilection to the contrary, you were automatically channeled into the stream which was deemed most academically prestigious, namely the classics stream (Latin and Greek languages and literature). Remarkably, my father, despite his own training and employment, never exerted the least pressure on me to take the science option (which was generally regarded, for reasons which are unclear to me, but were probably a reflection of a common attitude in Britain at that time, as the least prestigious of all). So I indeed ended up, at the age of thirteen, committed to spending most of my school years on the classics, with only a small fraction available for other subjects such as English, history, or mathematics. My memories of this classics education are mixed: on the one hand, I still recall the soul-destroying discipline of the "gender rhymes", jingles which we were required to memorize to remind us of the correct gender of Latin nouns, and fragments of which still 50 years later seem to take up kilobytes of storage space in my brain which I am sure could be put to better use ("... and common are to either sex, *artifex* and *opifex* . . ."). On the other hand, some of my teachers were enthusiastic and inspiring, and introduced me at an early stage to the poetry of Catullus and Horace and the historical analysis of Thucydides.

Although I by and large enjoyed the academic side of life at Beaumont and prospered at it, in most other respects I was a decided misfit: Almost all my fellow-pupils were fee-paying, which almost automatically meant that they enjoyed, at home, a lifestyle of which we had no experience; except in my last two years I was one of a small minority of "day-boys" (that is, I commuted daily from home rather than residing at the school throughout the term); and, perhaps most importantly, because of my academic precocity I was placed in classes with boys who were mostly a couple of years older than me. As a result, the five years I spent there, while not unhappy, do not stand out in retrospect as a particularly joyful period of my life.

However, at least two things happened during those years which were to have, in different ways, very happy long-term consequences. The first was that, for a reason I cannot now recall (I think it may have been to help my recovery from an illness which had kept me in bed for a month) my father decided to send me on a mountaineering course in Snowdonia. Unlike the rolling hills of Surrey and Berkshire which I had explored on my solitary hikes, the mountains around Llanberis were the real thing, craggy, mostly pathless and, in bad weather, decidedly not to be trifled with; I fell in love with them at first sight, and mountaineering in all its forms, from easy strolls to technically relatively demanding rock-climbing, has been a major passion of my life ever since.

The second serendipitous event actually occurred after I had competed for, and obtained, a scholarship to Balliol College, Oxford, in December 1954. I had to wait a few months before taking up the scholarship, and in those days there were no nationally organized programs to employ people in my position in voluntary service overseas, etc., so I somewhat reluctantly stayed at school. Also living at the school was a retired priest, Fr. Charles O'Hara, who had at one period of his career been a university teacher of mathematics and had in fact written a textbook on projective geometry. One day he ran into me in the corridor and said to me: "You seem to have a lot of time on your hands. Well, I have a lot of time on my hands. Why don't you come along to my room for a couple of hours a week, and I'll show you some interesting things in modern mathematics?" Now, at that point I had only the minimal mathematics required for the "O-level" examination (though I don't recall clearly, I think it did not even include differential calculus, and certainly not the integral variety), and I had absolutely no reason to think that I would ever in my career need anything more sophisticated. However, I indeed had time to kill, Fr. O'Hara seemed enthusiastic, and perhaps to humor him as much as anything, I agreed to his proposal. So for the rest of those two terms he gave me, for a couple of hours a week, a sort of Cook's tour of ideas in modern mathematics, involving concepts such as groups, rings, and fields which my O-level education had never even grazed and which I found quite fascinating. Even more importantly in retrospect, he actively encouraged me to do a few simple exercises involving those concepts, and to my initial surprise I found I could in fact do them without great difficulty. After the end of the school year I had other things on my mind and promptly forgot all the details of what he had taught me; but this was to be the first of a series of fortunate accidents that helped to shape my eventual career.

In early October of 1955 I went up to Oxford to take up my scholarship at Balliol, with the intention of reading (majoring in) the degree technically known as Literae Humaniores, and informally as Greats (on which more below). For me, coming as I did from a school background which even by the standards of those days was somewhat restrictive and conservative, entering university was like stepping through a door and finding myself in a completely new world. I still have idyllic memories of my first two weeks in Oxford, with the autumn colors in the parks at their peak and the scent in the air of infinite possibilities, both intellectual and social. Although of course that initial sense of excitement didn't last forever, by and large my nine years at Oxford, as undergraduate and graduate, were very happy indeed; not only did I thoroughly enjoy and prosper at my academic work, but I made a variety of friends from all over the world, especially from south and east Asia, with many of whom I have kept in lifelong contact. On the sports side, I had done some sculling at school, and initially rowed bow in the Balliol third eight. However, it was not long before the coach noticed that I was two stone lighter than the then coxswain ("cox") and interchanged us, and thereafter my destiny was fixed; I eventually rose to be cox of the first eight, and while I did not exactly steer them to victory (in fact, in the year of my coxswainship we did, formally, just about as badly as it is possible to do!), I nevertheless feel we put up a good fight. (Incidentally, I should have no such career possibility today: one consequence of the gender-integration of the Oxford colleges which took place in the 70's is that the coxes of college eights are now, for obvious reasons, almost invariably female).

The Oxford Greats degree is one of those many British institutions which need to be understood in historical rather than logical terms. It takes four years (12 trimesters) , and for the first five trimesters one studies Greek and Latin languages and literature, thus making it the obvious choice of degree for anyone who, like me, has specialized in the classics at school. For the last seven trimesters one studies in parallel "ancient", that is Greek and Roman, literature and philosophy; the philosophy side of the degree has only a relatively small "ancient" component (Plato and Aristotle), and is (or was in my day) mostly centered in the analytic and mostly Anglo-Saxon tradition (Descartes, Locke, Berkeley, Hume, Russell, Wittgenstein… and in more contemporary terms Ayer, Ryle, Austin…). I thoroughly enjoyed all the components of the course; I found particularly rewarding the individual (or sometimes two-person) tutorials characteristic of the "Oxbridge" system, in which I could discuss with my tutors, at least prima facie on equal terms, the most recent journal articles on the origins of coinage in the ancient Mediterranean or the concept of machine consciousness. (One disappointment I encountered on eventually switching to physics is that in that subject, as I suspect in most of the "hard" sciences, such cutting-edge discussions are not realistically possible at the undergraduate level).

I am often asked whether and how my Greats training has been useful to me in my subsequent career in physics. To that question I have a joking answer, namely that unlike (apparently) some of my physics colleagues I at least know the difference between the Greek letters φ and ψ! However, there is a serious answer: I certainly do feel the philosophy component of the degree, at least, has helped to shape the way at which I look at the world and in particular at the problems of physics. This is not something which is easy to quantify or make concrete; I have never undergone a course of psycho-analysis, but I imagine that anyone who does so can never again look at the world in quite the same light, and I think that the same is true for the kind of rigorous course of analytic philosophy which I went through at Oxford. Another analogy might be with the feeling I experienced later when I learned to speak Japanese with a reasonable degree of fluency; it is as if one has learned to use a muscle that one did not know one had. At any rate I have never for a moment regretted the years I spent on this degree.

Towards the end of my third year in Oxford it gradually began to dawn on me that I could not go on being a student for ever and must start looking for gainful employment. Not being endowed with a great deal of initiative, I first looked around to see what careers previous Greats students at Balliol had chosen. With few exceptions, it seemed that they had either gone into the British civil service or become teachers, at university or high-school level, of one of the subjects they had studied (classics, ancient history or philosophy). As regarded the first option, it did not take me long to figure out that even were I to pass the examination for the civil service (something which was by no means a foregone conclusion, since it tested qualities rather different from the ones which had advantaged me in a purely academic context), a career there was likely to be both uncongenial to me and of dubious benefit to the British public. So I started contemplating a career in academia; and since the subject which I had enjoyed most, and performed best in, was philosophy, this seemed the natural choice.

But the more I thought about the prospect of doing a doctorate in philosophy and eventually obtaining a university lectureship in that subject, the more I realized that in my bones I just did not want to do it. Why not? It never occurred to me – as it no doubt would have to someone with more imagination – that maybe I was not actually cut out for an academic career at all. Rather I asked myself what exactly it was about philosophy as such which deterred me from doing it professionally for life. And I eventually came up with the answer that it was because what counted as good or bad work in philosophy – at least as it was practiced in Oxford at that time – seemed to depend so strongly on the precise nuances of one's wording (something, incidentally, which I suspect may have subtly disadvantaged non-native English speakers in that discipline); there seemed to be no objective criterion of what was correct or not, or even what was good work or bad, and I felt in my bones that it was just such a criterion that I needed if I was going to pursue an academic career. I did indeed briefly consider the possibility of going into pure mathematics, but rejected it on the grounds that in mathematics, almost by the definition of the subject, to be wrong means you are stupid: I wanted the possibility of being wrong without being stupid – of being wrong, if you like, for interesting and nontrivial reasons. Physics seemed to fill that bill, and while I had zero formal training in that subject, the confidence which I had acquired from Fr. O'Hara in advanced mathematics led me to believe that that aspect of the subject, at least, would not give me major difficulty. So, in the early summer of 1958 I took my courage in both hands and applied to do a second Oxford undergraduate degree, in physics, following the anticipated completion in spring 1959 of my Greats degree.

In those days in Britain to do a second undergraduate degree in anything, let alone in a subject in which one had no secondary-school experience, was practically unheard of, and I immediately faced several practical obstacles. In the first place, I had to persuade some college to accept me; secondly, I had to find some way of financing two more years of undergraduate education, and finally, since 1959 was known to be the last year of conscription in Britain, I had to persuade my draft board that my application to do a second degree was not just a ruse to avoid military service for ever (a consequence it in fact had). At this point I had another piece of luck: in October of 1957 engineers in the Soviet Union had propelled into space the first Sputnik, thus stealing a march on the West in what not just governments, but every layman could see was an extraordinarily important area of technology. Immediately the cry went up from politicians and the press: how have the Soviets managed to get ahead of us in this crucially important enterprise? And the answer was not long coming: It was because we have encouraged all our best brains to study useless subjects (such as classics) rather than useful ones (such as science and engineering and particularly physics). Immediately all sorts of scholarships became available for students in the arts who wished to transfer to science; and while I did not in the end need to apply for any of these, I think that the general shift in cultural attitudes which they reflected was an enormous psychological boost to me in making the switch, and in particular may have been the crucial element in helping my prospective tutors to convince the draft board that they could dispense with my services on the parade ground.

The two people who played the most indispensable role in helping me make the transition are David Brink and Michael Baker. In those days most Oxford colleges did not have a tutor in theoretical physics as such, but Balliol had recently appointed David in this role, and since I initially applied to Balliol it was he who had to take a decision on whether I was or was not a reasonable prospect for a degree in physics, despite my lack of background. He asked me to read various sections of that beautiful book, "What is mathematics?" by Courant and Robbins in the summer vacation between my third and fourth years, and when I returned in October tested me on them; on the basis of my performance he recommended that Balliol should accept me for a second degree. However, by that time I had applied somewhat speculatively to Merton, where Michael Baker was a tutor in physics, for a Domus scholarship. These scholarships were normally given to students undertaking a postgraduate degree, but in my case the Fellows generously made an exception. Thus, I ended up doing my degree in physics at Merton; Michael was my principal tutor, but since he was an experimentalist and Merton did not at the time have a tutor in theoretical physics as such, I went to David for most of the more theoretically oriented parts of the course. In order to get a classified degree I had to complete the course in two years rather than the standard three, and this resulted in a certain degree of mental indigestion: I recall simultaneously struggling with old-fashioned problems involving rods and pulleys from Humphrey and Topping's "Intermediate Mechanics" and with the properties of Hilbert space as set out in Mandl's text on quantum mechanics.

The final examination at Oxford was in those days (maybe it still is) an ordeal unimaginable by those who have not experienced it: there was no continuous assessment at all, and one's whole academic fate depended on a succession of closed-book three-hour written papers (in the case of Greats fourteen of them), one after the other, two per day, with only the weekend as a short respite. Not surprisingly, the population of the psychiatric wards of the local hospitals used to peak markedly before and during those periods. Actually, in my Greats degree I had thrived on the stress and come out with a straight first class, so I was probably over-confident going into the physics final exam. In the event this nearly turned into a complete disaster: the first paper was, I think, on thermodynamics, an area where problems notoriously require just the right trick to solve them, and after an hour and a half of the three hours I had got nowhere with the first problem I had tackled. I remember sitting at my desk in the exam room, my head figuratively if not literally in my hands, and contemplating the prospect of having dashed the hopes of all those who had supported and shown faith in me; it was one of the blackest moments of my life. In the end I pulled myself together and was able to answer one or two questions on that paper, as well as coping reasonably with the remaining papers; but when I was told that I was on the borderline between two (unspecified) classes and would therefore have to undergo a viva (oral exam) to decide between them, I was convinced that the two classes in question were a second and a third. Imagine my incredulity and, eventually, my delight when a few days after the viva, a letter arrived informing me that I had a first (and would therefore qualify more or less automatically for public funding for an advanced degree should I so chose). If nothing else, I think that experience has convinced me that no examination system, and certainly not the one I went through, can be an infallible measure even of purely academic ability.

I was now in a position to proceed to postgraduate research in physics. I had specialized on the theoretical side as an undergraduate, and my memories of the (rather small) amount of lab work I had had to do were not particularly rosy, so it seemed natural to apply for research on the theoretical side. But where, in which area and to whom? The first question answered itself easily, since I felt (perhaps wrongly) that no university other than Oxford itself was likely to appreciate my peculiar academic background. As to the second, the main areas of theoretical work at Oxford in those days were particle theory and condensed-matter (or, as it was known in those days, "solid-state") physics. The time was early 1961, and the advice I got from the theorists I consulted was that the current state of particle theory was not very exciting (an opinion which had to be revised a few months later, when the concept of unitary symmetry burst on the scene); so I opted for the condensed-matter area. Finally, one person who was willing to overlook my unorthodox credentials was Dirk ter Haar, then a reader in theoretical physics and a fellow of Magdalen College; so I signed up for research under his supervision. As with all his students in that period, the tentatively assigned topic of my thesis was "Some Problems in the Theory of Many-Body Systems," which left me a considerable degree of latitude.

Even by the standards of British universities in those days, Dirk's supervisory style was somewhat unusual. He took a great interest in the personal welfare of his students and their families, and was meticulous in making sure they received adequate support; indeed, in the middle of my second year of research he encouraged me to apply for a Prize Fellowship (junior fellowship) at Magdalen. To my great surprise I was successful, I am sure in considerable measure thanks to his advocacy, and thereafter was able to enjoy a lifestyle rather more opulent than had been possible on the standard graduate studentship which had supported me earlier. On the other hand, Dirk's method of supervising graduate research was to throw his students in at the deep end: by a few months into our relationship I had got the message that it was up to me not only to solve my thesis problem, once posed, but to find a viable problem in the first place. (I try to encourage my own graduate students to do the same, although I do not take Dirk's extreme position of refusing, in effect, to make any suggestions at all). In the end my thesis work consisted of studies of two somewhat disconnected problems in the general area of liquid helium, one on higher-order phonon interaction processes in superfluid ^{4}He and the other on the properties of dilute solutions of ^{4}He in normal liquid ^{3}He (a system which unfortunately turned out to be much less experimentally accessible than the other side of the phase diagram, dilute solutions of ^{3}He in ^{4}He). Although both halves resulted in publications, neither made much impact, and the only feature which in retrospect distinguishes my D. Phil. thesis in any way is that a small part of it (the acknowledgements) is, as permitted by the Oxford examination statutes and as a result of what our forebears would no doubt have described as a "conceit", written in Latin.

One part of my graduate student experience which I have always felt was particularly valuable to me was the undergraduate teaching I did during this period. I had started this, I think as soon as I began graduate research, with a view to supplementing my rather subsistence-level studentship, but kept it up even after the Magdalen fellowship eliminated the need for this; as I recall, it was six hours per week of one-on-one or one-on-two individual tutorials in the undergraduate "theoretical option", and I found I enjoyed it immensely and learned at least as much from it as from my formal research activities. Nowadays I always insist, as a matter of principle rather than financial economy, that my own graduate students spend at least one semester per year as teaching assistants.

In the spring of 1964, as I approached the three-year deadline for submission of my D. Phil. thesis, I began to think about postdoctoral work. In applying I had two criteria: the group in question should be a world-class center of excellence in many-body theory, and the environment should be as different as possible from Oxford (where I had by now spent nine years, more than a third of my life). So posed, the problem had an essentially unique solution, namely the group of David Pines at the University of Illinois at Urbana-Champaign, and it was there that I applied; I don't think I even bothered applying anywhere else (in those pre-information revolution days, the preparation of multiple applications was quite a daunting and time-consuming task). Despite this, I was quite surprised and gratified to be accepted. (David tells me that the letter Dirk wrote for me said in effect that I had a good training on the arts side but didn't know any physics – a statement which in view of my highly compacted education in that subject was certainly true, but to which David's reaction was "well, that's something we can teach him!"). So I spent the period August 1964 – August 1965 at UIUC, and David and his colleagues (John Bardeen, Gordon Baym, Leo Kadanoff and others) did indeed teach me a great deal. And it was indeed very different from Oxford . . .

From the academic point of view that year at UIUC was a turning-point in my career; as recounted in my Nobel lecture, I not only became interested in the (at the time hypothetical) superfluid phase of liquid ^{3}He, but produced my first research (on Fermi-liquid effects in the superfluid phase) to make any real impact internationally. However, I did not find the physical environment congenial (it was only a few years since Dutch elm disease had wiped out more of the town's mature trees, something which is an essential ingredient of the quality of life in small Midwestern towns), and I remember swearing an oath to myself, as I left at the end of the year, that I would never come back and live permanently in Champaign-Urbana – an oath which I was to break rather spectacularly twenty years later.

My Magdalen fellowship still had a couple of years to run, and the natural thing would have been to return to Oxford and spend them there. However, I had become increasingly interested in north-east Asia, and anxious to spend some time there before committing myself to the standard British academic career; and the Fellows of the college eventually consented, with considerable generosity, to allow me to spend a year of my fellowship in the group of Professor Takeo Matsubara at Kyoto University in Japan. This year at Kyoto was a marvelous experience for me. It was my first real experience of living and working in a foreign culture, and I tried to exploit it to the fullest: I lived in a standard Japanese student room, put a considerable amount of effort into learning the language and tried as far as possible to avoid using English or mingling very much with other foreigners. At this time this was not at all the typical behavior of Western visitors, and many years later I was told that it had caused considerable interest and speculation among some of my Japanese colleagues, who finally hit on what had to be the only logical explanation – that I must be a trainee CIA agent! Fortunately this opinion did not prevent me making many good friends during that year, and I have very warm memories of all the parties, mountain hikes and home visits we shared. Although the academic work I did during this year was overall not particularly remarkable, it included the paper on two-band superconductors which was to play a crucial role in my research on superfluid ^{3}He a few years later.

After one more postdoctoral year which I spent in "roving" mode, spending time at Oxford, Harvard and Illinois, in the autumn of 1967 I took up a lectureship at the University of Sussex, where I was to spend the next fifteen years of my career. Sussex was one of the "new" universities founded in the fifties, and at that point was only a few years old, but had managed to attract among other things a lively group of theoretical physicists under the leadership of Roger Blin-Stoyle and a constellation of able low-temperature experimentalists headed by Douglas Brewer. However, what attracted me most was the liberal and collegial atmosphere both in the University as a whole and in the Physics Department, which actively encouraged people to explore their intellectual interests across traditional academic boundaries; quite a few of my colleagues, though trained in the traditional areas of physics, ended up spending much of their time on science policy, physics education and elsewhere. However, although I enjoyed this relaxed environment, I spent my first five years at Sussex mostly teaching the standard undergraduate physics courses and, in the time available for research, working on various problems in theoretical low-temperature physics, including some such as the possible "supersolid" phase of helium which appear to still be of interest nearly forty years later. Being without family attachments, I was able to spend a lot of time abroad during the vacations, and in particular had an extended stay at the Max-Planck-Institute in München and a couple of trips to winter schools in Karpacz, Poland (at the time something of a minor adventure, since the Iron Curtain was still firmly in place).

In the summer of 1972 there occurred the series of events which were to shape my research career for the next decade; I have recounted these in my Nobel lecture and will not repeat them here. At just about the same time an important event occurred on the personal front: I became engaged to Haruko Kinase, at that time an undergraduate student at Sussex, and we married in June 1972. (Since Haruko's nationality was and is Japanese, and her undergraduate major was international relations, the family joke is that at least she passed the practical!). We had a sort of extended honeymoon in Japan, living in central Tokyo with Haruko's parents while I worked in the group of Professor Yasushi Wada at the Hongo campus of Tokyo University. One spin-off from this stay was that I got to know Shin Takagi, who was in his final year as a graduate student in the Wada group, and was sufficiently impressed with him that I invited him to Sussex as a postdoc; he spent a couple of years there, and we collaborated extensively on superfluid ^{3}He as well as having many conversations ranging over the whole of physics. Incidentally, if my early years in the extreme outer reaches of the London conurbation are not counted, the only large cities anywhere in the world where I have ever lived for more than a few months are Kyoto and Tokyo; I can think of a lot worse.

As this note has to be of finite length, I will speed up at this point and just review the important events in my life since 1974. Soon after our return from Japan, in early 1975, Haruko and I acquired a small house that we still own in the warren of streets above London Road station in central Brighton, and it was there (or more accurately on the twelfth floor of the nearby Royal Sussex County Hospital) that our daughter Asako was born in September 1978, and there that she spent the first five years of her life. After returning from Japan I spent several more very enjoyable years at Sussex, with various excursions including two one-semester trips in 1976 and 1977 to the University of Science and Technology in Kumasi, Ghana, with which Sussex had a teaching exchange arrangement. This was another interesting cultural shock; one unexpected aspect was that because of the generally relaxed atmosphere in Ghana, I had for the first time in many years a surfeit of free time, and I used it among other things to write a paper on nonlocal hidden-variables theories which I eventually published a full quarter-century later.

In the spring of 1982 I received, out of the blue, an offer from the University of Illinois at Urbana-Champaign of the MacArthur Chair with which the university had recently been endowed, and after the three of us had been across for a visit (where I observed, inter alia, that the trees had recovered from the blight of the early 60's!) we decided to make the move. As I had already committed myself to an eight-month stay at Cornell in early 1983, we finally arrived in Urbana in the early fall of that year, and have been there ever since, so that Asako has in effect grown up as an American (she and I both eventually received U.S. citizenship, which we hold dual with that of the U.K., in the summer of 2001). Haruko eventually obtained a Ph.D. in cultural anthropology from the University of Illinois, and is currently doing research on the hospice system; Asako has graduated, also from UIUC, with a joint major in geography and chemistry. My own research interests have shifted away from superfluid ^{3}He since around 1980 (although I still maintain an interest in some problems involving violent departures from equilibrium); I have worked inter alia on the low-temperature properties of glasses, high-temperature superconductivity, the BEC atomic gases and above all on the theory of experiments to test whether the formation of quantum mechanics will continue to describe the physical world as we push it up from the atomic level towards that of everyday life (a program for which my shorthand is "building Schrödinger's Cat in the laboratory"). It is satisfying that this program, which when proposed twenty-five years ago met with considerable skepticism, seems in the last three or four years to have come to fruition, in the sense that several experimental groups have realized, in Josephson devices, quantum superpositions which can be legitimately regarded as of the "Schrödinger's Cat" type.

When I look back on the chain of events which led to the research which the Nobel Committee has recognized, I realize how incredibly fortunate I have been – not just because of the intervention in my career of seemingly irrelevant events such as the propulsion of Sputnik into orbit in the fall of 1957, but because so many people were prepared to put their faith in me, and in particular my ability to make a successful career in physics, at a time when the evidence in favor of that proposition was nonexistent or perhaps even negative. I shall remain forever grateful.

SUPERFLUID 3-He:THE EARLY DAYS AS SEEN BY A THEORIST

Nobel Lecture, December 8, 2003

by

ANTHONY J. LEGGETT

University of Illinois, Department of Physics, 1110 West Green Street, Urbana, Ijl 61801-3080, USA.

It is needless to say that I feel it a great honor and privilege to have been selected for the 2003 Nobel prize in physics for my theoretical work on superfluid ^{3}He; I am particularly pleased to be sharing the award with Professors Ginzburg and Abrikosov, whom I have always looked up to as giants of the closely related fleld of superconductivity. The story of how, in roughly the twelve-month period July 1972–July 1973, we came to a theoretical understanding of the experimental data on what we now know as superfluid ^{3}He is a sort of complex detective tale, involving many actors besides me; for reasons of time I will concentrate in this lecture on my own involvement, and will have to omit several important developments in which I had no direct role.

The element helium comes in two (stable) forms, ^{4}He and ^{3}He; at low temperatures and pressures both form liquids rather than solids. The liquid phase of the common isotope, ^{4}He, was realized nearly a century ago, and since 1938 has been known to show, at temperatures below about 2K, the property of superfluidity – the ability to flow through the narrowest capillaries without apparent friction. By contrast, the liquid form of the rare isotope, ^{3}He, has been available only since about 1950, when enough of it was produced by the decay of the tritium manufactured in nuclear reactors. However, it was soon recognized that liquid ^{3}He is in many ways similar to a system which has been known for much longer, namely the electrons in metals. Although there is one obvious difference (the electrons in metals are electrically charged whereas the ^{3}He atom is electrically neutral), both systems are dense systems of particles which have spin 1/2 and are therefore expected to obey Fermi-Dirac statistics. (By contrast, the atoms of ^{4}He have spin zero and should therefore obey Bose-Einstein statistics). If we consider a non-interacting gas of such particles in thermal equilibrium at a temperature $T \ll T_F = \varepsilon_F / k_B$ (where ε_F is the "Fermi energy", determined by the mass and density), then all states lying well below ε_F in energy are occupied by a single particle and all those well above ε_F are empty; rearrangement of the particles can take place only in an energy "shell" of width $\sim k_B T$ around ε_F, and all the thermal, transport and response properties are thus determined by the properties of the states in this shell. In a famous 1956 paper L. D. Landau [1] showed that under appropriate conditions this picture remains qualitatively valid even in the presence of strong inter-particle interactions; the system is then known as a "degenerate Fermi liquid". Experiments on liquid ^{3}He in the fifties and early sixties showed that this system indeed appeared to be behaving as a degenerate Fermi liquid below ~100 mK, down to the lowest temperatures then attainable (around 3mK).

Now, it has been known for nearly a century that the electrons in metals, which have a Fermi temperature of ~ 10^4–10^5K, may sometimes, at temperatures ≲ 20K, enter the so-called superconducting state, in which they can flow without apparent resistance; this is just the analog, for a charged system, of the superfluidity of liquid ^{4}He. Since for liquid ^{3}He the Fermi temperature is only a few K, it would have been reasonable to speculate that the atoms might undergo a similar transition at temperatures of the order of mK; since the atoms are electronically neutral, the result would be not superconductivity but rather superfluidity, as in ^{4}He. However, in the absence of a microscopic theory of superconductivity no quantitative approach to this question suggested itself.

Remarkable progress in the phenomenological description of superconductivity was made in the early fifties (as recognized in the awards to my colaureates), in particular by introducing the concept of a "macroscopic wave function" or order parameter. The microscopic underpinning [2] of this concept was provided by my late colleague John Bardeen and his collaborators Leon Cooper and Bob Schrieffer in 1957, in what is now universally recognized as the correct microscopic theory of superconductivity (at least as it was known at that time), the "BCS" theory. They postulated that in the superconducting state the electrons within a "shell" of width ~ $k_B T_c$ around the Fermi energy (where T_c is the temperature of the superconducting transition) tend to form "Cooper pairs", a sort of giant "di electronic molecule", whose radius is huge compared to the average distance between electrons (so that between any two electrons forming a Cooper pair there are billions of other electrons, each forming their own pairs). A very essential feature of the BCS theory of superconductivity is that the Cooper pairs, once formed, must all behave in exactly the same way, that is, they must have exactly the same wave function, as regards both the center of mass and the relative coordinate. In fact, the "macroscopic wave function" of Ginzburg and Landau turns out to be nothing but the common center-of-mass wave function of all the pairs. This wave function can have a very nontrivial spatial variation, and it is this variation which gives rise to the effects described by Professors Ginzburg and Abrikosov.

By contrast, in BCS theory the "internal" (relative) wave function of the pairs is rather boring; the two electrons have opposite spins, thus the pair has total spin zero, and the relative orbital angular momentum is also zero, so that in atomic notation the state of the "molecule" is $^1S_0$, and all properties of the system are completely isotropic; there are no "orientational" degrees of freedom.

When at the end of the fifties people started to extend the ideas of BCS to liquid ^{3}He, they soon realized an important difference between this system and the electrons in metals: The interaction potential between two He atoms is strongly repulsive at short distances, and becomes attractive only for interatomic separation $r \sim r_o \sim 3$ Å; thus, the atoms forming a Cooper pair cannot approach much more closely than this. Since they come from states close to the Fermi surface, however, their relative momentum must be of the order of $p_F \equiv (2m\varepsilon_F)^{1/2} \sim 1$ Å^{-1}. This then means that their dimensionless relative angular momentum l must be of order $p_F r_o/\hbar$ and thus must be nonzero (most probably 1 or 2), in contrast to the case of metallic superconductors where as we have seen the pairs have $l = 0$. (A more quantitative calculation bears out the conclusion of this intuitive argument). The Fermi statistics then imply that if l is even (so that the orbital state is symmetric under exchange of the two particles) then the spin state must be a singlet (S = 0) as in the original BCS case, while if l is odd the spin state must be a triplet (S = 1). It was then not unreasonable to expect that in either case (provided $l \neq 0$) the orbital properties might be anisotropic, and that in the case of spin triplet pairing (odd l) the spin properties might also be so; however, as we shall see, this question is less straightforward than it might seem.

In the early sixties there was a good deal of theoretical interest in the possible existence and properties of a Cooper-paired (hence presumably superfluid) phase of liquid ^{3}He; in the present context two developments in particular are worthy of note. In a seminal 1961 paper [3], Anderson and Morel made a systematic adaptation of the ideas of BCS to this system: that is, they explicitly assumed (as had others) that in the $l \neq 0$ case, just as in the $l =$ 0 one considered by BCS, all Cooper pairs form in exactly the same state, with respect not only to their center-of-mass motion but also with respect to their internal (relative) state. It is worth emphasizing that this assumption is not trivial; indeed there was at least one nearly contemporary paper [4] which made a quite different and perhaps prima facie more attractive assumption, namely that (in the $l = 2$ case) all the five Zeeman substates are equally populated, in such a way as to give a state of the system whose physical properties are totally isotropic.[1] By contrast, Anderson and Morel showed that with their assumption the physical properties would in general be anisotropic both in orbital and, in the odd-l (spin triplet) case, spin space. They studied in detail two cases: that of pairing in a relative d-state ($l = 2$), where they worked out various physical properties in considerable detail, and more briefly that of a p-state ($l = 1$). In the latter case they chose a particular state, namely one in which pairs form only in the $S_z = +1(\uparrow\uparrow)$ and $S_z = -1(\downarrow\downarrow)$ states (a so-called "equal-spin-pairing" (ESP) state), and furthermore these states are associated with the same orbital wave function, which intuitively corresponds to the pairs having an orbital angular momentum $\hbar$ along a direction conventionally denoted by the unit vector $\hat{\mathbf{l}}$; this state has subsequently, for reasons we shall see, acquired the name of the "Anderson-Brinkman-Morel" (ABM) state. As far as I know, there was no particular reason at the time to choose this state rather

[1] For arguments against this idea, see refs. [5] and [6]. Many years later the idea was revived [7] but again refuted [8].

than, for example, the state, which within the generalized BCS theory is degenerate with ABM, in which the $S_z = +1$ and $S_z = -1$ pairs have opposite angular momenta ("axial" state); so that the choice they actually made was serendipitous.

A second very important advance was made independently by Vdovin [9] in the former Soviet Union and Balian and Werthamer [10] (BW) in the West; they observed that in the odd-l case it is possible to form pairs simultaneously in all three Zeeman substates in such a way that the pair wave function is a superposition, i.e., schematically of the form (where **r** denotes the relative coordinate)

$$\Psi_{pair} = F_{\uparrow\uparrow}(\mathbf{r})\,|\uparrow\uparrow\rangle + F_{\downarrow\downarrow}(\mathbf{r})\,|\downarrow\downarrow\rangle + F_{\uparrow\downarrow}(\mathbf{r})(\frac{1}{\sqrt{2}}(\uparrow\downarrow + \downarrow\uparrow)) \tag{1}$$

It should be emphasized that all the Cooper pairs still occupy the single superposition(1) (we do not have one-third of the pairs occupying each Zeeman substate independently!). The "ESP" states considered in earlier work such as that of Anderson and Morel are special cases of (1) with $F_{\uparrow\downarrow}(\mathbf{r}) \equiv 0$; in particular, the ABM state has (up to a phase) $F_{\uparrow\uparrow}(\mathbf{r}) = F_{\downarrow\downarrow}(\mathbf{r})$. In the specific case $l = 1$, Vdovin and BW showed that it is possible to choose $F_{\uparrow\uparrow}(\mathbf{r})$ to correspond to (apparent) angular momentum $L_z = -1$ and $F_{\downarrow\downarrow}$ and $F_{\uparrow\downarrow}$ similarly to correspond to $L_z = +1$ and $L_z = 0$ respectively; in this way they constructed a state with $L = S = 1$ but $J \equiv |\mathbf{L}+\mathbf{S}| = 0$, i.e., in atomic notation a $^3P_0$ state. By the Wigner-Eckart theorem such a state should be isotropic in all its properties, whether involving orbital motion, spin or their combination; in particular, the spin susceptibility should be isotropic and equal to 2/3 of the normal-state value (reflecting the fact that a third of the spin state is $S_z = 0$ with respect to any axis). Vdovin and BW showed that within the generalized BCS calculation which they used this state (conventionally known as the BW state) should be more stable than any $l = 1$ ESP state.

As a result of these and other considerations, the general expectation in the theoretical community by around 1964 was that (a) liquid ^{3}He might enter a Cooper-paired state, which would then display inter alia the property of superfluidity, at a temperature which was difficult to predict, and (b) that the symmetry of such a state was most likely to be either $l = 2$, $S = 0$ or the BW ($^3P_0$) state; in either case the spin susceptibility should be isotropic and reduced from its normal-state value (to zero in the $l = 2$ case, by a factor of 1/3 in the BW case).

It was at this point that I became actively interested in the possible superfluid phase of ^{3}He. I had done a little work on normal ^{3}He in my D. Phil. thesis (using Landau Fermi-liquid theory which at that time was still something of a novelty in the West), and had subsequently gone to work as a postdoc with David Pines at the University of Illinois at Urbana-Champaign. One day, after I had been there for about a month, John Bardeen and Leo Kadanoff came into my office and said "Look, John Wheatley down in the basement is doing experiments on the spin diffusion coefficient of liquid ^{3}He, and is getting into the temperature regime where people think the superfluid transition might occur. Why don't you try to work out how this would be reflected in the spin diffusion behavior?" I started this calculation but never finished it; in retrospect this was probably just as well, since we now know that because of the anomalous NMR behavior of superfluid ^{3}He (on which much more below) the spin diffusion is an extraordinarily complicated problem, and it is unlikely in the extreme that I would have got it right in the absence of experimental clues.

The reason I never finished the spin diffusion calculation was that I got side-tracked on to a problem that struck me as much more interesting, namely how to combine the ideas of BCS on pairing in a weakly interacting Fermi system with Landau's "Fermi-liquid" theory of the normal state of ^{3}He. After some vicissitudes (see ref. [11]) I was able to reformulate the Landau theory in terms of a set of "molecular fields", a form in which it was then almost trivial to apply it to the hypothetical superfluid phase as well as to the normal one. I was able to show (for the $l = 0$ case) that while the spin susceptibility and normal density of a degenerate superfluid Fermi liquid, like that of a weakly interacting superfluid Fermi gas, both fall to zero in the limit $T \to 0$, their temperature dependence is in general quite different from that of the latter (and from one another)[2]. In the case of the spin susceptibility, I also applied the "molecular-field" idea to the BW state, and showed inter alia that in ^{3}He Fermi-liquid effects should depress the $T = 0$ susceptibility (relative to the normal-state value) from the "weakly interacting" value of 2/3 to about 1/3. In subsequent work I generalized Landau's calculation of the low-frequency collective excitations of a Fermi liquid, including zero sound, to the superfluid phase.

During the two years following my postdoctoral year at UIUC, I wandered about quite a bit, both geographically[3] and intellectually. Among other problems in the low-temperature area which I considered during this period was one concerning the possible collective excitations of a "two-band" superconductor, that is, a superconducting metal in which the Fermi surface interacts two different bands. I think I had probably read, and been influenced by, P. W. Anderson's elegant formulation of the theory of superfluidity in ^{4}He in terms of conjugate "number" and "phase" variables; at any rate, it occurred to me that a two-band superconductor should show a sort of "internal Josephson effect" corresponding to fluctuations of the relative number of electrons in the two bands and of the relative phase of the Cooper pairs in them, and in a paper published in Progress of Theoretical Physics [14] I discussed the appropriate microscopic definitions of the corresponding operators ΔN, $\Delta\phi$ and their commutation relations, namely, $[\Delta N, \Delta\phi] = i$. At the time this work sank more or less without trace, in part because by the time it was published

[2] Many years later I discovered that in both cases my conclusions had been to some extent anticipated, in refs. [12] and (arguably) [13] respectively. However, at the time they were a great boost to my morale!

[3] See the biographical note which precedes this lecture.

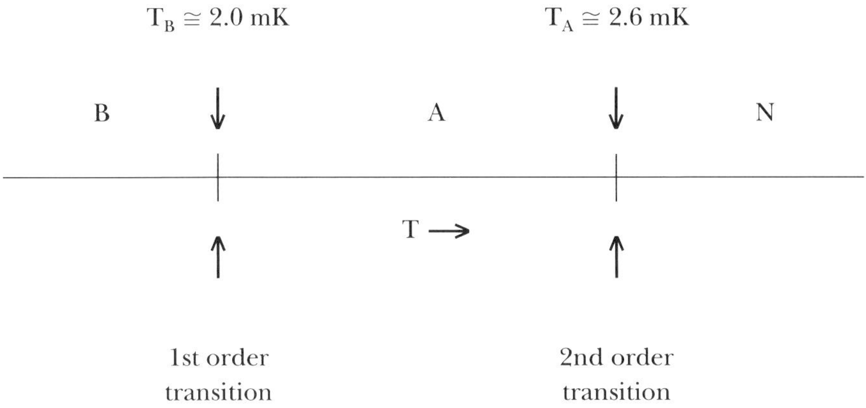

Figure 1: The phase diagram of liquid ^3He at melting pressure below 3mK.

it had already become clear that the experimental evidence for the existence of two-band superconductors in nature was dubious; however, it was to play a crucial role in the subsequent history.[4]

In the fall of 1967 I took up a lectureship at the University of Sussex, and for the next few years, in the intervals allowed by my teaching duties, I continued to work on various problems in low-temperature physics, including liquid ^3He. However, I found myself becoming increasingly bored with this area of research, and indeed with much of conventional physics; at the same time, thanks in part to a remarkable series of lectures delivered by my colleague Brian Easlea, I got more and more intrigued by the conceptual foundations of quantum mechanics, and by the summer of 1972 had made a firm decision that I would abandon the sort of physics that gets published in Phys. Rev. B and devote myself full-time to foundational studies. (Fortunately, in those days even lectureship positions in British universities carried tenure, so that it was possible to make such a switch without drastically affecting one's career prospects!)

In July 1972 I was on a climbing holiday in Scotland when I heard that Bob Richardson, whom I knew at a distance as an experimentalist at Cornell working on, among other things, liquid and solid ^3He, would be passing through Sussex for a day and would like to talk to me. I was certainly keen to meet Bob (I had had some correspondence with him concerning an effect in normal liquid ^3He on which he had done the experiment and I the theory), but I was enjoying my holiday, and I remember debating with myself whether to go home a day early to meet him or not. In the end I think the weather made the decision for me: it rained on the morning in question, I went home early and spent several hours talking to Bob. What he told me that day changed my

[4] In early 2003, Ponomarev *et al.* [15] reported experimental evidence for the existence in MgB_2 of the mode predicted in ref. [14].

whole research career and led, thirty years later, to my presence in Stockholm today.

In earlier work the Cornell group (at that time consisting of Doug Osheroff, Dave Lee and Bob himself) had studied the pressurization curve (that is, the graph of pressure versus time) of a mixture of liquid and solid ^{3}He, and had observed two small but reproducible anomalies, which with some hindsight indicate phase transitions in the liquid: a second-order transition at $T_A \sim 2.6$ mK and a first-order one at about $T_B \sim 2$ mK (see figure 1). Using the nomenclature which later became standard, the liquid phase which exists between 2 and 2.6 mK is the A phase and that below 2 mK is the B phase: thus the normal (N) – A transition is second order and the A–B one first order. The group had published their experimental results [16] in early 1972, but had erroneously interpreted the phase transitions as occurring in the solid rather than the liquid; I had vaguely known about this work but had not been particularly excited by it, in part because there was a general expectation that some kind of magnetic-ordering transition would occur in solid ^{3}He in the mK regime[5], and it did not seem that the resulting phase was likely to be particularly novel in its properties.

What Bob told me, however, was that the group (by now augmented by Willy Gully) had gone on to do nuclear magnetic resonance (NMR) experiments [18] on the solid-liquid mixture in the temperature regime in which the thermodynamic anomalies occurred. These were straightforward c.w. absorption measurements, with the *rf* field polarized perpendicular to the external *dc* field H_{ext}. The first, qualitative conclusion, which was drawn from an analysis of the spatial profile of the NMR, absorption, was quite unambiguously that the phase transitions were occurring in the liquid component of the mixture, not the solid. Now, what an NMR experiment in the above standard geometry measures is the *rf* power absorption as a function of the frequency ω of the *rf* field; provided a linear approximation is valid, from this one can directly infer the imaginary part of the (transverse) spin density response function of the system, $\chi(\omega)$, and application of a Kramers-Kronig relation to the latter (cf. below) then allows one to infer the *dc* susceptibility χ_0.

In the normal (N) phase the absorption profile has a very sharp resonance at the (temperature-independent) Larmor frequency $\omega_{res} = \gamma H_{ext}$, where γ is the gyromagnetic ratio of the free ^{3}He atom, and the *dc* susceptibility χ_0 is also temperature-independent and of the general order of magnitude of that predicted for a free Fermi gas with the mass and density of liquid ^{3}He. These results were already well established and not considered at all surprising: the discrepancy between the experimental and free-gas values of χ_0 is entirely consistent with the interpretation of the N phase as a degenerate Landau Fermi liquid, and the fact that the observed resonance frequency is (within experimental error) the free-atom value tells us that any effective magnetic field, other than those of "exchange" origin, arising from the system itself are

[5] A few years later, such a transition was indeed observed [17] at a temperature of about 1 mK.

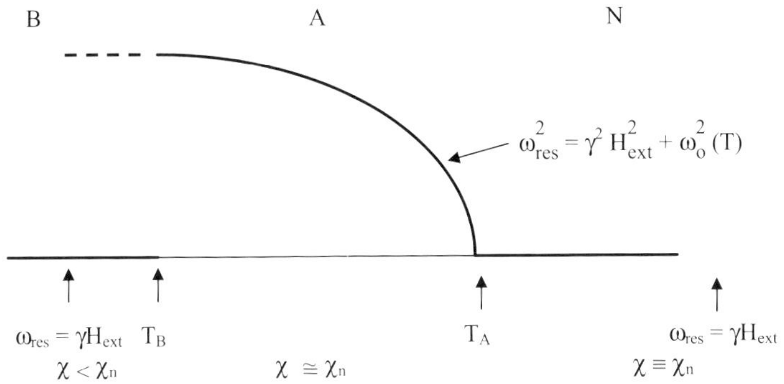

Figure 2: The behavior of the NMR resonance frequency ω_{res} in liquid ^3He below 3mK. (The qualitative behavior of the static susceptibility χ is also noted.).

negligible. (The "exchange" fields, which are in part responsible for the renormalization of the *dc* susceptibility, are automatically parallel to the total spin of the system; thus they cannot exert a torque on the latter and thus cannot shift the resonance frequency, which is determined by the external field alone: cf. below). This is exactly what is expected, since the only possible origin of such a non-exchange field is the tiny dipole interaction (on which more below).

By contrast, the NMR behavior in the *A* and *B* phases is highly unconventional: see figure 2. As the temperature falls below the temperature T_A of the second-order transition, the resonance absorption peak remains very sharp but shifts upwards in frequency. The shift is not proportional to the external field H_{ext} but rather obeys a "Pythagorean" law:

$$\omega^2_{res} = \gamma^2 \mathcal{H}^2_{ext} + \omega^2_o(T) \tag{2}$$

with ω_o^2 having the approximate temperature-dependence

$$\omega^2_o(T) \cong A(1 - T/T_A), \qquad A/(2\pi)^2 \cong 5 \times 10^{10} Hz^2 \tag{3}$$

When the system undergoes the first-order phase transition into the B phase, the resonance frequency drops back discontinuously to the Larmor (N-state) value $\omega_{res} = \gamma H_{ext} \equiv \omega_L$. Meanwhile the *dc* susceptibility, obtained as described from the absorption data using a Kramers-Kronig relation, remains constant in the *A* phase at its *N*–state value;[6] however it drops discontinuously at the *A–B* transition by a factor of ~ 50%, and thereafter decreases further with temperature in the *B* phase.

The behavior of the susceptibility struck me as surprising but not extraor-

[6] Subsequent work [19] has shown that there is in fact a very small jump (a fraction of a percent) in χ_o at the *N–A* transition.

the double commutator is nothing but the negative of $\partial^2 < \hat{H} >/\partial\theta_x^2$, the second derivative of the mean energy under such a rotation.

Suppose, first, that apart from the Zeeman energy in the external field, $\mu_n \Sigma_i S_{zi} H$, the only spin-dependent energies in the problem are spin-conserving, i.e., they depend only on the magnitude of the total spin **S**, not on its direction; this is the case, in particular, for energies of "exchange" origin. Such terms clearly cannot contribute to the double commutator, which then turns out to be simply equal to $\gamma^2 S_z H_{ext} \equiv \gamma^2 \chi_o H_{ext}^2$; thus in this case the resonance is uniquely constrained to lie exactly at the Larmor frequency $\omega_L \equiv \gamma H_{ext}$. Thus, as already stated, any shift of ω_{res} away from the Larmor value, such as occurs in the A phase, is unambiguous evidence for the operation of some spin-non-conserving energy. In liquid ^{3}He, as we have already seen, the only known such energy is the nuclear dipole-dipole interaction, which has the standard form

$$\hat{H}_D = \mu_n^2 \sum_{ij} (\hat{r}_{ij}^{-3}) \left\{ \hat{\boldsymbol{\sigma}}_i \cdot \hat{\boldsymbol{\sigma}}_j - \frac{3\hat{\boldsymbol{\sigma}}_i \cdot \mathbf{r}_{ij} \hat{\boldsymbol{\sigma}}_j \cdot \mathbf{r}_{ij}}{r_{ij}^2} \right\} \quad (r_{ij} \equiv | \mathbf{r}_i - \mathbf{r}_j |) \tag{7}$$

Evidently this expression is not invariant under rotation of the spins alone, without simultaneous rotation of the orbital coordinates, so in principle it can contribute to the double commutator in eqn. (6). In fact, this equation becomes explicitly:

$$\omega_{res}^2 = \gamma^2 \mathcal{H}_{ext}^2 + \chi_o^{-1} \partial^2 \langle H_D \rangle / \partial \theta_x^2 \tag{8}$$

Comparing this expression with the experimental result, eqns. (2) and (3), we find that in the A phase we must have

$$\frac{\partial^2 \langle H_D \rangle (T)}{\partial \theta_x^2} = K(1 - T/T_A) \quad K \cong 10^{-3} ergs/cm^3 \tag{9}$$

while in the N and B phases the quantity $\partial^2 \langle H_D \rangle / \partial \theta_x^2$ must be zero within the accuracy of the experiment.

Now, it is intuitively rather clear that the quantity $\partial^2 \langle H_D \rangle / \partial \theta_x^2$ cannot be larger in order of magnitude than $\langle H_D \rangle$ itself. Thus, the A-phase shift in the resonance frequency can be explained – but only if the expectation value of the nuclear dipole energy in that phase is of the order of magnitude (9). But how can this be? The maximum possible value (call it g_D) of the dipolar energy of a single pair of spins i, j, obtained by setting r_{ij} in (7) equal to the distance of closest approach (~ 2.5 Å) is only about $10^{-7}K$. Now it is true that if one multiplies g_D by the average density n (which should be a crude measure of the number of pairs which closely approach one another) one gets a quantity of the order of 1 erg/cm^3, more than enough to account for the value (9); in principle, there is enough dipole energy available! The problem, however, lies with the thermal disorder: Given that two ^{3}He nuclei have an orbital separation $\mathbf{r}_{ij}$, the energy advantage of a "right" spin orientation (spins parallel to $\mathbf{r}_{ij}$) over a "wrong" one (spins perpendicular to $\mathbf{r}_{ij}$) is at most only ~ g_D, so

TABLE I. Analogy between SBSOS and Ferromagnetism

FERROMAGNET	LIQUID ^{3}HE
$H = H_0 + H_Z$ ↑ invariant under simult. rotation of all spins	$H = H_0 + H_D$ ↑ invariant under relative rotation of spin + orbital coordinate systems
$H_Z = -\mu_B H \sum_i S_{Zi}$ (extl. field) breaks spin-rot^n symmetry	$H_D = g_D \sum_{ij} \left(\frac{\sigma_i \cdot \sigma_j - 3\sigma_i \cdot r_{ij}\, \sigma_j \cdot r_{ij}}{(r_{ij}^3 / r_0^3)} \right)$, $g_D \equiv \mu_0 \mu_n^2 r_0^3$ breaks relative spin-orbit rot^n symmetry
Paramagnetic phase ($T > T_c$): spins behave independently, kT competes with $\mu_B H \Rightarrow$ polarization $\sim \mu_B H / kT \ll 1 \Rightarrow$ $\langle H_Z \rangle \sim N(\mu_B H)^2/kT$	Normal phase ($T > T_A$): pairs of spins behave independently $\Rightarrow$ polarization $\sim g_D/kT \ll 1 \Rightarrow$ $\langle H_D \rangle \sim N g_D^2/kT$
Ferromagnetic phase ($T < T_c$): H_0 forces all spins to lie parallel $\Rightarrow k_B T$ competes with $N\mu_B H$ $\Rightarrow \langle S_Z \rangle \sim 1 \Rightarrow \langle H_Z \rangle \sim N\mu_B H$	Ordered phase ($T < T_A$): H_0 forces all pairs to behave similarly $\Rightarrow$ kT competes with Ng_D $\Rightarrow \langle H_D \rangle \sim Ng_D$ $\sim 10^{-3}$ ergs/cm^3 !

Table 1. Analogy between SBSOS and Ferromagnetism.

even in a classical model the preference for "right" orientations over "wrong" ones should only be a factor of order g_D/k_BT; this then gives an expectation value of $\langle H_D \rangle$ of order ng_D^2/k_BT, which is too small to account for (9). Actually, the situation is worse than this, since in a degenerate Fermi liquid such as ^{3}He the "polarizability" of the (pairs of) spins by the dipole interaction (as by an external field) is governed not by the thermal energy k_BT but by the much larger Fermi energy k_BT_F, so that the resulting value of $\langle H_D \rangle$ is only of order ng_D^2/k_BT_F. An explicit calculation of $\langle H_D \rangle$ in the Fermi-liquid model of the normal phase confirms this estimate, showing that the associated contribution of the dipole forces to the right-hand side of eqn. (8) is too small for the resulting shift to be seen experimentally, as is indeed the case.[8]

Where could this argument possibly go wrong? I worried over it day and night for a couple of weeks, and then the solution suddenly hit me: it is an idea which I called "spontaneously broken spin-orbit symmetry", or SBSOS for short. It is most easily explained by analogy with the more familiar case of a magnetic system described by the (isotropic) Heisenberg Hamiltonian, the explicit correspondence is given in Table I. The crucial point, in the magnet-

[8] Much more recently, a shift of the predicted order of magnitude in the N phase has in fact been seen, [21], although it is not clear that its detailed properties agree with theoretical expectations.

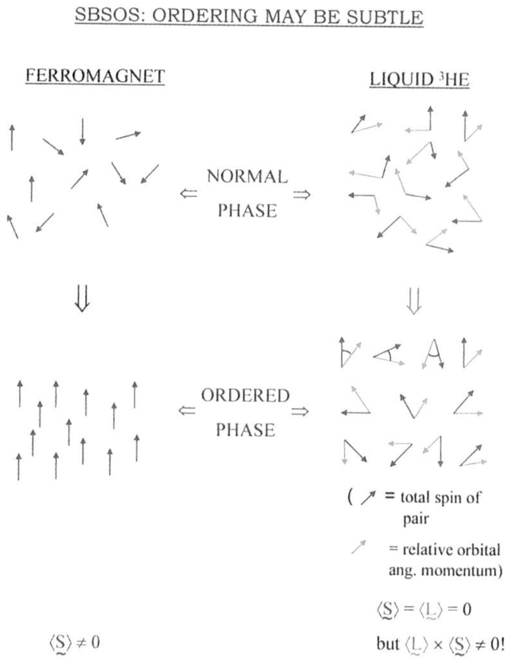

Figure 3: A comparison between ferromagnetism and SBSOS.

ic case, is that in the paramagnetic phase the competition of the thermal energy k_BT is with the single-spin Zeeman energy g_Z; since for reasonable fields we have $g_Z \ll k_BT$, the degree of polarization is small ($\sim g_Z/k_BT \ll 1$) and the expectation value of the total Zeeman energy is $\sim Ng_Z^2/k_BT$, i.e., second order in g_Z (N is the total number of spins). However, in the ferromagnetic phase, the isotropic (Heisenberg) part of the Hamiltonian, $\hat{H}_o$, while it cannot pick out a direction in space, nevertheless forces all (or most of) the spins to lie in the same direction. Now, instead of each spin having to choose individually to lie parallel or antiparallel to the field (corresponding to energies differing by g_Z) we need to choose (crudely speaking) between all spins lying parallel or all lying antiparallel; the energy difference between these two configurations is Ng_Z rather than g_Z, and it is with this large energy that the thermal energy k_BT has to compete. Since typically $Ng_Z \gg k_BT$, the polarization is almost 100%, and the resulting value of the total Zeeman energy is $\sim Ng_Z$, i.e., linear rather than quadratic in g_Z.

In a similar way, we may suppose that in the normal phase of liquid ^{3}He the different pairs of nuclear spins behave more or less independently; then, as sketched above, it is the single-pair energy dipole, g_D, which competes with the thermal energy k_BT, the degree of "polarization" is $\sim g_D/k_BT \ll 1$, and the resulting expectation value $\langle H_D \rangle$ is $\sim Ng_D^2/k_BT$. Now however suppose that in the A phase (on whose specific nature we do not at the moment speculate) the effect of the spin-conserving terms in the energy (kinetic energy, van der Waals potential energy etc.) is such that, while not favoring any particular configuration of the nuclear spins relative to the orbital coordinates, it forces all pairs to have the same configuration. Then, crudely speaking, we are forced to choose between a configuration in which all pairs have the "right" configuration, and one in which they all have the "wrong" one. The corresponding energy difference is now not g_D but Ng_D, so that provided that $Ng_D \gg k_BT$ (or actually $\gg k_BT_F$), as is certainly the case in practice, the "degree of polarization" of the pairs is of the order of 1 and the resulting value of $\langle H_D \rangle$ can in principle be of order Ng_D, as required by the experimental value of the constant K in eqn. (9).

It is worth emphasizing that the concept of SBSOS is in some ways more subtle than the analogous one of the breaking of rotational symmetry in a ferromagnet. In particular, as illustrated in figure 3, it is entirely compatible with the total "spin of the pairs" $\mathbf{S}_{pair}$ and their relative orbital angular momentum $\mathbf{L}_{pair}$ both having expectation value zero (although in general quantities like $\langle (\mathbf{L} \times \mathbf{S})_{pair} \rangle$ may be non-zero). Even in cases in which $\mathbf{S}_{pair}$ is nonzero,[9] this need not automatically of itself imply that the total spin $\mathbf{S}$ of the system is nonzero; $\mathbf{S}_{pair}$ is a measure of the spin correlations of those pairs of atoms which are close in space. It is this consideration, incidentally, which explains the failure of the simple argument, above, which purports to show that the nuclear dipole moments cannot possibly produce the required "Pythagorean" field to explain the A-phase resonance shift: Consider the equation of motion of the total spin of the system in the presence of the external field and the dipole interaction. This has the form

$$\frac{d\mathbf{S}}{dt} \equiv \frac{d}{dt}\sum_i \mathbf{S}_i = \gamma \sum_i \mathbf{S}_i \times \underset{\sim}{\mathcal{H}}_i \tag{10}$$

$$\underset{\sim}{\mathcal{H}}_i \equiv \underset{\sim}{\mathcal{H}}_{ext} + \underset{\sim}{\mathcal{H}}_d, \quad \mathcal{H}_{d\alpha} \equiv \sum_j f_{\alpha\beta}(\mathbf{r}_{ij}) S_{j\beta} \tag{11}$$

where α and β indicate Cartesian components of the spin values. Eqn. (10) is a quantum-mechanical (operator) equation of motion, and we need to take the expectation values of the quantities on the two sides:

$$\frac{d\langle \mathbf{S} \rangle}{dt} = \gamma \left\{ \langle \mathbf{S} \rangle \times \underset{\sim}{\mathcal{H}}_{ext} + \sum_i \langle \mathbf{S}_i \times \underset{\sim}{\mathcal{H}}_i \rangle \right\} \tag{12}$$

The argument given above implicitly assumes that we can make the approximation $\langle \mathbf{S}_i \times \underset{\sim}{H}_i \rangle \cong \langle \mathbf{S}_i \rangle \times \langle \underset{\sim}{H}_i \rangle$ in which case the maximum possible value of any precessional frequency associated with the dipole forces is indeed of the order of γ times the maximum value of $\langle \underset{\sim}{H}_i \rangle$, i.e., $\sim 1\,G$. In a state without SBSOS (and at temperatures such that $k_BT \gg g_D$) the "mean-field" approximation (13) is very good; however, in the presence of SBSOS it fails dramatically, and in fact the second term on the RHS of eqn. (12) can have a substantial value even when the total spin polarization $\mathbf{S} \equiv \Sigma_i \mathbf{S}_i$ is very small.

[9] e.g., the A_1 phase (see e.g., ref. [22] section XIII).

the Anderson-Brinkman-Morel (ABM) phase). This work constitutes a major qualitative leap beyond the simple BCS theory, and was in my view a crucial contribution to the solution of the puzzle of the new phases.

April 1973 was the spring vacation at Sussex, and at the invitation of Bob Richardson I was able to spend the whole month as a visitor at the Laboratory of Atomic and Solid State Physics at Cornell. This month was without doubt the single most exciting month of my more than 40-year academic career: I was alone, and was able to spend sixteen hours a day, seven days a week, thinking about the microscopic basis of the NMR data. I spent much of the time commuting between the basement of Clark Hall, where the experiments were ongoing, and the sixth-floor offices of the theory group, who were of course equally excited about the problem and eager to exchange ideas about it.

The theory I came up with [30] during this month can actually be viewed as a natural generalization of my long-forgotten work on the "internal Josephson effect" in a two-band superconductor. Consider an ESP triplet state: the two spin configurations (↑↑ and ↓↓) are then exactly analogous to the pairs formed, in the superconducting case, in the two bands, and the analog of the number difference ΔN is simply S_z , the z-component of the total (not the Cooper-pair!) spin of the system. What is the analog of the phase difference $\Delta\phi$ between the pairs in the two bands? If we consider for notational simplicity an ABM type state, so that the orbital wave function factors out, the spin state is of the form of a superposition:

$$\Psi_{pair} \sim a|\uparrow\uparrow\rangle + b|\downarrow\downarrow\rangle \equiv a|S_z = +1\rangle + b|S_z \equiv -1\rangle \tag{14}$$

so $\Delta\varphi \equiv arg(b/a)$. But for a spin-1 system, a change in the relative phase of the S_z = 1 and S_z = -1 components of the wave function is just equivalent to a rotation of the spin coordinates (or equivalently to the (inverted) rotation of the spin itself) around the z-axis. Hence, up to a factor of 2, $\Delta\varphi$ is just the angle θ_z of such a rotation. By the same arguments as used in ref. [14], one would then expect that S_z and θ_z satisfy the commutation relation

$$[S_z, \theta_z] = i \tag{15}$$

However, there is nothing special about the z-axis. One would therefore expect[12] that if **S** denotes the total spin vector and the vector $\boldsymbol{\theta}$ specifies the operation of rotation through an angle $|\theta|$ about the axis $\hat{\theta}$, then we should have the generalized commutation relation, valid for an arbitrary pairing state,

$$[\boldsymbol{S}_\alpha, \boldsymbol{\theta}_\beta] = \boldsymbol{i\delta}_{\alpha\beta} \tag{16}$$

[12] The derivation of the dynamics in the original paper uses the standard $\underline{d}$ vector notation (see e.g., ref. [22], section (XI. D.)) and is rigorous. Here I use a somewhat less precise but hopefully intuitive notation and argument.

Now comes the crunch: The commutation relations (16), while rigorous, will lead us to a useful dynamics only if the effective Hamiltonian for the latter can be expressed in terms of **S** and **θ** alone. Is this possible? I argued that it is, for the following reason: We know from experiment that the characteristic frequency associated with the dipole forces[13] (the quantity $\omega_0(T)$) is small compared to the other obvious characteristic frequencies in the problem, namely the "gap frequency"$\Delta(T)/\hbar$ and the (N-phase) quasiparticle relaxation rate τ^{-1}. Thus, during the NMR all the "microscopic" degrees of freedom (e.g., the distribution of normal quasiparticles, or the configuration of the Cooper pairs apart from their overall spin orientation) should follow the macroscopic degrees of freedom **S** and **θ** adiabatically, and one can make a sort of "Born-Oppenheimer" approximation in which the effective Hamiltonian is simply the minimum value of the (free) energy for the given values of those two variables.

Now the only energy, apart from the Zeeman term, which depends on the overall rotation angle of the nuclear spin system is the dipolar energy $H_d(\theta)$, which can be calculated explicitly as a function of θ for any given assumed Cooper-pair configuration (e.g., the ABM phase).[14] As to the **S**-dependent energy, it consists of the Zeeman energy plus a "polarization" energy which in a degenerate Fermi liquid has contributions from kinetic-energy (Pauli-principle) and "exchange" effects; in any event, for a given value of **S** the minimum value of this energy is simply $\frac{1}{2}\gamma^2\chi_0^{-1}\mathbf{S}^2$ where χ_0 is the static spin susceptibility (which is in general a function of temperature, etc.) Thus the total effective Hamiltonian in the Born-Oppenheimer (adiabatic) approximation takes the simple form

$$H(\mathrm{S},\theta) = \frac{1}{2}\gamma^2\chi_0^{-1}\mathrm{S}^2 - \gamma\mathrm{S}\cdot\mathbf{H}(\mathrm{t}) + \mathrm{H_D}(\boldsymbol{\theta}) \tag{17}$$

(where $\mathcal{H}$ is the total external magnetic field, that is in general the sum of the dc field and the rf one). From eqns. (16) and (17) we immediately obtain the equations of motion of **S** and θ:

$$\frac{d\mathrm{S}}{dt} = \mathrm{S}\times\mathcal{H}(\mathrm{t}) - \frac{\partial \mathrm{H_D}}{\partial\boldsymbol{\theta}} \tag{18a}$$

$$\frac{d\boldsymbol{\theta}}{dt} = \mathcal{H}(t) - \chi_0^{-1}\mathrm{S} \tag{18b}$$

Note that in thermal equilibrium $d\boldsymbol{\theta}/\mathrm{dt}$ is zero as it should be. Although strictly speaking the quantities **S, θ** appearing in eqns. (18) are quantum me-

[13] In high fields the Larmor frequency may exceed τ^{-1}, but this does not matter since the relevant motion is simply uniform rotation of the spin system.
[14] It is implicitly assumed that the reference (equilibrium) state is the one which minimizes the expectation value of H_D.

chanical operators, once one has got to them it is an excellent approximation, under all conditions realized to date in liquid ^{3}He, to treat them semiclassically, i.e., to take the expectation value of eqns. (11) and replace $\langle \partial \mathrm{H_D}/\partial \boldsymbol{\theta} \rangle$ by $(\partial/\partial \langle \boldsymbol{\theta} \rangle) \mathrm{H_D}(\langle \boldsymbol{\theta} \rangle)$, thereafter treating $\langle \mathbf{S} \rangle \equiv \mathbf{S}$ and $\langle \boldsymbol{\theta} \rangle \equiv \boldsymbol{\theta}$ as classical variables.[15]

The equations of motion (18) should be exact within the Born-Oppenheimer approximation; it is clear that they are conservative and thus cannot account for the finite linewidths observed experimentally, but in 1973 I left that question for future work.[16] In principle they can be solved for any motion of the spin system which respects this approximation, whether or not it is linear in the rf field. In the linear case, which was at the time the most urgent for the interpretation of the existing experiments, it turns out that in the standard "transverse" geometry the values of the resonance frequencies (there may be more than one of them) are determined by the eigenvalues of the tensor quantity,

$$\Omega_{ij}^2 \equiv \partial^2 \langle \mathrm{H}_D \rangle / \partial\theta_i \partial\theta_j \tag{19}$$

Since the form of this tensor is characteristic of the particular pairing state assumed, we see that the experimental NMR behavior is a "fingerprint" of the pairing state. In particular, let us assume for definiteness $\boldsymbol{l} = 1$ pairing and consider the three most frequently discussed states. For the ABM state, the form of the tensor $\hat{\Omega}$ is such that only a single resonance line is predicted in the NMR absorption, with a frequency which is correctly given by eqn. (31) as derived from the sum rules.[17] For the "axial" state, by contrast, the linearized versions of eqns. (18) predict two resonance lines, with a splitting which decreases rapidly with the dc external field and would have thus been invisible in the experiments existing in April 1973, but should be easily visible at lower fields. Although by this time it had already been shown [33],[26], that within the framework of a generalized Ginzburg-Landau theory the axial phase is never thermodynamically stable, the above observation suggested that it would be desirable to confirm, by NMR "fingerprinting" at lower dc fields, that the A phase shows no splitting of the resonance at any field, and such experiments were rapidly initiated both at Cornell and by Doug Osheroff at Bell Labs, to which he had by that time moved after completing his doctorate.

As regards the BW phase, the predictions made by the linearized versions of eqns. (18) are very intriguing. It turns out that the determination of the correct form (orientation) of the equilibrium state is quite delicate: The $^3P_0$

[15] For a discussion of the NMR behavior under conditions where the semiclassical approximation might fail, see ref. [31].

[16] In fact, in later work with S. Takagi [32] I was able to generalize eqns. (18) so to produce an account of the dissipation.

[17] Strictly speaking this is true provided we assume that the so-called *d*-vector of lies perpendicular to the direction of the rf field. (There is a subtlety connected with eqn. (4) which there is no space to discuss here.)

state originally considered by Vdovin and by BW does not minimize the nuclear dipole energy, and in fact to do this is necessary to rotate the spin coordinates through an angle of $\cos^{-1}(\frac{1}{4}) = 104°$ relative to the orbital one. However, in zero dc external field it is obvious from the isotropy of the $^3P_0$ state that the axis $\hat{\boldsymbol{\omega}}$ of this rotation is completely arbitrary. A finite external field along the z-axis breaks this degeneracy by depopulating, very slightly, the $S_z = 0$ Zeeman component of the pair state relative to the $S_z = \pm 1$ components, and it turns out that this means that it is energetically advantageous to choose $\hat{\boldsymbol{\omega}}$ to be along the (positive or negative) z-axis. The relevant energy, while it thus determines the equilibrium state, is sufficiently small that, at least at first sight, it does not have to be taken into account explicitly in the dynamics, which can thus be calculated by assuming a "pseudo-isotropic" state, that is, one obtained from the (truly isotropic) $^3P_0$ state by rotating the spins relative to the orbital coordinates through $104°$ around the z-axis. Now the structure of the tensor $\hat{\Omega}_{ij}$ in (eqn. (19) for this state is rather striking: A small rotation of the nuclear spin system away from the equilibrium configuration around any axis in the xy-plane corresponds, up to the relevant order, simply to a change in the axis of rotation while keeping the angle fixed, and therefore brings into play only the very small energy which stabilizes $\hat{\Omega}$ in the z-direction; at first sight, at least, it is consistent to neglect this,[18] and the transverse resonance therefore remains at the Larmor value as in the normal phase. On the other hand a rotation around the z-axis involves a change of the angle of the spin-orbit rotation away from the equilibrium value of $104°$, and therefore should bring into play a dipole energy of the same order of magnitude as seen in the A-phase transverse resonance.

But how to see this effect? The answer is to look for a "longitudinal" resonance, that is, a finite-frequency resonance in the absorption spectrum of an rf field polarized parallel rather than perpendicular to the dc field. (In the normal phase, as in any phase lacking SBSOS, this absorption spectrum is simply a diffusive peak centered on zero frequency). The presence of such a resonance, at a frequency which could be estimated with some confidence on the basis of the "calibration" of the dipole energy from the A-phase shift, should when coupled with the absence of a transverse shift identify the B phase almost uniquely as the BW state. (A similar longitudinal resonance is predicted to occur in both the ABM and axial phases).

Although initially neither the Anderson-Brinkman theory of the stability of the ABM phase nor the spin dynamics which I had developed during my stay at Cornell found universal acceptance (cf. ref. [11]), by the late summer of 1973 the smoke had cleared suffciently that most researchers (including me) were convinced that both were correct and that the existing experimental data were entirely consistent with the identification of the A phase as ABM and the B phase as BW. The clinching NMR observations, of a lack of splitting of the A-phase transverse resonance at low fields [35] and of a longitudinal res-

[18] Though cf. ref. [35].

onance at roughly the predicted frequency in the B phase [34], came in within a year or so, and from then on these identifications, while occasionally challenged in the literature, appear to have stood the test of time, and are by now standard textbook material. It should of course be added that the hypothesis of Cooper pairing leads to many other predictions, above all that of superfluid behavior, and that this and the vast majority of other predictions are by now very well confirmed experimentally (see for example the comprehensive text [36] by Vollhardt and Wölfle).

Thirty years later, following the award of the Nobel prize for my participation in these events, one question which I am frequently and legitimately asked by journalists and others is "What is superfluid ^{3}He good for?" Particularly when standing next to my UIUC colleague and co-laureate Paul Lauterbur, whose MRI research has obviously been of such direct and immediate benefit to mankind, I find this question somewhat embarrassing, since in the most direct and practical sense the most honest answer is: nothing whatever! Helium is the most chemically inert of the elements; its application best known to the general public is probably the filling of balloons at children's parties, and its main scientific application is in cryogenics, an area in which the (rare and expensive) light isotope ^{3}He cannot compete with its much more readily available cousin ^{4}He. If this were not enough, the superfluid phases of ^{3}He occur only at less than one hundred-thousandth of room temperature, a regime which is hardly conducive to most kinds of practical application. So while it is conceivable that the "superfluid amplification" property to be discussed below may in the future be put to use in metrology (the determination of the fundamental constants), from all other points of view superfluid ^{3}He may well be the most practically useless system ever discovered.

If we take a broader view, however, and content ourselves with indirect applications, the picture is much rosier. With the arguable exception of the "fractional quantum Hall" systems discovered ten years later, the superfluid phases of liquid ^{3}He are probably the most sophisticated physical systems of which we can claim a quantitative understanding, showing a subtlety of correlation unprecedented in all of known physics; and the lessons learned from them have been very widely applied elsewhere, both in condensed matter physics (for example to the cuprate superconductors, which like ^{3}He are believed to form Cooper pairs in an "exotic " (non-s-wave) pairing state), and in particle physics and cosmology; indeed, whole books(e.g., ref. [37]) have been written on the analogies between various phenomena known experimentally to occur in superfluid ^{3}He and some postulated in particle physics and/or the cosmology of the early universe. A second area in which the uniquely rich structure of the order parameter (pair wave function) of superfluid ^{3}He has had fruitful consequences is in studies of chaos and turbulence, and particularly of the way in which topological defects in the order parameter are generated in quenching through a phase transition (a process which is in fact frequently regarded as a model for processes believed to occur in the early universe).

However, to my mind the single most exciting property characteristic of the superfluid phases of ^{3}He is the "superfluid amplification" associated with the existence of SBSOS. By "superfluid (or Bose) amplification" I mean that the fact that all the Cooper pairs have to behave identically has the consequence that tiny effects which in a normal system would be completely quenched by thermal disorder may become spectacularly visible. In itself this property is of course not peculiar to ^{3}He; in a BCS superconductor the Josephson effect and, less obviously, the Meissner effect itself can be regarded as illustrations of this behavior. What is new and unique[19] about the superfluid ^{3}He system is the fact that the Cooper pairs have one or more nontrivial internal (orientational) degrees of freedom, and must all behave identically with respect to these as well as to the center-of-mass motion (i.e., most show the property of SBSOS). As a result, the phenomenon of superfluid amplification occurs also in properties associated with the internal (relative) motion. We already saw one striking example of this in the NMR behavior; I would like to close with a second potential application of this principle which, if it can be realized experimentally, should be even more spectacular.

Unlike the gravitational and electromagnetic interactions, which have direct and obvious manifestations at the everyday (macroscopic) level, the weak interaction postulated in particle physics has so far manifested itself directly only at the microscopic level. One striking property of this interaction which distinguishes it from the other known interactions (gravitational, electromagnetic and strong) is that it violates the principle of invariance under spatial inversion (P); the consequences of this violation have been seen in scattering experiments, and more recently in the optical behavior of heavy atoms. It is interesting to ask: Is it possible to see the effects of this P-violation at the macroscopic level? In discussing this question it is important to bear in mind that, except for a very small "CP-violating" component (of order 10^{-3} of the total), the weak interaction respects the principle of invariance under time reversal (T). As a result, if we ignore for present purposes the small CP-violating component, the weak interaction cannot give rise to an electric dipole moment (EDM) on any "simple" quantum system (elementary particle, atom or molecule). The reason is as follows: by the Wigner-Eckart theorem, such an EDM (**d**) would have to be proportional to the only vector quantity characterizing this system, namely the total angular momentum **J**. But the ansatz **d** = const. **J** evidently violates T as well as P, and hence (in the absence of CP-violation) is unviable.

Consider however the Cooper pairs in the (pseudo-isotropic) B phase of superfluid ^{3}He. Because of the nonlinearity of the generalized BCS equation which describes the formation of pairs, the pseudo-isotropic state (as distinct from the original $^3P_0$ state of BW and Vdovin) is not characterized by a well-defined angular momentum $\mathbf{L}+\mathbf{S} \equiv \mathbf{J}$. However, it does possess a characteris-

[19] At least until the recent advent of BEC alkali gases with a hyperfine degree of freedom.

tic vector which has a finite expectation value in the ground state, namely the vector $\langle \hat{\mathbf{L}} \times \hat{\mathbf{S}} \rangle$, which turns out to be directed along the "spin-orbit rotation axis" $\hat{\omega}$ (see above). Since both **L** and **S** are axial vectors and odd under time reversal, the ansatz

$$\mathbf{d} = \text{const.}\ \langle (\mathbf{L} \times \mathbf{S}) \rangle$$

violates P but not T, and is therefore allowed by symmetry even in the absence of CP-violation.

A calculation [38] of the order of magnitude of the EDM expected to be produced by the weak interaction in ^{3}He-B indicates that it involves a number of dimensionless factors which are each individually small in this system, and that as a result the EDM on a single Cooper pair is something like ten orders of magnitude smaller than the smallest EDM which would be visible (if realized on, e.g., the neutron) in current beam experiments. However, now comes the crunch: because of the property of SBSOS, all the Cooper pairs must have exactly the <u>same</u> value of $\langle \mathbf{L} \times \mathbf{S} \rangle$ and hence of **d**, so that the total EDM of the system is in principle macroscopic! Although calculations indicate it is still extremely small, it is not obvious beyond the reach of existing measurement techniques, and Doug Osheroff at Stanford is currently building an experiment to look for it. If it can be seen, it will be the first ever example of a <u>direct</u> manifestation of the weak interaction, and in particular of its characteristic property of P-violation, at the level of everyday life.

In conclusion I would first like to thank those who did the hard work involved in nominating me for the Nobel prize, and the Physics Committee of the Royal Swedish Academy of Sciences for the even harder work which eventually resulted in my selection; I am very conscious that I am only one of a number of researchers whose work both before and after the experimental discovery helped to clarify the nature of the new phases, and in this context it is impossible not to mention one name in particular, that of Phil Anderson, who with various collaborators contributed so many vital insights during these years and later. Secondly, I would like to thank all those who in one way or another helped me along the path which has led me to stand here today. In particular I am grateful to the late Fr. Charles O'Hara for introducing me to the marvels of modern mathematics and giving me the confidence that I could do it if I ever had to; to David Brink and Michael Baker for accepting me, a classics graduate with zero formal training in physics, to do a second undergraduate degree in that subject, and to the Fellows of Merton College, Oxford, for providing the means for me to do it: to the late Dirk ter Haar for accepting me–still the proverbial dark horse–to do an advanced degree in theoretical physics, and to the faculty of Magdalen College, Oxford, for electing me to a fellowship which considerably more than kept body and soul together while I did it and thereafter; and to David Pines for choosing me as his postdoc at University of Illinois at Urbana-Champaign, and for his encouragement in my first ventures into the area of superfluid ^{3}He. I owe special thanks to Bob Richardson, Doug Osheroff, Dave Lee and Willy Gully for their generosity in sharing their data with me long before publication, and for innumerable, fruitful and inspiring discussions; and to David Mermin, Vinay Ambegaokar, Joe Serene and other members of the Cornell theoretical group for their help, particularly (but not only) with the technical details of the calculation. In this context I should also mention Shin Takagi, who while not around for the period I have reviewed became an invaluable collaborator in my later work on superfluid ^{3}He.

Finally, I owe a huge debt to the various people who have supported me in my research over the years, whether, like my former colleagues at Sussex and my current ones at Illinois, by providing a congenial academic environment, or in less direct but equally important ways through their friendship and encouragement. In this last category I am especially grateful to my wife, Haruko, to whom I was getting engaged just as all the excitement I have described was happening, and who nevertheless tolerated and even encouraged my going off by myself to Cornell for that vital month–and who, with my daughter Asako, has been equally supportive of my research, and tolerant of its sometimes unreasonable-seeming demands, over the last 30 years. To all of you, my sincere and heartfelt thanks.

REFERENCES

[1] L.D. Landau, Zh. Eksp. Teor. Fiz. **30**, 1058 (1956): translation. *Sov. Phys. JETP* **3**, 920 (1957).

[2] J. Bardeen, L.N. Cooper and J.R. Schrieffer, *Phys. Rev.* **108**, 1175 (1957).

[3] P.W. Anderson and P. Morel, *Phys. Rev.* **123**, 1911 (1961).

[4] L.P. Gor'kov and V.M. Galitskii, *Zh. Eksp. Teor. Fiz.* **40**, 1124 (1961): translation, *Sov. Phys. JETP* **13**, 792 (1962).

[5] D. Hone, *Phys. Rev. Letters* **8**, 370 (1962).

[6] R. Balian, L.H. Nosanow and N.R. Werthamer, *Phys. Rev. Letters* **8**, 372 (1962).

[7] V.R. Chechetkin, *Fiz. Nizk. Temp.* **8**, 41 (1982): translation, *Sov. J. Low Temp. Phys.* **7**, 1982.

[8] S. Yip, *Phys. Letters* **105A** 66 (1984).

[9] Yu. A. Vdovin, in Application of Methods of Quantum Field Theory to Problems of Many Particles, ed. A.I. Alekseyeva (Moscow, Gosatomizdat 1963) (in Russian).

[10] R. Balian and N.R. Werthamer, *Phys. Rev.* **131**, 1553 (1963).

[11] A.J. Leggett, *Physica* 109 B & C. 1393 (1982).

[12] P. Fulde and R.A. Ferrell, *Phys. Rev.* **135A**, 550 (1964).

[13] A. I. Larkin, *Zh. Eksp. Teor. Fiz.* **46**, 2188 (1964): translation, *Sov. Phys. JETP* **19**, 1478 (1964).

[14] A.J. Leggett, *Prog. Theor. Phys.* **36**, 901 (1966).

[15] Ya. G. Ponomarev *et al.*, cond-mat/0303640

[16] D.D. Osheroff, R.C. Richardson and D.M. Lee, *Phys. Rev. Letters* **28**, 885 (1972).

[17] W.P. Halperin, C.N. Archie, F.B. Rasmussen, R.A. Buhrman and R.C. Richardson, *Phys. Rev. Letters* **32**, 927 (1974).

[18] D.D. Osheroff, W.J. Gully, R.C. Richardson and D.M. Lee, *Phys. Rev. Letters* **29**, 920 (1973).

[19] D.N. Paulson, H. Kojima and J.C. Wheatley, *Phys. Lett.* **47**, 457 (1974).

[20] D. Pines and P. Nozières, Theory of Quantum Liquids, vol. I, Benjamin, New York 1966, section 2.2.

[21] T.M. Haard, J.B. Kycia, M.R. Rand, H.H. Hensley, Y. Lee, P.J. Hamot, D.T. Sprague and W.P. Halperin, *Phys. Rev. Letters* **72**, 860 (1994).

[22] A.J. Leggett, *Revs. Mod. Phys.* **47**, 331 (1975).
[23] A.J. Leggett, *Phys. Rev. Letters* **29**, 1227 (1972).
[24] A.J. Leggett, *J. Phys. C* **6**, 2187 (1973).
[25] P.W. Anderson, *Phys. Rev. Letters* **30**, 368 (1973).
[26] P.W. Anderson and W.F. Brinkman, *Phys. Rev. Letters* **30**, 1108 (1973).
[27] A. Layzer and D. Fay, *Int. J. Magn.* **1**, 135 (1971).
[28] W.F. Brinkman, J. Serene and P.W. Anderson, *Phys. Rev. A* **10**, 2386 (1974).
[29] D.N. Paulson, R.T. Johnson and J.C. Wheatley, *Phys. Rev. Letters* **31**, 746 (1973).
[30] A.J. Leggett, *Phys. Rev. Letters* **31**, 352 (1973).
[31] A.J. Leggett, Synthetic Metals, in press.
[32] A.J. Leggett and S. Takagi, *Ann. Phys.* **106**, 79 (1977).
[33] N.D. Mermin and G. Stare, *Phys. Rev. Letters* **30**, 1135 (1973).
[34] H.M. Bozler, M.E.R. Bernier, W.J. Gully, R.C. Richardson and D.M. Lee, *Phys Rev. Letters* **32**, 875 (1974).
[35] D.D. Osheroff, *Phys. Rev. Letters* **33**, 1009 (1974).
[36] D. Vollhardt and P. Wölfle, The Superfluid Phases of Helium -3, Taylor and Francis, London 1990.
[37] G.E. Volovik, The Universe in a Helium Droplet, Clarendon Press, Oxford, 2003.
[38] A.J. Leggett, *Phys. Rev. Letters* **39**, 587 (1977).

Physics 2004

DAVID J. GROSS, H. DAVID POLITZER and FRANK WILCZEK

"for the discovery of asymptotic freedom in the theory of the strong interaction"

THE NOBEL PRIZE IN PHYSICS

Speech by Professor Lars Brink of the Royal Swedish Academy of Sciences.
Translation of the Swedish text.

Your Majesties, Your Royal Highnesses, Ladies and Gentlemen,

When Isaac Newton saw the apple fall he understood how gravity works and he was able to formulate a law for it. He applied this law to explain the orbit of the moon travelling around the earth. At the same time he could explain how the force of gravity between two bodies diminishes when the distance between them increases. Later, it was discovered that there is a similar decrease of the electric force between two bodies carrying electric charges as they are separated from each other. But what is a force? How do forces arise? How can they act in empty space – in a vacuum? Almost exactly a hundred years ago, Albert Einstein realized that light can be described as a bunch of particles, called photons, which move from one electrically charged body to another and mediate the electromagnetic force between the bodies. They mediate a force in the same way as the football that the goalkeeper cannot stop or as the cannon ball that can destroy a tower of a fortress. We can also remind ourselves of how Baron von Münchhausen threw himself and grabbed a cannon ball to travel over a wall.

Why does the force diminish with increasing distance? The cannon ball meets resistance from the air but the force due to the motion of photons also decreases in vacuum. This is because according to quantum theory a beam of photon particles can also be thought of as a wave (as in the standard description of light). The further away from the source you are, the smaller the section of the wave that will hit you. When the wave that appeared when Krakatoa exploded in 1883 hit the coastline of Sumatra it caused enormous damage, but when it hit the coast of Africa much further away it was just a swell in the water. We can also understand the force of gravity in the same way as the electromagnetic force. We have not been able to discover the force particles here, but we are convinced that they exist.

At the same time as the physicist wants to understand the fundamental forces that act in Nature, he or she wants to understand what the fundamental building blocks are. We have divided up matter into atoms, which have been further subdivided into electrons and the nucleus, which is made of protons and neutrons. It became clear very early on that within the nuclei there must be different interactions, the weak nuclear force responsible for the radioactive decays, and the strong nuclear force holding the nucleus together even when the repulsive electric force between the protons tries to separate them. These forces act over very short ranges, ranges as improbably small as the nucleus itself. To understand these forces and to understand the fundamental building blocks of Nature have been the great task of particle physics for the last fifty years. This year's Nobel Prize completes the picture that the work behind several earlier prizes initiated and as a result we now know the fundamental building blocks and we have a description of the four fundamental forces.

One of these earlier discoveries was the understanding that protons and neutrons are composite objects, which are made of even more fundamental particles called quarks. It was found that quarks come in various species that have different types of charges, which came to be called colour charges. These charges are in some sense analogous to electric charges, so it was natural to believe that the quarks should behave like the electrons. However, unlike electrons, no free quarks were ever discovered. Perversely, it seemed as if the force increased as quarks were separated. Conversely, plenty of evidence accrued that when two quarks got close to each other they could hardly feel each other. This behaviour is called "asymptotic freedom." Could a theory of the quarks behave like that? This behaviour of the force between quarks seemed to be outside the realm of any theory of the kind that had so successfully explained the electromagnetic force. So around 1970 particle physics stood in front of a great dilemma. Common sense, and all calculations that were attempted told us that the force between quarks should behave in a manner that contradicted the experimental facts. In the end the issue came down to one specific question. Does any theory predict a minus sign in the right place? All theories that were tested gave the incorrect positive sign.

In 1973, David Gross and Frank Wilczek and David Politzer considered a novel class of theories. To the surprise of the world and to their own great astonishment they found the result –11/3 that signalled that these theories are asymptotically free. Seldom has a negative result had such a positive effect! A theory for the strong force between the quarks could now quickly be formulated and a detailed comparison with experiments could be performed. During the last fifteen years experiments at large accelerators have confirmed the theory with great accuracy. The theory of Gross, Politzer and Wilczek successfully describes the physics of quarks, the matter from which we are to a very large extent built.

Since the discovery, further research has shown that these theories are unique. No other theories can account for the experimental picture and it is wonderful to know that Nature has chosen the only theory that we have found to be possible.

Professor Gross, Professor Politzer, Professor Wilczek,

You have been awarded the 2004 Nobel Prize in Physics for your discovery of asymptotic freedom in the theory of the strong interactions. It is an honour for me to convey the warmest congratulations of the Royal Swedish Academy of Sciences. I now ask you to step forward to receive your Nobel Prizes from the hands of His Majesty the King.

David Gross

DAVID J. GROSS

I was born in Washington, D.C., on February 19, 1941, the eldest of four sons. My father, Bertram Meyer, born in Philadelphia, son of immigrant Jewish parents from Czechoslovakia-Hungary, had attended the University of Pennsylvania as an English major. From 1941 a staff member of U.S. Sen. James E. Murray of Montana, he helped after the war to write the Employment Act of 1946. This law established a commitment to maintain high levels of employment and a charter for the President's Council of Economic Advisers (CEA). He served as the first assistant to the chairman after the CEA was established in 1948. My mother, Nora (Faine), was born in the Ukraine. In 1914 her father immigrated to the United States and was joined by his wife and two children only after the end of World War I. Remarkably, both children made their way to college; my uncle completed Harvard Law School, and my mother graduated from Barnard. She majored in chemistry, but never pursued a career in science, an unimaginable goal for women in the midst of the Depression.

My childhood in Arlington, Va., a middle class suburb of Washington, was uneventful. Ours was a very intellectual family, and we were encouraged to read at a very early age. My father would always try to interest his children (and everyone around him) in his intellectual passions. At the age of 11, I was recruited to proofread his book, The Legislative Struggle: A Study in Social Combat, which he later used to launch an academic career. I was paid 10 cents per page, and by the time I finished, I hated the word "committee". Dinner conversation – always intense – dealt with adult issues, mostly politics and issues of social justice. My father and mother treated us children as intellectual equals, thus greatly bolstering our self-confidence and our interest in ideas of all kinds.

In 1953 my father joined a group of Democratic exiles from the Eisenhower administration who were sent to Israel as an advisory team together with the first US aid package. Israel, then struggling to absorb half a million refugees from Europe and construct a new state, accepted the money and the advice (but was wise enough to disregard most of the advice). After the team returned in 1955, my father remained and joined the Hebrew University, where he set up the School of Business Administration. Jerusalem in the 1950's, a small, divided city, the seat of the government and the University, contained a remarkable mixture of religious fanatics and refugees from all over the world. It was a far cry from Arlington, Va.

I was thrust into school with no knowledge of Hebrew, and for the first time in my life had to struggle to compete with my peers. One of the great advantages of growing up in Jerusalem at that time was the absence of television and the many other distractions present then and now in the United States. Thrown back onto my own resources, I became an avid reader. My brother Larry and I rapidly went through the library run by the United States Information Agency at the American Consulate in Jerusalem. Then, when my father became a professor at the Hebrew University, we had access to the University Library on Mount Scopos. From the age of 13, I was attracted to physics and mathematics. My interest in these subjects derived mostly from popular science books that I read avidly. Early on I was fascinated by theoretical physics and determined to become a theoretical physicist. I had no real idea what that meant, but it seemed incredibly exciting to spend one's life attempting to find the secrets of the universe by using one's mind. I studied mathematics on my own and soon, exceeding the knowledge of my teachers, was excused from mathematics classes. My high school education was excellent, with many highly qualified teachers, who today would be employed in higher education or in industry.

Determined to become a theoretical physicist, upon graduating from High School, I entered the Hebrew University and majored in physics and mathematics. The physics education was reasonably good, and the mathematics education was excellent. After receiving a B.Sc., I applied to graduate schools in the United States. At that time (but soon to change) students from Israel were unheard of, and I was only accepted at the University of California at Berkeley. In retrospect this was fortunate. Berkeley was, for many reasons, at that time a very exciting place. It was at the center of the remarkable social and political developments of the 1960's – the fight for social justice and the opposition to the war in Vietnam. But more importantly for me, Berkeley was the center of elementary particle physics, with new discoveries appearing monthly at, what was then called, the "Rad Lab" (short for "Berkeley Radiation Laboratory"), before "radiation" became a dirty word, and the name was changed to the "Lawrence Berkeley Laboratory." In 1964, I started to do research under the supervision of Geoffrey Chew, the charismatic leader of the S-Matrix "bootstrap" approach to the strong interactions. I found this revolutionary new theory very exciting at first, but gradually became disillusioned. I rapidly finished a thesis and spent most of my last year at Berkeley in thoughts of new directions.

Concluding my graduate studies in 1966, I was nominated to the Harvard Society of Fellows, a wonderful institution that appointed eight fellows each year in many different fields. The only responsibilities were to dine together with the senior fellows once each week. Harvard was a bit of a shock after Berkeley. Schwinger had long dominated the theoretical scene, but now a new generation of theorists was just appearing. In 1966, Shelley Glashow and Sidney Coleman had returned to Harvard and Steve Weinberg was visiting. After Berkeley I found Harvard rather formal and unfriendly, but the Society of Fellows was a wonderful experience. I was fortunate to have as junior colleagues Curtis Callan and Roman Jackiw, with whom I had very productive collaborations. This is where I started on the path that led to the discovery of asymptotic freedom and QCD (the story told in my Nobel lecture.)

In 1969, I had many offers for faculty positions. I decided to go to Princeton, as an assistant professor, where I was to stay for 27 years. Murph Goldberger and Sam Treiman were my senior theoretical colleagues at Princeton. They created a marvelous atmosphere to do research and were enormously supportive of me. They promoted me to tenure in 1971, before I had even begun to think of the possibility. Thus I was able to pursue my research free of the anxiety that often plagues young scientists today.

Between the University and the Institute for Advanced Study, there was a remarkable group of young theorists. I collaborated with John Schwarz and Andre Neveu on string theory for a while, but mostly with Curtis Callan and with Roger Dashen on gauge theory. At Princeton I was lucky to have had a remarkable group of graduate students, especially Frank Wilczek and Edward Witten. This was fortunate since I have always enjoyed collaborating with other physicists and students rather than working in solitude. I find that my best ideas emerge in heated exchanges with others. My way of dealing with students, then and now, was to involve them closely with my current work and very often to work with them directly. I always found it difficult to come up with trivial or easy problems that I could assign to a student and preferred to involve students in the much more ambitious projects that I myself was working on. Such collaborations were no doubt good training for my students, but were also very beneficial for me.

After the discovery of asymptotic freedom and the emergence of QCD, I spent many years on the dynamics of gauge theories in the attempt to solve QCD. Much progress was made, but the goal of a controllable analytic solution of the theory was not realized. By the beginning of the 1980's, I along with many others had shifted my attention to more speculative physics – the challenge of unifying all the forces of nature and confronting quantum gravity. In the early 1980's, I began to work on string theory again and in 1984, together with a group of younger collaborators, discovered the heterotic string, which at the time seemed to offer the possibility of explaining the Standard Model from string theory. Since then I have been working largely on string theory. Although remarkable advances have been made, the potential of this theory to unify all the forces of nature and to make contact with experiment has not yet been realized.

In 1996, I was offered the directorship of the Institute for Theoretical Physics in Santa Barbara (for the second time). I was ready for new challenges and attracted by the opportunity to come to the renowned ITP. Having spent thirty years teaching and doing research, I was ready to take on other responsibilities. The directorship of the ITP was a wonderful opportunity since most of the work consisted of scientific leadership and not administration, and I could continue a vigorous research activity as well. I have not regretted this decision and have greatly enjoyed directing the Institute and participating in the remarkable span of science that takes place at its programs.

I have two children with my first wife, Shulamith (Toaff): Ariela Gross, who is an historian and professor of law at the University of Southern California and the mother of my grandchildren, Raphaela and Sophia; and Elisheva

Gross, who is completing her doctorate in psychology at the University of California at Los Angeles. I now live with my second wife, Jacquelyn Savani, and my stepdaughter, Miranda Savani, in Santa Barbara, California.

HONORS

Alfred P. Sloan Foundation Fellow, 1970–74
Fellow, American Physical Society, elected 1974
Fellow, American Academy of Arts and Sciences, elected 1985
Member, National Academy of Sciences, elected 1986
Fellow, American Association for the Advancement of Science, elected 1987
Recipient, J. J. Sakurai Prize of the American Physical Society, 1986
Recipient, MacArthur Foundation Fellowship Prize, 1987
Recipient, Dirac Medal, 1988
Docteur Philosophiae Honoris Causa, University of Montpellier, 2000
Recipient, Oskar Klein Medal, Stockholm University, 2000
Recipient, Harvey Prize, Technion-Israel Institute of Technology, 2000
Doctor Philosophiae Honoris Causa, Hebrew University, Jerusalem, 2001
Recipient, European Physical Society Prize in Elementary Particle Physics, 2003
Recipient, Grande Medaille D'Or, Academie des Sciences, France, 2004
Recipient, Nobel Prize in Physics, 2004

THE DISCOVERY OF ASYMPTOTIC FREEDOM AND THE EMERGENCE OF QCD

Nobel Lecture, December 8, 2004

by

DAVID J. GROSS

Kavli Institute for Theoretical Physics, UCSB, Santa Barbara, California, USA.

INTRODUCTION

The progress of science is much more muddled than is depicted in most history books. This is especially true of theoretical physics, partly because history is written by the victorious. Consequently, historians of science often ignore the many alternate paths that people wandered down, the many false clues they followed, the many misconceptions they had. These alternate points of view are less clearly developed than the final theories, harder to understand and easier to forget, especially as these are viewed years later, when it all really does make sense. Thus reading history one rarely gets the feeling of the true nature of scientific development, in which the element of farce is as great as the element of triumph.

The emergence of QCD is a wonderful example of the evolution from farce to triumph. During a very short period, a transition occurred from experimental discovery and theoretical confusion to theoretical triumph and experimental confirmation. In this Nobel lecture I shall describe the turn of events that led to the discovery of asymptotic freedom, which in turn led to the formulation of QCD, the final element of the remarkably comprehensive theory of elementary particle physics – the *Standard Model*. I shall then briefly describe the experimental tests of the theory and the implications of asymptotic freedom.

PARTICLE PHYSCIS IN THE 1960'S

The early 1960's, when I started my graduate studies at UC Berkeley, were a period of experimental supremacy and theoretical impotence. The construction and utilization of major accelerators were proceeding at full steam. Experimental discoveries and surprises appeared every few months. There was hardly any theory to speak of. The emphasis was on phenomenology, and there were only small islands of theoretical advances here and there. Field theory was in disgrace; S-Matrix theory was in full bloom. Symmetries were all the rage. Of the four forces observed in nature, only gravity and electromagnetism were well understood. The other two forces, the weak force responsible for radioactivity and the strong nuclear force that operated within the nucleus, were largely mysterious. Particle physics was divided into the study of the weak and the strong interactions, the two mysterious forces that operate within the nucleus. In the case of the weak interactions, there was a rather successful phenomenological theory, but not much new data. The strong interactions were where the experimental and theoretical action was, particularly at Berkeley. They were regarded as especially unfathomable. In hindsight this was not surprising since nature was hiding her secrets. The basic constituents of hadrons (strongly interacting particles) were invisible. We now know that these are quarks, but no one had ever seen a quark, no matter how hard protons were smashed into protons. Furthermore, the "color" charges we now know are the source of the Chromodynamic fields, the analogs of the electric charge, were equally invisible. The prevalent feeling was that it would take a very long time to understand the nuclear force and that it would require revolutionary concepts. Freeman Dyson had asserted that "*the correct theory will not be found in the next hundred years*". For a young graduate student, such as myself, this was clearly the biggest challenge.

QUANTUM FIELD THEORY

Quantum field theory was originally developed for the treatment of Electrodynamics, immediately after the completion of quantum mechanics and the discovery of the Dirac equation. It seemed to be the natural tool for describing the dynamics of elementary particles. The application of quantum field theory to the nuclear forces had important early success. Fermi formulated a powerful and accurate phenomenological theory of beta decay, which (though deficient at high energy) was to serve as a framework for exploring the weak interactions for three decades. Yukawa proposed a field theory to describe the nuclear force and predicted the existence of heavy mesons, which were soon discovered. On the other hand quantum field theory was confronted from the beginning with severe difficulties. These included the infinities that appeared as soon as one went beyond lowest order perturbation theory, as well as the lack of any non-perturbative tools. By the 1950's the suspicion of field theory had deepened to the point that a powerful dogma emerged – that field theory was fundamentally wrong, especially in its application to the strong interactions.

The renormalization procedure, developed by R. Feynman, J. Schwinger, S. Tomanaga and F. Dyson, which had eliminated the ubiquitous infinities that occurred in calculations by expressing physical observables in terms of physical parameters, was spectacularly successful in Quantum Electrodynamics. However, the physical meaning of renormalization was not truly understood. The feeling of most was that renormalization was a trick. This was especially the case for the pioneering inventors of quantum field theory. They were prepared at the first appearance of divergences to renounce their belief in quantum field theory and to brace for the next revolution. However it was also the feeling of the younger leaders of the field, who had laid the foundations of perturbative quantum field theory and renormalization in the late 1940's.

The prevalent feeling was that renormalization simply swept the infinities under the rug, but that they were still there and rendered the notion of local fields meaningless. To quote Feynman, speaking at the 1961 Solvay conference [1], *"I still hold to this belief and do not subscribe to the philosophy of renormalization"*.

Field theory was almost totally perturbative at that time; all nonperturbative techniques that had been tried in the 1950's had failed. The path integral, developed by Feynman in the late 1940's, which later proved so valuable for a nonperturbative formulation of quantum field theory as well as a tool for semiclassical expansions and numerical approximations, was almost completely forgotten. In a sense, the Feynman rules were too successful. They were an immensely useful, picturesque and intuitive way of performing perturbation theory. However, these alluring qualities also convinced many that all that was needed from field theory were these rules. They diverted attention from the non-perturbative dynamical issues facing field theory. In my first course on quantum field theory at Berkeley in 1965, I was taught that Field Theory = Feynman Rules. Today we know that there are many phenomena, especially confinement in QCD, that cannot be understood perturbatively.

In the United States, the main reason for the abandonment of field theory for the strong interactions was simply that one could not calculate. American physicists are inveterate pragmatists. Quantum field theory had not proved to be a useful tool with which to make contact with the explosion of experimental discoveries. The early attempts in the 1950's to construct field theories of the strong interactions were total failures. In hindsight this was not surprising since a field theory of the strong interactions faced two enormous problems. First, which fields to use? Following Yukawa, the first attempts employed pion and nucleon fields. Soon, with the rapid proliferation of particles, it became evident that nothing was special about the nucleon or the pion. All the hadrons, the strange baryons and mesons as well as the higher spin recurrences of these, appeared to be equally fundamental. The obvious conclusion that all hadrons were composites of more fundamental constituents was thwarted by the fact that, no matter how hard hadrons were smashed into one another, one had not been able to liberate these hypothetical constituents. This was not analogous to the paradigm of atoms made of nucleons and electrons or of nuclei composed of nucleons. The idea of permanently bound, confined, constituents was unimaginable at the time. Second, since the pion-nucleon coupling was so large, perturbative expansions were useless, and all attempts at non-perturbative analysis were unsuccessful.

In the case of the weak interactions, the situation was somewhat better. Here one had an adequate effective theory, the four-fermion Fermi interaction, which could be usefully employed, using perturbation theory to lowest order, to organize and understand the emerging experimental picture of the weak interactions. The fact that this theory was non-renormalizable meant that beyond the Born approximation it lost all predictive value. This disease increased the suspicion of field theory. Yang-Mills theory, which had appeared in the mid 1950's, was not taken seriously. Attempts to apply Yang-Mills

strings in 1972. I worked on this theory for two years, at CERN and then at Princeton with Schwarz and Neveu. At first I felt that this model, which captured many of the features of hadronic scattering, might provide the long sought alternative to a field theory of the strong interactions. However by 1971 I realized that there was no way that this model could explain scaling, and I felt strongly that scaling was the paramount feature of the strong interactions. In fact the dual resonance model lead to incredibly soft behavior at large momentum transfer, quite the opposite of the hard scaling observed. Also, it required for consistency many features that were totally unrealistic for the strong interactions – massless vector and tensor particles. These features later became the motivation for the hope that string theory may provide a comprehensive and unified theory of all the forces of nature. This hope remains strong. However the relevant energy scale is not 1 GeV but rather 10^{19} GeV!

The data on deep inelastic scattering were getting better. No violations of scaling were observed, and the free-field-theory sum rules worked. I remember well the 1970 Kiev conference on high-energy physics. There I met S. Polyakov and S. Migdal, uninvited but already impressive participants at the meeting. Polyakov, Migdal and I had long discussions about deep inelastic scattering. Polyakov knew all about the renormalization group and explained to me that naive scaling cannot be right. Because of renormalization the dimensions of operators change with the scale of the physics being probed. Not only that, dimensionless couplings also change with scale. They approach at small distances fixed point values that are generically those of a strongly coupled theory, resulting in large anomalous scaling behavior quite different from free field theory behavior. I retorted that the experiments showed otherwise. He responded that this behavior contradicts field theory. We departed; he convinced, as many were, that experiments at higher energies would change, I that the theory would have to be changed. The view that the scaling observed at SLAC was not a truly asymptotic phenomenon was rather widespread. The fact that scaling set in at rather low momentum transfers, "precocious scaling," reinforced this view. Thus the cognoscenti of the renormalization group (Wilson, Polyakov, and others) believed that the non-canonical scaling indicative of a non-trivial fixed point of the renormalization group would appear at higher energies.

Much happened during the next two years. Gerhard 't Hooft's spectacular work on the renormalizability of Yang-Mills theory reintroduced non-Abelian gauge theories to the community [27]. The electroweak theory of S. Glashow, S. Weinberg and A. Salam was revived. Field theory became popular again, at least in application to the weak interactions. The path integral reemerged from obscurity. Kenneth Wilson's development of the operator product expansion [28] provided a tool that could be applied to the analysis of deep inelastic scattering. The Callan-Symanzik equations simplified the renormalization group analysis, which was then applied to the Wilson expansion [29]. The operator product analysis was extended to the light cone, the relevant region for deep-inelastic scattering [30]. Most important from my point of view was the revival of the renormalization group by Wilson [31]. The renormalization group stems from the fundamental work of Gell-Mann and Low, Stueckelberg and Petermann, and Bogoliubov and Shirkov [32]. This work was neglected for many years, partly because it seemed to provide only information about physics for large space-like momenta, which are of no direct physical interest. Also, before the discovery of asymptotic freedom, the ultraviolet behavior was not calculable using perturbative methods, and there were no others. Thus it appeared that the renormalization group provided a framework in which one could discuss, but not calculate, the asymptotic behavior of amplitudes in a physically uninteresting region.

THE PLAN

By the end of 1972, I had learned enough field theory, especially renormalization group methods, to tackle the problem of scaling head on. I decided, quite deliberately, to prove that local field theory could not explain the experimental fact of scaling and thus was not an appropriate framework for the description of the strong interactions. Thus, deep inelastic scattering would finally settle the issue as to the validity of quantum field theory. The plan of the attack was twofold. First, I would prove that "Ultraviolet Stability," the vanishing of the effective coupling at short distances, later called asymptotic freedom, was necessary to explain scaling. Second, I would show that there existed no asymptotically free field theories. The latter was to be expected. After all the paradigm of quantum field theory, Quantum Electrodynamics (QED), was infrared stable; the effective charge grew larger at short distances and no one had ever constructed a theory in which the opposite occurred. If the effective coupling were, contrary to QED, to decrease at short distances, one might explain how the strong interactions turn off in this regime and produce scaling. Indeed, one might suspect that this is the only way to get point-like behavior at short distances. It was well understood, due to Wilson's work and its application to deep inelastic scattering, that one might expect to get scaling in a quantum field theory at a fixed point of the renormalization group. However this scaling would not have canonical, free field theory, behavior. Such behavior would mean that the scaling dimensions of the operators that appear in the product of electromagnetic currents at light-like distances had canonical, free field dimensions. This seemed unlikely. I knew that if the fields themselves had canonical dimensions, then for many theories this implied that the theory was trivial, i.e., free. Surely this was also true if the composite operators that dominated the amplitudes for deep-inelastic scattering had canonical dimensions.

By the spring of 1973, Callan and I had completed a proof of this argument, extending an idea of G. Parisi [33] to all renormalizable field theories, with the exception of non-Abelian gauge theories. The essential idea was to prove that the vanishing anomalous dimensions of the composite operators, at an assumed fixed point of the renormalization group, implied the vanishing anomalous dimensions of the fields. This then implied that the theory was free at this fixed point. The conclusion was that naïve scaling could be explained only if the assumed fixed point of the renormalization group was at the origin of coupling space, i.e., the theory must be asymptotically free [34]. Non-Abelian gauge theories were not included in the argument since both arguments broke down for these theories. The discovery of asymptotic freedom made this omission irrelevant.

The second part of the argument was to show that there were no asymptotically free theories at all. I had set up the formalism to analyze the most general renormalizable field theory of fermions and scalars – again excluding non-Abelian gauge theories. This was not difficult, since to investigate asymptotic freedom it suffices to study the behavior of the β-functions in the vicinity of the origin of coupling constant space, i.e., in lowest order perturbation theory (one-loop approximation). I almost had a complete proof but was stuck on my inability to prove a necessary inequality. I discussed the issue with Sidney Coleman, who was spending the spring semester in Princeton. He came up with the missing ingredient, and added some other crucial points – and we had a proof that no renormalizable field theory that consisted of theories with arbitrary Yukawa, scalar or Abelian gauge interactions could be asymptotically free [35]. A. Zee had also been studying this. He too was well aware of the advantages of an asymptotically free theory and was searching for one. He derived, at the same time, a partial result, indicating the lack of asymptotic freedom in theories with SU(N) invariant Yukawa couplings [36].

THE DISCOVERY OF ASYMPTOTIC FREEDOM

Frank Wilczek started work with me in the fall of 1972. He had come to Princeton as a mathematics student, but soon discovered that he was really interested in particle physics. He switched to the physics department, after taking my field theory course in 1971, and started to work with me. My way of working with students, then and now, was to involve them closely with my current work and very often to work with them directly. This was certainly the case with Frank, who functioned more as a collaborator than a student from the beginning. I told him about my program to determine whether quantum field theory could account for scaling. We decided that we would calculate the β-function for Yang-Mills theory. This was the one hole in the line of argument I was pursuing. It had not been filled largely because Yang-Mills theory still seemed strange and difficult. Few calculations beyond the Born approximation had ever been done. Frank was interested in this calculation for other reasons as well. Yang-Mills theory was already in use for the electro-weak interactions, and he was interested in understanding how these behaved at high energy.

Coleman, who was visiting in Princeton, asked me at one point whether anyone had ever calculated the β-function for Yang-Mills theory. I told him that we were working on this. He expressed interest because he had asked his student, H. David Politzer, to generalize the mechanism he had explored with Eric Weinberg, that of dynamical symmetry breaking of an Abelian gauge theory, to the non-Abelian case. An important ingredient was the knowledge of the renormalization flow, to decide whether lowest order perturbation theory could be a reliable guide to the behavior of the energy functional. Indeed, Politzer went ahead with his own calculation of the β-function for Yang-Mills theory.

Our calculation proceeded slowly. I was involved in the other parts of my program and there were some tough issues to resolve. We first tried to prove on general grounds, using spectral representations and unitarity, that the theory could not be asymptotically free, generalizing the arguments of Coleman and me to this case. This did not work, so we proceeded to calculate the β-function for a Yang-Mills theory. Today this calculation is regarded as quite simple and even assigned as a homework problem in quantum field theory courses. At the time it was not so easy. This change in attitude is the analogue, in theoretical physics, of the familiar phenomenon in experimental physics whereby yesterday's great discovery becomes today's background. It is always easier to do a calculation when you know what the result is and you are sure that the methods make sense. One problem we had to face was that of gauge invariance. Unlike QED, where the charge renormalization was trivially gauge invariant (since the photon is neutral), the renormalization constants in QCD were all gauge dependent. However, the physics could not depend on the gauge. Another issue was the choice of regularization. Dimensional regularization had not really been developed yet, and we had to convince ourselves that the one-loop β-function was insensitive to the regularization used. We did the calculation in an arbitrary gauge. Since we knew that the answer had to be gauge invariant, we could use gauge invariance as a check on our arithmetic. This was good since we both kept on making mistakes. In February the pace picked up, and we completed the calculation in a spurt of activity. At one point a sign error in one term convinced us that the theory was, as expected, non-asymptotically free. As I sat down to put it all together and to write up our results, I caught the error. At almost the same time Politzer finished his calculation and we compared our results. The agreement was satisfying [37,38].

Why are non-Abelian gauge theories asymptotically free? Today we can understand this in a very physical fashion, although it was certainly not so clear in 1973. It is instructive to interrupt the historical narrative and explain, in modern terms, why QCD is asymptotically free. The easiest way to understand this is by considering the magnetic screening properties of the vacuum [39]. In a relativistic theory one can calculate the dielectric constant, ε, in terms of the magnetic permeability, μ, since $\varepsilon\mu=1$ (in units where c=velocity of light=1). In classical physics all media are diamagnetic. This is because, classsically, all magnets arise from electric currents and the response of a system to an applied magnetic field is to set up currents that act to decrease the field (Lenz's law). Thus $\mu< 1$, a situation that corresponds to electric screening or $\varepsilon>1$. However, in quantum mechanical systems paramagnetism is possible. This is the case in non-Abelian gauge theories where the gluons are charged particles of spin one. They behave as permanent color magnetic dipoles that align themselves parallel to an applied external field increasing its magnitude and producing $\mu>1$. We can therefore regard the anti-screening of the Yang-Mills vacuum as paramagnetism. QCD is asymptotically free because the anti-screening of the gluons overcomes the screening due to the quarks. The arithmetic works as follows. The contribution to ε (in some units) from a particle of charge q is $-q^2/3$, arising from ordinary dielectric (or diamagnetic) screening. If the particle has spin s (and thus a permanent dipole moment γs), it contributes $(\gamma s)^2$ to μ. Thus a spin one gluon (with $\gamma = 2$, as in Yang-Mills theory) gives a contribution to μ of $\delta\mu = (-1/3+2^2)q^2 = 11/3\ q^2$; whereas a spin one-half quark contributes, $\delta\mu = -(-1/3+(2/2)^2)\ q^2 = -2/3\ q^2$ (the extra minus arises because quarks are fermions). In any case, the upshot is that as long as there are not too many quarks the anti-screening of the gluons wins out over the screening of the quarks. The formula for the β-function of a non-Abelian gauge theory is given by:

$$\beta(\alpha) \equiv \mu\frac{d\alpha(\mu)}{d\mu} = \frac{\alpha^2}{\pi}b_1 + \frac{\alpha^3}{\pi^2}b_2 + \ldots \quad \text{, where} \quad \alpha = \frac{g^2}{4\pi} \tag{1}$$

Our result was that [37,38]

$$b_1 = -\left[\frac{11}{6}C_A - \frac{2}{3}\Sigma_R n_R T_R\right] \tag{2}$$

Here C_R is the eigenvalue of the quadratic Casimir operator in the representation R of SU(N) (for the adjoint representation, C_A=N), T_R is the trace of the square of the generators for the representation R of SU(N) (T_A=N and for the fundamental representation, $T_F = 1/2$), and n_R is the number of fermions in the representation R. In the case of SU(3), as in QCD, C_A=N, T_F=1/2, and thus

$$b_1 = -11/2 + n_F/3. \tag{3}$$

Thus one can tolerate as many as 16 triplets of quarks before losing asymptotic freedom.

NON-ABELIAN GAUGE THEORIES OF THE STRONG INTERACTIONS

For me the discovery of asymptotic freedom was totally unexpected. Like an atheist who has just received a message from a burning bush, I became an immediate true believer. Field theory was not wrong – instead scaling must be explained by an asymptotically free gauge theory of the strong interactions. Our first paper contained, in addition to the report of the asymptotic freedom of Yang-Mills theory, the hypothesis that this could offer an explanation for scaling, a remark that there would be logarithmic violations of scaling and most important of all the suggestion that the strong interactions must be based on a color gauge theory [37].

Our abstract reads: *"It is shown that a wide class of non-Abelian gauge theories have, up to calculable logarithmic corrections, free-field asymptotic behavior. It is suggested that Bjorken scaling may be obtained from strong-interaction dynamics based on non-Abelian gauge symmetry."* The first paragraph reads: *"Non-Abelian gauge theories have received much attention recently as a means of constructing unified and renormalizable theories of the weak and electromagnetic interactions. In this note we report on an investigation of the ultraviolet asymptotic behavior of such theories. We have found that they possess the remarkable feature, perhaps unique among renormalizable theories, of asymptotically approaching free-field theory. Such asymptotically free theories will exhibit, for matrix elements of currents between on-mass-shell states, Bjorken scaling. We therefore suggest that one should look to a non-Abelian gauge theory of the strong interactions to provide the explanation for Bjorken scaling, which has so far eluded field theoretic understanding."*

We had a specific theory in mind. Since the deep-inelastic experiments indicated that the charged constituents of the nucleon were quarks, the gluons had to be flavor neutral. Thus the gluons could not couple to flavor. We were very aware of the growing arguments for the color quantum number. Not just the quark model spectroscopy that was the original motivation of Han, Nambu and Greenberg, but the counting factor (of three) that went into the evaluation of the $\pi \rightarrow 2\gamma$ decay rate from the axial anomaly (this had been recently emphasized by W. Bardeen, H. Fritzsch and Gell-Mann [40]), and the factor of three that color provided in the total annihilation cross section. Thus the gluons could couple to color and all would be well. Thus we proposed [37]: *"One particularly appealing model is based on three triplets of fermions, with Gell-Mann's SU(3)xSU(3)as a global symmetry and a SU(3) "color" gauge group to provide the strong interactions. That is, the generators of the strong interaction gauge group commute with ordinary SU(3)x SU(3) currents and mix quarks with the same isospin and hypercharge but different 'color'. In such a model the vector mesons are (flavor) neutral, and the structure of the operator product expansion of electromagnetic or weak currents is essentially that of the free quark model (up to calculable logarithmic corrections)."* Thus we proposed that the strong interactions be described by the theory we now call QCD!

Callan and I had already discussed the appearance of logarithmic corrections to scaling in asymptotically free theories [34]. We analyzed deep inelastic scattering in an asymptotically free theory and discovered *"that in such asymptotically free theories naive scaling is violated by calculable logarithmic terms."* Thus we were well aware what the form of the scaling deviations would be in such a theory, Wilczek and I immediately started to calculate the logarithmic deviations from scaling. We were tremendously excited by the possibility of deriving exact experimental predictions from first principles that could conclusively test our asymptotically free theories of the strong interactions. We had already evaluated the asymptotic form of the flavor non-singlet structure functions, which were the easiest to calculate, at the time our Physical Review Letter was written, but did not have room to include the results. We immediately started to write a longer paper in which the structure of the theory would be spelled out in more detail and the dynamical issues would be addressed, especially the issue of confinement. In our letter we were rather noncommittal on this issue. We had tentatively concluded that Higgs mesons would destroy asymptotic freedom, but had only begun to explore the dynamical consequences of unbroken color symmetry. The only thing we were sure of was that *"perturbation theory is not trustworthy with respect to the stability of the symmetric theory nor to its particle content"* [37]. Politizer's paper appeared just after ours. He pointed out the asymptotic freedom of Yang-Mills theory and speculated on its implications for the dynamical symmetry breaking of these theories. His abstract reads; *"An explicit calculation shows perturbation theory to be arbitrarily good for the deep Euclidean Green's functions of any Yang-Mills theory and of many Yang-Mills theories with fermions. Under the hypothesis that spontaneous symmetry breakdown is of dynamical origin, these symmetric Green's functions are the asymptotic forms of the physically significant spontaneously broken solution, whose coupling could be strong."* No mention is made of either Bjorken scaling or of the strong interactions [38].

In our second paper, written a few months later, we outlined in much greater detail the structure of asymptotically free gauge theories of the strong interactions and the predictions for the scaling violations in deep-inelastic scattering [41]. The paper was delayed for about two months because we had problems with the singlet structure functions – due to the operator mixing of physical operators with ghost operators. This problem was similar to the issue of gauge invariance that had plagued us before. Here the problem was more severe. Physical operators, whose matrix elements were measurable in deep-inelastic scattering experiments, mixed under renormalization with ghost operators that could have no physical meaning. Finally we deferred the analysis of the singlet structure functions to a third paper, in which we resolved this issue [42]. We showed that, even though this mixing was real and unavoidable, the ghost operators decoupled from physical measurements. In the second paper we discussed in detail the choice between symmetry breaking and unbroken symmetry and noted that *"another possibility is that the gauge symmetry is exact. At first sight this would appear to be ridiculous since it would imply the existence of massless, strongly coupled vector mesons. However, in asymptotically free theories these naïve expectations might be wrong. There may be little connection between the 'free' Lagrangian and the spectrum of states... The infrared behavior of Green's functions in this case is determined by the strong-coupling limit of the theory. It may be very well that this infrared behavior is such so as to suppress all but color singlet states, and that the colored gauge fields as well as the quarks could be 'seen' in the large-Euclidean momentum region but never produced as real asymptotic states."* [41].

Steve Weinberg reacted immediately to asymptotic freedom. He wrote a paper in which he pointed out that in an asymptotically free gauge theory of the strong interactions the order α interactions produced by electroweak interactions can be calculated ignoring the strong force; and found that these effects do not violate conservation of parity and strangeness, in agreement with observation, as long as there were no colored scalars [43]. This led him to suggest that a theory with unbroken color symmetry could explain why we do not see quarks and gluons. There is a slight difference between our respective conjectures. Weinberg argued that perhaps the infrared divergences, caused by the masslessness of the gluons in an unbroken color gauge theory, would make the rate of production of non-singlet states vanish. Today we believe in the existence of non-confining, Coulomb phases, with unbroken color symmetry, for some supersymmetric non-Abelian gauge theories. We argued that perhaps the growth of the effective coupling at large distances, the infrared behavior of the coupling caused by the flip side of asymptotic freedom (later dubbed infrared slavery by Georgi and Glashow) would confine the quarks and gluons in color singlet states.

In October 1973, Fritzsch, Gell-Mann and H. Leutwyler submitted a paper in which they discussed the *"advantages of color octet gluon picture"*. Here they discussed the advantages of *"abstracting properties of hadrons and their currents from a Yang-Mills gauge model based on colored quarks and color octet gluons"* [44]. They discussed various models and pointed out the advantages of each. The first point was already discussed at the NAL high-energy physics conference in August 1972. There Gell-Mann and Fritzsch had discussed their program of *"abstracting results from the quark-gluon model"* [45]. They discussed various models and asked, *"shall treat the vector gluon, for convenience, as a color singlet"*. In October 1973, Fritzsch, Gell-Mann and Leutwyler also noted that in the non-relativistic quark model with a Coulomb potential mediated by vector gluons the potential is attractive in color singlet channels, which might explain why these are light, a point that had been made previously by H. Lipkin [46]. They also noted the asymptotic freedom of such theories, but did not regard this as an argument for scaling since *"we conjecture that there might be a modification at high energies that produces true scaling"*. Finally they noted that the axial U(1) anomaly in a non-Abelian gauge theory might explain the notorious U(1) problem, although they could not explain how, since the anomaly itself could be written as a total divergence.(It required the discovery of instantons to find the explanation of the U(1) problem.)

THE EMERGENCE AND ACCEPTANCE OF QCD

Although it was clear to me that the strong interactions must be described by non-Abelian gauge theories, there were many problems. The experimental situation was far from clear, and the issue of confinement remained open. However, within a small community of physicists the acceptance of the theory was very rapid. New ideas in physics sometimes take years to percolate into the collective consciousness. However, in rare cases such as this there is a change of perception analogous to a phase transition. Before asymptotic freedom it seemed that we were still far from a dynamical theory of hadrons; afterwards it seemed clear that QCD was such a theory. (The name QCD first appeared in a review by W. Marciano and H. Pagels [47], where it was attributed to Gell-Mann. It was such an appropriate name that no one could complain.) Asymptotic freedom explained scaling at short distances and offered a mechanism for confinement at large distance. Suddenly it was clear that a non-Abelian gauge theory was consistent with everything we knew about the strong interactions. It could encompass all the successful strong interaction phenomenology of the past decade. Since the gluons were flavor neutral, the global flavor symmetries of the strong interactions, SU(3)x SU(3), were immediate consequences of the theory, as long as the masses of the quarks (the mass parameters of the quarks in the Lagrangian, not the physical masses that are effectively infinite due to confinement) are small enough.

Even more alluring was the fact that one could calculate. Since perturbation theory was trustworthy at short distances many problems could be tackled. Some theorists were immediately convinced, among them Altarelli, Appelquist, Callan, Coleman, Gaillard, R. Gatto, Georgi, Glashow, Kogut, Ben Lee, Maiani, Migdal, Polyakov, Politzer, Susskind, Weinberg, Zee. At large distances however perturbation theory was useless. In fact, even today after 31 years of study we still lack reliable, analytic tools for treating this region of QCD. This remains one of the most important areas of theoretical particle physics. However, at the time the most important thing was to convince oneself that the idea of confinement was not inconsistent. One of the first steps in that direction was provided by lattice gauge theory. I first heard of Wilson's lattice gauge theory [48] when I gave a lecture at Cornell in the late spring of 1973. Wilson had started to think of this approach soon after asymptotic freedom was discovered. The lattice formulation of gauge theory (independently proposed by Polyakov) had the enormous advantage, as Wilson pointed out in the fall of 1973, that the strong coupling limit was particularly simple and exhibited confinement. Thus one had at least a crude approximation in which confinement was exact. It is a very crude approximation, since to arrive at the continuum theory from the lattice theory one must take the weak-coupling limit. However, one could imagine that the property of confinement was not lost as one went continuously from strong to weak lattice coupling, i.e., there was no phase transition. Moreover one could, as advocated by Wilson, study this possibility numerically using Monte Carlo methods to construct the lattice partition function. However, the first quantitative results of this program did not emerge until 1981. By now the program of calculating the hadronic mass spectrum has come close to its goal, achieving now reliable results that fit the low-lying spectrum to a few percent!

Personally I derived much solace in the coming year from two examples of soluble two-dimensional field theories. One was the $(\bar{\psi}\psi)^2$ theory that Neveu and I analyzed and solved for large N [49]. This provided a soluble example of an asymptotically free theory that underwent dimensional transmutation, solving its infrared problems by generating a dynamical fermion mass through spontaneous symmetry breaking. This provided a model of an asymptotically free theory, with no built in mass parameters. We could solve this model and check that it was consistent and physical. The other soluble model was two-dimensional QCD, analyzed by 't Hooft in the large N limit [50]. Two dimensional gauge theories trivially confine color. This was realized quite early and discussed for Abelian gauge theory, the Schwinger model, by A. Casher, Kogut and Susskind, as a model for confinement in the fall of 1973 [51]. However, QCD_2 is a much better example. It has a spectrum of confined quarks which in many ways resembles the four dimensional world. These examples gave many of us total confidence in the consistency of the concept of confinement. It clearly was possible to have a theory whose basic fields do not correspond to asymptotic states, to particles that one can observe directly in the laboratory. Applications of the theory also began to appear. Two calculations of the β-function to two loop order were performed, with the result that, in the notation of (2), $b_2 = -\left[\frac{17}{12}C_A^{\ 2} - \frac{1}{2}C_F T_F n - \frac{5}{6}C_A T_F n\right]$ [52]. Appelquist and Georgi, and Zee calculated the corrections to the scaling of the e^+-e^- annihilation cross-section[53]; Gaillard, and Lee, and independently Altarelli and Maiani, calculated the enhancement of the $\Delta I=1/2$ non-leptonic decay matrix elements [54]. The analysis of scaling violations for deep-inelastic scattering continued [55], and the application of asymptotic freedom, what is now called perturbative QCD, was extended to many new processes.

The experimental situation developed slowly, and initially looked rather bad. I remember in the spring of 1974 attending a meeting in Trieste. There I met Burt Richter who was gloating over the fact that $R = \sigma_{e^+e^- \to hadrons}/\sigma_{e^+e^- \to \mu^+\mu^-}$ was increasing with energy, instead of approaching the expected constant value. This was the most firm of all the scaling predictions. R must approach a constant in any scaling theory. In most theories one cannot predict the value of the constant. However, in an asymptotically free theory the constant is predicted to equal the sum of the squares of the charges of the constituents. Therefore, if there were only the three observed quarks, one would expect that R -> $3[(1/3)^2 + (1/3)^2 + (2/3)^2] = 2$. However, Richter reported that R was increasing, passing through 2, with no sign of flattening out. Now many of us knew that charmed particles had to exist. Not only were they required, indeed invented, for the GIM mechanism to work, but as C. Bouchiat, J. Illiopoulos and L. Maini, and independently R. Jackiw and I [56] showed, if the charmed quark were absent the electro-weak theory would be anomalous and non-renormalizable. Gaillard, Lee and Rosner had written an important and insightful paper on the phenomenology of charm [57]. Thus, many of us thought that since R was increasing probably charm was being produced. In 1974 the charmed mesons, much narrower than anyone imagined (except for Appelquist and Politzer [58]), were discovered, looking very much like positronium, and easily interpreted as Coulomb bound states of quarks. This clinched the matter for many of the remaining skeptics. The rest were probably convinced once experiments at higher energy began to see quark and gluon jets.

The precision tests of the theory, the logarithmic deviations from scaling, took quite a while to observe. I remember very well a remark made to me by a senior colleague, in April of 1973 when I was very excited, right after the discovery of asymptotic freedom. He remarked that it was unfortunate that our new predictions regarding deep-inelastic scattering were logarithmic effects, since it was unlikely that we would see them verified, even if true, in our lifetime. This was an exaggeration, but the tests did take a long time to appear. Confirmation only started to trickle in 1975–78 at a slow pace. By now the predictions are indeed verified, in many cases to better than 1%. Nowadays, when you listen to experimentalists talk about their results they point to their Lego plots and say, "Here we see a quark, here a gluon." Believing is seeing,

seeing is believing. We now believe in the physical reality of quarks and gluons, we now believe in asymptotic simplicity of their interactions at high energies; so we can see quarks and gluons. The way in which we see quarks and gluons, indirectly through the effects they have on our measuring instruments, is not much different from the way we see electrons.

IMPLICATIONS OF ASYMPTOTIC FREEDOM

The most important implication of asymptotic freedom is QCD itself with point like behavior of quarks at short distance and the strong confining force at large distance. But in addition asymptotic freedom greatly increased our confidence in the consistency of quantum field theory, produced the first example of a theory with no adjustable parameters, enabled us to probe the very early history of the universe and allowed us to extrapolate the standard model to high energy.

a. Consistency of quantum field theory.

Traditionally, fundamental theories of nature have had a tendency to break down at short distances. This often signals the appearance of new physics that is discovered once one has experimental instruments of high enough resolution (energy) to explore the higher energy regime. Before asymptotic freedom it was expected that any quantum field theory would fail at sufficiently high energy, where the flaws of the renormalization procedure would appear. To deal with this, one would have to invoke some kind of fundamental length. In an asymptotically free theory this is not necessarily the case; the decrease of the effective coupling for large energy means that no new physics need arise at short distances. There are no infinities at all, the bare coupling is finite, and in fact it vanishes. The only divergences that arise are an illusion that appears when one tries to compare, in perturbation theory, the finite effective coupling at finite distances with the vanishing effective coupling at infinitely short distances.

Thus the discovery of asymptotic freedom greatly reassured us of the consistency of four-dimensional quantum field theory. We can trust renormalization theory asymptotically free theories, even though perturbation theory is only an asymptotic expansion, since it gets simpler in the regime of short distances. We are very close to having a rigorous mathematical proof of the existence of asymptotically free gauge theories in four dimensions – at least when placed into a finite box to tame the infrared dynamics that produces confinement.

b. No adjustable parameters.

At first glance QCD has only one parameter, the dimensionless number that specifies the strength of the force (if we neglect the quark masses, an excellent approximation for ordinary hadrons since the light quarks are so light). But through the dependence of the charge on distance or energy, the theory produces a dynamical mass scale. One defines the mass scale of QCD to be the energy at which the charge equals some value, say 1. Then, via this phenomenon of *dimensional transmutation*, all masses, indeed all observables, are calculable in terms of the dynamically generated mass scale. It is sometimes claimed that the origin of mass is the Higgs mechanism that is responsible for the breaking of the electroweak symmetry that unbroken would forbid quark masses. This is incorrect. Most, 99%, of the proton mass is due to the kinetic and potential energy of the massless gluons and the essentially massless quarks, confined within the proton.

Thus, QCD provides the first example of a complete theory, with no adjustable parameters and with no indication within the theory of a distance scale at which it must break down. Indeed, were it not for the electro-weak interactions and gravity, we might be satisfied with QCD as it stands. It is the best example we possess of a perfect, complete theory.

c. The early history of the universe.

The universe has been expanding since the big bang, thus early on it was hot and dense. To trace the history of the universe we must understand the dynamics that operates when the universe was hot and particles were very energetic. Before the standard model we could not go back further than 200,000 years after the big bang. Today, especially since QCD simplifies at high energy, we can extrapolate to very early times, where nucleons melt and quarks and gluons are liberated to form a quark-gluon plasma.

d. Unification.

One the most important implication of asymptotic freedom is the insight it gave into the unification of all the forces of nature. Almost immediately after the discovery of asymptotic freedom and the proposal of the non-Abelian gauge theories of the strong interactions, the first attempts were made to unify all the interactions. This was natural, given that one was using very similar theories to describe all the known interactions. Furthermore, the apparently insurmountable barrier to unification – namely the large difference in the strength of the strong interactions and the electro-weak interactions – was seen to be a low energy phenomenon. Since the strong interactions decrease in strength with increasing energy these forces could have a common origin at very high energy. H. Georgi, H. Quinn and S. Weinberg showed that the couplings run in such a way as to merge somewhere around 10^{14} to 10^{16} GeV [59]. This is our most direct clue as to where the next threshold of fundamental physics lies, and hints that at this immense energy all the forces of nature, including gravity, are unified.

ACKNOWLEDGEMENTS

As I end I would like to thank not only the Nobel Foundation, but nature itself, who has given us the opportunity to explore her secrets and the fortune to have revealed one of her most mysterious and beautiful aspects – the strong force.

REFERENCES

1. Feynman, R., in *The Quantum Theory of Fields* – The 12th Solvay Conference, 1961 (Interscience, New York).
2. Landau, L.D. and I.Pomeranchuk, *Dokl. Akad. Nauk SSSR* **102**, 489 (1955).
3. Landau, L.D., in *Niels Bohr and the Development of Physics,* (ed. W.Pauli), p. 52. (Pergamon Press, London, 1955).
4. Landau, L.D., Fundamental Problems, in *Pauli Memorial Volume,* pg. 245, (Interscience, New York, 1960).
5. VanTerent'ev, M.V. and Vanyashin, V.S., *Soviet Physics JETP* **21**, 380 (1965).
6. Chew, G. in *S-Matrix Theory,* (W.A. Benjamin Inc, 1963).
7. Gell-Mann, M., Current Algebra: Quarks and What Else?, in *Proceedings of the XVI Int'l Conference on High Energy Physics,* Vol. **4**, 135 (1972).
8. Gell-Mann, M. and Neeman, Y. *The Eightfold Way* (W.A. Benjamin Inc.) (1964).
9. Gell-Mann, M., The Symmetry Group of Vector and Axial Vector Currents, *Physics,* **1**, 63 (1964).
10. Gell-Mann, M., A Schematic Model of Baryons and Mesons, *Phys. Lett.* **8**, 214 (1964); Zweig, G., *CERN Report No. TH401,* 4R12 (1964) (unpublished).
11. Greenberg, O.W., Spin and Unitary-Spin Independence in a Paraquark Model of Baryons and Mesons, *Phys. Rev. Lett.* **13**, 598 (1964).
12. Nambu,Y., Dynamical Symmetries & Fundamental Fields, *Proceedings of 2nd Coral Gables Conf. on Symmetry Principles at High Energy,* 133(1965); Systematics of Hadrons in Subnuclear Physics, *Preludes in Theoretical Physics,* (1965).
13. Han, M.Y. and Nambu,Y., Three-Triplet Model with Double SU(3) Symmetry, *Phys. Rev.* **139 B**, 1006 (1965).
14. Low, F., in *Proceedings of the International Conference on High Energy Physics 1966,* pg. 249 (1966).
15. Adler, S., Sum Rules for the Axial Vector Coupling Constant Renormalization in Beta Decay, *Phys. Rev.* **140 B**, 736 (1965); Weisberger,W.I. Unsubtracted Dispersion Relations and the Renormalization of the Weak Axial-Vector Coupling Constants, *Phys. Rev.* **143 B**, 1302 (1966).
16. Dashen, R. and Gell-Mann, M., Representations of Local Current Algebra at Infinite Momentum, *Phys. Rev. Lett.* **17**, 340 (1966).
17. Bjorken, J.D. Applications of the Chiral U(6)XU(6)Algebra of Current Densities, *Phys. Rev.* **148 B**, 1467 (1966); Current Algebra at Small Distances, *Varenna School Lectures,* Course XLI, Varenna, Italy (1967).
18. Callan, C.G. and Gross, D. J., Crucial Test of a Theory of Currents, *Phys. Rev. Lett* **21**, 311 (1968).
19. Bloom, E.D. *et al.*, High-Energy Inelastic e-p Scattering at 6° and 10°, *Phys. Rev. Lett.* **23**, 930 (1969).
20. Callan, C. G. and Gross, D. J., High-Energy Electroproduction and the Constitution of the Electric Current, *Phys. Rev. Lett.* **22**, 156 (1968).
21. Gross, D.J. and Llewelyn-Smith,C., High Energy Neutrino-Nucleon Scattering, Current Algebra and Partons, *Nucl. Phys.* **B14**, 337 (1969).
22. Feynman, R.P., High-Energy Collisions of Hadrons, *Phys.Rev.Lett.* **23**, 1415 (1969).
23. Drell, S.D. and Yan, T.M., Partons and Their Applications at High Energies, *Ann. Phys.* (NY) **66**, 578 (1971).
24. Gross, D.J. and Wess, J., Scale Invariance, Conformal Invariance and the High Energy Behavior of Scattering Amplitudes, *Phys. Rev.* **D2**, 753 (1970).
25. Callan, C.G., Broken Scale Invariance in Scalar Field Theory, *Phys. Rev.* **D2**, 1541 (1970); Symanzik, K., Small Distance Behavior in Field Theory and Power Counting, *Comm. Math. Phys.* **18**, 227 (1970).
26. Veneziano, G., Construction of a Crossing-Symmetric Regge-Behaved Amplitude for Linearly Rising Trajectories, *Nuovo Cimento,* **57 A**, 190 (1968).

27. 't Hooft, G., Renormalizable Lagrangians for Massive Yang-Mills Fields, *Nucl. Phys.* **35**, 167 (1967).
28. Wilson, K., Renormalization Group and Strong Interactions, *Phys. Rev.* **D3**, 1818 (1971). 29.
29. Callan, C.G., Broken Scale Invariance and Asymptotic Behavior, *Phys. Rev.* **D5**, 3202 (1972); Symanzik, K., Small-Distance-Behavior Analysis and Wilson Expansions, *Comm. Math. Phys.* **23**, 49 (1971); Christ, N., Hasslacher, B. and Mueller, A., Light-Cone Behavior of Perturbation Theory, *Phys. Rev.* **D6**, 3543 (1972); Callan, C. G .and Gross, D. J., Bjorken Scaling in Quantum Field Theory, *Phys. Rev.* **D8**, 4383 (1973).
30. Jackiw, R. Van Royen, R. and West, G., Measuring Light-Cone Singularities, *Phys. Rev.* **D2**, 2473 (1970); Frishman, Y., Operator Products at Almost Light Like Distances, *Ann. Phys.* **66**, 373 (1971); Leutwyler, H. and Stern, J., Singularities of Current Commutators on the Light Cone, *Nucl. Phys.* **B20**, 77 (1970); Gross, D. J., Light Cone Structure of Current Commutators in the Gluon-Quark Model, *Phys. Rev.* **D4**, 1059 (1971); Christ, N., Hasacher, B. and Mueller, A., Light-Cone Behavior of Perturbation Theory, *Phys. Rev.* **D6**, 3543 (1972).
31. Wilson, K. and Kogut, J., The Renormalization Group and the ε expansion. *Phys. Rep.* **31**, 792 (1973).
32. Gell-Mann, M. and Low, F., Quantum Electrodynamics at Small Distances, *Phys. Rev.* **95**, 1300 (1954). Stueckelberg, E. and Petermann, A., La Normalisation des Constantes dans la Theorie des Quanta, *Helv. Phys. Acta* **26**, 499 (1953). N. N. Bogoliubov and D. V. Shirkov, *Introduction to the Theory of Quantized Fields,* Interscience, New York, (1959).
33. Parisi, G., Deep Inelastic Scattering in a Field Theory with Computable Large-Momenta Behavior, *Lett. al Nuovo Cimento* **7**, 84 (1973).
34. Callan, C.G. and Gross, D. J., Bjorken Scaling in Quantum Field Theory, *Phys. Rev.* **D8**, 4383 (1973).
35. Coleman, S. and Gross, D. J., Price of Asymptotic Freedom, *Phys. Rev. Lett.* **31**, 851 (1973).
36. Zee, A., Study of the Renormalization Group for Small Coupling Constants, *Phys. Rev.* **D7**, 3630 (1973).
37. Gross, D. J. and Wilczek,F., Ultraviolet Behavior of Non-Abelian Gauge Theories, *Phys. Rev. Lett.* **30**, 1343 (1973).
38. Politzer, H.D., Reliable Perturbative Results for Strong Interactions?, *Phys. Rev. Lett.* **30**, 1346 (1973).
39. Nielsen, N.K., Asymptotic Freedom as a Spin Effect, *Am.J. Phys.* **49**, 1171(1981).
40. Bardeen, W.A., Fritzsch, H. and Gell-Mann, M., Light-Cone Current Algebra, π^0 Decay, and e^+-e^- Annihilation, in *Scale and Conformal Symmetry in Hadron Physics*, ed. R. Gatto (Wiley & Sons, 1973).
41. Gross, D. J. and Wilczek, F., Asymptotically Free Gauge Theories. I, *Phys. Rev.* **D8**, 3633 (1973).
42. Gross, D. J. and Wilczek, F., Asymptotically Free Gauge Theories. II, *Phys. Rev.* **D9**, 980 (1974).
43. Weinberg, S., Non-Abelian Gauge Theories of the Strong Interactions, *Phys. Rev. Lett.* **31**, 494 (1973).
44. Fritzsch, H., Gell-Mann, M. and Leutwyler, H., Advantages of the Color Octet Gluon Picture, *Phys. Lett.* **47 B**, 368 (1973).
45. Fritzsch,H. and Gell-Mann, M., Current Algebra: Quarks and What Else?, *Proceedings of the XVI Int. Conf. on High Energy Physics*,Vol.2,164 (1972).
46. Lipkin, H., Triality, Exotics and the Dynamical Basis of the Quark Model, *Phys. Lett.* **45 B**, 267 (1973).
47. Marciano,W. and Pagels, H., Quantum Chromodynamics, *Phys. Rep.* **C36**, 137 (1978).
48. Wilson, K., Confinement of Quarks, *Phys. Rev.* **D10**, 2445 (1974).
49. Gross, D. J. and Neveu, A., Dynamical Symmetry Breaking in an Asymptotically Free Theory, *Phys. Rev.* **D10**, 3235 (1974).
50. t' Hooft, G., A Planar Diagram Theory for Strong Interactions, *Nucl. Phys.* **B72**, 461 (1974).

51. A. Casher, J. Kogut and L. Susskind, Vacuum Polarization and the Quark-Parton puzzle, *Phys. Rev. Lett.* **31**, 792 (1973); Vacuum Polarization and the Absence of Free Quarks, *Phys. Rev.* **D10**, 732 (1974).
52. Caswell, W., Asymptotic Behavior of Non-Abelian Gauge Theories to Two-Loop Order, *Phys. Rev. Lett.* **33**, 244 (1974); Jones, D., Two-Loop Diagrams in Yang-Mills Theory, *Nucl. Phys.* **B75**, 531, (1974).
53. Appelquist, T. and Georgi, H., e^+-e^- Annihilation in Gauge Theories of Strong Interactions, *Phys. Rev.* **D8**, 4000 (1973); Zee, A., Electron-Positron Annihilation in Stagnant Field Theories, *Phys. Rev.* **D8**, 4038, (1973).
54. Gaillard, M.K. and Lee, B.W. $\Delta I=1/2$ Rule for Non-Leptonic Decays in Asymptotically Free Field Theories, *Phys. Rev. Lett.* **33**, 108 (1974); Altarelli, G. and Maiani, L., Octet Enhancement of Non-Leptonic Weak Interactions in Asymptotically Free Gauge Theories *Phys. Lett.* **52 B**, 351 (1974).
55. Gross, D. J., How to Test Scaling in Asymptotically Free Theories, *Phys. Rev. Lett.* **32**, 1071 (1974).
56. Bouchiat, C., Iliopoulpos, J. and Meyer, Ph., An Anomaly Free Version of Weinberg's Model, *Phys. Lett.* **38 B**, 519 (1972); Gross, D.J. and Jackiw, R., The Effect of Anomalies on Quasi-Renormalizable Theories, *Phys. Rev.* **D6**, 477 (1972).
57. Gaillard, M.K., Lee, B.W. and Rosner, J.L., Search for Charm, *Rev. Mod. Phys.* **47**, 277 (1975).
58. Appelquist, T. and Politzer, H. D., Heavy Quarks and e^+-e^- Annihilation, *Phys. Rev. Lett.* **34**, 43 (1975).
59. Georgi, H., Quinn, H.R. and Weinberg, S., Hierarchy of Interactions in Unified Gauge Theories, *Phys. Rev. Lett.* **33**, 451 (1974).

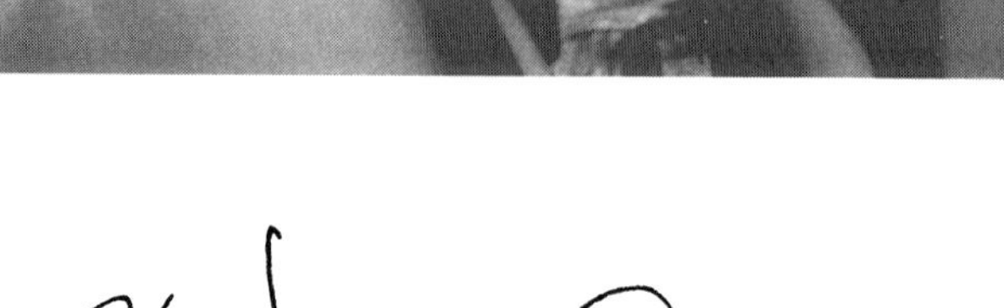

THE DILEMMA OF ATTRIBUTION

Nobel Lecture, December 8, 2004

by

H. David Politzer

California Institute of Technology (Caltech), Pasadena, USA.

I suspected that there were some members of the live audience who were somewhat apprehensive about sitting through the morning's physics lectures. After all, there were three guys there to talk about one minus sign. If it were just two people and a plus sign, +, one could talk about the | and the other about the –. However, to my mind, this year's awards represent or symbolize not just a minus sign but a large body of significant advances in our understanding of fundamental physics and are the work of not just three people but a great many scientists, stretching out over many years and many countries. This is really a prize for that whole community.

Sidney Coleman, my beloved teacher from graduate school, referred to this community as *i fratelli fisici,* by which he meant the brotherhood of physicists. Most of us spoke at least a bit of broken Italian, a legacy of the grand and highly influential summer schools organized by Nino Zichichi in Erice, Sicily. Indeed, one of my fondest reflections on my particle physics career is having been able to arrive at a train station, virtually anywhere in the world, and be greeted by a total stranger who immediately treated me like an old friend.

I'd love to tell you all their stories, but I certainly don't know them all, nor do I have time (or space) even for those that I do. So I've chosen a few of the people and a few of the stories with which to make a particular point. You can judge for yourself at the end how well I've succeeded. And I'll deal mostly with theorists because I know them best – although I must say that I do regard theoretical physics as a fundamentally parasitic profession, living off the labors of the real physicists.

I'd like to address one particular aspect of the impact of these prizes. To a considerable extent they have come to represent milestones in the progress of science. And it is a testament to the care and wisdom exercised in the selection process just how important the prizes have become. To the public, they spark continued interest in science's most important advances. But even within the world of the scientific experts, the prizes likewise serve as markers of this progress. The use of history in science education may be a contributing factor to why this is so and how it works. As teachers of the next generation of scientists, we always seek to compress and simplify all the developments that have come before. We want to bring our students as quickly as possible to the frontier of current understanding. From this perspective, the actual history, which involves many variants and many missteps, is only a hindrance. And the neat, linear progress, as outlined by the sequence of gleaming gems recognized by Nobel Prizes, is a useful fiction. But a fiction it is. The truth is often far more complicated. Of course, there are the oft-told priority disputes, bickering over who is responsible for some particular idea. But those questions are not only often unresolvable, they are often rather meaningless. Genuinely independent discovery is not only possible, it occurs all the time. Sometimes a yet harder problem in the prize selection process is to identify what is the essential or most important idea in some particular, broader context. So it's not just a question of who did it, i.e., who is responsible for the work, but what "it" is. I.e., what is the significant "it" that should stand as a symbol for a particularly important advance.

I've no interest in recounting my whole life's story or even my physics career. Rather, I want to focus on the context of the particular work cited in this year's awards. So I begin this saga with a trip I took with Erick Weinberg, a fellow graduate student, friend, and something of a mentor (he was a year ahead of me) from Cambridge, Massachusetts to Hoboken, New Jersey (I think it was 1970) to a conference to hear our teacher, Sidney Coleman, speak. He was delivering a paper titled "Why Dilatation Generators Don't Generate Dilatations." We had read a written version but hoped that his talk would help us understand it better. It was a several hour drive. Somewhere along the way, I asked Erick to explain to me a bit about what were called Yang-Mills or non-Abelian gauge theories. I had heard the name but was otherwise ignorant. They'd been invented in 1954 and were the last and least understood entry in a short list of what came to be considered the only possible descriptions of fundamental particle interactions. Erick explained the defining basics but told me that nothing was known about their consequences and that many of the most famous senior particle theorists had gotten seriously confused about them. (The list of such notables included Dick Feynman, Shelly Glashow, Abdus Salam, and Steve Weinberg.) And now it seemed that no senior physicist wanted to discuss them; their ignorance and confusion were too embarrassing.

(While delivering my talk live in Stockholm, it occurred to me I should have had a little light or a bell that went off when I mentioned a Nobel Laureate – because part of my point is to try to understand who is and who isn't. The relevant names are already familiar to the physicist segment of the audience. But for the sake of the general audience, I just raised my finger discretely. Here I'll use a superscript $\mathcal{N}$. So far, there's Yang$^{\mathcal{N}}$, Feynman$^{\mathcal{N}}$, Glashow$^{\mathcal{N}}$, Salam$^{\mathcal{N}}$, Steve Weinberg$^{\mathcal{N}}$, but not Coleman or Erick Weinberg.)

It turns out there was one brave soul, Tini Veltman$^{\mathcal{N}}$, who never gave up on Yang-Mills theory, and, with his best-ever grad student, Gerard 't Hooft$^{\mathcal{N}}$, cracked the case in 1971. I think it worth noting that I, personally, know of no one who claimed to understand the details of 't Hooft 's paper. Rather we all learned it from Ben Lee, who combined insights from his own work (that renormalization constants are independent of the choice of ground state in such theories), from hitherto unnoticed work from Russia (Fadde'ev and Popov on quantization and Feynman rules), and from the simple encouragement from 't Hooft 's paper that it was possible. (It is amazing how much easier it can be to solve a problem once you are assured that a solution exists!)

The bit of physics I remember best from the Hoboken conference was from a talk by T.D. Lee[N]. He spoke with confidence that the weak interactions were mediated by a heavy bosonic particle that carried the force, and he gave its mass. (Several years later he was proven right.) The clearest version of that theory had been written down by Steve Weinberg[N] in 1967. But no one in that period ever referred to Weinberg's paper. For example, I don't think that Weinberg's paper had any influence on T.D. Lee's thinking. In fact, when what is now known as the Weinberg-Salam model was recognized by the Nobel Foundation, Sidney Coleman published in *Science* magazine in 1979 a citation search he did documenting that essentially no one paid any attention to Weinberg's Nobel Prize winning paper until the work of 't Hooft[N] (as explicated by Ben Lee). In 1971 interest in Weinberg's paper exploded. I had a parallel personal experience: I took a one-year course on weak interactions from Shelly Glashow[N] in 1970, and he never even mentioned the Weinberg-Salam model or his own contributions to the theory (for which he shared that Nobel Prize; by the way, his contribution to that theory was largely his PhD thesis work, done under the direction of Julian Schwinger[N], who had already published papers on the non-Abelian gauge bosons as carriers of the weak force in the mid- 1960's.) I note again that I also don't personally know anyone who ever read Salam's work on the subject either, except for John Ward, and he was actually the co-author on the relevant papers. – He is not a Nobel Laureate.

A further aside on the work of 't Hooft[N] and Veltman[N], whose contributions were enormously profound and influential, albeit really rather difficult to characterize for a lay audience. One of their many contributions (called, in the business, dimensional regularization) is a tool of essential significance, both for settling issues of principle and for doing explicit calculations. Dimensional regularization also was invented independently for the same purposes and appeared in an earlier paper, now mostly forgotten, by Bollini and Giambiaggi.

Coleman's talk in Hoboken was about his then, early understanding of what came to be known as the renormalization group. His thinking was very much influenced by the independent work of Kurt Symanzik and Curt Callan. However, the undisputed champion of the renormalization group was Ken Wilson[N], (one of my all-time, absolute heroes) for which he received Nobel recognition. That a prize was given to Wilson[N] and Wilson[N] alone in 1982 perhaps reflects the depth of his understanding, the precision of his detailed, physical predictions, and his evangelical zeal. We should remember, however, that the renormalization group work that led to experimentally confirmed predictions, which were in the field of phase transitions and are the substance of the citation for that prize, was all done in collaboration with Michael Fisher; we should remember that the basic, formal work was done independently and published earlier by Wegner and Houghton; and we should remember that the essential physical ideas were articulated independently and earlier by Leo Kadanoff. Furthermore, the renormalization group was actually invented in 1954 by Murray Gell-Mann[N] and Francis Low. But even that formulation of the renormalization group appeared in an earlier, independent paper of Stueckelberg and Petermann.

In the early days following the triumph of the Weinberg-Salam model, at one point Glashow[N] asked Coleman a practical question that came up in his own work. (The specific technical question was "What happens if the whole theory has less symmetry than the classical scalar (spin zero) sector?") And Coleman answered the question, but he also recognized that the answer was worthy of a deeper, clearer understanding. So, he embarked on its study, in the simplest possible contexts, with my buddy Erick Weinberg. I tagged along in this effort and occasionally made some contribution.

(Here's an anecdote of my first meeting with Niccolo Cabbibo, a charming man, responsible for a monumental contribution to our understanding of the weak interactions and their relation to the strong interactions, which is now largely overlooked because of the telescoping of history into a compact introduction to the present. We were both visiting the University of Chicago, staying at the Windemere Hotel. We chatted over dinner and after as rats scurried between our feet. He is the only person who ever mentioned to me noticing my name in the acknowledgments of Coleman and Weinberg's classic paper.)

During this work with Coleman and Weinberg, one day I wondered and then asked Coleman, "What happens if there are no scalar fields (spin zero particles) in the first place?" It was an innocent but inordinately profound question which occupied us both quite intensively for the next several months. I learned an enormous amount just working on it. And I benefited from far closer and more extensive interactions with Coleman than he awarded to most of his students – because he was actively working on the problem with me. However, I never made what Coleman considered substantial "progress" as measured by his standards. On the other hand, I did many things that, in retrospect, would have been publishable on their own. For example, I was very proud of a trick I invented (only to be told later that it was first done by Heisenberg[N]) for solving (at least in the simplest approximation [what's called 1/N]) what came later to be known as the Gross-Neveu model.

Coleman took a leave of absence from Harvard, taking his sabbatical at Princeton. At that point, I decided I needed a research program on which I could proceed on my own – something that might not meet Coleman's high standards but on which I might have some chance of success. I decided to look into whether the renormalization group had anything to say about the low energy (or ground state) behavior of Yang-Mills theory. An analogous analysis for electrodynamics appeared in the classic textbook of Bogoliubov and Shirkov, though Coleman characterized the relevant chapter there as "mysterious". This was a possible approach to the question I articulated regarding no scalar fields, but I thought I might be able to follow the steps of Bogoliubov and Shirkov explicitly as a guide.

A key first step was to know the Yang-Mills beta function. (I assumed [correctly] in my live talk that its definition had been made clear in the earlier remarks of my co-recipients; it is, after all, the minus sign to which I first alluded.) By the way, Erick Weinberg was supposed to compute it for an appendix of his thesis, to carry out a generalization of a renormalization group flow argument that appears in the Coleman-Weinberg paper, except for a realistic, non-Abelian weak interaction theory. But, in the end, I guess he figured he had enough stuff to get his degree, and it was time for him to move on to something new. I had actually hoped we'd compare notes, but he never attempted the calculation.

I visited Coleman a couple of times in Princeton. When I described to him my new, specific research program, I asked if he knew whether the beta function had already been computed. He thought not but said we should ask David Gross, who was down the hall. David said no, and we discussed briefly then that, while the calculation may have seemed to some to be daunting, it would, in fact, be straightforward.

Fortunately for both of us – and for Frank, too – he was probably wrong, though this episode is fraught with ambiguity. To my knowledge, there are no relevant printed records of the crucial bits of the story, which have been handed down only as folklore, existing in a variety of variant versions.

At a major particle physics meeting in Marseilles the previous year, attended by many particle physics luminaries, Symanzik gave a talk precisely about what came to be known as asymptotic freedom. He described how it could account for the otherwise mysterious results from SLAC on electron-proton scattering. Symanzik knew that the beta functions for other theories were all positive. In fact, many wise people thought there was a general, model-independent argument for positivity. For example, Schwinger[N] later asked me after hearing me speak on the subject, "What about the positivity of the spectral function?" I.e., intermediate, physical states come with positive probabilities. (This refers to an argument that is, indeed, relevant to other theories.) Symanzik said it would be interesting to know the answer for Yang-Mills theory, and then 't Hooft[N] announced it was negative. In some versions of the story, 't Hooft[N] spoke up at the question and answer period following Symanzik's talk. However, there are attendees of the meeting who have no such recollection. In other versions, it was a private exchange between 't Hooft[N] and Symanzik.

There are a variety of first-, second-, and third-hand accounts of why nothing further was heard on this subject from 't Hooft[N]. I won't repeat them here.* But I'd like to speculate a complementary perspective as to why no

* I will add one conjecture to the list, though it is not something I ever confirmed with 't Hooft[N]. It is possible that at that time 't Hooft[N] knew the sign of the beta function but not its coefficient. His calculations employed dimensional regularization and dimensional subtraction. From these he would have known the sign of the renormalization constants. However, the fundamental definition of the beta function makes reference to the response of the theory to scale transformations. Dimensional regularization introduces a scale in a subtle way – when one analytically continues away from the superficially scale invariant dimensions. How the traditional renormalization group is represented in this context is something that was worked out only a couple of years later.

one else at the meeting got wind of it or otherwise took any notice. (Admittedly, I wasn't there. So this is pure speculation.) Most theorists' attention then was on weak interactions, and this is a strong interaction issue. But that's not a good enough excuse. People did, after all, talk a lot about the scaling of the Stanford electron-proton experiments. Rather, I think Symanzik's speaking style played a crucial role. He was a charming, intense, sweet, and brilliant man. But his live delivery left something to be desired. I remember a different talk I heard him give on a somewhat related subject. He used hand-written slides for an overhead projector (which were the industry standard at the time for technical presentations). However, he obviously wrote out his slides with lined paper underneath as a guide, using every line. So he ended up with over twenty five lines of equations and text per page. His handwriting was typical German: undecipherable, at least to Americans, looking like endless up-down-up-down-up-down. The clincher, though, was when an equation on one page referred to an equation on another. He'd slap the second slide on top of the first, off-set the two by half a line, and point to both.

I slowly and carefully completed a calculation of the Yang-Mills beta function. I happen to be ambidextrous and mildly dyslexic. So I have trouble with left/right, in/out, forward/backward, etc. Hence, I derived each partial result from scratch, paying special attention to signs and conventions. It did not take long to go from dismay over the final minus sign (it was indeed useless for studying low energy phenomena) to excitement over the possibilities. I phoned Sidney Coleman. He listened patiently and said it was interesting. But, according to Coleman, I had apparently made an error because David Gross and his student had completed the same calculation, and they found it was plus. Coleman seemed to have more faith in the reliability of a team of two, which included a seasoned theorist, than in a single, young student. I said I'd check it yet once more. I called again about a week later to say I could find nothing wrong with my first calculation. Coleman said yes, he knew because the Princeton team had found a mistake, corrected it, and already submitted a paper to *Physical Review Letters*.

On learning of the Gross-Wilczek-Politzer result, Ken Wilson[N], who might have thought of its impossibility along the same lines as I attributed to Schwinger[N], above, knew who to call to check the result. He realized that there were actually several people around the world who had done the calculation, *en passant* as it were, as part of their work on radiative corrections to weak interactions in the newly-popular Weinberg-Salam model. They just never thought to focus particularly on this aspect. But they could quickly confirm for Wilson[N] by looking in their notebooks that the claimed result was, indeed, correct.

Steve Weinberg[N] and Murray Gell-Mann[N] were among those to instantly embrace non-Abelian color SU(3) gauge theory as the theory of the strong interactions. In Gell-Mann's case, it was in no small part because he had already invented it (!) with Harald Fritzsch and christened it QCD. He had previously articulated three solid arguments for choosing this particular theory. (For the physicists, those arguments were: baryon statistics, $\pi \rightarrow 2\gamma$, and the electron-positron annihilation cross section). And asymptotic freedom, i.e., the negative beta function, was the clincher. I'd only heard of Gell-Mann[N] and Fritzsch's work second hand, from Shelly Glashow[N], and he seemed think it shouldn't be taken too seriously. I only later realized it was more Glashow's mode of communication than his serious assessment of the plausibility of the proposal. In any case, I had completely lost track of Gell-Mann[N] and Fritzsch's QCD.

After the first seminar I ever gave on this subject (it was at MIT), I was approached by Ken Johnson (who, himself had done pioneering work on the renormalization group years earlier) and Vicki Weisskopf. "Very nice," they said. "Too bad that it is in glaring contradiction to at least two important classes of experiments." One problem was the electron-positron cross section, which had only gotten much worse since Gell-Mann[N] and Fritzsch's proposal of QCD, and the other was the issue of large angle products in proton-proton collisions. There were many more energetic particles produced than expected (naively) from QCD. By the way, this second issue attracted Dick Feynman's attention. And it wasn't until a couple of years later and his careful analysis with Rick Field that QCD was reconciled with those experiments. Only then did Feynman[N] join the ranks of the believers.

The experimentally measured electron-positron cross section (as a function of increasing collision energy) had leveled off – instead of continuing to drop steeply, which was thought to be a QCD prediction. In Aspen, Colorado, in the summer of 1974, I crossed paths with Ken Wilson[N], who, characteristically succinct, said, "It's charm, and it's not short distance." Tom Appelquist and I made it our task to understand those oracular comments and flesh out their consequences. By the end of the summer, the reconciliation of QCD with the experimental measurements was pretty clear to us. Tom toured the country explaining our work. His seminars included a sketch of what the cross section really was as opposed to what the experimentalists reported and an estimate, albeit technically an upper bound, on the astoundingly long lifetime of a particle that was being produced and decaying as yet unnoticed. Many people heard those talks and remember them, and there is at least one, objective written record of their existence: Sid Drell gave an account in a piece he wrote subsequently for *Scientific American* about charm. At the time, there were already many what-proved-to-be wrong papers trying to interpret the electron-positron experiments, and the SLAC experiment leader, Burt Richter[N], was touring the country explaining that he had made the monumental discovery that the electron was actually a little hadron, i.e., a strongly interacting particle like the proton, only much smaller in diameter. (This discovery, or at least the same experimental results, had been observed a few years earlier at the Cambridge Electron Accelerator, a joint Harvard-MIT venture. But no one believed it, and the machine was decommissioned.) Appelquist and I were drafting a paper. But I was the conservative one, perhaps overly influenced, I later realized, by a talk that I had heard by Steve Adler as to how large the discrepancy between naive QCD calculation and experimental measurements could be before the theory was in definite trouble. I focused on the things we could most reliably compute and did not appreciate the correctness of Tom's more general arguments.

In November that Fall came the experimental announcements. SLAC observed a particle (they called it the ψ) and ultimately observed a whole cross section just as predicted by Appelquist. And observation was simultaneously announced by Sam Ting$^{\mathcal{N}}$, in an experiment that identified a pimple, which Ting$^{\mathcal{N}}$ eponymously titled the $\mathcal{J}^*$, on what had been known as Lederman's Shoulder. (That's Leon Lederman$^{\mathcal{N}}$.) That is to say that Ting's experiment had actually been done earlier by Lederman$^{\mathcal{N}}$. The earlier experiment had cruder resolution, but it clearly indicated that there was something anomalous at just that energy.

Appelquist and I hurriedly dashed off a short version of our work to *Physical Review Letters*, where it was immediately and unequivocally rejected by senior editor Sy Pasternak. It was against that journal's policy to let authors engage in the coining of frivolous, new terminology. In the case at hand, our friend and colleague Alvaro De Rújula, on hearing of our work, had coined the term "charmonium", which in a single word was able to transmit the central new idea of the paper to any serious particle physics reader. Ultimately, Shelly Glashow$^{\mathcal{N}}$ brokered a compromise with Pasternak. We could use "charmonium" in the text but not in the title. The negotiations caused a delay of a couple of weeks – a long time in those heady days. As a consequence, publication came along side several other long-since-forgotten papers, instead of being hard on the heals of the experimental discovery.

That our explanation was correct was soon widely appreciated, and it convinced almost all of the remaining skeptics of the validity of QCD. I suspect that the consensus on this issue was a major contributing factor to the Royal Swedish Academy of Sciences' recognition within just a couple of years of Richter$^{\mathcal{N}}$ and Ting's discovery.

I hope you all now understand why I owe Tom Appelquist a huge, profound, and public apology. We certainly could have submitted for publication in September substantially the same paper we ultimately wrote two months later.

Now, somewhat out of chronological order, I'd like to express my thanks to my old friend and collaborator Howard Georgi. After the calculation of the beta function, it was fairly obvious what should be done next. One had to redo some calculations that had been done earlier by Norman Christ, Brosl Hasslacher, and Al Mueller but in the context of what was now, obviously, the right theory. Here, again, a missing name from that collaboration but who had a major impact was Georgio Parisi. Well, Howard Georgi checked up periodically on my progress, and I admitted having some technical trouble. So he volunteered to help, and we went on to do an enormous number of clever things together.

Apropos clever, there are some advances that require considerable mental struggle and lengthy argumentation, only to virtually disappear as non-issues,

*a reasonable approximation to the relevant Chinese character.

because they're simply obvious from a newer perspective. For example, the fact that quarks could have a mass, something unambiguously quantifiable and measured in grams – in spite of their never existing as isolated particles – was one such issue on which I battled with many physicists, including, for example, Gell-Mann[N] and Steve Weinberg[N]. The heavy charm quark gave impetus to those considerations, but there was a conceptual battle that had to be fought against older prejudices formed in the limited context of the "light" quarks. Younger physicists today can't even imagine that there was ever an issue.

Heavy quarks appeared once again in my research life. Joe Polchinski asked Mark Wise, a colleague of mine at Caltech, a question about heavy quark calculations, which Mark and I proceeded to answer. It was again a case where, unbeknownst to us, the work had already been done, this time by Misha Shifman and Mike Voloshin in the Soviet Union. Furthermore, I again missed the most important phenomenological consequences of that line of thought. Those had to wait for the collaboration of Mark Wise with Nathan Isgur. That heavy quark physics depends only trivially on the actual value of the heavy quark mass was obvious to me and probably most anyone else who gave it a thought. What Isgur and Wise noted was that in a world with more than one type of heavy quark, this gives rise to symmetries of monumentally useful importance. (The second heavy quark, the so-called bottom quark, was identified only several years after the first, i.e., the charmed quark.)

The establishment by the mid-1970's of QCD as the correct theory of the strong interactions completed what is now known prosaically as the Standard Model. It offers a description of all known fundamental physics except for gravity, and gravity is something that has no discernible effect when particles are studied a few at a time. However, the situation is a bit like the way that the Navier-Stokes equation accounts for the flow of water. The equations are at some level obviously correct, but there are only a few, limited circumstances in which their consequences can be worked out in any detail. Nevertheless, many leading physicists were inclined to conclude in the late 1970's that the task of basic physics was nearly complete, and we'd soon be out of jobs. A famous example was the inaugural lecture of Stephen Hawking as Lucasian Professor of Mathematics, a chair first held by Isaac Barrow at Cambridge University. Hawking titled his lecture, "Is the End in Sight for Theoretical Physics?" And he argued strongly for "Yes".

But more recent observations of astronomers have turned things on their heads. Recall, if you will, that among the many stupendous insights of Isaac Newton, the second Lucasian Professor, was the idea that the stuff of the heavens was the same stuff as matter here on Earth. This was revolutionary. And he asserted that the laws that governed the motion of stuff in the heavens were the same laws as applied to matter on Earth. (That there are laws at all may be his most profound insight. It is certainly what came to define the whole discipline of physics.) For three centuries we accumulated stunning detailed confirmation of these of Newton's assertions. But in a very fundamental way both of these ideas now appear to be about as wrong as they possibly could be

– at least that's the simplest interpretation of our current large-scale astrophysical observations. It turns out that we haven't a clue what virtually all of the matter in the universe consists of – except that it's not made of the particles that make up matter on Earth or in the stars. Furthermore, the force which governs the largest scale motions in the universe has nothing to do with the forces of the Standard Model or with gravity as it is familiar here on Earth.

There is a very active field of theoretical research which seeks to go beyond the Standard Model. Success in these endeavors would mean explaining the apparently arbitrary aspects of the Standard Model; success would mean bringing an account of gravity into the picture; and success would mean illuminating the previously mentioned issues in astrophysics. However, we now face a very serious problem in advancing the experimental frontier, a problem which few people like to discuss. It seems to me that ever since Leeuwenhoek, advances in the resolving power of our "microscopes" have come with similar investments of capital and manpower. I.e., an increase by an order of magnitude in the one required an increase by roughly an order of magnitude of the other – at least once we average over fits and starts and brilliant insights. The last big machine planned and canceled in the U.S. was to cost about $10 billion. (That's $\$10^{10}$.) That would have allowed us to reach distances small enough to study the interactions of weak bosons directly. The realm of the conjectured "unification" of the forces of the Standard model, the realm of their possible unification with gravity, and the basic physics of String Theory, the most widely pursued approach to a physics more fundamental than the Standard Model, are all more than a dozen orders of magnitude further away. However, $\$10^{22}$ is simply not available for this line of research (or anything else for that matter).

The question of the benefit of this work incurred on mankind, an aspect stipulated in Mr. Nobel's will, is a whole other topic. But, as I said at the outset, I certainly appreciate the care and wisdom invested by the Royal Swedish Academy of Sciences in identifying noteworthy advances in fundamental physics – and in identifying the particular advance that we celebrate today. The reality of the actual progress of science is, however, often very complicated, as I hope I have conveyed from my few examples. The committees of the Academy know this full well, but their deliberations are confidential. I felt strongly that more of the public should contemplate these matters if they wish to understand not just the ideas of science but also how they have developed. I also hope that more of the scientific community would remember them, too.

My presentation in Stockholm ended at this point, but, in the days that followed, it prompted a variety of comments, questions, and exchanges. I'd like to add here a brief version of one of them. I was asked, point blank, what I actually thought of the 2004 Nobel Prizes in Physics, aside from the obvious personal considerations. And this is a distilled version of my reply. Recognition of the theory of the strong interactions is an obvious choice – for all the reasons that have been discussed in my co-recipients' lectures, in the presentation speech in Stockholm by Lars Brink, in the assembled material of the Nobel Foundation, and in the wide coverage elsewhere. However, in my view, getting to our current level of understanding has been a rich and complex story. Nevertheless, I believe that it is the overwhelming consensus (but by no means unanimous) opinion of researchers in the field – and I personally agree – that the discovery of asymptotic freedom was a genuinely crucial event. For some, it made everything clear. For others, it was only the beginning. And for yet others, it was the beginning of the final chapter. But in any case, it was key.

REFERENCES

The two books I would recommend first which give excellent accounts of this epoch in particle physics (and more) are:

1. *The Second Creation,* by Robert Crease and Charles Mann.
2. *Constructing Quarks,* by Andrew Pickering.

The first is colorful in its rendition of the personalities, rather accurate in its physics, and totally accessible to the interested layman. The second is a more scholarly endeavor. Pickering began his career as a particle theorist, a contemporary of mine. He includes considerable scientific detail but still aims at a non-technical audience.

Two marvelous books on twentieth century physics for the interested layman which focus more on scientific substance rather than historical process are:

3. *The Cosmic Code*
4. *Perfect Symmetry*

both by Heinz Pagels, something of a self-styled New York City dandy, but as charming a person as could be. He died young but just as he dreamed it would be.

The Cosmic Code is about quantum mechanics, and *Perfect Symmetry* covers more of the sweep of particle physics and cosmology.

I have not sought out the actual published references for the relevant points in my narrative. They're not hard to find, but this is not a refereed journal. And yet, there is a potentially enlightening aspect to my having put this together purely from memory in October and November of 2004. While standard references are unequivocally available in the published record, what actually transpired, leading to those publications, is not. We rely on people's personal accounts. And now we enter the interesting realm where participants in the same event may have very different and mutually contradictory perceptions of what transpired, *and* those perceptions may shift as time passes. While intentional deception is not an unheard of phenomenon, these phenomena effect the reports of people with the highest integrity. Although evaluating the accuracy of my personal recollections may be very difficult, at least it would be possible to see how good my memory is with respect to items that can be confirmed or refuted.

Frank Wilczek

FRANK A. WILCZEK

The most deeply formative events of my scientific career long preceded my first contact with the research community; indeed, some of them preceded my birth.

My grandparents emigrated from Europe in the aftermath of World War I, as young teenagers; on my father's side they came from Poland and on my mother's side from Italy, near Naples. My grandparents arrived with nothing, and no knowledge of English. My grandfathers were a blacksmith and a mason, respectively. Both my parents were born on Long Island, in 1926, and they have lived there ever since. I was born in 1951, and grew up in a place called Glen Oaks, which is in the northeast corner of Queens, barely within the city limits of New York City.

I've always loved all kinds of puzzles, games, and mysteries. Some of my earliest memories are about the questions I "worked on" even before I went to school. When I was learning about money, I spent a lot of time trying out various schemes of exchanging different kinds of money (e.g., pennies, nickels, and dimes) in complicated ways back and forth, hoping to discover a way to come out ahead. Another project was to find ways of getting very big numbers in a few steps. I discovered simple forms of repeated exponentiation and recursion for myself. Generating large numbers made me feel powerful.

With these inclinations, I suspect I was destined for some kind of intellectual work. A few special circumstances led me to science, and eventually to theoretical physics.

My parents were children during the time of the Great Depression, and their families struggled to get by. This experience shaped many of their attitudes, and especially their aspirations for me. They put very great stock in education, and in the security that technical skill could bring. When I did well in school they were very pleased, and I was encouraged to think about becoming a doctor or an engineer. As I was growing up my father, who worked in electronics, was taking night classes. Our little apartment was full of old radios and early-model televisions, and with the books he was studying. It was the time of the Cold War. Space exploration was a new and exciting prospect, nuclear war a frightening one; both were ever-present in newspapers, TV, and movies. At school, we had regular air raid drills. All this made a big impression on me. I got the idea that there was secret knowledge that, when mastered, would allow Mind to control Matter in seemingly magical ways.

Another thing that shaped my thinking was religious training. I was brought up as a Roman Catholic. I loved the idea that there was a great drama

Betsy from around the time I met her.

Me, speaking at Fermilab in 1977.

and a grand plan behind existence. Later, under the influence of Bertrand Russell's writings and my increasing awareness of scientific knowledge, I lost faith in conventional religion. A big part of my later quest has been trying to regain some of the sense of purpose and meaning that was lost. I'm still trying.

I went to public schools in Queens, and was fortunate to have excellent teachers. Because the schools were big, they could support specialized and advanced classes. At Martin van Buren High School there was a group of thirty or so of us who went to many such classes together, and both supported and competed with one another. More than half of us went on to successful scientific or medical careers.

I arrived at the University of Chicago with large but amorphous ambitions. I flirted with brain science, but soon decided that the central questions were not ready for mathematical treatment, and that I lacked the patience for laboratory work. I read voraciously in many subjects, but I wound up majoring in mathematics, largely because doing that gave me the most freedom. During my last term at Chicago, I took a course about the use of symmetry and group theory in physics from Peter Freund. He was an extremely enthusiastic and inspiring teacher, and I felt an instinctive resonance with the material. I went to Princeton University as a graduate student in the math department, but kept a close eye on what was going on in physics. I became aware that deep ideas involving mathematical symmetry were turning up at

My hero Feynman giving me grief at Murph Goldberger's birthday party in 1983; the devil horns are from Sam Treiman.

Me and my family in 2001.

the frontiers of physics; specifically, the gauge theory of electroweak interactions, and the scaling symmetry in Wilson's theory of phase transitions. I started to talk with a young professor named David Gross, and my proper career as a physicist began.

The great event of my early career was to help discover the basic theory of the strong force, QCD. That is the subject of the following lecture. The equations of QCD are based on gauge symmetry principles, and we make progress with them using (approximate) scaling symmetry. It was very gratifying to find that the ideas I admired as a student could be used to get a powerful and accurate theory for an important part of fundamental physics. I continue to apply these ideas in new ways, and I am certain that they have a great future.

An aspect of my later work that is not much reflected in the lecture, has been to use insights and methods from "fundamental" physics to address "applied" questions, and vice versa. I'm not sure that fractional quantum numbers, transmuted quantum statistics, exotic superfluidities, or the gauge theory of swimming at low Reynolds number have really arrived as applied physics (yet?), but I've derived a lot of joy from my discoveries in these areas.

To me, the unity of knowledge is a living ideal and goal. I continue, as in my student days, to read voraciously in many subjects, and to think about them. I hope to further expand the horizons of my writing and work in the future.

I've been blessed with a wife, Betsy Devine, and two daughters, Amity and Mira, who've been an inexhaustible source of joy and entertainment.

ASYMPTOTIC FREEDOM: FROM PARADOX TO PARADIGM

Nobel Lecture, December 8, 2004

by

FRANK A. WILCZEK

Massachusetts Institute of Technology (MIT), Cambridge, USA.

1 A PAIR OF PARADOXES

In theoretical physics, paradoxes are good. That's paradoxical, since a paradox appears to be a contradiction, and contradictions imply serious error. But Nature cannot realize contradictions. When our physical theories lead to paradox we must find a way out. Paradoxes focus our attention, and we think harder.

When David Gross and I began the work that led to this Nobel Prize [1, 2, 3, 4], in 1972, we were driven by paradoxes. In resolving the paradoxes we were led to discover a new dynamical principle, asymptotic freedom. This principle in turn has led to an expanded conception of fundamental particles, a new understanding of how matter gets its mass, a new and much clearer picture of the early universe, and new ideas about the unity of Nature's forces. Today I'd like to share with you the story of these ideas.

1.1 *Paradox 1: Quarks are Born Free, but Everywhere They are in Chains*

The first paradox was phenomenological.

Near the beginning of the twentieth century, after pioneering experiments by Rutherford, Geiger and Marsden, physicists discovered that most of the mass and all of the positive charge inside an atom is concentrated in a tiny central nucleus. In 1932, Chadwick discovered neutrons, which together with protons could be considered as the ingredients out of which atomic nuclei could be constructed. But the known forces, gravity and electromagnetism, were insufficient to bind protons and neutrons tightly together into objects as small as the observed nuclei. Physicists were confronted with a new force, the most powerful in Nature. It became a major challenge in fundamental physics, to understand this new force.

For many years physicists gathered data to address that challenge, basically by bashing protons and neutrons together and studying what came out. The results that emerged from these studies, however, were complicated and hard to interpret.

What you would expect, if the particles were really fundamental (indestructible), would be the same particles you started with, coming out with just their trajectories changed. Instead, the outcome of the collisions was often many particles. The final state might contain several copies of the originals, or different particles altogether. A plethora of new particles was discovered in this way. Although these particles, generically called hadrons, are unstable, they otherwise behave in ways that broadly resemble the way protons and neutrons behave. So the character of the subject changed. It was no longer natural to think of it as simply as the study of a new force that binds protons and neutrons into atomic nuclei. Rather, a new world of phenomena had come into view. This world contained many unexpected new particles, that could transform into one another in a bewildering variety of ways. Reflecting this change in perspective, there was a change in terminology. Instead of the nuclear force, physicists came to speak of the strong interaction.

In the early 1960s, Murray Gell-Mann and George Zweig made a great advance in the theory of the strong interaction, by proposing the concept of quarks. If you imagined that hadrons were not fundamental particles, but rather that they were assembled from a few more basic types, the quarks, patterns clicked into place. The dozens of observed hadrons could be understood, at least roughly, as different ways of putting together just three kinds ("flavors") of quarks. You can have a given set of quarks in different spatial orbits, or with their spins aligned in different ways. The energy of the configuration will depend on these things, and so there will be a number of states with different energies, giving rise to particles with different masses, according to $m = E/c^2$. It is analogous to the way we understand the spectrum of excited states of an atom, as arising from different orbits and spin alignments of electrons. (For electrons in atoms the interaction energies are relatively small, however, and the effect of these energies on the overall *mass* of the atoms is insignificant.)

The rules for using quarks to model reality seemed quite weird, however.

Quarks were supposed to hardly notice one another when they were close together, but if you tried to isolate one, you found that you could not. People looked very hard for individual quarks, but without success. Only bound states of a quark and an antiquark – mesons – or bound states of three quarks – baryons – are observed. This experimental regularity was elevated into The Principle of Confinement. But giving it a dignified name didn't make it less weird.

There were other peculiar things about quarks. They were supposed to have electric charges whose magnitudes are fractions ($\frac{2}{3}$ or $\frac{1}{3}$) of what appears to be the basic unit, namely the magnitude of charge carried by an electron or proton. All other observed electric charges are known, with great accuracy, to be whole-number multiples of this unit. Also, identical quarks did not appear to obey the normal rules of quantum statistics. These rules would require that, as spin $\frac{1}{2}$ particles, quarks should be fermions, with antisymmetric wave functions. The pattern of observed baryons cannot be understood using antisymmetric wave functions; it requires symmetric wave functions.

The atmosphere of weirdness and peculiarity surrounding quarks thickened into paradox when J. Friedman, H. Kendall, R. Taylor and their collaborators at the Stanford Linear Accelerator (SLAC) used energetic photons to poke into the inside of protons [5]. They discovered that there are indeed entities that look like quarks inside protons. Surprisingly, though, they found that when quarks are hit hard they seem to move (more accurately: to transport energy and momentum) as if they were free particles. Before the experiment, most physicists had expected that whatever caused the strong interaction of quarks would also cause quarks to radiate energy abundantly, and thus rapidly to dissipate their motion, when they got violently accelerated.

At a certain level of sophistication, that association of radiation with forces appears inevitable, and profound. Indeed, the connection between forces and radiation is associated with some of the most glorious episodes in the history of physics. In 1864, Maxwell predicted the existence of electromagnetic radiation – including, but not limited to, ordinary light – as a consequence of his consistent and comprehensive formulation of electric and magnetic forces. Maxwell's new radiation was subsequently generated and detected by Hertz, in 1883 (and over the twentieth century its development has revolutionized the way we manipulate matter and communicate with one another). Much later, in 1935, Yukawa predicted the existence of pions based on his analysis of nuclear forces, and they were subsequently discovered in the late 1940s; the existences of many other hadrons were predicted successfully using a generalization of these ideas. (For experts: I have in mind the many resonances that were first seen in partial wave analyses, and then later in production.) More recently the existence of *W* and *Z* bosons, and of color gluons, and their properties, was inferred before their experimental discovery. Those discoveries were, in 1972, still ahead of us, but they serve to confirm, retroactively, that our concerns were worthy ones. Powerful interactions ought to be associated with powerful radiation. When the most powerful interaction in nature, the strong interaction, did not obey this rule, it posed a sharp paradox.

1.2 *Paradox 2: Special Relativity and Quantum Mechanics Both Work*

The second paradox is more conceptual. Quantum mechanics and special relativity are two great theories of twentieth-century physics. Both are very successful. But these two theories are based on entirely different ideas, which are not easy to reconcile. In particular, special relativity puts space and time on the same footing, but quantum mechanics treats them very differently. This leads to a creative tension, whose resolution has led to three previous Nobel Prizes (and ours is another).

The first of these prizes went to P. A. M. Dirac (1933). Imagine a particle moving on average at very nearly the speed of light, but with an uncertainty in position, as required by quantum theory. Evidently it there will be some probability for observing this particle to move a little faster than average, and therefore faster than light, which special relativity won't permit. The only known way to resolve this tension involves introducing the idea of antiparticles. Very roughly speaking, the required uncertainty in position is accommodated by allowing for the possibility that the act of measurement can involve the creation of several particles, each indistinguishable from the original, with different positions. To maintain the balance of conserved quantum numbers, the extra particles must be accompanied by an equal number of antiparticles. (Dirac was led to predict the existence of antiparticles through a sequence of ingenious interpretations and re-interpretations of the elegant relativistic wave equation he invented, rather than by heuristic reasoning of the sort I've presented. The inevitability and generality of his conclusions, and their direct relationship to basic principles of quantum mechanics and special relativity, are only clear in retrospect).

The second and third of these prizes were to R. Feynman, J. Schwinger, and S.-I. Tomonaga (1965) and to G. 't Hooft and M. Veltman (1999) respectively. The main problem that all these authors in one way or another addressed is the problem of ultraviolet divergences.

When special relativity is taken into account, quantum theory must allow for fluctuations in energy over brief intervals of time. This is a generalization of the complementarity between momentum and position that is fundamental for ordinary, non-relativistic quantum mechanics. Loosely speaking, energy can be borrowed to make evanescent virtual particles, including particle-antiparticle pairs. Each pair passes away soon after it comes into being, but new pairs are constantly boiling up, to establish an equilibrium distribution. In this way the wave function of (superficially) empty space becomes densely populated with virtual particles, and empty space comes to behave as a dynamical medium.

The virtual particles with very high energy create special problems. If you calculate how much the properties of real particles and their interactions are changed by their interaction with virtual particles, you tend to get divergent answers, due to the contributions from virtual particles of very high energy.

This problem is a direct descendant of the problem that triggered the introduction of quantum theory in the first place, i.e. the "ultraviolet catastrophe" of black body radiation theory, addressed by Planck. There the problem was that high-energy modes of the electromagnetic field are predicted, classically, to occur as thermal fluctuations, to such an extent that equilibrium at any finite temperature requires that there is an infinite amount of energy in these modes. The difficulty came from the possibility of small-amplitude fluctuations with rapid variations in space and time. The element of discreteness introduced by quantum theory eliminates the possibility of very small-amplitude fluctuations, because it imposes a lower bound on their size. The (relatively) large-amplitude fluctuations that remain are predicted to occur very rarely in thermal equilibrium, and cause no problem. But quantum fluctuations are much more efficient than are thermal fluctuations at exciting the high-energy modes, in the form of virtual particles, and so those modes come back to haunt us. For example, they give a divergent contribution to the energy of empty space, the so-called zero-point energy.

Renormalization theory was developed to deal with this sort of difficulty. The central observation that is exploited in renormalization theory is that although interactions with high-energy virtual particles appear to produce divergent corrections, they do so in a very structured way. That is, the same corrections appear over and over again in the calculations of many different physical processes. For example in quantum electrodynamics (QED) exactly two independent divergent expressions appear, one of which occurs when we calculate the correction to the mass of the electron, the other of which occurs when we calculate the correction to its charge. To make the calculation mathematically well-defined, we must artificially exclude the highest energy modes, or dampen their interactions, a procedure called applying a cut-off, or regularization. In the end we want to remove the cut-off, but at intermediate stages we need to leave it in, so as to have well-defined (finite) mathematical expressions. If we are willing to take the mass and charge of the electron from experiment, we can identify the formal expressions for these quantities, including the potentially divergent corrections, with their measured values. Having made this identification, we can remove the cutoff. We thereby obtain well-defined answers, in terms of the measured mass and charge, for everything else of interest in QED.

Feynman, Schwinger, and Tomonoga developed the technique for writing down the corrections due to interactions with any finite number of virtual particles in QED, and showed that renormalization theory worked in the simplest cases. (I'm being a little sloppy in my terminology; instead of saying the number of virtual particles, it would be more proper to speak of the number of internal loops in a Feynman graph.) Freeman Dyson supplied a general proof. This was intricate work, that required new mathematical techniques. 't Hooft and Veltman showed that renormalization theory applied to a much wider class of theories, including the sort of spontaneously broken gauge theories that had been used by Glashow, Salam, and Weinberg to construct the (now) standard model of electroweak interactions. Again, this was intricate and highly innovative work.

This brilliant work, however, still did not eliminate all the difficulties. A very profound problem was identified by Landau [6]. Landau argued that virtual particles would tend to accumulate around a real particle as long as there was any uncancelled influence. This is called screening. The only way for this screening process to terminate is for the source plus its cloud of virtual particles to cease to be of interest to additional virtual particles. But then, in the end, no uncancelled influence would remain – and no interaction!

Thus all the brilliant work in QED and more general field theories represented, according to Landau, no more than a temporary fix. You could get finite results for the effect of any particular number of virtual particles, but when you tried to sum the whole thing up, to allow for the possibility of an arbitrary number of virtual particles, you would get nonsense – either infinite answers, or no interaction at all.

Landau and his school backed up this intuition with calculations in many different quantum field theories. They showed, in all the cases they calculated, that screening in fact occurred, and that it doomed any straightforward attempt to perform a complete, consistent calculation by adding up the contributions of more and more virtual particles. We can sweep this problem under the rug in QED or in electroweak theory, because the answers including only a small finite number of virtual particles provide an excellent fit to experiment, and we make a virtue of necessity by stopping there. But for the strong interaction that pragmatic approach seemed highly questionable, because there is no reason to expect that lots of virtual particles won't come into play, when they interact strongly.

Landau thought that he had destroyed quantum field theory as a way of reconciling quantum mechanics and special relativity. Something would have to give. Either quantum mechanics or special relativity might ultimately fail, or else essentially new methods would have to be invented, beyond quantum field theory, to reconcile them. Landau was not displeased with this conclusion, because in practice quantum field theory had not been very helpful in understanding the strong interaction, even though a lot of effort had been put into it. But neither he, nor anyone else, proposed a useful alternative.

So we had the paradox, that combining quantum mechanics and special relativity seemed to lead inevitably to quantum field theory; but quantum field theory, despite substantial pragmatic success, self-destructed logically due to catastrophic screening.

2 PARADOX LOST: ANTISCREENING, OR ASYMPTOTIC FREEDOM

These paradoxes were resolved by our discovery of asymptotic freedom.

We found that some very special quantum field theories actually have antiscreening. We called this property asymptotic freedom, for reasons that will soon be clear. Before describing the specifics of the theories, I'd like to indicate in a rough, general way how the phenomenon of antiscreening allows us to resolve our paradoxes.

Antiscreening turns Landau's problem on its head. In the case of screening, a source of influence – let us call it charge, understanding that it can represent something quite different from electric charge – induces a canceling cloud of virtual particles. From a large charge, at the center, you get a small observable influence far away. Antiscreening, or asymptotic freedom, implies instead that a charge of intrinsically small magnitude catalyzes a cloud of virtual particles that enhances its power. I like to think of it as a thundercloud that grows thicker and thicker as you move away from the source.

Since the virtual particles themselves carry charge, this growth is a self-reinforcing, runaway process. The situation appears to be out of control. In particular, energy is required to build up the thundercloud, and the required energy threatens to diverge to infinity. If that is the case, then the source could never be produced in the first place. We've discovered a way to avoid Landau's disease – by banishing the patients!

At this point our first paradox, the confinement of quarks, makes a virtue of theoretical necessity. For it suggests that there *are* in fact sources – specifically, quarks – that cannot exist on their own. Nevertheless, Nature teaches us, these confined particles can play a role as building-blocks. If we have, nearby to a source particle, its antiparticle (for example, quark and antiquark), then the catastrophic growth of the antiscreening thundercloud is no longer inevitable. For where they overlap, the cloud of the source can be canceled by the anticloud of the antisource. Quarks and antiquarks, bound together, can be accommodated with finite energy, though either in isolation would cause an infinite disturbance.

Because it was closely tied to detailed, quantitative experiments, the sharpest problem we needed to address was the paradoxical failure of quarks to radiate when Friedman, Kendall, and Taylor subjected them to violent acceleration. This too can be understood from the physics of antiscreening. According to this mechanism, the color charge of a quark, viewed up close, is small. It builds up its power to drive the strong interaction by accumulating a growing cloud at larger distances. Since the power of its intrinsic color charge is small, the quark is actually only loosely attached to its cloud. We can jerk it away from its cloud, and it will – for a short while – behave almost as if it had no color charge, and no strong interaction. As the virtual particles in space respond to the altered situation they rebuild a new cloud, moving along with the quark, but this process does not involve significant radiation of energy and momentum. That, according to us, was why you could analyze the most salient aspects of the SLAC experiments – the inclusive cross-sections, which only keep track of overall energy-momentum flow – as if the quarks were free particles, though in fact they are strongly interacting and ultimately confined.

Thus both our paradoxes, nicely dovetailed, get resolved together through antiscreening.

The theories that we found to display asymptotic freedom are called nonabelian gauge theories, or Yang-Mills theories [7]. They form a vast generalization of electrodynamics. They postulate the existence of several different kinds of charge, with complete symmetry among them. So instead of one entity, "charge", we have several "colors". Also, instead of one photon, we have a family of color gluons.

The color gluons themselves carry color charges. In this respect the nonabelian theories differ from electrodynamics, where the photon is electrically neutral. Thus gluons in nonabelian theories play a much more active role in the dynamics of these theories than do photons in electrodynamics. Indeed, it is the effect of virtual gluons that is responsible for antiscreening, which does not occur in QED.

It became evident to us very early on that one particular asymptotically free theory was uniquely suited as a candidate to provide the theory of the strong interaction. On phenomenological grounds, we wanted to have the possibility to accommodate baryons, based on three quarks, as well as mesons, based on quark and antiquark. In light of the preceding discussion, this requires that the color charges of three different quarks can cancel, when you add them up. That can oocur if the three colors exhaust all possibilities; so we arrived at the gauge group $SU(3)$, with three colors, and eight gluons. To be fair, several physicists had, with various motivations, suggested the existence of a three-valued internal color label for quarks years before [8]. It did not require a great leap of imagination to see how we could adapt those ideas to our tight requirements.

By using elaborate technical machinery of quantum field theory (including the renormalization group, operator product expansions, and appropriate dispersion relations) we were able to be much more specific and quantitative about the implications our theory than my loose pictorial language suggests. In particular, the strong interaction does not simply turn off abruptly, and there is a non-zero probability that quarks will radiate when poked. It is only asymptotically, as energies involved go to infinity, that the probability for radiation vanishes. We could calculate in great detail the observable effects of the radiation at finite energy, and make experimental predictions based on these calculations. At the time, and for several years later, the data was not accurate enough to test these particular predictions, but by the end of the 1970s they began to look good, and by now they're beautiful.

Our discovery of asymptotic freedom, and its essentially unique realization in quantum field theory, led us to a new attitude towards the problem of the strong interaction. In place of the broad research programs and fragmentary insights that had characterized earlier work, we now had a single, specific candidate theory – a theory that could be tested, and perhaps falsified, but which could not be fudged. Even now, when I re-read our declaration [3]

> Finally let us recall that the proposed theories appear to be uniquely singled out by nature, if one takes both the SLAC results and the renormalization-group approach to quantum field theory at face value.

I re-live the mixture of exhilaration and anxiety that I felt at the time.

3 A FOURSOME OF PARADIGMS

Our resolution of the paradoxes that drove us had ramifications in unanticipated directions, and extending far beyond their initial scope.

3.1 *Paradigm 1: The Hard Reality of Quarks and Gluons*

Because, in order to fit the facts, you had to ascribe several bizarre properties to quarks – paradoxical dynamics, peculiar charge, and anomalous statistics – their "reality" was, in 1972, still very much in question. This despite the fact that they were helpful in organizing the hadrons, and even though Friedman, Kendall, and Taylor had "observed" them! The experimental facts wouldn't go away, of course, but their ultimate significance remained doubtful. Were quarks basic particles, with simple properties, that could be used to in formulating a profound theory – or just a curious intermediate device, that would need to be replaced by deeper conceptions?

Now we know how the story played out, and it requires an act of imagination to conceive how it might have been different. But Nature is imaginative, as are theoretical physicists, and so it's not impossible to fantasize alternative

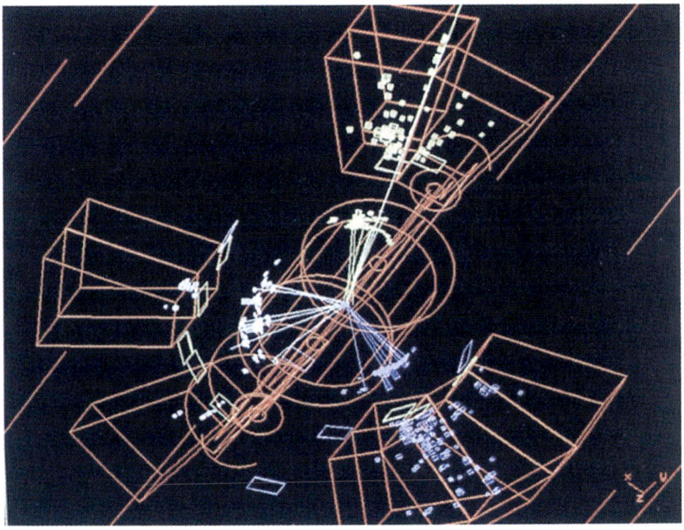

Figure 1: A photograph from the L3 collaboration, showing three jets emerging from electron-positron annihilation at high energy [9]. These jets are the materialization of a quark, antiquark, and gluon.

histories. For example, the quasiparticles of the fractional quantum Hall effect, which are not basic but rather emerge as collective excitations involving ordinary electrons, also cannot exist in isolation, and they have fractional charge and anomalous statistics! Related things happen in the Skyrme model, where nucleons emerge as collective excitations of pions. One might have fantasized that quarks would follow a similar script, emerging somehow as collective excitations of hadrons, or of more fundamental preons, or of strings.

Together with the new attitude toward the strong interaction problem, that I just mentioned, came a new attitude toward quarks and gluons. These words were no longer just names attached to empirical patterns, or to notional building blocks within rough phenomenological models. Quarks and (especially) gluons had become ideally simple entities, whose properties are fully defined by mathematically precise algorithms.

You can even see them! Here's a picture, which I'll now explain.

Asymptotic freedom is a great boon for experimental physics, because it leads to the beautiful phenomenon of jets. As I remarked before, an important part of the atmosphere of mystery surrounding quarks arose from the fact that they could not be isolated. But if we change our focus, to follow flows of energy and momentum rather than individual hadrons, then quarks and gluons come into view, as I'll now explain.

There is a contrast between two different kinds of radiation, which expresses the essence of asymptotic freedom. Hard radiation, capable of significantly re-directing the flow of energy and momentum, is rare. But soft radiation, that produces additional particles moving in the same direction, without deflecting the overall flow, is common. Indeed, soft radiation is associated

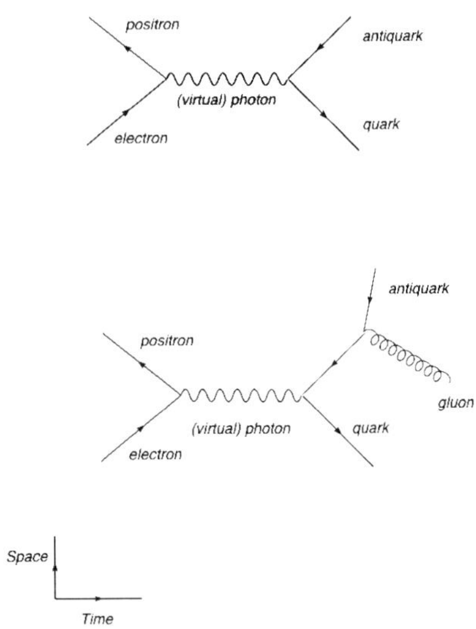

Figure 2: These Feynman graphs are schematic representations of the fundamental processes in electron-positron annihilation, as they take place in space and time. They show the origin of two-jet and three-jet events.

with the build-up of the clouds I discussed before, as it occurs in time. Let's consider what it means for experiments, say to be concrete the sort of experiment done at the Large Electron Positron collider (LEP) at CERN during the 1990s, and contemplated for the International Linear Collider (ILC) in the future. At these facilities, one studies what emerges from the annihilation of electrons and positrons that collide at high energies. By well-understood processes that belong to QED or electroweak theory, the annihilation proceeds through a virtual photon or Z boson into a quark and an antiquark. Conservation and energy and momentum dictate that the quark and antiquark will be moving at high speed in opposite directions. If there is no hard radiation, then the effect of soft radiation will be to convert the quark into a spray of hadrons moving in a common direction: a jet. Similarly, the antiquark becomes a jet moving in the opposite direction. The observed result is then a 2-jet event. Occasionally (about 10% of the time, at LEP) there will be hard radiation, with the quark (or antiquark) emitting a gluon in a significantly new direction. From that point on the same logic applies, and we have a 3-jet event, like the one shown in Figure 1. The theory of the underlying space-time process is depicted in Figure 2. And roughly 1% of the time 4 jets will occur, and so forth. The relative probability of different numbers of jets, how it varies with the overall energy, the relative frequency of different angles at which the jets emerge and the total energy in each – all these detailed aspects of the "antenna pattern" can be predicted quantitatively. These predictions reflect the basic couplings among quarks and gluons, which define QCD, quite directly.

The predictions agree well with very comprehensive experimental measurements. So we can conclude with confidence that QCD is right, and that what

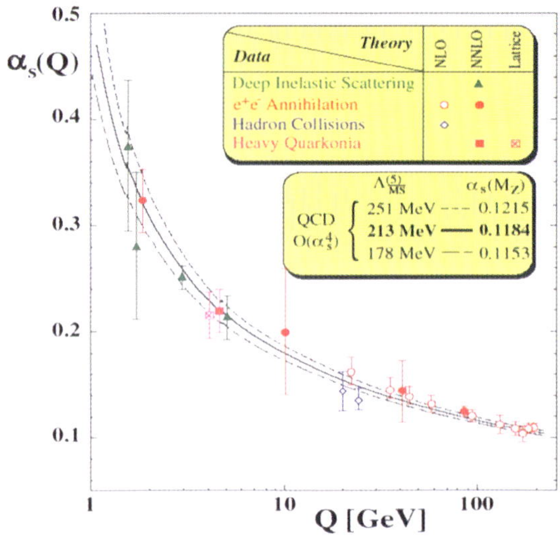

Figure 3: Many quite different experiments, performed at different energies, have been successfully analyzed using QCD. Each fits a large quantity of data to a single parameter, the strong coupling α_s. By comparing the values they report, we obtain direct confirmation that the coupling evolves as predicted [10].

you are seeing, in Figure 1, is a quark, an antiquark, and a gluon – although, since the predictions are statistical, we can't say for sure which is which!

By exploiting the idea that hard radiation processes, reflecting fundamental quark and gluon interactions, control the overall flow of energy and momentum in high-energy processes, one can analyze and predict the behavior of many different kinds of experiments. In most of these applications, including the original one to deep inelastic scattering, the analysis necessary to separate out hard and soft radiation is much more involved and harder to visualize than in the case of electron-positron annihilation. A lot of ingenuity has gone, and continues to go, into this subject, known as perturbative QCD. The results have been quite successful and gratifying. Figure 3 shows one aspect of the success. Many different kinds of experiments, performed at many different energies, have been successfully described by QCD predictions, each in terms of the one relevant parameter of the theory, the overall coupling strength. Not only must each experiment, which may involve hundreds of independent measurements, be fit consistently, but one can then check whether the values of the coupling change with the energy scale in the way we predicted. As you can see, it does. A remarkable tribute to the success of the theory, which I've been amused to watch evolve, is that a lot of the same activity that used to be called *testing* QCD is now called *calculating backgrounds*.

As a result of all this success, a new paradigm has emerged for the operational meaning of the concept of a fundamental particle. Physicists designing and interpreting high-energy experiments now routinely describe their results in terms of producing and detecting quarks and gluons: what they mean, of course, is the corresponding jets.

3.2 *Paradigm 2: Mass Comes from Energy*

My friend and mentor Sam Treiman liked to relate his experience of how, during World War II, the U.S. Navy responded to the challenge of training a large number of radio engineers starting with very different levels of preparation, ranging down to near zero. They designed a crash course for it, which Sam took. In the training manual, the first chapter was devoted to Ohm's three laws. Ohm's first law is $V = IR$. Ohm's second law is $I = V/R$. I'll leave it to you to reconstruct Ohm's third law.

Similarly, as a companion to Einstein's famous equation $E = mc^2$ we have his second law, $m = E/c^2$.

All this isn't quite as silly as it may seem, because different forms of the same equation can suggest very different things. The usual way of writing the equation, $E = mc^2$, suggests the possibility of obtaining large amounts of energy by converting small amounts of mass. It brings to mind the possibilities of nuclear reactors, or bombs. Stated as $m = E/c^2$, Einstein's law suggests the possibility of explaining mass in terms of energy. That is a good thing to do, because in modern physics energy is a more basic concept than mass. Actually, Einstein's original paper does not contain the equation $E = mc^2$, but rather $m = E/c^2$. In fact, the title is a question: "Does the Inertia of a Body Depend Upon its Energy Content?" From the beginning, Einstein was thinking about the origin of mass, not about making bombs.

Modern QCD answers Einstein's question with a resounding "Yes!" Indeed, the mass of ordinary matter derives almost entirely from energy – the energy of massless gluons and nearly massless quarks, which are the ingredients from which protons, neutrons, and atomic nuclei are made.

The runaway build-up of antiscreening clouds, which I described before, cannot continue indefinitely. The resulting color fields would carry infinite energy, which is not available. The color charge that threatens to induce this runaway must be cancelled. The color charge of a quark can be cancelled either with an antiquark of the opposite color (making a meson), or with two quarks of the complementary colors (making a baryon). In either case, perfect cancellation would occur only if the particles doing the canceling were located right on top of the original quark – then there would be no uncanceled source of color charge anywhere in space, and hence no color field. Quantum mechanics does not permit this perfect cancellation, however. The quarks and antiquarks are described by wave functions, and spatial gradients in these wave function cost energy, and so there is a high price to pay for localizing the wave function within a small region of space. Thus, in seeking to minimize the energy, there are two conflicting considerations: to minimize the field energy, you want to cancel the sources accurately; but to minimize the wave-

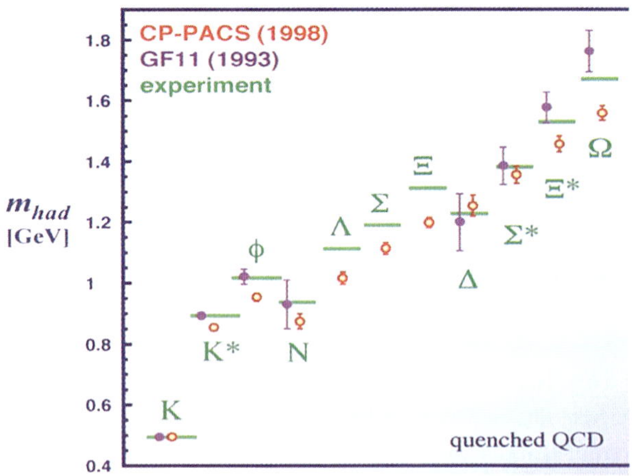

Figure 4: Comparison of observed hadron masses to the energy spectrum predicted by QCD, upon direct numerical integration of the equations, exploiting immense computer power [11]. The small remaining discrepancies are consistent with what is expected given the approximations that were necessary to make the calculation practical.

function localization energy, you want to keep the sources fuzzy. The stable configurations will be based on different ways of compromising between those two considerations. In each such configuration, there will be both field energy and localization energy. This gives rise to mass, according to $m = E/c^2$, even if the gluons and quarks started out without any non-zero mass of their own. So the different stable compromises will be associated with particles that we can observe, with different masses; and metastable compromises will be associated with observable particles that have finite lifetimes.

To determine the stable compromises concretely, and so to predict the masses of mesons and baryons, is hard work. It requires difficult calculations that continue to push the frontiers of massively parallel processing. I find it quite ironical that if we want to compute the mass of a proton, we need to deploy something like 10^{30} protons and neutrons, doing trillions of multiplications per second, working for months, to do what one proton does in 10^{-24} seconds, namely figure out its mass. Maybe it qualifies as a paradox. At the least, it suggests that there may be much more efficient ways to calculate than the ones we're using.

In any case, the results that emerge from these calculations are very gratifying. They are displayed in Figure 4. The observed masses of prominent mesons and baryons are reproduced quite well, stating from an extremely tight and rigid theory. Now is the time to notice also that one of the data points in Figure 3, the one labeled "Lattice", is of a quite different character from the others. It is based not on the perturbative physics of hard radiation, but rather on the comparison of a direct integration of the full equations of QCD with experiment, using the techniques of lattice gauge theory.

The success of these calculations represents the ultimate triumph over our two paradoxes:

- The calculated spectrum does not contain anything with the charges or other quantum numbers of quarks; nor of course does it contain massless gluons. The observed particles do not map in a straightforward way to the primary fields from which they ultimately arise.
- Lattice discretization of the quantum field theory provides a cutoff procedure that is independent of any expansion in the number of virtual particle loops. The renormalization procedure must be, and is, carried out without reference to perturbation theory, as one takes the lattice spacing to zero. Asymptotic freedom is crucial for this, as I discussed – it saves us from Landau's catastrophe.

By fitting some fine details of the pattern of masses, one can get an estimate of what the quark masses are, and how much their masses are contributing to the mass of the proton and neutron. It turns out that what I call QCD Lite – the version in which you put the u and d quark masses to zero, and ignore the other quarks entirely – provides a remarkably good approximation to reality. Since QCD Lite is a theory whose basic building-blocks have zero mass, this result quantifies and makes precise the idea that most of the mass of ordinary matter – 90 % or more – arises from pure energy, via $m = E/c^2$.

The calculations make beautiful images, if we work to put them in eye-friendly form. Derek Leinweber has made some striking animations of QCD fields as they fluctuate in empty space. Figure 5 is a snapshot from one of his animations. Figure 6 from Greg Kilcup, displays the (average) color fields, over and above the fluctuations, that are associated with a very simple hadron, the pion, moving through space-time. Insertion of a quark-antiquark pair, which we subsequently remove, produces this disturbance in the fields.

These pictures make it clear and tangible that the quantum vacuum is a dynamic medium, whose properties and responses largely determine the behavior of matter. In quantum mechanics, energies are associated with frequencies, according to the Planck relation $E = h\nu$. The masses of hadrons, then, are uniquely associated to tones emitted by the dynamic medium of space when it is disturbed in various ways, according to

$$\nu = mc^2/h \tag{1}$$

We thereby discover, in the reality of masses, an algorithmic, precise Music of the Void. It is a modern embodiment of the ancients' elusive, mystical "Music of the Spheres".

3.3 *Paradigm 3: The Early Universe was Simple*

In 1972 the early universe seemed hopelessly opaque. In conditions of ultra-high temperatures, as occurred close to the Big Bang singularity, one would have lots of hadrons and antihadrons, each one an extended entity that inter-

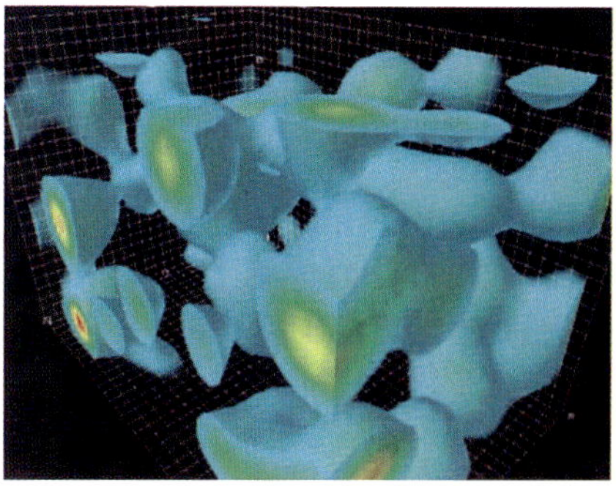

Figure 5: A snapshot of spontaneous quantum fluctuations in the gluon fields [12]. For experts: what is shown is the topological charge density in a typical contribution to the functional integral, with high frequency modes filtered out.

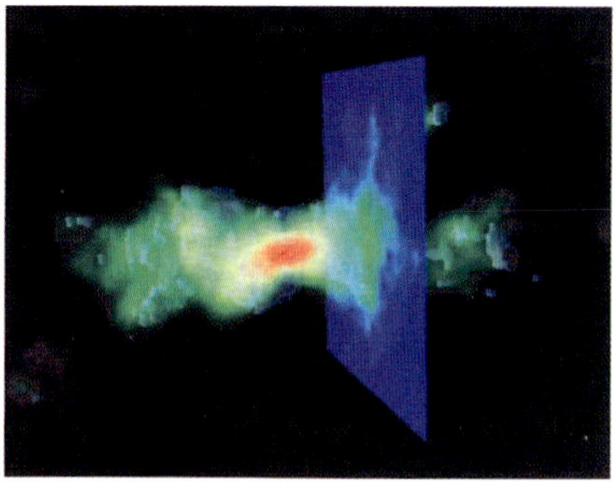

Figure 6: The calculated net distribution of field energy caused by injecting and removing a quark-antiquark pair [13]. By calculating the energy in these fields, and the energy in analogous fields produced by other disturbances, we predict the masses of hadrons. In a profound sense, these fields are the hadrons.

acts strongly and in complicated ways with its neighbors. They'd start to overlap with one another, and thereby produce a theoretically intractable mess.

But asymptotic freedom renders ultra-high temperatures friendly to theorists. It says that if we switch from a description based on hadrons to a description based on quark and gluon variables, and focus on quantities like total energy, that are not sensitive to soft radiation, then the treatment of the strong interaction, which was the great difficulty, becomes simple. We can calculate to a first approximate by pretending that the quarks, antiquarks and gluons behave as free particles, then add in the effects of rare hard interactions. This makes it quite practical to formulate a precise description of the properties of ultra-high temperature matter that are relevant to cosmology.

We can even, over an extremely limited volume of space and time, reproduce Big Bang conditions in terrestrial laboratories. When heavy ions are caused to collide at high energy, they produce a fireball that briefly attains temperatures as high as 200 MeV. "Simple" may not be the word that occurs to you in describing the explosive outcome of this event, as displayed in Figure 7, but in fact detailed study does permit us to reconstruct aspects of the initial fireball, and to check that it was a plasma of quarks and gluons.

3.4 *Paradigm 4: Symmetry Rules*

Over the course of the twentieth century, symmetry has been immensely fruitful as a source of insight into Nature's basic operating principles. QCD, in particular, is constructed as the unique embodiment of a huge symmetry group, local $SU(3)$ color gauge symmetry (working together with special relativity, in the

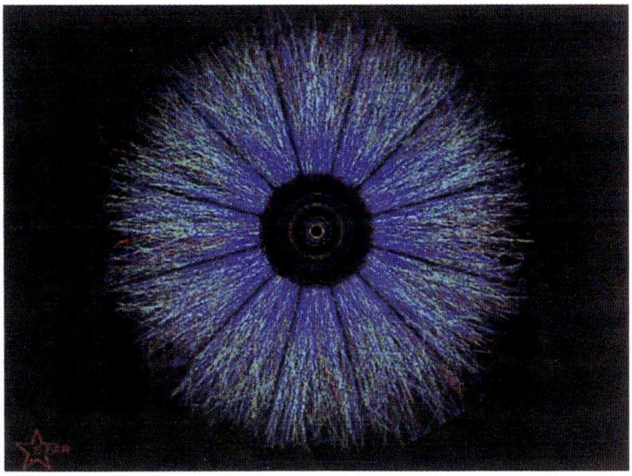

Figure 7: A picture of particle tracks emerging from the collision of two gold ions at high energy. The resulting fireball and its subsequent expansion recreate, on a small scale and briefly, physical conditions that last occurred during the Big Bang [14].

context of quantum field theory). As we try to discover new laws, that improve on what we know, it seems good strategy to continue to use symmetry as our guide. This strategy has led physicists to several compelling suggestions, which I'm sure you'll be hearing more about in future years! QCD plays an important role in all of them – either directly, as their inspiration, or as an essential tool in devising strategies for experimental exploration.

I will discuss one of these suggestions schematically, and mention three others telegraphically.

3.4.1 *Unified Field Theories*

Both QCD and the standard electroweak standard model are founded on gauge symmetries. This combination of theories gives a wonderfully economical and powerful account of an astonishing range of phenomena. Just because it is so concrete and so successful, this rendering of Nature can and should be closely scrutinized for its aesthetic flaws and possibilities. Indeed, the structure of the gauge system gives powerful suggestions for its further fruitful development. Its product structure $SU(3) \times SU(2) \times U(1)$, the reducibility of the fermion representation (that is, the fact that the symmetry does not make connections linking all the fermions), and the peculiar values of the quantum number hypercharge assigned to the known particles all suggest the desirability of a larger symmetry.

The devil is in the details, and it is not at all automatic that the superficially complex and messy observed pattern of matter will fit neatly into a simple mathematical structure. But, to a remarkable extent, it does.

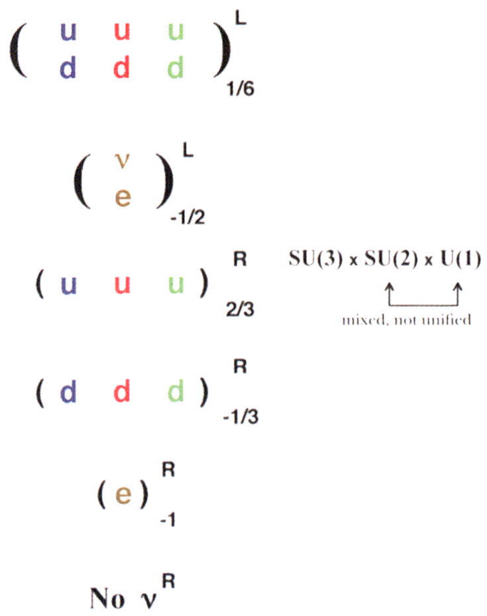

	R	W	B	G	P
u	+	-	-	+	-
u	-	+	-	+	-
u	-	-	+	+	-
d	+	-	-	-	+
d	-	+	-	-	+
d	-	-	+	-	+
u^c	-	+	+	-	-
u^c	+	-	+	-	-
u^c	+	+	-	-	-
d^c	-	+	+	+	+
d^c	+	-	+	+	+
d^c	+	+	-	+	+
ν	+	+	+	+	-
e	+	+	+	-	+
e^c	-	-	-	+	+
N	-	-	-	-	-

Hypercharge Y = -1/6 (R+W+B) + 1/4 (G+P)

Figure 8: A schematic representation of the symmetry structure of the standard model. There are three independent symmetry transformations, under which the known fermions fall into five independent units (or fifteen, after threefold family repetition). The color gauge group $SU(3)$ of QCD acts horizontally, the weak interaction gauge group $SU(2)$ acts vertically, and the hypercharge $U(1)$ acts with the relative strengths indicated by the subscripts. Right-handed neutrinos do not participate in any of these symmetries.

Figure 9: The hypothetical enlarged symmetry $SO(10)$ [15] accommodates all the symmetries of the standard model, and more, into a unified mathematical structure. The fermions, including a right-handed neutrino that plays an important role in understanding observed neutrino phenomena, now form an irreducible unit (neglecting family repetition). The allowed color charges, both strong and weak, form a perfect match to what is observed. The phenomenologically required hypercharges, which appear so peculiar in the standard model, are now theoretically determined by the color and weak charges, according to the formula displayed.

Most of what we know about the strong, electromagnetic, and weak interactions is summarized (rather schematically!) in Figure 8. QCD connects particles horizontally in groups of 3 ($SU(3)$), the weak interaction connects particles vertically in groups of 2 ($SU(2)$) in the horizontal direction and hypercharge ($U(1)$) senses the little subscript numbers. Neither the different interactions, nor the different particles, are unified. There are three different interaction symmetries, and five disconnected sets of particles (actually fifteen sets, taking into account the threefold repetition of families).

We can do much better by having more symmetry, implemented by additional gluons that also change strong into weak colors. Then everything clicks into place quite beautifully, as displayed in Figure 9.

There seems to be a problem, however. The different interactions, as observed, do not have the same overall strength, as would be required by the ex-

Unification of gauge couplings

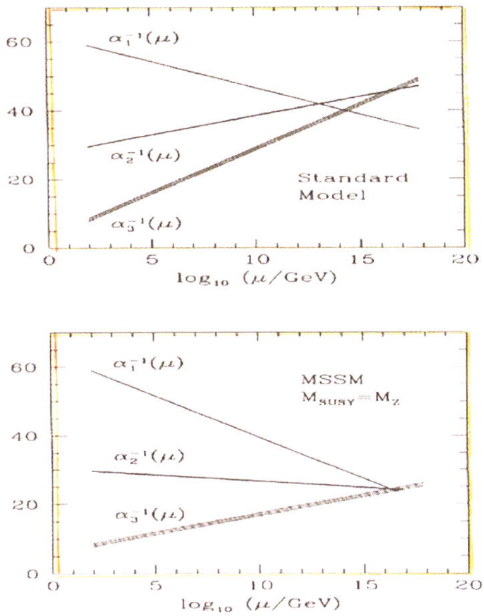

Gravity fits too!
(roughly)

Figure 10: We can test the hypothesis that the disparate coupling strengths of the different gauge interactions derive a common value at short distances, by doing calculations to take into account the effect of virtual particle clouds [16]. These are the same sort of calculations that go into Figure 3, but extrapolated to much higher energies, or equivalently shorter distances. Top panel: using known virtual particles. Bottom panel: including also the virtual particles required by low-energy supersymmetry [17].

tended symmetry. Fortunately, asymptotic freedom informs us that the observed interaction strengths at a large distance can be different from the basic strengths of the seed couplings viewed at short distance. To see if the basic theory might have the full symmetry, we have to look inside the clouds of virtual particles, and to track the evolution of the couplings. We can do this, using the same sort of calculations that underlie Figure 3, extended to include the electroweak interactions, and extrapolated to much shorter distances (or equivalently, larger energy scales). It is convenient to display inverse couplings and work on a logarithmic scale, for then the evolution is (approximately) linear. When we do the calculation using only the virtual particles for which we have convincing evidence, we find that the couplings do approach each other in a promising way, though ultimately they don't quite meet. This is shown in the top panel of Figure 10.

Interpreting things optimistically, we might surmise from this near-success that the general idea of unification is on the right track, as is our continued reliance on quantum field theory to calculate the evolution of couplings. After all, it is hardly shocking that extrapolation of the equations for evolution of the couplings beyond their observational foundation by many orders of magnitude is missing some quantitatively significant ingredient. In a moment I'll mention an attractive hypothesis for what's missing.

A very general consequence of this line of thought is that an enormously large energy scale, of order 10^{15} GeV or more, emerges naturally as the scale of unification. This is a profound and welcome result. It is profound, because the large energy scale – which is far beyond any energy we can access directly – emerges from careful consideration of experimental realities at energies more than ten orders of magnitude smaller! The underlying logic that gives us such leverage is a synergy of unification and asymptotic freedom, as follows. If evolution of couplings is to be responsible for their observed gross inequality then, since this evolution is only logarithmic in energy, it must act over a very wide range.

The emergence of a large mass scale for unification is welcome, first, because many effects we might expect to be associated with unification are observed to be highly suppressed. Symmetries that unify $SU(3) \times SU(2) \times U(1)$ will almost inevitably involve wide possibilities for transformation among quarks, leptons, and their antiparticles. These extended possibilities of transformation, mediated by the corresponding gauge bosons, undermine conservation laws including lepton and baryon number conservation. Violation of lepton number is closely associated with neutrino oscillations. Violation of baryon number is closely associated with proton instability. In recent years neutrino oscillations have been observed; they correspond to miniscule neutrino masses, indicating a very feeble violation of lepton number. Proton instability has not yet been observed, despite heroic efforts to do so. In order to keep these processes sufficiently small, so as to be consistent with observation, a high scale for unification, which suppresses the occurrence of the transformative gauge bosons as virtual particles, is most welcome. In fact, the unification scale we infer from the evolution of couplings is broadly consistent with the observed value of neutrino masses, and that fact encourages further vigorous pursuit of the quest to observe proton decay.

The emergence of a large mass scale for unification is welcome, secondly, because it opens up possibilities for making quantitative connections to the remaining fundamental interaction in Nature: gravity. It is notorious that gravity is absurdly feebler than the other interactions, when they are compared acting between fundamental particles at accessible energies. The gravitational force between proton and electron, at any macroscopic distance, is about $Gm_em_p/\alpha \sim 10^{-40}$ of the electric force. On the face of it, this fact poses a severe challenge to the idea that these forces are different manifestations of a common source – and an even more severe challenge to the idea that gravity, because of its deep connection to space-time dynamics, is the primary force.

By extending our consideration of the evolution of couplings to include gravity, we can begin to meet these challenges.

- Whereas the evolution of gauge theory couplings with energy is a subtle quantum mechanical effect, the gravitational coupling evolves even classically, and much more rapidly. For gravity responds directly to energy-momentum, and so it appears stronger when viewed with high-energy probes. In moving from the small energies where we ordinarily measure to unification energy scales, the ratio GE^2/α ascends to values that are no longer absurdly small.
- If gravity is the primary force, and special relativity and quantum mechanics frame the discussion, then Planck's system of physical units, based on Newton's constant G, the speed of light c, and Planck's quantum of action h, is privileged. Dimensional analysis then suggests that the value of naturally defined quantities, measured in these units, should be of order unity. But when we measure the proton mass in Planck units, we discover

$$m_p \sim 10^{-18}\sqrt{\frac{hc}{G}} \tag{2}$$

On this hypothesis, it makes no sense to ask "Why is gravity so feeble?". Gravity, as the primary force, just is what it is. The right question is the one we confront here: "Why is the proton so light?". Given our new, profound understanding of the origin of the proton's mass, which I've sketched for you today, we can formulate a tentative answer. The proton's mass is set by the scale at which the strong coupling, evolved down from its primary value at the Planck energy, comes to be of order unity. It is then that it becomes worthwhile to cancel off the growing color fields of quarks, absorbing the cost of quantum localization energy. In this way, we find, quantitatively, that the tiny value of the proton mass in Planck units arises from the fact that the basic unit of color coupling strength, g, is of order $\frac{1}{2}$ at the Planck scale! Thus dimensional reasoning is no longer mocked. The apparent feebleness of gravity results from our partiality toward the perspective supplied by matter made from protons and neutrons.

3.4.2 *Supersymmetry*

As I mentioned a moment ago, the approach of couplings to a unified value is suggested, but not accurately realized, if we infer their evolution by including the effect of known virtual particles. There is one particular proposal to expand the world of virtual particles, which is well motivated on several independent grounds. It is known as low-energy supersymmetry [18].

As the name suggests, supersymmetry involves expanding the symmetry of the basic equations of physics. This proposed expansion of symmetry goes in a different direction from the enlargement of gauge symmetry. Supersymmetry makes transformations between particles having the same color charges and different spins, whereas expanded gauge symmetry changes the color charges while leaving spin untouched. Supersymmetry expands the space-time symmetry of special relativity.

In order to implement low-energy supersymmetry, we must postulate the existence of a whole new world of heavy particles, none of which has yet been observed directly. There is, however, a most intriguing indirect hint that this idea may be on the right track: If we include the particles needed for low-energy supersymmetry, in their virtual form, in the calculation of how couplings evolve with energy, then accurate unification is achieved! This is shown in the bottom panel of Figure 10.

By ascending a tower of speculation, involving now both extended gauge symmetry and extended space-time symmetry, we seem to break though the clouds, into clarity and breathtaking vision. Is it an illusion, or reality? This question creates a most exciting situation for the Large Hadron Collider (LHC), due to begin operating at CERN in 2007, for this great accelerator will achieve the energies necessary to access the new world of heavy particles, if it exists. How the story will play out, only time will tell. But in any case I think it is fair to say that the pursuit of unified field theories, which in past (and many present) incarnations has been vague and not fruitful of testable consequences, has in the circle of ideas I've been describing here attained entirely new levels of concreteness and fecundity.

3.4.3 *Axions* [19]

As I have emphasized repeatedly, QCD is in a profound and literal sense constructed as the embodiment of symmetry. There is an almost perfect match between the observed properties of quarks and gluons and the most general properties allowed by color gauge symmetry, in the framework of special relativity and quantum mechanics. The exception is that the established symmetries of QCD fail to forbid one sort of behavior that is not observed to occur. The established symmetries permit a sort of interaction among gluons – the so-called θ term – that violates the invariance of the equations of QCD under a change in the direction of time. Experiments provide extremely severe limits on the strength of this interaction, much more severe than might be expected to arise accidentally.

By postulating a new symmetry, we can explain the absence of the undesired interaction. The required symmetry is called Peccei-Quinn symmetry after the physicists who first proposed it. If it is present, this symmetry has remarkable consequences. It leads us to predict the existence of new very light, very weakly interacting particles, *axions*. (I named them after a laundry detergent, since they clean up a problem with an axial current.) In principle axions might be observed in a variety of ways, though none is easy. They have interesting implications for cosmology, and they are a leading candidate to provide cosmological dark matter.

3.4.4 *In Search of Symmetry Lost* [20]

It has been almost four decades since our current, wonderfully successful theory of the electroweak interaction was formulated. Central to that theory is the concept of spontaneously broken gauge symmetry. According to this concept, the fundamental equations of physics have more symmetry than the actual physical world does. Although its specific use in electroweak theory involves exotic hypothetical substances and some sophisticated mathematics, the underlying theme of broken symmetry is quite old. It goes back at least to the dawn of modern physics, when Newton postulated that the basic laws of mechanics exhibit full symmetry in three dimensions of space despite the fact that everyday experience clearly distinguishes 'up and down' from 'sideways' directions in our local environment. Newton, of course, traced that asymmetry to the influence of Earth's gravity. In the framework of electroweak theory, modern physicists similarly postulate that the physical world is described by a solution wherein all space, throughout the currently observed Universe, is permeated by one or more (quantum) fields that spoil the full symmetry of the primary equations.

Fortunately this hypothesis, which might at first hearing sound quite extravagant, has testable implications. The symmetry-breaking fields, when suitably excited, must bring forth characteristic particles: their quanta. Using the most economical implementation of the required symmetry breaking, one predicts the existence of a remarkable new particle, the so-called Higgs particle. More ambitious speculations suggest that there should be not just a single Higgs particle, but rather a complex of related particles. Low-energy supersymmetry, for example, requires at least five "Higgs particles".

Elucidation of the Higgs complex will be another major task for the LHC. In planning this endeavor, QCD and asymptotic freedom play a vital supporting role. The strong interaction will be responsible for most of what occurs in collisions at the LHC. To discern the new effects, which will be manifest only in a small proportion of the events, we must understand the dominant backgrounds very well. Also, the production and decay of the Higgs particles themselves usually involves quarks and gluons. To anticipate their signatures, and eventually to interpret the observations, we must use our understanding of how protons – the projectiles at LHC – are assembled from quarks and gluons, and how quarks and gluons show themselves as jets.

4 THE GREATEST LESSON

Evidently asymptotic freedom, besides resolving the paradoxes that originally concerned us, provides a conceptual foundation for several major insights into Nature's fundamental workings, and a versatile instrument for further investigation.

The greatest lesson, however, is a moral and philosophical one. It is truly awesome to discover, by example, that we humans can come to comprehend Nature's deepest principles, even when they are hidden in remote and alien realms. Our minds were not created for this task, nor were appropriate tools ready at hand. Understanding was achieved through a vast international effort involving thousands of people working hard for decades, competing in the small but cooperating in the large, abiding by rules of openness and honesty. Using these methods – which do not come to us effortlessly, but require nurture and vigilance – we can accomplish wonders.

5 POSTCRIPT: REFLECTIONS

That was the conclusion of the lecture as I gave it. I'd like to add, in this written version, a few personal reflections.

5.1 *Thanks*

Before concluding I'd like to distribute thanks.

First I'd like to thank my parents, who cared for my human needs and encouraged my curiosity from the beginning. They were children of immigrants from Poland and Italy, and grew up in difficult circumstances during the Great Depression, but managed to emerge as generous souls with an inspiring admiration for science and learning. I'd like to thank the people of New York, for supporting a public school system that served me extremely well. I also got a superb undergraduate education, at the University of Chicago. In this connection I'd especially like to mention the inspiring influence of Peter Freund, whose tremendous enthusiasm and clarity in teaching a course on group theory in physics was a major influence in nudging me from pure mathematics toward physics.

Next I'd like to thank the people around Princeton who contributed in crucial ways to the circumstances that made my development and major work in the 1970s possible. On the personal side, this includes especially my wife Betsy Devine. I don't think it's any coincidence that the beginning of my scientific maturity, and a special surge of energy, happened at the same time as I was falling in love with her. Also Robert Shrock and Bill Caswell, my fellow graduate students, from whom I learned a lot, and who made our extremely intense life-style seem natural and even fun. On the scientific side, I must of course thank David Gross above all. He swept me up in his drive to know and to calculate, and through both his generous guidance and his personal example started and inspired my whole career in physics. The environment for theoretical physics in Princeton in the 1970s was superb. There was an atmosphere of passion for understanding, intellectual toughness, and inner confidence whose creation was a great achievement. Murph Goldberger, Sam Treiman, and Curt Callan especially deserve enormous credit for this. Also Sidney Coleman, who was visiting Princeton at the time, was very actively interested in our work. Such interest from a physicist I regarded as uniquely brilliant was inspiring in itself; Sidney also asked many challenging specific questions that helped us come to grips with our results as they developed. Ken Wilson had visited and lectured a little earlier, and his renormalization group ideas were reverberating in our heads.

Fundamental understanding of the strong interaction was the outcome of decades of research involving thousands of talented people. I'd like to thank my fellow physicists more generally. My theoretical efforts have been inspired by, and of course informed by, the ingenious persistence of my experimental colleagues. Thanks, and congratulations, to all. Beyond that generic thanks I'd like to mention specifically a trio of physicists whose work was particularly important in leading to ours, and who have not (yet?) received a Nobel Prize for it. These are Yoichiro Nambu, Stephen Adler, and James Bjorken. Those heroes advanced the cause of trying to understand hadronic physics by taking the concepts of quantum field theory seriously, and embodying them in specific mechanistic models, when doing so was difficult and unfashionable. I'd like to thank Murray Gell-Mann and Gerard 't Hooft for not quite inventing everything, and so leaving us something to do. And finally I'd like to thank Mother Nature for her extraordinarily good taste, which gave us such a beautiful and powerful theory to discover.

This work is supported in part by funds provided by the U.S. Department of Energy (D.O.E.) under cooperative research agreement DE-FC02-94ER40818.

5.2 *A Note to Historians*

I have not, here, given an extensive account of my personal experiences in discovery. In general, I don't believe that such accounts, composed well after the fact, are reliable as history. I urge historians of science instead to focus on the contemporary documents; and especially the original papers, which by definition accurately reflect the understanding that the authors had at the time, as they could best articulate it. From this literature, it is I think not difficult to identify where the watershed changes in attitude I mentioned earlier occurred, and where the outstanding paradoxes of strong interaction physics and quantum field theory were resolved into modern paradigms for our understanding of Nature.

REFERENCES

[1] In view of the nature and scope of this write-up, its footnoting will be light. Our major original papers [2, 3, 4] are carefully referenced.

[2] D. Gross and F. Wilczek, *Phys. Rev. Lett.* **30**, 1343 (1973).

[3] D. Gross and F. Wilczek, *Phys. Rev.* **D8**, 3633 (1973).

[4] D. Gross and F. Wilczek, *Phys. Rev.* **D9**, 980 (1974).

[5] J. Friedman, H. Kendall, and R. Taylor received the Nobel Prize for this work in 1990.

[6] L. Landau, in *Niels Bohr and the Development of Physics*, ed. W. Pauli (McGraw-Hill, New York 1955).

[7] C.-N. Yang and R. Mills, *Phys. Rev.* **96**, 191 (1954).

[8] An especially clear and insightful early paper, in which a dynamical role for color was proposed, is Y. Nambu, in *Preludes in Theoretical Physics* ed. A. De-Shalit, H. Feshbach, L. van Hove (North-Holland, Amsterdam 1966).

[9] Figure courtesy L3 collaboration, CERN.

[10] Figure courtesy S. Bethke, hep-ex/0211012.

[11] Figure courtesy Center for Computational Physics, University of Tsukuba.

[12] Figure courtesy D. Leinweber, http://www.physics.adelaide.edu.au/theory/staff/leinweber/VisualQCD/Nobel/.

[13] Figure courtesy G. Kilcup, http://www.physics.ohio-state.edu/~kilcup.
[14] Figure courtesy STAR collaboration, Brookhaven National Laboratory.
[15] Unification based on $SO(10)$ symmetry was first outlined in H. Georgi, in *Particles and Fields - 1974,* ed. C. Carlson (AIP, New York, 1975).
[16] H. Georgi, H. Quinn, and S. Weinberg, *Phys. Rev. Lett.* **33**, 451 (1974).
[17] S. Dimopoulos, S. Raby, and F. Wilczek, *Phys. Rev.* **D24**, 1681 (1981).
[18] A standard review is H. P. Nilles, *Phys. Rep.* **110**, 1 (1984).
[19] A standard review is J. Kim, *Phys. Rep.* **150**, 1 (1987). I also recommend F. Wilczek, hep-ph/0408167.
[20] I treat this topic more amply in F. Wilczek, *Nature* **433**, 239 (2005).

Physics 2005

ROY J. GLAUBER

"for his contribution to the quantum theory of optical coherence"

and the other half jointly to

JOHN L. HALL and THEODOR W. HÄNSCH

"for their contributions to the development of laser-based precision spectroscopy, including the optical frequency comb technique"

THE NOBEL PRIZE IN PHYSICS

Speech by Professor Stig Stenholm of the Royal Swedish Academy of Sciences.
Translation of the Swedish text.

Your Majesties, Your Royal Highnesses, Ladies and Gentlemen,

We live in a world of light. We experience our surroundings by sight. Likewise, light from the most distant galaxies gives us knowledge about the Universe. This year's Prize is concerned with light.

Light displays a characteristic dual nature: It appears both as a wave motion but also as a stream of discrete particles of light, as photons. When a photon hits a material, it can emit one and only one electron. Light may be described by classical optics, but observing it is always based on the absorption of one quantum of energy.

Optics has been an important part of physics for a long time, and we live with its technical applications every day. Hence, physicists were slow in recognizing the need for a quantum theory of optics. The emergence of the laser, however, made it essential to distinguish its special light from the more disordered radiation emitted by hot bodies. This induced Roy Glauber to utilize the quantum theory to describe the properties of light and how these can be observed. His work laid the foundations for the field of research today called Quantum Optics. This makes it possible to test our conception of the fundamental features of reality.

These results are based on utilization of the quantum character of light. At the same time, light is a wave motion, where a precise color corresponds to a precise distance between the crests of the waves. Because the velocity of light is constant, the distance between two crests always corresponds to a definite time interval; they occur with a definite frequency. Physicists have long struggled to measure time with extreme accuracy. We need good references for time. A good reference has to be available to everybody, and hence it is chosen from the world of atoms; we have so-called atomic clocks. But a reference is not sufficient; one must also be able to establish the measure of an unknown period of time as compared with the reference. We need a measuring rod to compare two intervals of time.

John Hall and Theodor Hänsch have worked on ever-improved standard references for frequency measurements. In order to compare an unknown period of light with the reference, they have developed the frequency comb technique. This gives a sequence of exactly separated frequencies and a method to set this measuring rod against an unknown frequency. Thus one obtains an extremely accurate number for the unknown period. This allows spectroscopic measurements with extremely high precision. Today this technique is as exact as the methods developed for this purpose during earlier decades, but it promises many times improved accuracy.

The history of physics shows that, when the accuracy of measurements is improved, new physics may be discovered and explored. The work honored today facilitates tests of our basic theories in physics. The character of time and space may be clarified, and the limitations of the laws of physics may be established.

This year's Nobel Prize explores two of the ultimate modes of behavior of light. Its lumpy quanta determine the smallest packet of energy that can be utilized in measurements. At the other extreme, its role as a perfect wave motion may act as a reference for time periods.

Professor Glauber has shown how to apply quantum considerations to all light sources. Thus he initiated the field of Quantum Optics, which today offers both challenging tasks for the scientists and promising techniques for future quantum engineering.

Professor Hall and Professor Hänsch have developed highly precise methods for spectroscopy and provided the measuring rod to compare optical signals, the frequency comb. Thus the determination of the frequency of an unknown light becomes both simpler and more exact. The present state of the art is highly precise, but future developments promise progress far beyond the achievements of today.

Professor Glauber, Professor Hall and Professor Hänsch,

You have been awarded the 2005 Nobel Prize in Physics for your research into the properties of light, extending our knowledge and technology into the extremes of its dual nature. On behalf of the Royal Swedish Academy of Sciences, I convey to you the warmest congratulations. I now ask you to step forward to receive your Nobel Prizes from the hands of His Majesty the King.

Roy J. Glauber

ROY J. GLAUBER

What is it that makes a dedicated scientist out of a kid with an everyday background? Is it the ungovernable forces that seem to shape all our lives, or is it the development of our own curiosity and tastes that tips the balance of randomness? I've always been puzzled by those questions and can't claim to have found serious answers. Perhaps these recollections will reveal one, even if it escapes me as I write.

To be a traveling salesman in the 1920's gave one possession of a company-owned car, acquaintance with a potentially vast area of the country, and a slightly better income than one would earn within the tight confines of New York City. My father, having enjoyed some experience with that life before getting married in 1924, couldn't wait to get back on the road, a possibility he had to postpone for about three years until his wife and new-born son, at last aged two, were ready to travel. The itinerant life was a restless one and quite disconnected. After long hours spent driving through endless farmlands we would stay overnight at the houses of farmers who had hung the sign "Tourists-Vacancy" near the road – never two successive nights in the same house. That was long before the days of roadside motels. Even hotels were scarce in some of the small towns we visited. Rural electrification was not yet a reality and I became quite accustomed to the smells of illumination by kerosene and acetylene lamps, as well as all the odors of barnyards and outdoor plumbing.

The periods in which my father was visiting customers, in whatever small town we were passing through posed a problem for my mother. Trained as an elementary school teacher, but pregnant before she had begun teaching, she was determined to make these passages as instructive for me as she could. The most interesting place in each town, as well as I could make out, was the fire department. We received guided tours of their living quarters and fire engines all over the Midwest. Where fire departments were lacking, visits to assorted courtrooms, police departments and even local lockups would do for my introduction to civics.

In one Cleveland hotel room when I was four, we actually had a radio. It occupied a wooden cabinet about the size of a steamer trunk. I remember insisting there must be a man inside it. He had given his name as Maurice Chevalier. Discovering that the cabinet top was hinged, I opened it and can feel still my bafflement at discovering within it only a few glowing radio tubes.

The 1929 market crash had an immediate impact for me. The company my

Life on the road. Standing on the bumper of my father's car, Ohio, 1930.

father was working for failed and the car we had been using was repossessed. The result was my first ride on a train, an exciting experience that there was no occasion to repeat, once my father had another job and a most imposing new car, a Marmon, a kind of Cadillac of its day, and one of many brands destined for early extinction.

The arrival of a baby sister in 1931 and my need to begin school meant that we had somehow to settle down. My folks decided we would return to New York, but the only way to do it, under the circumstances, was for our family to move into a crowded apartment in upper Manhattan with my father's mother and aunt. It was quite a shock, moving there from the wide-open expanses of the middle west, to sit in the crowded classrooms of an ancient school building. I had had very little experience playing with other kids in the small towns we had visited, and had no idea how to deal with the crowds of kids who managed somehow to play on the concrete sidewalks and in the adjoining gutters.

My mother was talented at crafts of various sorts. She sewed and embroidered well and, though untrained, she sketched and painted quite skillfully. She encouraged me to draw as soon as I could hold pencils or crayons steadily. That was the beginning of my career as a creative artist, specializing in speeding trains, airplanes and the occasional dirigible. It was a necessary release from the need to get fresh air by playing on the sidewalks.

After one school year in Manhattan, we found an apartment in Sunnyside, an attractive area of Queens, and moved there in 1932. It was a kind of deliverance. The neighborhood was spacious and not yet fully built up. It consisted largely of modest single family houses that all had at least a small area devoted to yards or gardens. All the blocks of individual houses and even our apartment building had central areas of lawn with space for children to play – not withstanding all the "Please keep off the grass" signs. There were even vacant lots with tiny hills and semi-permanent puddles that lent excitement to the four daily treks to and from school.

Sunnyside's residents consisted mostly of young families, quite a few contending with the unemployment so widespread in those depression years. The building for Public School 150 however was clean, well lit, and quite new. Its teachers were mostly young and optimistic, a vigorous contrast to the atmosphere that prevailed more generally in the country. The school's annual Christmas play, written by the fifth and sixth graders in 1932 was entitled, "Santa's Depression." It depicted Santa Claus as being broke, and unable to afford to make his usual rounds, until everyone pitched in to help him out. In the same period there were demonstrations against foreclosures on home mortgages going on in the neighborhood streets. Those depression years cast a long shadow over the lives of children no less than their parents.

My earlier years had left me with no experience in sports of any kind so it wasn't easy discovering how to engage in outdoor exercise. Unlike Manhattan, however, Sunnyside had many residential streets with little traffic. The best solution to my exercise problem was roller skating – with steel-wheeled skates that clamped to one's everyday shoes. Those steel wheels were quite noisy rolling or scraping on concrete, notwithstanding their good ball bearings, and I wore down to those very bearings many a set of skate wheels, cruising the neighborhood streets.

Electricity mystified me throughout childhood and I vividly remember once at age seven trying to see what it was all about. Plugging lamp cords into wall sockets must lead to the flow of something through those wires, but whatever it was, one never got to see it before it was swallowed up by the lamp. One morning I awoke early, determined to catch sight of it. I screwed the wires of a short length of lamp cord into a male plug and inserted it into a wall socket, leaving free the frayed wires at the other end. There was a bright blue flash at that end, accompanied by a muffled bang. That was followed by silence, till my parents awoke and began wondering why none of the light switches seemed to be working. The fuse was easily replaced, but I never overcame my surprise that what passed so silently through slender wires could behave so aggressively.

My most interesting projects were the ones I could pursue indoors like building models of contemporary airplanes of all shapes and of ships and locomotives. I had no cash allowance to spend on such projects and so was wholly dependent on gifts of construction kits from uncles and aunts. When those were scarce, as they sometimes were, I ventured into other areas, attempting to use crate wood to construct the projects suggested in various instruction books written for young boys. The most interesting of these projects usually failed, and I began to conclude the authors could never themselves have really built the exciting things they were describing. Their version of a guitar, for example, fashioned from a cigar box and some cheese box wood never had the rigidity to permit stretching a guitar string tightly enough. My guitar looked a bit like one but couldn't sound a single note.

There were other failures, many of them, but each brought new experience in the use of hand tools. An uncle, to encourage this construction bent, presented me with a three-year subscription to Popular Mechanics magazine. That magazine, besides celebrating all of the mechanical wonders of the age, included brief plans for all sorts of home projects: door chimes, folding tables, towel racks, bookends and knife sharpeners. The subscription did a great deal to keep my interest in mechanical things alive, but I can't say I ever succeeded in building any of those worthy projects. And I doubt that anyone who didn't have a machine shop at his disposal ever did either.

All the sawing, drilling and sanding I was doing at home left little time for drawing and painting, but there was ample opportunity to pursue those interests in school. Tempera paints were available there, a certain amount of free time, and a good deal of encouragement from the teachers, who felt a need to keep the backs of their classrooms decorated with mural paintings executed by the kids. They were painted over large areas of brown wrapping paper that covered the rear blackboards. I enjoyed designing those huge works and loved the freedom painting them gave me from sitting at my classroom desk.

The school produced a magazine every term and when my design for the cover of the Christmas 1935 issue was accepted, I felt like the Michelangelo of the fifth grade. In fact I did have some involvement with sculpture as well. Small carvings in soap, greatly encouraged by the Procter and Gamble Corporation, were a medium of the day, and I made many of them, mostly of musicians playing instruments. But a more conventional medium was Plasticine clay, which remains permanently soft. Those sculptures tended not to last long, but we managed, with a teacher, to take a few to a real sculptor's studio, and I was fascinated there to learn to make plaster molds and permanent castings.

If I was fully determined in the fifth and sixth grades to become an artist of one sort or another, it was not without a certain note of caution. My uncle, Sam Adler, was a gifted artist who had not yet succeeded in selling any of his work, nor did those years seem to promise that he ever would. His advice to me was that becoming an artist was an excellent idea, provided my motivation was so strong as to leave no alternative. I began then to feel that my artwork was not spontaneous enough, that if I were a true artist I shouldn't have to think so hard before even starting drawings; they should just pour out more instinctively. My involvement with art receded to a hobby.

The years in which the depression lingered must have been difficult ones for the owners of apartment buildings. Faced with many vacancies, they offered rent-free months and other incentives to new tenants, so there was

My cover for the PS 150, Queens, NY School Magazine, December 1935.

always a certain degree of restlessness among the city's apartment dwellers. In 1936, when I was ten, my parents decided that the higher ground of the Bronx – and the top floor of a six story apartment house – would be a better place to live than the flat sea-level expanse of Long Island. A precipitous increase of the local population density went with that move, and it became once again impossible for me or my sister to spend much time outdoors, in the streets. My first salvation was reading. I visited the local public library regularly and began reading the great adventure stories of Jules Verne, Alexander Dumas and Walter Scott. The junior high school I went to seemed mired in a curriculum too timid to do anything serious, and altogether flat-footed at what it did undertake. Mathematics, I remember, consisted of memorizing the decimal equivalents of the familiar "business fractions" and doing compound interest calculations out longhand. I was so put off by those lessons I occasionally got failing grades. Our premature introduction to French required our memorizing a list of proverbs which didn't literally translate into their English counterparts.

That junior high school experience was typified by what was called "music appreciation" in the auditorium assemblies. The principal, a Mr. Snyder, had himself written words to accompany several dozen themes of the great works of music. Singing his rhyming words he evidently felt, set to the themes of the great composers, should imprint those masterpieces on our young memories. Indeed they did, but it was at the expense of burdening those themes permanently with his infernal doggerel.

But that school did offer my first exposure to science and it was exciting. We were shown how to coil wire around nails and make them into electromagnets with the current from dry cells. Those 6 volt dry cells, widely used to power doorbells, cost 25 cents in the local 5 and 10. From that time on I was never without them.

My ambition to be an artist was further dampened by an art appreciation course largely devoted to biographies of the less scandalous painters, and punctuated by black and white lantern slides of their masterpieces. That course also had a creative element devised to avoid, at all costs, creating any sort of untidiness in the classroom. I was encouraged to draw with pastel crayons, again on a large sheet of wrapping paper hung at the back of the classroom, while the other kids who felt less inclined toward art, were set to copying mounted cartoon panels on drawing paper. Neither the pastels for me, which were intended evidently to make the room look like an art class, nor the cartoon exercises for the other kids, seemed to have any instructive value.

My lingering interests in art presently became centered on puppetry and marionettes. The instructions I saw for making them in some magazine articles and a handbook seemed to offer an interesting combination of sculpture and construction. After fashioning several puppet heads of papier maché and painting them, I set about constructing their marionette bodies and string controls. When it came the turn of our class to present a play in the school auditorium I volunteered to produce a small troupe of marionettes and a stage appropriate for the class presentation. Our decision to stage the fairy tale "Rumplestiltskin" turned out to set a more imposing task than I had imagined. Fortunately my mother came to my aid, offering not only to costume the marionettes but to help in constructing several. The task kept both of us busy for a solid month. The eventual presentation by the class, speaking for and operating the marionettes, must have been some sort of success since we had to repeat it several times. But I was ultimately embarrassed by the fact that so much of the work had been visibly my mother's, and resolved that any further projects of mine would be wholly independent.

When did my interest in science become more serious? It really wasn't too serious, I'd have to admit, until still another change of location. My parents, realizing in 1937 that the move to the Bronx had not been a success, decided to move to an apartment at the north end of Manhattan. We lived in a more spacious neighborhood there and across a peaceful street from Inwood Park – the only uncultivated area left in Manhattan. The school I attended there for the ninth grade, which was nominally the first year of high school, was materially less boring than the prior year's. Algebra was an altogether new beginning and even a redemption for mathematics. It was finally freed of all that dismal arithmetic. That was a joy more than sufficient to overcome the uselessness of so many of the procedures that the curriculum did include. Who, even in those days, could imagine seriously needing to carry out long division of lengthy polynomials, or see any need to teach that procedure to children? No one with experience beyond teacher training could have been responsible for that curriculum.

But general science was another of the subjects we studied, and the energy and enthusiasm of its young teachers more than made up for its attenuated subject matter. I had read an elementary book on astronomy by that time and had been taken by my Aunt Sarah on an exciting visit to the Hayden Planetarium. I found that I could easily visualize the diurnal motions of the stars, the monthly motion of the moon, and somewhat more sketchily the motions of the readily visible planets, Jupiter, Saturn and Venus. The images associated with astronomy quickly captured my imagination, and I began to read everything about it I could find.

The encyclopedia had some simplistic diagrams of how a telescope works, and they seemed to assure me that I could build one from some ordinary magnifying glasses I had accumulated. I did that and was amazed by the rainbow colored edges I saw on the image of the moon and presently dismayed by its overall fuzzy quality. It took a bit of reading to discover what the trouble was – chromatic aberration, endemic to all such primitive refractors. The cure – the only one accessible to me – would be to build a reflecting telescope. But that would be a long-term project, fortunately one that had already been pioneered by quite a few adult amateur astronomers. There was a book, in two volumes, in fact, that drew together the experiences of several amateurs and gave a good deal of guidance, if not detailed instruction, for grinding, polishing and figuring the mirror, and for constructing the remaining optical elements and the mounting. Going through the entire procedure required nine months of work. Coarse and fine grinding of the mirror took only a couple of weeks, but polishing and figuring it to its final shape consumed months. Building a stable yet flexible mounting for the telescope, one that would permit me to follow objects in the sky, compensating for the earth's rotation was another matter entirely. I had only a few hand tools appropriate to woodworking, and still no access to machine tools of any sort. Constructing the wooden cell to house the mirror involved strenuous use of a coping saw and a wood file for several days on end.

The steel polar axis for the equatorial mounting was originally the steering shaft of a Ford car. The proprietor of the junkyard I found it in was happy to give it to me free. But I had somehow to put a 41° bend into that shaft, to equal the latitude of New York. I drew an outline of the bend I needed on a sheet of asbestos and took it off together with the three foot shaft to a garage that I knew refashioned truck housings. The owner was tickled by the project, heated the shaft in his forge till it glowed brightly and pounded it into the precise shape I needed. It must have taken him a good three-quarters of an hour altogether, and I felt I owed him payment for his time. He thought the matter over, and I recall his smile as he said that would come to 25 cents. In fact I got a good deal of aid over those months from people who were pleased to help an ambitious kid with virtually no money to spend on his projects. My accumulated savings of $10 were no more than half spent during those nine months.

Stability of the telescope mounting demanded that it be fairly massive but not too heavy to be carried by hand. The only way I could use it, after all, was to carry it upstairs to the roof of the apartment building. A weight of 40 or 50 pounds seemed appropriate for the base of the mounting, and I would have to make it of cast concrete. A schoolmate kindly brought me a sack of cement and a bag of sand contributed by his father, a local contractor. I fashioned a mold of the appropriate shape from recycled box wood and filled it with the concrete mixture called for by the instructions on the cement sack. The only place available to me for the casting operation was the wooden floor of my bedroom, between my bed and work desk. I had taken some precautions against the leakage of a little water by covering the floor first with waxed paper and a layer of newspaper. My understanding of the setting of concrete was that some miracle of chemistry would incorporate all of the water into the finally hardened product, with none left to leak out onto the floor. That is how plaster of Paris had hardened. But the result was a memorable lesson. I couldn't say what fraction of the hardening was eventually due to drainage, but it must have been appreciable. I had to spend an entire day mopping up pools of water around the hardening mass. I suppose when concrete sidewalks harden their leakage just seeps into the ground below. In my case it would have been the apartment downstairs.

Observing with the telescope wasn't too easy either. In winter the apartment house roof was cold and often windy. Because of the city lights the sky was rarely dark enough to permit seeing the fainter objects, usually diffuse nebulae or distant galaxies. Still there were the thrilling topography of the moon, frequent views of the major planets and countless planetary nebulae, double stars, and clusters of all sorts. Lacking the means to find the fainter objects mechanically, I had to go about tracking them down by locating their positions on star maps relative to the brighter stars or objects easier to find. By putting in at least a little time on most clear nights I managed over the next year or two to find most of the hundred or so extended objects catalogued by the Italian astronomer Messier. I even managed to fashion a film holder and cardboard shutter for the telescope so that I took through it a sequence of moon pictures during the lunar eclipse of November 8, 1938.

The possibility of performing optical tests as exquisitely sensitive as the Foucault test of the telescope mirror's figure with even the most primitive sorts of equipment convinced me that optics was full of miracles. Some other miracles I had seen involved the mysteries of light polarization. The Polaroid Corporation was sponsoring an exhibit I had visited at the Museum of Science and Industry at Rockefeller Center that showed, among other things, the remarkable colors that appeared in transparent materials like cellophane when seen between crossed sheets of Polaroid film. How could

I procure any of the magical Polaroid film? That seemed hopeless for a 12-year old, but I had heard of the possibility of light polarization by reflection. The best reflectors for the purpose would be smooth and black – to avoid the complications of transmission. My father, who at that time in 1937 was selling jewelry displays made of just such black glass, found me several rectangular pieces of the right size. I was able then to mount all the optical elements, including a 25 watt light bulb within a cigar box and use the device to reveal the same sorts of polarization phenomena I had seen at the museum. Seeing the unseen in that way turned out to be as much of a thrill as any I had with the telescope.

In the late 30's an organization with the imposing name The American Institute of the City of New York began organizing activities for young people interested in science. They held science congresses during the Christmas vacations and science fairs during the spring school break, both at the Museum of Natural History. The science congresses were patterned after professional scientific meetings, and split just as incoherently into many sessions, according to fields and specialties. Each session had several ten minute talks presented by the kids as contributed papers. One of those presentations in 1937 was my own description of the plans for the forthcoming 200-inch telescope at Mt. Palomar. It was a visionary image that kept my spirits up while I was having troubles of my own building my 6-inch diameter telescope. The sponsoring Institute saw to it that our talks were attended by at least a sprinkling of mature scientists whom they could somehow persuade to volunteer. I was flattered that my own talk was attended, if only briefly, by Dr. Clyde Fisher, the curator of the Hayden Planetarium. One of his assistant lecturers, Dorothy Bennett, stayed for the whole ten minutes and dropped a suggestion to me that added immensely to my experience over the next four years.

Dorothy Bennett was something of a wonder. Seeking a career in New York, she had arrived there as a fresh graduate of the University of Minnesota just in time for the economic debacle of 1929. With boundless energy and no prior acquaintance with astronomy she found a position working on the plans for the City's new planetarium. One of her many inspirations was to begin in 1930 a citywide astronomy club for kids of high school age. It met on Saturday evenings biweekly, in an imposing auditorium on the top floor of the Roosevelt Memorial building, adjacent to the Planetarium. There the kids, who came in by subway from the far reaches of the city, heard invited lectures by real astronomers. It was that club that Miss Bennett suggested I try attending. I was indeed excited by it and caught up in it from the first meeting I went to. It then formed a large part of my life till I went off to college.

Watched over by Dorothy in a kind of godmotherly role, the Junior Astronomy Club actually had a permanent office in a former watchman's apartment in the basement of the Roosevelt Memorial. There it held committee meetings, originated large mailings to the membership and ground out its monthly mimeographed publication, the Junior Astronomy News. I rushed to take part in all of those programs, ceaselessly amazed that the club could manage all of its activities on dues that only came to 25 cents per

Taken at December 1940 Science Congress talk. At left is a 6 inch richest-field telescope f4, much more portable than my original f8, in the center is the diffraction grating spectroscope, at right a photomicrographic camera, with odd bits of paraphernalia in the foreground.

to visit their "atom smasher", a vertically mounted Van de Graaf generator, and to talk briefly with a couple of real scientists, including a well known theorist, E. U. Condon. Then, as a climax to the trip, I was ushered into the office of the president of the corporation. He promptly drew from his top desk drawer a tattered old pocket notebook. It was his official record, he explained, of the hours he had worked for the company at the turn of the century for a wage of only a few cents an hour.

The junior year of high school meant starting to think about going to college. The teacher assigned as my guidance counselor, thinking perhaps of the experience of his colleagues, assured me that there were too few positions available anywhere for astronomers or physicists, and that I would be best off going to an engineering school. He felt I should apply to a range of them, but he saw Rensselaer Polytech as the ideal compromise. The father of my best friend, a Harvard graduate himself, gave me rather different advice. Disappointed at the unlikelihood of his own son's admission to Harvard, he guessed that I might make it. More to the point, he suggested that scholarship support could be available. Neither my parents nor I would otherwise

SPECIAL JOINT MEETING

NEW YORK ELECTRICAL SOCIETY
and
The American Institute Science & Engineering Clubs

TO-MORROW'S SCIENTISTS

Chairman: IRWIN ARIAS, *American Institute Science Laboratory*

Speakers:

PATRICK CARNER
Haaren High School

ROY GLAUBER
Bronx High School of Science

BARUCH BLUMBERG
Far Rockaway High School

WILLIAM C. DIEFENBACH
Stuyvesant High School

By cooperation with The American Institute of New York City, all our speakers are doing special work at the Science Laboratory.

A demonstration meeting to see and hear how youth is preparing for the problems confronting the fields of science and engineering.

Wednesday
May 28, 1941
8:00 p.m.

The speakers are representative of young people and illustrate how they are preparing themselves to formulate attacks on problems, how to design, construct, operate and evaluate results from laboratory research equipment. Also the development of sound techniques and the ability to demonstrate and explain the use and value of their studies.

May 1941 announcement of talks at the Electrical Engineering Auditorium by four high school students. My talk was about photographs I had taken with the instruments I had shown at the 1940 Science Congress. The talk by Baruch Blumberg dealt with a model refrigerator he had constructed. He didn't continue with physics, however, and switched to medicine, winning the Nobel Prize in 1976 for the discovery of the hepatitis B virus. The chairman, Irwin Arias, also turned out to be a distinguished hepatologist.

have been so presumptuous as to imagine that large a leap in social status. I did fill out the lengthy Harvard applications, however, and take the several required examinations. The application for the scholarship awarded by the New York Harvard Club involved a searching interview conducted in a large, oak-paneled room by a dean and half a dozen club member contributors. I was eventually admitted to a number of colleges, including Rensselaer, but without scholarship aid. Harvard, on the other hand, granted me a Harvard Club scholarship, while making it clear that there were many more exams to take before I would be declared admitted.

Beginning at Harvard in the fall of 1941 meant suddenly being treated like a member of the gentry. We had waiter service at our dining tables and daily printed menus listing alternative dishes. Of course, some fraction of the waiters were fellow classmates, working for board. Our society was stratified in many other ways as well. The rents for the dormitory rooms were graded according to their location, with the result that the scholarship students were clustered in the less desirable areas. They never got to meet the occupants of the higher priced real estate. I scarcely minded any of that. I had come from a different world than those normal Harvard students. College was for them primarily a social experience, overlaid by a burden of course work. For me, on the other hand, having skipped a couple of grades along the way, and some two years younger than most of my classmates, it was the other way around. I enjoyed a few social contacts, but worked hard at my studies, finding them demanding at times, but on the whole well planned and satisfying.

That freshman year was punctuated on December 7 by the Japanese attack on Pearl Harbor and by the entry the next day of the U.S. into the war in Europe as well as the Pacific. The next few months saw significant changes in our lives as students. The rather searching physics course I was taking had been planned as the first half of a two year cycle. Because faculty members were departing for war work the remaining year-long course would not be offered as planned. It would instead be packed into the second semester of the first-year course. That proved to be quite a tall order but a fast way of learning.

The entire school then began operating during the summer and accelerating its course programs with the thought of providing as much education as possible before the young men left for the armed forces. In the meantime Harvard's dining halls lost their graciousness and were transformed into the cafeteria-style mess halls they have been ever since. The draft age, then 21, was presently lowered to 18 and the university began losing students in large numbers. With its faculty depleted the Physics Department announced that its graduate courses were shortly to be given for the last time "for the duration." That announcement made it a good idea to jump directly into the graduate courses, skipping the intermediate ones which had looked neither demanding nor very interesting anyway. It was with the war thus cracking the whip that I managed to assimilate most of the courses of a graduate school education by the time I turned 18 in September 1943. At that point I felt ready for war work myself and filled out a questionnaire sent out by an agency called the National Roster of Scientific Personnel. Its purpose was to ascertain scientific training and try to place people accordingly.

The armed forces by that time had become vastly larger than the country's immediate needs. The Army developed what it called a Specialized Training Program, in effect for storage of its legions in the universities for a year or so, until they would be needed in the invasion of Europe. The program exposed a large population of draftees to college courses for the first time and was a productive experiment in education. I was given a position teaching elementary physics in the program and had my hands full doing that along with taking a full program of courses of my own.

Then one day in October 1943, a stranger in a dark suit appeared in the Physics Department office evidently asking for me. He introduced himself as a Mr. Trytten from Washington, D.C. and asked to speak privately to me. We withdrew to a faculty meeting room in which the blinds were never raised. Closing the door, he asked if I would be interested in joining a new project that was engaged in interesting work. That it was "out west" was the most he would tell me about either the location of the place or what it was doing. It sounded fascinating nonetheless, and I found the security questionnaires he put before me easy to fill out. Having so little prior history helped. His seeking me out seemed to relate to my having filled out the National Roster blanks. It was then a matter of several weeks before my security clearance had been completed and I was instructed to send whatever belongings I needed to the now famous Post Office Box 1663, Santa Fe, New Mexico. In my case it was a trunk sent not by mail but by Railway Express. There were many occasions, then and later to imagine what a capacious P.O. box that one must be.

I could find many tiny hints at what was going on out there, all of them questionable and several, as it later became clear, completely wrong. The most solid hint was in fact a negative one. For about two years after the discovery of uranium fission in 1939 there had been occasional notes in the New York Times speculating on the possibility of starting a chain reaction. They had stopped appearing, it was hard to say just when, but at least two years earlier. So I had no idea whether it had become a dead issue, or my offer of a position implied some real progress toward a chain reaction.

The train ride from New York to Lamy, New Mexico, the stop for Santa Fe, consumed two and a half days. A driver from Los Alamos had come to the station principally to meet a short man in a black overcoat, but took me along, stopping first at an unassuming project office in Santa Fe, where I learned that the man in the overcoat was John Von Neumann, a legendary mathematician.

The ride from Santa Fe up to "the Hill" was an experience I shall never forget. First there was the breathtaking scenery of the canyons of the Pajarito Plateau. Then there was my fellow passenger, John Von Neumann, who engaged in a lively conversation for most of the trip with the driver, whom I learned only later was a mathematician who had worked with "Johnny" earlier. With a thought perhaps of maintaining security, they discussed some calculations underway using the most outlandish mathematical terminology, and describing mathematical errors in physical terms that I knew represented physical impossibilities. The ride was an incredible mixture of visual thrills and intellectual enigmas.

I was astonished, shortly after arrival at the project, to be told that the chain reaction had long since been achieved in Chicago and the present intention was to construct a reaction fast enough to be a bomb. It was disturbing news and I recoiled from it at first, but the challenges and uncertainties involved helped reconcile me to it. More importantly, I felt, as everyone else on the project did, that whatever these uncertainties might be, the Germans, possessing the same understanding we had, were likely to be working on the bomb as well. And if they reached that goal before we did they would not be sentimental about using it to stave off eventual defeat. That fear applied only to the known expertise of the Germans. The conflict with Japan didn't appear to motivate anyone's involvement in the project.

The project was only a few months old when I joined it but most of its eventual leaders were already there. Not many were yet well-known. They were remarkably youthful. Oppenheimer in his late thirties was one of the oldest. He had a universal understanding of the work and an eloquence in describing it that kept us spellbound. Hans Bethe, the leader of the theoretical division, had a penetrating understanding that seemed capable of formulating absolutely anything quantitatively and evaluating it effortlessly, an aura he maintained even many years later. Feynman was there as leader of a small theoretical group. He was often cantankerously teasing the security people. His lectures were always offbeat performances demonstrating novel approaches to problems in ways devoted as much to entertainment as to the technical message. There were others, too many to mention, and among them as an occasional visitor, Niels Bohr, whom we called Nicholas Baker for obvious reasons, together with his son Aage.

Overwhelmed by these giants, my own position in the Theory Division at age 18 was a modest one. There were many problems in neutron diffusion such as finding the critical mass that required more careful formulations than had been carried out in the earliest projections. I worked on those for the better part of the two years I spent at Los Alamos and wrote three lengthy secret papers on those subjects.

There were many delays before the Trinity Test of the bomb in July 1945 and with them the uncertainty of how well it would work grew steadily. Unable to secure a position among the experimenters at Alomogordo, I had to be content with watching for the flash from the top of Sandia Peak near Albuquerque. I saw the flash indeed and some of the glow that followed from a distance of over a hundred miles. The test was followed by some tense days, leading up to August 6, when the use of the bomb at Hiroshima was announced. One thing the portentous announcement meant was a certain release from secrecy. We could now resume contact with the outside world. We could say, if only in general terms what we had been working on. But there were no celebrations of any sort until the war was over a few days later.

Resuming the life of an undergraduate at Harvard early in 1946 proved surprisingly difficult, even though I needed only a few credits to graduate. Having had a team of assistants to do calculations for me at Los Alamos didn't make it any easier, I found, to do my own homework back at school, particularly when I felt I had moved beyond all that. Fortunately that time was brief, and then I became a graduate student. But I had already taken most of the graduate courses on offer, and so was largely left on my own, being allowed to register, in effect, for independent reading and research. The principal reason for my remaining at Harvard was the addition of Julian Schwinger to the faculty. I had met him during a brief appearance he made at Los Alamos, late in 1945, and was immediately so impressed with his knowledge and his incredibly informative lecturing style that I felt he was unique among teachers and would be the ideal thesis advisor as well. I became friendly with Julian over the next three years and was never less than amazed by his ability to construct elegant mathematical structures that would permit him to see further than any of his contemporaries. There were times in those postwar years when it seemed he was responsible for most of the progress in theoretical physics, and very likely would be for years to come. His lectures were

Life with Wolfgang Pauli; a Spring 1950 outing. Prepared to photograph Pauli kicking the ball into the lake, as he had done earlier, I stood to one side, carefully aiming the camera at him. Pauli indeed kicked the ball, and I managed to snap the shutter just before the camera hit me squarely in the face.

brilliantly delivered and notes on them were highly prized and reproduced wherever they could be found. Many students crowded in to work with him, however, and he limited the time he spent with them, so they didn't always produce great theses. Though nominally registered to work with Julian, I actually worked by myself and produced in 1949 a quantum field theoretical thesis that was useful to my later development but scarcely much better than the others of the day.

Robert Oppenheimer, who seemed to know more of me than I had imagined, invited me to spend my first postdoctoral year in Princeton at the Institute for Advanced Study. The group of 20 or so postdocs who gathered there included quite a few eventual leaders of the postwar generation of theorists. None had a stable position anywhere else and so the atmosphere was quite competitive. In the first term of 1950, Wolfgang Pauli was scheduled to visit the Institute. Following the advice of friends who had worked in Zürich, I arranged with Pauli to return with him to Zürich and work with

At the Les Houches Summer School, July 1954. Don Hughes spoke about neutron physics, Fermi about pion scattering, and I lectured on particle collision theory.

him until the fall of that year when I would return to Princeton and the Institute. Having a few months to live in Zürich and to travel over Europe was the principal experience of that encounter. Pauli at age 50 had relaxed into the role of a critic and was no longer inspiring much research. He did retain a mordant sense of humor, however, and was forever doing his best to tease me. Teasing others as well, if not insulting them outright, he was always interesting to be around.

After another year I spent at the Institute, Oppy found me a teaching position. It was only a temporary one, replacing Feynman at Caltech. Feynman was to spend the year in Brazil, where by his own account, he worked hard on the bongo drums, and Caltech needed someone to teach quantum mechanics. The chemistry department out there, under Linus Pauling, seemed to be an exceptionally active one. My research for the year was devoted to resolving a puzzle they had encountered in studying electron diffraction by molecules. Solving the problem didn't interest me in molecules very much, but it did involve me deeply in scattering problems in which the incident particles were of wavelength much smaller than the ranges of interactions. Those problems, I understood, would become steadily more important in nuclear physics as accelerator energies were increased. I continued studying those problems then when I was invited back to Harvard in the fall of 1952 and for some years after that. The result was a species of nuclear diffraction theory analogous in some ways to optical diffraction theory, but generalized to include inelastic collisions between incident particles and complex nuclear systems. The theory is even used these days to treat the high-energy collisions of pairs of heavy nuclei.

Once I was back at Harvard I began to climb the academic ladder of professorial positions and was able to direct the thesis work of a number of gifted students. Theoretical physicists weren't nearly as specialized in those days as they are now. All of theory was considered one's province and so those theses ranged over half a dozen fields, as did my own work.

The late 50's proved to be an exciting time for many reasons. A radically new light source, the laser, was being developed and there were questions in the air regarding the quantum structure of its output. That was particularly so in view of the surprising discovery of quantum correlations in ordinary light by Hanbury Brown and Twiss. A second source of excitement, all my own, was that I had met the young woman I was to marry, Cynthia Rich, and had been going out with her since 1957. We married in July 1960, bought a contemporary house a year later, and settled into quite a happy life together. That was the period in which I began to work on quantum optics with a surmise that the Hanbury Brown-Twiss correlation would be found absent from a stable laser beam, and then followed it with a sequence of more general papers on photon statistics and the meaning of coherence.

Our first child, a son Jeffrey, arrived in 1963. I remember feeling his arrival was a kind of redemption, a species of renewal for which I was more than grateful at age 38. I was doing a good deal of traveling in those days, particularly during vacations, and it always amazed me how transportable the baby was. We had no trouble taking him on short domestic trips anywhere, but thinking back on my own experience perhaps, waited till he was nearly 4 before taking him on a longer trip to Geneva for a sabbatical at CERN. My work in this period gravitated back to high-energy collision theory, since experiments had begun to reveal many of the results my diffractive multiple scattering theory had predicted.

Our second child, a daughter Valerie, didn't arrive until 1970, and by that time our placid and comfortable academic life had been roiled up in many ways. Years of demonstrations against the Vietnam War, the anguish of the black liberation movement, and finally the bitter recriminations of militant feminism had left the world of our university seriously fragmented. My wife, joining with the militants, decided that the days of traditional marriage were over, and that her own should be one of the first to go. The law, she found, would permit her to end it, of her own choice, while retaining custody of the children. Devastated by her decision, I simply couldn't believe she would hew to it, and it took some time to try to reach a settlement. By that time, indeed, she no longer sought active custody of the children, and having taken care of them earlier, I proceeded thereafter to raise them as a single father. It was a time-consuming occupation, but an immensely rewarding one, and I managed fortunately to remain involved and reasonably productive in my

Spreading the gospel at a July 1977 summer school organized by F. T. Arecchi at Villa le Pianore, Versilia, Italy. The message considerably outlasted the moustache.

work. I'm sure there is some number of papers I never got to write as a result, but raising those children and seeing them succeed was not an experience I would trade for the missing papers or any sort of recognition. Both Jeff and Val have families of their own now and are busy raising my grandchildren. I envy them that privilege, and wish I had the opportunity to be raising them all myself.

ONE HUNDRED YEARS OF LIGHT QUANTA

Nobel Lecture, December 8, 2005

by

ROY J. GLAUBER

Harvard University, 17 Oxford Street, Lyman 331, Cambridge, MA 02138, USA.

It can't have escaped you, after so many recent reminders, that this year marks the one hundredth birthday of the light quantum. I thought I would tell you this morning a few things about its century long biography. Of course we have had light quanta on earth for eons, in fact ever since the good Lord said "let there be quantum electrodynamics" – which is a modern translation, of course, from the biblical Aramaic. So in this talk I'll try to tell you what quantum optics is about, but there will hardly be enough time to tell you of the many new directions in which it has led us. Several of those are directions that we would scarcely have anticipated as all of this work started.

My own involvement in this subject began somewhere around the middle of the last century, but I would like to describe some of the background of the scene I entered at that point as a student. Let's begin, for a moment, even before the quantum theory was set in motion by Planck. It is important to recall some of the remarkable things that were found in the 19th century, thanks principally to the work of Thomas Young and Augustin Fresnel. They established within the first 20 years of the 19th century that light is a wave phenomenon, and that these waves, of whatever sort they might be, interpenetrate one another like waves on the surface of a pond. The wave displacements, in other words, add up algebraically. That's called superposition, and it was found thus that if you have two waves that remain lastingly in step with one another, they can add up constructively, and thereby reinforce one another in some places, or they can even oscillate oppositely to one another, and thereby cancel one another out locally. That would be what we call destructive interference.

Interference phenomena were very well understood by about 1820. On the other hand it wasn't at all understood what made up the underlying waves until the fundamental laws of electricity and magnetism were gathered together and augmented in a remarkable way by James Clerk Maxwell, who developed thereby the electrodynamics we know today. Maxwell's theory showed that light waves consist of oscillating electric and magnetic fields. The theory has been so perfect in describing the dynamics of electricity and magnetism over laboratory scale distances, that it has remained precisely intact. It has needed no fundamental additions in the years since the 1860's, apart from those concerning the quantum theory. It serves still, in fact, as the basis for the discussion and analysis of virtually all the optical instrumentation we have ever developed. That overwhelming and continuing success may eventually have led to a certain complacency. It seemed to imply that the field of optics, by the middle of the 20th century, scarcely needed to take any notice of the granular nature of light. Studying the behavior of light quanta was then left to the atomic and elementary particle physicists – whose interests were largely directed toward other phenomena.

The story of the quantum theory, of course, really begins with Max Planck. Planck in 1900 was confronted with many measurements of the spectral distribution of the thermal radiation that is given off by a hot object. It was known that under ideally defined conditions, that is for complete (or black) absorbers and correspondingly perfect emitters this is a unique radiation distribution. The intensities of its color distribution, under such ideal conditions, should depend only on temperature and not at all on the character of the materials that are doing the radiating. That defines the so-called black-body distribution. Planck, following others, tried finding a formula that expresses the shape of that black-body color spectrum. Something of its shape was known at low frequencies, and there was a good guess present for its shape at high frequencies.

The remarkable thing that Planck did first was simply to devise an empirical formula that interpolates between those two extremes. It was a relatively simple formula and it involved one constant which he had to adjust in order to fit the data at a single temperature. Then having done that, he found his formula worked at other temperatures. He presented the formula to the Germany Physical Society[1] on October 19, 1900 and it turned out to be successful in describing still newer data. Within a few weeks the formula seemed to be established as a uniquely correct expression for the spectral distribution of thermal radiation.

The next question obviously was: did this formula have a logical derivation of any sort? There Planck, who was a sophisticated theorist, ran into a bit of trouble. First of all he understood from his thermodynamic background that he could base his discussion on nearly any model of matter, however oversimplified it might be, as long as it absorbed and emitted light efficiently. So he based his model on the mechanical system he understood best, a collection of one-dimensional harmonic oscillators, each of them oscillating rather like a weight at the end of a spring. They had to be electrically charged. He knew from Maxwell exactly how these charged oscillators interact with the electromagnetic field. They both radiate and absorb in a way he could calculate. So then he ought to be able to find the equilibrium between these oscillators and the radiation field, which acted as a kind of thermal reservoir – and which he never made any claim to discuss in detail.

He found that he could not secure a derivation for his magic formula for the radiation distribution unless he made an assumption which, from a philosophical standpoint, he found all but unacceptable. The assumption was that the harmonic oscillators he was discussing had to possess energies that were distributed, not as the continuous variables one expected, but confined instead to discrete and regularly spaced values. The oscillators of frequency ν

would have to be restricted to energy values that were integer multiples, i.e. n-fold multiples (with $n = 0, 1, 2, 3...$) of something he called the quantum of energy, $h\nu$.

That number h was, in effect, the single number that he had to introduce in order to fit his magic formula to the observed data at a single temperature. So he was saying, in effect, that these hypothetical harmonic oscillators representing a simplified image of matter could have only a sequence amounting to a "ladder" of energy states. That assumption permits us to see immediately why the thermal radiation distribution must fall off rapidly with rising frequency. The energy steps between the oscillator states grow larger, according to his assumption, as you raise the frequency, but thermal excitation energies, on the other hand, are quite restricted in magnitude at any fixed temperature. High frequency oscillators at thermal equilibrium would never even reach the first step of excitation. Hence there tends to be very little high frequency radiation present at thermal equilibrium. Planck presented this revolutionary suggestion[2] to the Physical Society on December 14, 1900, although he could scarcely believe it himself.

The next great innovation came in 1905 from the young Albert Einstein, employed still at the Bern Patent Office. Einstein first observed that Planck's formula for the entropy of the radiation distribution, when he examined its high frequency contributions, looked like the entropy of a perfect gas of free particles of energy $h\nu$. That was a suggestion that light itself might be discrete in nature, but hardly a conclusive one.

To reach a stronger conclusion he turned to an examination of the photoelectric effect, which had first been observed in 1887 by Heinrich Hertz. Shining monochromatic light on metal surfaces drives electrons out of the metals, but only if the frequency of the light exceeds a certain threshold value characteristic of each metal. It would have been most reasonable to expect that as you shine more intense light on those metals the electrons would come out faster, that is with higher velocities in response to the stronger oscillating electric fields – but they don't. They come out always with the same velocities, provided that the incident light is of a frequency higher than the threshold frequency. If it were below that frequency there would be no photoelectrons at all.

The only response that the metals make to increasing the intensity of light lies in producing more photoelectrons. Einstein had a naively simple explanation for that.[3] The light itself, he assumed, consists of localized energy packets and each possesses one quantum of energy. When light strikes the metal, each packet is absorbed by a single electron. That electron then flies off with a unique energy, an energy which is just the packet energy $h\nu$ minus whatever energy the electron needs to expend in order to escape the metal.

It took until about 1914–16 to secure an adequate verification of Einstein's law for the energies of the photoelectrons. When Millikan succeeded in doing that, it seemed clear that Einstein was right, and that light does indeed consist of quantized energy packets. It was thus Einstein who fathered the light quantum, in one of the several seminal papers he wrote in the year 1905.

To follow the history a bit further, Einstein began to realize in 1909 that his energy packets would have a momentum which, according to Maxwell, should be their energy divided by the velocity of light. These presumably localized packets would have to be emitted in single directions if they were to remain localized, or to constitute "Nadelstrahlung" (needle radiation), very different in behavior from the broadly continuous angular distribution of radiation that would spread from harmonic oscillators according to the Maxwell theory. A random distribution of these needle radiations could look appropriately continuous, but what was disturbing about that was the randomness with which these needle radiations would have to appear. That was evidently the first of the random variables in the quantum theory that began disturbing Einstein and kept nettling him for the rest of his life.

In 1916 Einstein found another and very much more congenial way of deriving Planck's distribution by discussing the rate at which atoms radiate. Very little was known about atoms at that stage save that they must be capable of absorbing and giving off radiation. An atom lodged in a radiation field would surely have its constituent charges shaken by the field oscillations, and that shaking could lead either to the absorption of radiation or to the emission of still more radiation. Those were the processes of absorption or emission induced by the prior presence of radiation. But Einstein found that thermal equilibrium between matter and radiation could only be reached if, in addition to these induced processes, there exists also a spontaneous process, one in which an excited atom emits radiation even in the absence of any prior radiation field. It would be analogous to radioactive decays discovered by Rutherford. The rates at which these processes take place were governed by Einstein's famous B and A coefficients respectively. The existence of spontaneous radiation turned out to be an important guide to the construction of quantum electrodynamics.

Some doubts about the quantized nature of light inevitably persisted, but many of them were dispelled by Compton's discovery in 1922 that x-ray quanta are scattered by electrons according to the same rules as govern the collisions of billiard balls. They both obey the conservation rules for energy and momentum in much the same way. It became clear that the particle picture of light quanta, whatever were the dilemmas that accompanied it, was here to stay.

The next dramatic developments of the quantum theory, of course, took place between the years 1924 and 1926. They had the effect of ascribing to material particles such as electrons much of the same wave-like behavior as had long since been understood to characterize light. In those developments de Broglie, Heisenberg, Schrödinger and others accomplished literal miracles in explaining the structure of atoms. But however much this invention of modern quantum mechanics succeeded in laying the groundwork for a more general theory of the structure of matter, it seemed at first to have little new bearing on the understanding of electromagnetic phenomena. The spontaneous emission of light persisted as an outstanding puzzle.

Thus there remained a period of a couple of years more in which we described radiation processes in terms that have usually been called "semiclassical." Now the term "classical" is an interesting one – because, as you know, every field of study has its classics. Many of the classics that we are familiar with go back two or three thousand years in history. Some are less old, but all share an antique if not an ancient character. In physics we are a great deal more precise, as well as contemporary. Anything that we understood or could have understood prior to the date of Planck's paper, December 14, 1900, is to us "classical." Those understandings are our classics. It is the introduction of Planck's constant that marks the transition from the classical era to our modern one.

The true "semiclassical era," on the other hand, lasted only about two years. It ended formally with the discovery[4] by Paul Dirac that one must treat the vacuum, that is to say empty space, as a dynamical system. The energy distributed through space in an electromagnetic field had been shown by Maxwell to be a quadratic expression in the electric and magnetic field strengths. Those quadratic expressions are formally identical in their structure to the mathematical expressions for the energies of mechanical harmonic oscillators. Dirac observed that even though there may not seem to be any organized fields present in the vacuum, those mathematically defined oscillators that described the field energy would make contributions that could not be overlooked. The quantum mechanical nature of the oscillators would add an important but hitherto neglected correction to the argument of Planck.

Planck had said the energies of harmonic oscillators are restricted to values n times the quantum energy, $h\nu$, and the fully developed quantum mechanics had shown in fact that those energies are not $nh\nu$ but $(n + \frac{1}{2})\,h\nu$. All of the intervals between energy levels remained unchanged, but the quantum mechanical uncertainty principle required that additional $\frac{1}{2}\,h\nu$ to be present. We can never have a harmonic oscillator completely empty of energy because that would require its position coordinate and its momentum simultaneously to have the precise values zero.

So, according to Dirac, the electromagnetic field is made up of field amplitudes that can oscillate harmonically. But these amplitudes, because of the ever-present half quantum of energy $\frac{1}{2}\,h\nu$, can never be permanently at rest. They must always have their fundamental excitations, the so-called "zero-point fluctuations" going on. The vacuum then is an active dynamical system. It is not empty. It is forever buzzing with weak electromagnetic fields. They are part of the ground state of emptiness. We can withdraw no energy at all from those fluctuating electromagnetic fields. We have to regard them nonetheless as real and present even though we are denied any way of perceiving them directly.

An immediate consequence of this picture was the unification of the notions of spontaneous and induced emission. Spontaneous emission is emission induced by those zero-point oscillations of the electromagnetic field. Furthermore it furnishes, in a sense, an indirect way of perceiving the zero-point fluctuations by amplifying them. Quantum amplifiers tend to generate background noise that consists of radiation induced by those vacuum fluctuations.

It is worth pointing out a small shift in terminology that took place in the late 1920's. Once material particles were found to exhibit some of the wave-like behavior of light quanta, it seemed appropriate to acknowledge that the light quanta themselves might be elementary particles, and to call them "photons" as suggested by G. N. Lewis in 1926. They seemed every bit as discrete as material particles, even if their existence was more transitory, and they were at times freely created or annihilated.

The countless optical experiments that had been performed by the middle of the 20th century were in one or another way based on detecting only the intensity of light. It may even have seemed there wasn't anything else worth measuring. Furthermore those measurements were generally made with steady light beams traversing passive media. It proved quite easy therefore to describe those measurements in simple and essentially classical terms. A characteristic first mathematical step was to split the expression for the oscillating electric field E into two complex conjugate terms

$$E = E^{(+)} + E^{(-)}, \qquad (1)$$

$$E^{(-)} = (E^{(+)})^*, \qquad (2)$$

with the understanding that $E^{(+)}$ contains only positive frequency terms, i.e. those varying as $e^{-i\omega t}$ for all $\omega > 0$, and $E^{(-)}$ contains only negative frequency terms $e^{i\omega t}$. This is a separation familiar to electrical engineers and motivated entirely by the mathematical convenience of dealing with exponential functions. It has no physical motivation in the context of classical theory, since the two complex fields $E^{(\pm)}$ are physically equivalent. They furnish identical descriptions of classical theory.

Each of the fields $E^{(\pm)}(\boldsymbol{r}t)$ depends in general on both the space coordinate $\boldsymbol{r}$ and time t. The instantaneous field intensity at $\boldsymbol{r},t$ would then be

$$|E^{(+)}(\boldsymbol{r},t)|^2 = E^{(-)}(\boldsymbol{r},t)E^{(+)}(\boldsymbol{r},t). \qquad (3)$$

In practice it was always an average intensity that was measured, usually a time average.

The truly ingenious element of many optical experiments, going all the way back to Young's double-pinhole experiment, was the means their design afforded to superpose the fields arriving at different space-time points before the intensity observations were made. Thus in Young's experiment, shown in Figure 1, light penetrating a single pinhole in the first screen passes through two pinholes in the second screen and then is detected as it falls on a third screen. The field $E^{(+)}(\boldsymbol{r}t)$ at any point on the latter screen is the superposition of two waves radiated from the two prior pinholes with a slight difference in their arrival times at the third screen, due to the slightly different distances they have to travel.

If we wanted to discuss the resulting light intensities in detail we would find it most convenient to do that in terms of a field correlation function which we shall define as

$$G^{(1)}(\boldsymbol{r}_1 t_1, \boldsymbol{r}_2 t_2) = \left\langle E^{(-)}(\boldsymbol{r}_1 t_1)E^{(+)}(\boldsymbol{r}_2 t_2) \right\rangle. \qquad (4)$$

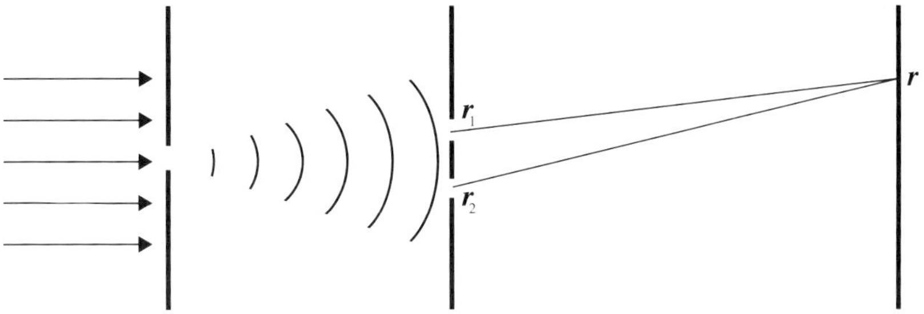

Figure 1. Young's experiment. Light passing through a pinhole in the first screen falls on two closely spaced pinholes in a second screen. The superposition of the waves radiated by those pinholes at r_1 and r_2 leads to interference fringes seen at points r on the third screen.

This is a complex-valued function that depends, in general, on both space-time points $\boldsymbol{r}_1 t_1$ and $\boldsymbol{r}_2 t_2$. The angular brackets $\langle ... \rangle$ indicate that an average value is somehow taken, as we have noted. The average intensity of the field at the single point $\boldsymbol{r}t$ is then just $G^{(1)}(\boldsymbol{r}t, \boldsymbol{r}t)$.

If the field $E^{(+)}(\boldsymbol{r}t)$ at any point on the third screen is given by the sum of two fields, i.e. proportional to $E^{(+)}(\boldsymbol{r}_1 t_1) + E^{(+)}(\boldsymbol{r}_2 t_2)$, then it is easy to see that the average intensity at $\boldsymbol{r}t$ on the screen is given by a sum of four correlation functions,

$$G^{(1)}(\boldsymbol{r}_1 t_1 \boldsymbol{r}_1 t_1) + G^{(1)}(\boldsymbol{r}_2 t_2 \boldsymbol{r}_2 t_2) + G^{(1)}(\boldsymbol{r}_1 t_1 \boldsymbol{r}_2 t_2) + G^{(1)}(\boldsymbol{r}_2 t_2 \boldsymbol{r}_1 t_1). \tag{5}$$

The first two of these terms are the separate contributions of the two pinholes in the second screen, that is, the intensities they would contribute individually if each alone were present. Those smooth intensity distributions are supplemented however by the latter two terms of the sum, which represent the characteristic interference effect of the superposed waves. They are the terms that lead to the intensity fringes observed by Young.

Intensity fringes of that sort assume the greatest possible contrast or visibility when the cross correlation terms like $G^{(1)}(\boldsymbol{r}_1 t_1 \boldsymbol{r}_2 t_2)$ are as large in magnitude as possible. But there is a simple limitation imposed on the magnitude of such correlations by a familiar inequality. There is a formal sense in which cross correlation functions like $G^{(1)}(\boldsymbol{r}_1 t_1 \boldsymbol{r}_2 t_2)$ are analogous to the scalar products of two vectors and are thus subject to a Schwarz inequality. The squared absolute value of that correlation function can then never exceed the product of the two intensities. If we let x abbreviate a coordinate pair r,t, we must have

$$|\, G^{(1)}(x_1 x_2)\,|^2 \leq G^{(1)}(x_1 x_1)\, G^{(1)}(x_2 x_2). \tag{6}$$

The upper bound to the cross-correlation is attained if we have

$$|\, G^{(1)}(x_1 x_2)\,|^2 = G^{(1)}(x_1 x_1)\, G^{(1)}(x_2 x_2), \tag{7}$$

and with it we achieve maximum fringe contrast. We shall then speak of the fields at x_1 and x_2 as being optically coherent with one another. That is the definition of relative coherence that optics has traditionally used.[5]

There is another way of stating the condition for optical coherence that is also quite useful, particularly when we are discussing coherence at pairs of points extending over some specified region in space-time. Let us assume that it is possible to find a positive frequency field $\mathcal{E}(\boldsymbol{r}t)$ satisfying the appropriate Maxwell equations and such that the correlation function (4) factorizes into the form

$$G^{(1)}(\boldsymbol{r}_1 t_1, \boldsymbol{r}_2 t_2) = \mathcal{E}^*(\boldsymbol{r}_1 t_1)\mathcal{E}(\boldsymbol{r}_2 t_2). \tag{8}$$

While the necessity of this factorization property requires a bit of proof,[6] it is at least clear that it does bring about the optical coherence that we have defined by means of the upper bound in the inequality (6) since in that case we have

$$|G^{(1)}(\boldsymbol{r}_1 t_1 \boldsymbol{r}_2 t_2)|^2 = |\mathcal{E}(\boldsymbol{r}_1 t_1)|^2 |\mathcal{E}(\boldsymbol{r}_2 t_2)|^2. \tag{9}$$

In the quantum theory, physical variables such as $E^{(\pm)}(\boldsymbol{r}t)$ are associated, not with simple complex numbers, but with operators on the Hilbert-space vectors $|\ \rangle$ that represent the state of the system, which in the present case is the electromagnetic field. Multiplication of the operators $E^{(+)}(\boldsymbol{r}_1 t_1)$ and $E^{(-)}(\boldsymbol{r}_2 t_2)$ is not in general commutative, and the two operators can be demonstrated to act in altogether different ways on the vectors $|\ \rangle$ that represent the state of the field. The operator $E^{(+)}$, in particular, can be shown to be an annihilation operator. It lowers by one the number of quanta present in the field. Applied to an n-photon state, $|n\rangle$, it reduces it to an $n-1$ photon state, $|n-1\rangle$. Further applications of $E^{(+)}(\boldsymbol{r}t)$ keep reducing the number of quanta present still further, but the sequence must end with the $n = 0$ or vacuum state, $|vac\rangle$, in which there are no quanta left. In that state we must finally have

$$E^{(+)}(\boldsymbol{r}t)\,|vac\rangle = 0. \tag{10}$$

The operator adjoint to $E^{(+)}$, which is $E^{(-)}$ must have the property of raising an n-photon state to an $n + 1$ photon state, so we may be sure, for example, that the state $E^{(-)}(\boldsymbol{r}t)\,|vac\rangle$ is a one-photon state. Since the vacuum state can not be reached by raising the number of photons, we must also require the relation

$$\langle vac|\,E^{(-)}(\boldsymbol{r}t) = 0, \tag{11}$$

which is adjoint to Eq. (10).

The results of quantum measurements often depend on the way in which the measurements are carried out. The most useful and informative ways of discussing such experiments are usually those based on the physics of the measurement process itself. To discuss measurements of the intensity of light then we should understand the functioning of the device that detects or counts photons.

Such devices generally work by absorbing quanta and registering each such absorption process, for example, by the detection of an emitted photo-electron. We need not go into any of the details of the photoabsorption process to see the general nature of the expression for the photon counting probability. All we need to assume is that our idealized detector at the point $\boldsymbol{r}$ has negligibly small size and has a photo-absorption probability that is independent of frequency so that it can be regarded as probing the field at a well-defined time t. Then if the field makes a transition from an initial state $|i\rangle$ to a final state $|f\rangle$ in which there is one photon fewer, the probability amplitude for that particular transition is given by the scalar product – or matrix element

$$\langle f|\, E^{(+)}(\boldsymbol{r}t)\,|i\rangle\,. \tag{12}$$

To find the total transition probability we must find the squared modulus of this amplitude and sum it over the complete set of possible final states $|f\rangle$ for the field. The expression for the completeness of the set of states $|f\rangle$ is

$$\sum_f \; |f\rangle\,\langle f| = 1,$$

so that we then have a total transition probability proportional to

$$\sum_f \; |\langle f|\, E^{(+)}(\boldsymbol{r}t)\,|i\rangle\,|^2 = \sum_f \; \langle i|\, E^{(-)}(\boldsymbol{r}t)\,|f\rangle\,\langle f|\, E^{(+)}(\boldsymbol{r}t)\,|i\rangle$$

$$= \langle i|\, E^{(-)}(\boldsymbol{r}t)E^{(+)}(\boldsymbol{r}t)\;|i\rangle\,. \tag{13}$$

It is worth repeating here that in the quantum theory the fields $E^{(\pm)}$ are non-commuting operators rather than simple numbers. Thus one could not reverse their ordering in the expression (13) while preserving its meaning. In the classical theory we discussed earlier $E^{(+)}$ and $E^{(-)}$ are simple numbers that convey equivalent information. There is no physical distinction between photo-absorption and emission since there are no classical photons. The fact that the quantum energy $h\nu$ vanishes for $h \rightarrow 0$ removes any distinction between positive and negative values of the frequency variable.

The initial state of the field in our photon counting experiment depends, of course, on the output of whatever light source we use, and very few sources produce pure quantum states of any sort. We must thus regard the state $|i\rangle$ as depending in general on some set of random and uncontrollable parameters characteristic of the source. The counting statistics we actually measure then may vary from one repetition of the experiment to another. The figure we would quote must be regarded as the average taken over these repetitions. The neatest way of specifying the random properties of the state $|i\rangle$ is to define what von Neumann called the density operator

$$\rho = \{|i\rangle\,\langle i|\,\}_{\text{av}}, \tag{14}$$

which is the statistical average of the outer product of the vector $|i\rangle$ with itself. That expression permits us to write the average of the counting probability as

$$\left\{ \langle i | \, E^{(-)}(\boldsymbol{r}t) E^{(+)}(\boldsymbol{r}t) \, | i \rangle \right\}_{\text{av}} = \text{Trace}\left\{ \rho E^{(-)}(\boldsymbol{r}t) E^{(+)} \boldsymbol{r}t) \right\}. \tag{15}$$

Interference experiments like those of Young and Michelson, as we have noted earlier, often proceed by measuring the intensities of linear combinations of the fields characteristic of two different space-time points. To find the counting probability in a field like $E^{(+)}(\boldsymbol{r}_1 t_1) + E^{(+)}(\boldsymbol{r}_2 t_2)$, for example, we will need to know expressions like that of Eq. (15) but with two different space-time arguments $\boldsymbol{r}_1 t_1$ and $\boldsymbol{r}_2 t_2$. It is convenient then to define the quantum theoretical form of the correlation functions (4) as

$$G^{(1)}(\boldsymbol{r}_1 t_1 \boldsymbol{r}_2 \mathrm{t}_2) = \text{Trace}\left\{ \rho E^{(-)}(\boldsymbol{r}_1 t_1) E^{(+)}(\boldsymbol{r}_2 t_2) \right\}. \tag{16}$$

This function has the same scalar product structure as the classical function (4) and can be shown likewise to obey the inequality (6). Once again we can take the upper bound of the modulus of this cross-correlation function or equivalently the factorization condition (8) to define optical coherence.

It is worth noting at this point that optical experiments aimed at achieving a high degree of coherence have almost always accomplished it by using the most nearly monochromatic light attainable. The reason for that is made clear by the factorization condition (8). These experiments were always based on steady or statistically stationary light sources. What we mean by a steady state is that the function $G^{(1)}$ with two different time arguments, t_1 and t_2, can in fact only depend on their difference $t_1 - t_2$. Optical coherence then requires

$$G^{(1)}(t_1 - t_2) = \mathcal{E}^*(t_1)\mathcal{E}(t_2). \tag{17}$$

The only possible solution of such a functional equation for the function $\mathcal{E}(t)$ is one that oscillates with a single positive frequency. The requirement of monochromaticity thus follows from the limitation to steady sources. The factorization condition (8), on the other hand, defines optical coherence more generally for non-steady sources as well.

Although the energies of visible light quanta are quite small on the everyday scale, techniques for detecting them individually have existed for many decades. The simplest methods are based on the photoelectric effect and the use of electron photomultipliers to produce well defined current pulses. The possibility of detecting individual quanta raises interesting questions concerning their statistical distributions, distributions that should in principle be quite accessible to measurement. We might imagine, for example, putting a quantum counter in a given light beam and asking for the distribution of time intervals between successive counts. Statistical problems of that sort were never, to my knowledge, addressed until the importance of quantum correlations began to become clear in the 1950's. Until that time virtually all optical experiments measured only average intensities or quantum counting rates, and the correlation function $G^{(1)}$ was all we needed to describe them. It was in that decade, however, that several new sorts of experiments requiring a more general approach were begun. That period seemed to mark the beginning of quantum optics as a relatively new or rejuvenated field.

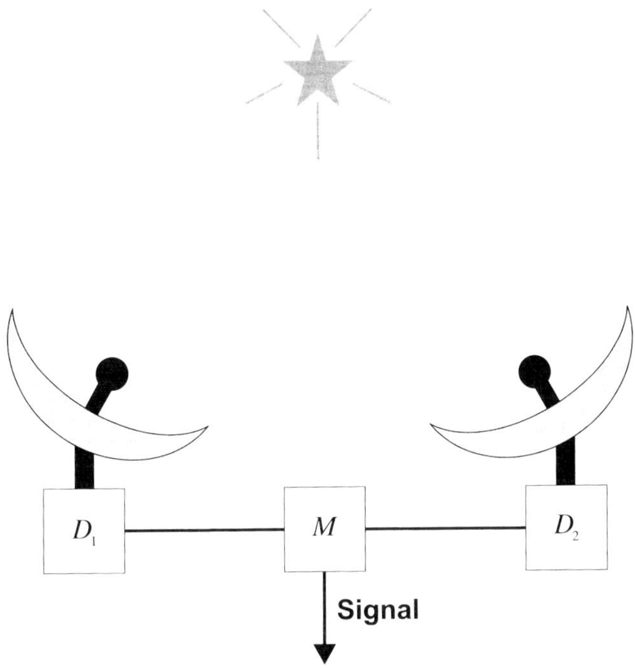

Figure 2. The intensity interferometry scheme of Hanbury Brown and Twiss. Radio frequency waves are received and detected at two antennas. The filtered low-frequency signals that result are sent to a device that furnishes an output proportional to their product.

In the experiment I found most interesting, R. Hanbury Brown and R. Q. Twiss developed a new form of interferometry.[7] They were interested at first in measuring the angular sizes of radio wave sources in the sky and found they could accomplish that by using two antennas, as shown in Figure 2, with a detector attached to each of them to remove the high-frequency oscillations of the field. The noisy low frequency signals that were left were then sent to a central device that multiplied them together and recorded their time-averaged values. Each of the two detectors then was producing an output proportional to the square of the incident field, and the central device was recording a quantity that was quartic in the field strengths.

It is easy to show, by using classical expressions for the field strengths, that the quartic expression contains a measurable interference term, and by exploiting it Hanbury Brown and Twiss did measure the angular sizes of many radio sources. They then asked themselves whether they couldn't perform the same sort of "intensity interferometry" with visible light, and thereby measure the angular diameters of visible stars. Although it seemed altogether logical that they could do that, the interference effect would have to involve the detection of pairs of photons and they were evidently inhibited in imagining the required interference effect by a statement Dirac makes in the first chapter of his famous textbook on quantum mechanics.[8] In it he is discussing why one sees intensity fringes in the Michelson interferometer,

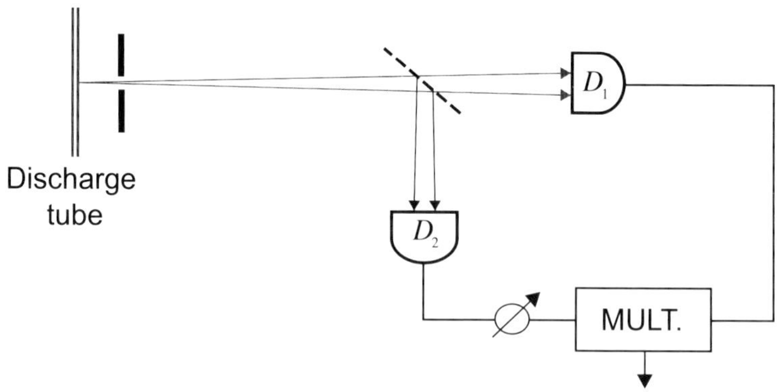

Figure 3. The Hanbury Brown-Twiss photon correlation experiment. Light from an extremely monochromatic discharge tube falls on a half-silvered mirror which sends the split beam to two separate photo-detectors. The random photocurrents from the two detectors are multiplied together and then averaged. The variable time delay indicated is actually achieved by varying the distance of the detector D_2 from the mirror.

and says in ringingly clear terms "Each photon then interferes only with itself. Interference between two different photons never occurs."

It is worth recalling at this point that interference simply means that the probability amplitudes for alternative and indistinguishable histories must be added together algebraically. It is not the photons that interfere physically, it is their probability amplitudes that interfere – and probability amplitudes can be defined equally well for arbitrary numbers of photons.

Evidently Hanbury Brown and Twiss remained uncertain on this point and undertook an experiment[9] to determine whether pairs of photons can indeed interfere. Their experimental arrangement is shown in Figure 3. The light source is an extremely monochromatic discharge tube. The light from that source is collimated and sent to a half-silvered mirror which sends the separated beams to two separate photo-detectors. The more or less random output signals of those two detectors are multiplied together, as they were in the radiofrequency experiments, and then averaged. The resulting averages showed a slight tendency for both of the photo-detectors to register photons simultaneously (Figure 4). The effect could be removed by displacing one of the counters and thus introducing an effective time delay between them. The coincidence effect thus observed was greatly weakened by the poor time resolution of the detectors, but it raised considerable surprise nonetheless. The observation of temporal correlations between photons in a steady beam was something altogether new. The experiment has been repeated several times, with better resolution, and the correlation effect has emerged in each case more clearly.[10]

The correlation effect was enough of a surprise to call for a clear explanation. The closest it came to that was a clever argument[11] by Purcell who used the semi-classical form of the radiation theory in conjunction with a formula for the relaxation time of radiofrequency noise developed in wartime radar

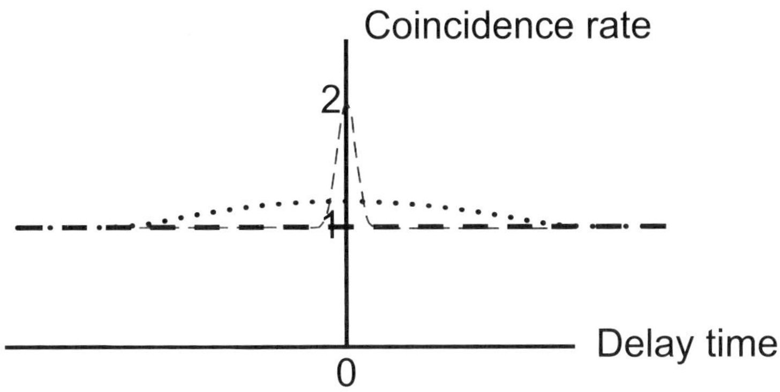

Figure 4. The photon coincidence rate measured rises slightly above the constant background of accidental coincidences for sufficiently small time delays. The observed rise was actually weakened in magnitude and extended over longer time delays by the relatively slow response of the photo-detectors. With ideal detectors it would take the more sharply peaked form shown.

studies. It seemed to indicate that the photon correlation time would be increased by just using a more monochromatic light source.

The late 1950's were, of course, the time in which the laser was being developed, but it was not until 1960 that the helium-neon laser[12] was on the scene with its extremely monochromatic and stable beams. The question then arose: what are the correlations of the photons in a laser beam? Would they extend, as one might guess, over much longer time intervals as the beam became more monochromatic? I puzzled over the question for some time, I must admit, since it seemed to me, even without any detailed theory of the laser mechanism, that there would not be any such extended correlation.

The oscillating electric current that radiates light in a laser tube is not a current of free charges. It is a polarization current of bound charges oscillating in a direction perpendicular to the axis of the tube (Figure 5). If it is sufficiently strong it can be regarded as a predetermined classical current, one that suffers negligible recoil when individual photons are emitted. Such currents, I knew,[13] emitted Poisson distributions of photons, which indicated clearly that the photons were statistically independent of one another. It seemed then that a laser beam would show no Hanbury Brown-Twiss photon correlations at all.

How then would one describe the delayed-coincidence counting measurement of Hanbury Brown and Twiss? If the two photon counters are sensitive at the space-time points $\boldsymbol{r}_1 t_1$ and $\boldsymbol{r}_2 t_2$ we will need to make use of the annihilation operators $E^{(+)}(\boldsymbol{r}_1 t_1)$ and $E^{(+)}(\boldsymbol{r}_2 t_2)$ (which do, in fact commute). The amplitude for the field to go from the state $|i\rangle$ to a state $|f\rangle$ with two quanta fewer is

$$\langle f|\ E^{(+)}(\boldsymbol{r}_2 t_2) E^{(+)}(\boldsymbol{r}_1 t_1)\,|i\rangle\,. \tag{18}$$

When this expression is squared, summed over final states $|f\rangle$ and averaged

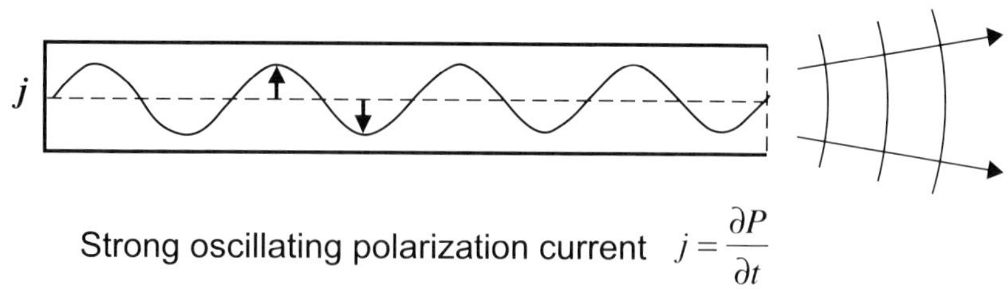

Figure 5. Schematic picture of a gas laser. The standing light wave in the discharge tube generates an intense transverse polarization current in the gas. Its oscillation sustains the standing wave and generates the radiated beam.

over the initial states $|i\rangle$ we have a new kind of correlation function that we can write as

$$G^{(2)}(\boldsymbol{r}_1 t_1 \boldsymbol{r}_2 t_2 \boldsymbol{r}_2 t_2 \boldsymbol{r}_1 t_1) = \text{Trace}\{\rho E^{(-)}(\boldsymbol{r}_1 t_1) E^{(-)}(\boldsymbol{r}_2 t_2) E^{(+)}(\boldsymbol{r}_2 t_2) E^{(+)}(\boldsymbol{r}_1 t_1)\}. \quad (19)$$

This is a special case of a somewhat more general second order correlation function that we can write (with the abbreviation $x_j = \boldsymbol{r}_j t_j$) as

$$G^{(2)}(x_1 x_2 x_3 x_4) = \text{Trace}\{\rho E^{(-)}(x_1) E^{(-)}(x_2) E^{(+)}(x_3) E^{(+)}(x_4)\}. \quad (20)$$

Now Hanbury Brown and Twiss had seen to it that the beams falling on their two detectors were as coherent as possible in the usual optical sense. The function $G^{(1)}$ should thus have satisfied the factorization condition (8), but that statement doesn't at all imply any corresponding factorization property of the functions $G^{(2)}$ given by Eqs. (19) or (20).

We are free to define a kind of second-order coherence by requiring a parallel factorization of $G^{(2)}$,

$$G^{(2)}(x_1 x_2 x_3 x_4) = \mathcal{E}^*(x_1)\mathcal{E}^*(x_2)\mathcal{E}(x_3)\mathcal{E}(x_4), \quad (21)$$

and the definition can be a useful one even though the Hanbury Brown-Twiss correlation assures us that no such factorization is present in their experiment. If it were present the coincidence rate according to Eq. (21) would be proportional to

$$G^{(2)}(x_1 x_2 x_2 x_1) = G^{(1)}(x_1 x_1)\ G^{(1)}(x_2 x_2), \quad (22)$$

that is, to the product of the two average intensities measured separately – and that is what was not found. Ordinary light beams, that is, light from ordinary sources, even extremely monochromatic ones as in the Hanbury Brown-Twiss experiment, do not have any such second order coherence.

We can go on defining still higher-order forms of coherence by defining n-th order correlation functions $G^{(n)}$ that depend on $2n$ space-time coordinates. The usefulness of such functions may not be clear since carrying out the n-fold delayed coincidence counting experiments that measure them would be quite difficult in practice. It is nonetheless useful to discuss such functions since they do turn out to play an essential role in most calculations of the statistical distributions of photons. If we turn on a photon counter for any given length of time, for example, the number of photons it records will be a random integer. Repeating the experiment many times will lead us to a distribution function for that number. To predict those distributions[14] we need, in general, to know the correlation functions $G^{(n)}$ of arbitrary orders.

Once we are defining higher order forms of coherence, it is worth asking whether we can find fields that lead to factorization of the complete set of correlation functions $G^{(n)}$. If so, we could speak of those as possessing full coherence. Now, are there any such states of the field? In fact there are lots of them, and some can describe precisely the fields generated by predetermined classical current distributions. These fields have the remarkable property that annihilating a single quantum in them by means of the operator $E^{(+)}$ leaves the field essentially unchanged. It just multiplies the state vector by an ordinary number. That is a relation we can write as

$$E^{(+)}(\boldsymbol{r}t)\,\big|\,\big\rangle = \mathcal{E}(\boldsymbol{r}t)\,\big|\,\big\rangle, \tag{23}$$

where $\mathcal{E}(\boldsymbol{r}t)$ is a positive frequency function of the space-time point $\boldsymbol{r}t$. It is immediately clear that such states must have indefinite numbers of quanta present. Only in that way can they remain unchanged when one quantum is removed. This remarkable relation does in fact hold for all of the quantum states radiated by a classical current distribution, and in that case the function $\mathcal{E}(\boldsymbol{r}t)$ happens to be the classical solution for the electric field.

Any state vector that obeys the relation (23) will also obey the adjoint relation

$$\big\langle\,\big|\,E^{(-)}(\boldsymbol{r}t) = \mathcal{E}^*(\boldsymbol{r}t)\,\big\langle\,\big|\,. \tag{24}$$

Hence the n-th order correlation function will indeed factorize into the form

$$G^{(n)}(x_1 \ldots . x_{2n}) = \mathcal{E}^*(x_1)\ldots.\mathcal{E}^*(x_n)\mathcal{E}(x_{n+1})\ldots.\mathcal{E}(x_{2n}) \tag{25}$$

that we require for n-th order coherence. Such states represent fully coherent fields, and delayed coincidence counting measurements carried out in them will reveal no photon correlations at all. To explain, for example, the Hanbury Brown-Twiss correlations we must use not pure coherent states but mixtures of them, for which the factorization conditions like Eq. (25) no longer hold. To see how these mixtures arise, it helps to discuss the modes of oscillation of the field individually.

The electromagnetic field in free space has a continuum of possible frequencies of oscillation, and a continuum of available modes of spatial oscillation at any given frequency. It is often simpler, instead of discussing all these modes at once, to isolate a single mode and discuss the behavior of that one alone. The field as a whole is then a sum of the contributions of the individual modes. In fact when experiments are carried out within reflecting enclosures the field modes form a discrete set, and their contributions are often physically separable.

The oscillations of a single mode of the field, as we have noted earlier, are essentially the same as those of a harmonic oscillator. The n-th excitation state of the oscillator represents the presence of exactly n light quanta in that mode. The operator that decreases the quantum number of the oscillator is usually written as a, and the adjoint operator – which raises the quantum number by one unit as $a^\dagger$. These operators then obey the relation

$$a\,a^\dagger - a^\dagger a = 1, \tag{26}$$

which shows that their multiplication is not commutative. We can take the field operator $E^{(+)}(\boldsymbol{r}t)$ for the mode we are studying to be proportional to the operator a. Then any state vector for the mode that obeys the relation (23) will have the property

$$a \left|\ \right\rangle = \alpha \left|\ \right\rangle \tag{27}$$

where α is some complex number. It is not difficult to solve for the state vectors that satisfy the relation (27) for any given value of α. They can be expressed as a sum taken over all possible quantum-number states $|n\rangle$, $n = 0, 1, 2\ldots.$ that takes the form

$$|\alpha\rangle = e^{-\frac{1}{2}|\alpha|^2} \sum_{n=0}^{\infty} \frac{\alpha^n}{\sqrt{n!}} |n\rangle, \tag{28}$$

in which we have chosen to label the state with the arbitrary complex parameter α. The states $|\alpha\rangle$ are fully coherent states of the field mode.

The squared moduli of the coefficients of the states $|n\rangle$ in Eq. (28) tell us the probability for the presence of n quanta in the mode, and those numbers do indeed form a Poisson distribution, one with the mean value of n equal to $|\alpha|^2$. The coherent states form a complete set of states in the sense that any state of the mode can be expressed as a suitable sum taken over them. As we have defined them they are equivalent to certain oscillator states defined by Schrödinger[15] in his earliest discussions of wave functions. Known thus from the very beginning of wave mechanics, they seemed not to have found any important role in the earlier development of the theory.

Coherent excitations of fields have a particularly simple way of combining. Let us suppose that one excitation mechanism brings a field mode from its empty state $|0\rangle$ to the coherent state $|\alpha_1\rangle$. A second mechanism could bring it, for example, from the state $|0\rangle$ to the state $|\alpha_2\rangle$. If the two mechanisms act simultaneously the resulting state can be written as $e^{i\varphi}|\alpha_1 + \alpha_2\rangle$ where $e^{i\varphi}$ is a phase factor that depends on α_1 and α_2, but we don't need to know it since it cancels out of the expression for the density operator

$$\rho = |\alpha_1 + \alpha_2\rangle\langle\alpha_1 + \alpha_2|. \tag{29}$$

This relation embodies the superposition principle for field excitations and tells us all about the resulting quantum statistics. It is easily generalized to treat the superposition of many excitations. If, say, j coherent excitations were present, we should have a density operator

$$\rho = |\alpha_1 + \ldots + \alpha_j\rangle\ \langle\alpha_1 + \ldots + \alpha_j|. \tag{30}$$

Let us suppose now that the individual excitation amplitudes α_j are in one

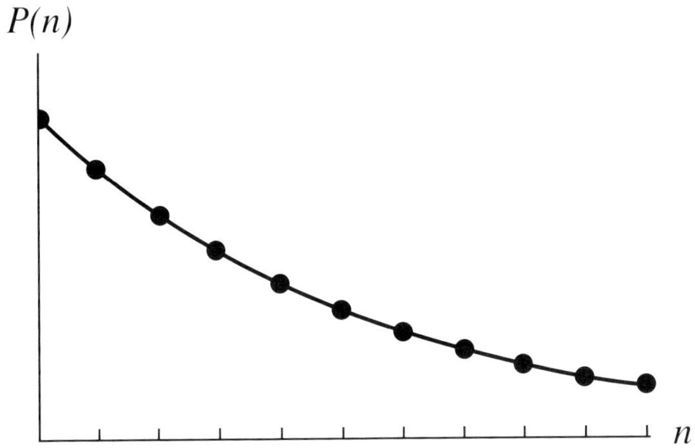

Figure 6. Geometrical or fixed-ratio sequence of probabilities for the presence of *n* quanta in a mode that is excited chaotically.

or another sense random complex numbers. Then the sum $\alpha_1 + \ldots + \alpha_j$ will describe a suitably defined random walk in the complex plane. In the limit $j \to \infty$ the probability distribution for the sum $\alpha = \alpha_1 + \ldots + \alpha_j$ will be given by a Gaussian distribution which we can write as

$$P(\alpha) = \frac{1}{\pi \langle n \rangle} e^{-\frac{|\alpha|^2}{\langle n \rangle}}, \tag{31}$$

in which the mean value of $|\alpha|^2$ which has been written as $\langle n \rangle$, is the mean number of quanta in the mode.

The density operator that describes this sort of random excitation is a probabilistic mixture of coherent states,

$$\rho = \frac{1}{\pi \langle n \rangle} \int e^{-\frac{|\alpha|^2}{\langle n \rangle}} |\alpha\rangle \langle\alpha| \, d^2\alpha, \tag{32}$$

where $d^2\alpha$ is an element of area in the complex plane. When we express ρ in terms of *m*-quantum states by using the expansion (28) we find

$$\rho = \frac{1}{1+\langle n \rangle} \sum_{m=0}^{\infty} \left(\frac{\langle n \rangle}{1+\langle n \rangle} \right)^m |m\rangle\langle m| \,. \tag{33}$$

This kind of random excitation mechanism is thus always associated with a geometrical or fixed-ratio distribution of quantum numbers (Figure 6). In the best known example of the latter, the Planck distribution, we have $\langle n \rangle = \left(e^{\frac{h\nu}{kT}} - 1 \right)^{-1}$, and the density operator (33) then contains the familiar thermal weights $e^{-\frac{mh\nu}{kT}}$.

There is something remarkably universal about the geometrical sequence of n-quantum probabilities. The image of chaotic excitation we have derived it from, on the other hand, excitation in effect by a random collection of lasers, may well seem rather specialized. It may be useful therefore to have a more general way of characterizing the same distribution. If a quantum state is specified by the density operator ρ, we may associate with it an entropy S given by

$$S = -\mathrm{Trace}(\rho \log \rho), \tag{34}$$

which is a measure, roughly speaking, of the disorder characteristic of the state. The most disordered, or chaotic, state is reached by maximizing S, but in finding the maximum we must observe two constraints. The first is

$$\mathrm{Trace}\ \rho = 1, \tag{35}$$

which says simply that all probabilities add up to one. The second is

$$\mathrm{Trace}\ (\rho\ a^{\dagger}a) = \langle n \rangle, \tag{36}$$

which fixes the average occupation number of the mode.

When S is maximized, subject to these two constraints, we find indeed that the density operator ρ takes the form given by Eq. (33). The geometrical distribution is thus uniquely representative of chaotic excitation. Most ordinary light sources consist of vast numbers of atoms radiating as nearly independently of one another as the field equations will permit. It should be no surprise then that these are largely maximum entropy or chaotic sources. When many modes are excited, the light they radiate is, in effect, colored noise and indistinguishable from appropriately filtered black body radiation.

For chaotic sources, the density operator (32) permits us to evaluate all of the higher order correlation functions $G^{(n)}(x_1 \ldots x_{2n})$. In fact they can all be reduced[14] to sums of products of first order correlation functions $G^{(1)}(x_i x_j)$. In particular, for example, the Hanbury Brown-Twiss coincidence rate corresponding to the two space time points x_1 and x_2 can be written as

$$G^{(2)}(x_1 x_2 x_2 x_1) = G^{(1)}(x_1 x_1)\, G^{(1)}(x_2 x_2) + G^{(1)}(x_1 x_2)\, G^{(1)}(x_2 x_1). \tag{37}$$

The first of the two terms on the right side of this equation is simply the product of the two counting rates that would be measured at x_1 and x_2 independently. The second term is the additional delayed coincidence rate detected first by Hanbury Brown and Twiss, and it is indeed contributed by a two-photon interference effect. If we let $x_1 = x_2$, which corresponds to zero time delay in their experiment, we see that

$$G^{(2)}(x_1 x_1 x_1 x_1) = 2\{G^{(1)}(x_1 x_1)\}^2, \tag{38}$$

or the coincidence rate for vanishing time delay should be double the background or accidental rate.

The Gaussian representation of the density operator in terms of coherent states is an example of a broader class of so-called “diagonal representations” that are quite convenient to use – when they are available. If the density operator for a single mode, for example, can be written in the form

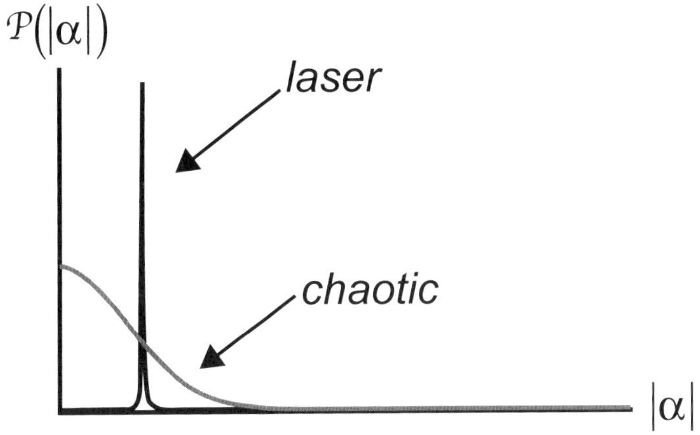

Figure 7. The quasiprobability function $\mathcal{P}(|\alpha|)$ for a chaotic excitation is Gaussian in form, while for a stable laser beam it takes on non-zero values only near a fixed value of $|\alpha|$.

$$\rho = \int \mathcal{P}(\alpha)\,|\alpha\rangle\langle\alpha|\,d^2\alpha \tag{39}$$

then the expectation values of operator products like $a^{\dagger n}a^m$ can be evaluated as simple integrals over the function $\mathcal{P}$ such as

$$\{a^{\dagger n}a^m\}_{\text{av}} = \int \mathcal{P}(\alpha)\,\alpha^{*n}\alpha^m d^2\alpha. \tag{40}$$

The function $\mathcal{P}(\alpha)$ then takes on some of the role of a probability density, but that can be a bit misleading since the condition that the probabilities derived from ρ all be positive or zero does not require $\mathcal{P}(\alpha)$ to be positive definite. It can and sometimes does take on negative values over limited areas of the α-plane in certain physical examples, and it may also be singular. It is a member, as we shall see, of a broader class of quasi-probability densities. The representation (39), the $\mathcal{P}$-representation, unfortunately is not always available.[16,17] It can not be defined, for example, for the familiar "squeezed" states of the field in which one or the other of the complementary uncertainties is smaller than that of the coherent states.

The difference between a monochromatic laser beam and a chaotic beam is most easily expressed in terms of the function $\mathcal{P}(\alpha)$. For a stationary laser beam the function $\mathcal{P}$ depends only on the magnitude of α and vanishes unless $|\alpha|$ assumes some fixed value. A graph of that function $\mathcal{P}$ is shown in Figure 7, where it can be compared with the Gaussian function for the same mean occupation number $\langle n\rangle$ given by Eq. (31).

How do we measure the statistical properties of photon distributions? A relatively simple way is to place a photon counter in a light beam behind either a mechanical or an electrical shutter. If we open the shutter for a given length of time t, the counter will register some random number n of photons. By repeating that measurement sufficiently many times we can establish a statistical distribution for those random integers n. The analysis necessary to derive this distribution mathematically can be a bit complicated since it re-

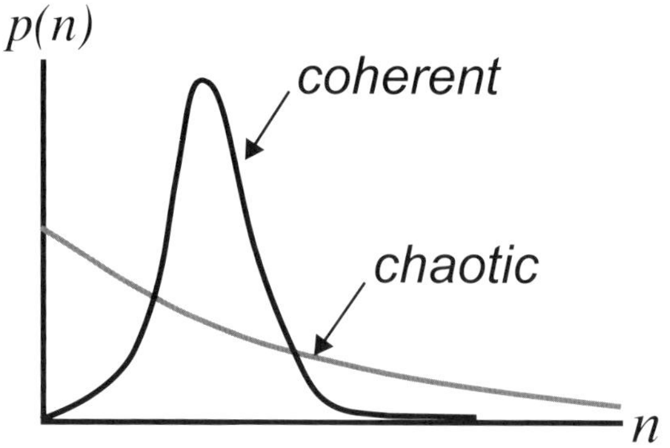

Figure 8. The two $\mathcal{P}(|\alpha|)$ distributions of Fig. 7 lead to different photon occupation number distributions $p(n)$: for chaotic excitation a geometric distribution, for coherent excitation a Poisson distribution.

quires, in general, a knowledge of all the higher order correlation functions. Experimental measurements of the distribution, conversely, can tell us about those correlation functions.

For the two cases in which we already know all the correlation functions, it is particularly easy to find the photocount distributions. If the average rate at which photons are recorded is w, then the mean number recorded in time t is

$$\langle n \rangle = wt .$$

In a coherent beam the result for the probability of n photocounts is just the Poisson distribution

$$p_n(t) = \frac{(wt)^n}{n!} e^{-wt}. \tag{41}$$

In a chaotic beam, on the other hand, the probability of counting n quanta is given by the rather more spread-out distribution

$$p_n(t) = \frac{1}{1+wt} \left(\frac{wt}{1+wt} \right)^n. \tag{42}$$

These results, which are fairly obvious from the occupation number probabilities implicit in Eqs. (28) and (33), are illustrated in Figure 8.

Here is a closely related question that can also be investigated experimentally without much difficulty. If we open the shutter in front of the counter at an arbitrary moment, some random interval of time will pass before the first photon is counted. What is the distribution of those random times? In a steady coherent beam, in fact, it is just an exponential distribution

$$\mathcal{W}_{\text{coh}} = we^{-wt} , \tag{43}$$

while in a chaotic beam it assumes the less obvious form

$$\mathcal{W}_{\text{ch}}(\text{t}) = \frac{w}{(1+wt)^2} . \tag{44}$$

There is an alternative way of finding a distribution of time intervals. Instead of simply opening a shutter at an arbitrary moment, we can begin the

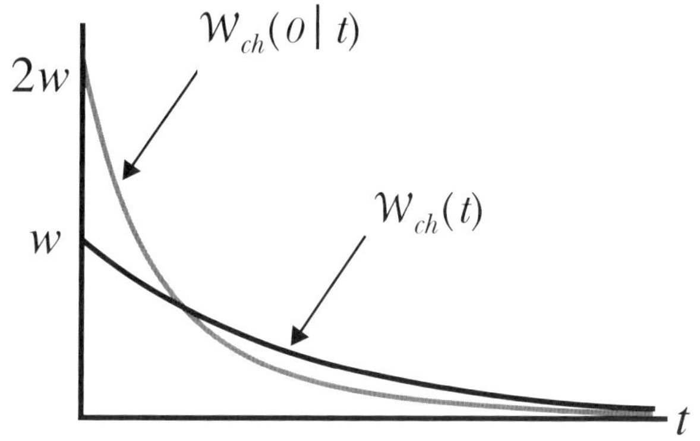

Figure 9. Time interval distributions for counting experiments in a chaotically excited mode: $\mathcal{W}_{ch}(t)$ is the distribution of intervals from an arbitrary moment until the first photocount. $\mathcal{W}_{ch}(0\,|\,t)$ is the distribution of intervals between two successive photocounts.

measurement with the registration of a given photocount at time zero and then ask what is the distribution, of time intervals until the next photocount. This distribution, which we may write as $\mathcal{W}(0\,|\,t)$, takes the same form for a coherent beam as it does for the measurement described earlier, which starts at arbitrary moments,

$$\mathcal{W}_{\text{coh}}(0\,|\,t) = we^{-wt} = \mathcal{W}_{\text{coh}}(t). \tag{45}$$

This identity is simply a restatement of the statistically independent or uncorrelated quality of all photons in a coherent beam.

For a chaotic beam, on the other hand, the distribution $\mathcal{W}_{\text{ch}}(o\,|\,t)$ takes a form quite different from $\mathcal{W}_{\text{ch}}(t)$. It is

$$\mathcal{W}_{\text{ch}}(0\,|\,t) = \frac{2w}{(1+wt)^3}, \tag{46}$$

an expression which exceeds $\mathcal{W}_{\text{ch}}(t)$ for times for which $wt < 1$, and is in fact twice as large as $\mathcal{W}_{\text{ch}}(t)$ for $t = 0$ (Figure 9). The reason for that lies in the Gaussian distribution of amplitudes implicit in Eqs. (31) and (32). The very fact that we have counted a photon at $t = 0$ makes it more probable that the field amplitude α has fluctuated to a large value at that moment, and hence the probability for counting a second photon remains larger than average for some time later. The functions $\mathcal{W}_{\text{ch}}(t)$ and $\mathcal{W}_{\text{ch}}(0\,|\,t)$ are compared in Figure 8.

All of the experiments we have discussed thus far are based on the procedure of photon counting, whether with individual counters or with several of them arranged to be sensitive in delayed coincidence. The functions they measure, the correlation functions $G^{(n)}$, are all expectation values of products of field operators written in a particular order. If one reads from right to left, the annihilation operator always precedes the creation operators in our correlation functions, as they do, for example, in Eq. (19) for $G^{(2)}$. It is that so-called "normal ordering" that gives the coherent states, and the quasiprobability density $\mathcal{P}(\alpha)$ the special roles they occupy in discussing this class of experiments.

But there are other kinds of expectation values that one sometimes needs in order to discuss other classes of experiments. These could, for example, involve symmetrically ordered sums of operator products, or even anti-normally ordered products which are opposite to the normally ordered ones. The commutation relations for the multiplication of field operators will ultimately relate all these expectation values to one another, but it is often possible to find much simpler ways of evaluating them. There exists a quasiprobability density that plays much the same role for symmetrized products as the function $\mathcal{P}$ does for the normally ordered ones. It is, in fact, the function Wigner[18] devised in 1932 as a kind of quantum mechanical replacement for the classical phase space density. For anti-normally ordered operator products, the role of the quasiprobability density is taken over by the expectation value which for a single mode is $\frac{1}{\pi}\langle\alpha|\,\rho\,|\alpha\rangle$. The three quasiprobability densities associated with the three operator orderings and whatever experiments they describe are all members of a larger family that can be shown to have many properties in common.[17]

The developments I have described to you were all relatively early ones in the development of the field we now call quantum optics. The further developments that have come in rapid succession in recent years are too numerous to recount here. Let me just mention a few. A great variety of careful measurements of photon counting distributions and correlations of the type we have discussed have been carried out[19] and furnish clear agreement with the theory. They have furthermore shown in detail how the properties of laser beams change as they rise in power from below threshold to above it.

The fully quantum mechanical theory of the laser was difficult to develop[20] since the laser is an intrinsically nonlinear device, but only through such a theory can its quantum noise properties be understood. The theories of a considerable assortment of other kinds of oscillators and amplifiers have now been worked out.

Nonlinear optics has furnished us with new classes of quantum phenomena such as parametric down conversion in which a single photon is split into a pair of highly correlated or entangled photons. Entanglement has been a rich source of the quantum phenomena that are perhaps most interesting – and baffling – in everyday terms.

It is worth emphasizing that the mathematical tools we have developed for dealing with light quanta can be applied equally well to the much broader class of particles obeying Bose-Einstein statistics. These include atoms of He^4, Na^{23}, Rb^{87}, and all of the others which have recently been Bose-condensed by optical means. When proper account is taken of the atomic interactions and the non-vanishing atomic masses, the coherent state formalism is found to furnish useful descriptions of the behavior of these bosonic gases.

The formalism seems likewise to apply to subatomic particles, to bosons that are only short-lived. The pions that emerge by hundreds or even thousands from the high-energy collisions of heavy ions are also bosons. The pions of similar charge have a clearly noticeable tendency to be emitted with closely correlated momenta, an effect which is evidently analogous to the Hanbury Brown-Twiss correlation of photons, and invites the same sort of analysis.[21]

Figure 10. Professor Einstein, encountered in the spring of 1951 in Princeton, NJ.

Particles obeying Fermi-Dirac statistics, of course, behave quite differently from photons or pions. No more than a single one of them ever occupies any given quantum state. This kind of reckoning associated with fermion fields is radically different therefore from the sort we have associated with bosons, like photons. It has proved possible, nonetheless, to develop an algebraic scheme[22] for calculating expectation values of products of fermion fields that is remarkably parallel to the one we have described for photon fields. There is a one-to-one correspondence between the mathematical operations and expressions for boson fields on the one hand and fermion fields on the other. That correspondence has promise of proving useful in describing the dynamics of degenerate fermion gases.

I'd like, as a final note, to share with you an experience I had in 1951, while I was a postdoc at the Institute for Advanced Study in Princeton. Possessed by the habit of working late at night – in fact on photon statistics[13] at the time – I didn't often appear at my Institute desk early in the day. Occasionally I walked out to the Institute around noon, and that was closer to the end of the work day for Professor Einstein. Our paths thus crossed quite a few times, and on one of those occasions I had ventured to bring my camera. He seemed more than willing to let me take his picture as if acknowledging his role as a local landmark, and he stood for me just as rigidly still. Here, in Figure 10, is the hitherto unpublished result. I shall always treasure that image, and harbor the enduring wish I had been able to ask him just a few questions about that remarkable year, 1905.

REFERENCES

1. M. Planck, Ann. d. Phys. **1**, 69 (1900).
2. M. Planck, Ann. d. Phys. **1**, 719 (1900).
3. A. Einstein, Ann. d. Phys. **17**, 132 (1905).
4. P. A. M. Dirac, Proc. Roy. Soc. **114**, 243, 710 (1927).
5. M. Born and E. Wolf, *Principles of Optics* (London: Pergamon Press, 1959), Chap. X.
6. U. M. Titulaer and R. J. Glauber, Phys. Rev. **140B**, 676 (1965), **145B**, 1041 (1966).
7. R. Hanbury Brown and R. Q. Twiss, Phil. Mag. Ser. 7, **45**, 663 (1954).
8. P. A. M. Dirac, *The Principles of Quantum Mechanics*, Fourth ed. (Oxford University Press, 1958), p. 9.
9. R. Hanbury Brown and R. Q. Twiss, Nature **177**, 27 (1956); Proc. Roy. Soc. (London) **A242**, 300 (1957), **A243**, 291 (1957).
10. G. A. Rebka and R. V. Pound, Nature **180**, 1035 (1957), D. B. Scarl, Phys.Rev. **175**, 1661 (1968).
11. E. M. Purcell, Nature **178**, 1449 (1956).
12. A. Javan, W. R. Bennett and D. R. Herriot, Phys. Rev. Lett. **6**, 106 (1961).
13. R. J. Glauber, Phys. Rev. **84**, 395 (1951).
14. R. J. Glauber, in *Quantum Optics and Electronics* (Les Houches 1964), eds. C. de Witt, A. Blandin and C. Cohen-Tannoudji (New York: Gordon and Breach Science Publishers, Inc., 1965), p. 63.
15. E. Schrödinger, Naturwissenschaften **14**, 664 (1926).
16. E. C. G. Sudarshan, Phys. Rev. Letters **10**, 277 (1963).
17. K. E. Cahill and R. J. Glauber, Phys. Rev. **177**, 1857 (1969), **177**, 1882 (1969).
18. E. Wigner, Phys. Rev. **40**, 749 (1932).
19. See for example, F. T. Arecchi, in: *Quantum Optics*, Course XLII, Enrico Fermi School, Varenna, 1969, ed. R. J. Glauber (New York: Academic Press, 1969); E. Jakeman and E. R. Pike, J. Phys. **A2**, 411 (1969).
20. M. O. Scully and W. E. Lamb, Jr., Phys. Rev. **159**, 208 (1967); M. Lax, in: *Brandeis University Summer Institute Lectures* (1966), Vol. II, eds. M. Cretien et al (New York: Gordon and Breach, 1968); H. Haken, *Laser Theory*, in: *Encyclopedia of Physics* XXV/2c, ed. S. Flügge (Heidelberg: Springer-Verlag, 1970).
21. R. J. Glauber, *Quantum Optics and Heavy Ion Physics*, http://arxiv.org/nucl-th/0604021, 2006.
22. K. E. Cahill and R. J. Glauber, Phys. Rev. **A59**, 1538 (1999).

Portrait photo of Roy J.Glauber by photographer J. Reed.

John L Hall

JOHN L. HALL

John Lewis Hall (known by many as Jan) was born on August 21, 1934 in Denver, Colorado. His father (John Ernest Hall) was trained as an electrical engineer and worked for the U. S. Bureau of Reclamation on many hydroelectric projects here and overseas. His mother, Rae (Long) Hall was an elementary school teacher and singer. Jan's interests in electricity and radio were supported by parental interest, but his research on black-powder rockets was gradually replaced by social activities such as scouts and a church youth group. After completing public schooling in Denver, he received a Westinghouse Scholarship to Carnegie Institute of Technology (now Carnegie Mellon University) and earned a B.S. (1956), M.S. (1958) and Ph.D. (1961) in Physics from the Carnegie Institute of Technology. He did his thesis work with Professor Robert T. Schumacher using a self-made electron microwave spin resonance spectrometer to study the hyperfine spectrum of interstitial hydrogen atoms in CaF_2 crystals. While considering various postgraduate opportunities, he chanced on a call for applications to be a National Research Council Fellow at the (then) National Bureau of Standards in Washington, D. C. Toward the end of that postdoc year, he was invited to join others from NBS to start a new research group at the University of Colorado Boulder campus. Since 1962, he has been responsible for a number of major innovations and developments in laser technology at JILA (formerly called the Joint Institute for Laboratory Astrophysics), operated by the University of Colorado and the National Institute of Standards and Technology (NIST).

He has authored more than 230 articles in refereed journals, and holds ten U.S. patents. He is a member of the National Academy of Sciences, and is a Fellow of the Optical Society of America and the American Physical Society. He is also a Senior Fellow Emeritus of the National Institute of Standards and Technology, and a Fellow Adjoint of JILA. He has trained numerous graduate students and post docs in the Physics Department at the University of Colorado, Boulder. He retired from NIST in November, 2004, but is still active in JILA and his consulting company (Hall Stable Lasers, LLC). His electronics hobbies and electro-optical systems design provide satisfying outlets for him, often leading to specialized instruments and other technical innovations, and occasional patents.

His JILA work has been recognized through a number of awards from NIST, the Department of Commerce, and the U. S. Office of Personnel Management. He has been awarded many professional, peer-generated honors by the Optical Society and the American Physical Society. In 2004, he received the IEEE Rabi Award and became a member of the French Légion d'Honneur. In 2005, he shared the Nobel Prize in Physics with T.W. Hänsch and Roy Glauber for "contributions to the development of laser-based precision spectroscopy, including the optical frequency comb technique."

Also a graduate of Carnegie Mellon, his wife, Lindy, (the former Marilyn Robinson) holds a Bachelor's degree from the Margaret Morrison Carnegie College (1957), and a Master's degree from Carnegie Library School (1959). She taught 7th grade English in a Boulder junior high for 17 years, followed by 5 years as media specialist there. She remains active as a consultant in curriculum development. They have three children, two in Colorado: Carey, who teaches in a middle school of the Jefferson County School District, and Jon, who specializes in automotive computer and engine troubleshooting. Thomas is a field engineer for a computer company near Boston. With his wife Margaret, and their two daughters (aged six and eight) and son (now two and a half), they often accept visiting grandparents at their home in Quincy, MA. A favorite gathering place for all is their mountain cabin at Marble, Colorado, in the mountains west of Aspen, which was built by the family during the 1980's. Jan is very proud of having done the electrical wiring and plumbing entirely by himself, and receiving no red flags or violation notices from the State inspectors.

"I am truly fortunate to have somehow always been in the right place at the right time, to capitalize on technical advances that made increasingly precise laser measurements and applications possible. I have worked with amazing colleagues and students, whose enthusiasm and vision made for excellent physics, great fun and the building of important programs. It was a great 45 years and I have really loved it. But nowadays Lindy and I often discuss my in-process transition on to personal plans and activities, leaving the next dazzling science to my younger friends."

DEFINING AND MEASURING OPTICAL FREQUENCIES: THE OPTICAL CLOCK OPPORTUNITY – AND MORE

Nobel Lecture, December 8, 2005

by

JOHN L. HALL

JILA, NIST, and University of Colorado, Boulder, CO 80309-0440, USA.

ABSTRACT

Four long-running currents in laser technology met and merged in 1999–2000. Two of these were the quest toward a stable repetitive sequence of ever-shorter optical pulses and, on the other hand, the quest for the most time-stable, unvarying optical frequency possible. The marriage of UltraFast- and UltraStable lasers was brokered mainly by two international teams and became exciting when a special "designer" microstructure optical fiber was shown to be nonlinear enough to produce "white light" from the femto-second laser pulses, such that the output spectrum embraced a full optical octave. Then, for the first time, one could realize an optical frequency interval equal to the comb's lowest frequency, and count out this interval as a multiple of the repetition rate of the femtosecond pulse laser. This "gear-box" connection between the radio frequency standard and any/all optical frequency standards came just as Sensitivity-Enhancing ideas were maturing. The four-way Union empowered an explosion of accurate frequency measurement results in the standards field and prepares the way for refined tests of some of our cherished physical principles, such as the time-stability of some of the basic numbers in physics (eg., the "fine-structure" constant, the speed of light, certain atomic mass ratios ...), and the equivalence of time-keeping by clocks based on different physics. The stable laser technology also allows time-synchronization between two independent femto-second lasers so exact they can be made to appear as if the source were a single laser. By improving pump/probe experiments, one important application will be in bond-specific spatial scanning of biological samples. This next decade in optical physics should be a blast!

OVERVIEW AND SUMMARY

The view backward over some momentous developments often suggests a kind of certainty and inevitability that may not have been evident, even in the slightest form, when the story was going on. One modern trend is to focus on some particular research project – one which is so simple and transparent that the Manager can expect to be successful in the chosen research task. But such a project will likely have modest consequences: Surely its consequences were at least dimly visible from the beginning. By contrast, this "Optical Frequency Comb" capability has come "out of the blue" from a remarkable synthesis of independent "state-of-the-art" developments in four distinct fields: UltraStable Lasers, UltraFast Pulse Lasers, Ultra-NonLinear Materials and Responses, and UltraSensitive Laser Spectroscopy. These separate fields were alike in their shared – but independent – pursuit of advancing simple and effective technology for using electromagnetic signals for their own spectroscopic and other optical physics interests in the visible domain. After the **Great Laser Technology Synthesis of 1999–2000**, celebrated by the brief name of "Optical Frequency Comb," the Optical Toolbox has really blossomed. In respecting our Patron, Dr. Nobel, we may be more expansive and clear: **the field of optics has blossomed explosively**!

The resulting new capabilities are unbelievably rich in terms of the tools and capabilities that have been created, and these in turn are reinforcing progress in these related contributing fields. For example, after the frenzy of the first generation frequency measurements, some of the Generation II comb applications now include: low-jitter time synchronization between ultrafast laser sources, coherent stitching-together of the spectra of separate fs laser sources so as to spectrally broaden and temporally shorten the composite pulse, optical waveform synthesis for Coherent Control experiments, precision measurement of optical nonlinearities using the phase measurement sensitivity of rf techniques, coherently storing a few hundred sequential pulses and then extracting their combined energy to generate correspondingly more intense pulses at a lower repetition rate… . Attractive topics of research for Generation III applications include precise remote synchronization of accelerator cavity fields and the stable reference oscillators for Large Array Microwave Telescopes; and potential reduction of the relative phase-noise of the oscillator references used for deep space telescopes. (NASA, VLBI …) That's just part of the first five years.

So in the precision metrology field, what exactly could one say is different now? In the same way we have enjoyed for the last half-century powerful spectroscopy methods with radiofrequency signals (consider Magnetic Resonance Imaging as one of its useful forms!), we now can use frequency-control methods for optical spectroscopy. But there is a really important difference: the number of cycles per second in the optical domain is roughly 10-million-fold greater than in the rf domain, even as the rf processes themselves are still a few million-fold faster than human perception scales. In essence, these large factors map into a corresponding improvement in resolution – our measurement capability. See the discussion below. With human senses we can perceive halves and quarters and tenths, and perhaps a little better. These capabilities are enhanced approximately by the product of these two large numbers, bringing us immediately into the garden of a few parts in 10^{14} metrology. We can do even better by averaging independent measurements.

METROLOGICAL STANDARDS AND SCIENCE

A Close and bi-directional Connection

On occasion, accumulation of progress in the details of some scientific enquiry leads us to a glorious new vision of some parts of our experience: basically a new insight or organizing principle becomes available. But behind this revelation normally is a huge amount of painstaking work, quantitatively stating experimental results, which normally are expressed in absolute units. Sometimes an experiment can provide its own internal calibration, but in the main we really need to have **practical standards** to reference the measurements against. Of course the Standards must themselves be reproduced and distributed before the scientific results can be confirmed by several labs. The best case is that the needed Standard is based on some fundamental physical effect, ideally a quantum effect, so it can be *independently realized* by different laboratories at the same accuracy. This standards-realization process is in a revolution itself! [1]

The Length Standard and its Relationship to Frequency/Time

It's useful to discuss a bit about metrological Standards, which we can initially take to be the seven base quantities of the **Système International d'Unités** (International System of Units), or more briefly the SI, or "the Metric System" [2]. From these seven base units, another ~30 useful derived units can be defined. For our purposes of *stretching measurement precision to the ultimate limits*, clearly **Time** and **Length** are the two quantities offering the highest potential precision. For eons the day was a natural unit for Time, but standards for Length have seemed artificial and arbitrary. In 1791 the Metric System was first discussed but, lacking serious metrology experience, these Age of Enlightenment gentlemen of the French Academy of Sciences decided that the Metre would be defined as some small fraction ($\frac{1}{4}$ x10^{-7}) of the earth's circumference on a great circle passing through the poles and France. Of course, having the standard based on surveying had some limitations in practical lab work, but at least the unit of length was finally a definite and basically absolute distance. This was welcome change since public exhibits in places such as Braunschweig, Germany and on Santorini Island, Greece show there was a succession of length standards in sequential use, as a new Duke of different personal arm length came into power. But by 1875, with the Treaty of the Metre Convention, a stable metal bar began to look like a good idea. While not fully universal and independently realizable, the factory could make many of these prototype Metre bars, and could confirm their equivalence.

The community of Metric countries in 1889 welcomed the improved X-cross-section meter bars known as the "International Prototype Metre" length standard. This design used graduations (lines) engraved onto a platinum-iridium bar, with a Meter defined as the separation between two graduation lines at 0 °C, measured with a specified mounting arrangement, and under atmospheric pressure. The 30 new bars were calibrated using an optical comparator technique, before dissemination of two to each country.

By 1890 A. A. Michelson had identified the exceptional coherence of the Cd red line, and by 1892 had used it with his new interferometer to determine the length of the International Prototype Metre. His measurements showed the defined Metre contained 1,553,164.13 units of the wavelength of the cadmium red line, measured in air at 760 mm of atmospheric pressure at 15 °C. For this and other contributions, Michelson was awarded the Nobel Prize in 1907. Of course thermal expansion was a limiting problem, such that when the low-expansion steel alloy <<Invar>> was invented, the creator (and Director of the BIPM), C. D. Guillaume, was awarded the Nobel Prize for 1920. However, the SI Metre definition was unchanged for 85 years: the Meter Bars worked well and optical comparitors got fatigueless photo-electric eyes.

Spectroscopic experiments and supporting Quantum Theory led to improved understanding and improved light sources. The metrological needs of the World Wars changed the Science climate, and transportation disruptions emphasized the advantage of having *independently-reproducible standards* based on quantum physics. Eventually, in 1960 the Eleventh General Conference on Weights and Measures was able to redefine the International Standard of Length as 1,650,763.73 vacuum wavelengths of orange light resulting from transitions between specified atomic energy levels of the krypton isotope of atomic weight of 86. Going forward with a new definition, one would say the Kr wavelength is $\lambda = 1\text{ m} / 1{,}650{,}763.73 = 0.605{,}780{,}211\ \mu\text{m}$. While the adopted Definition speaks about unperturbed atoms, in fact several shifts were observed in light from the discharge lamp used for realizing this Metre in practice. Pressure shifts and discharge operating conditions were stabilized by operating the lamp at a specified discharge current and at a fixed pressure and temperature (using the triple-point of liquid nitrogen). A field-induced gas flow of Kr^+ led to a wavelength difference of light viewed from the two cell ends. When laser comparisons with this standard were performed, the additional problem of radially-dependent Doppler shifts of the emitted light was discovered.

The 1960's and 1970's saw a number of different stabilized lasers systems introduced, refined, and the wavelengths measured and compared between various national labs. Basically, all these laser systems were entered into the competition to be the next International Length Standard. There were then 48 nations involved in the Metre Convention, so politically speaking, choosing one out of the many offered candidate lasers would be difficult. In addition, none of these approaches were overwhelmingly superior, when performance, cost and complexity were all considered. And scientifically, it seemed attractive for the new Length Standard definition to be based on the Speed of Light, introduced as a defined quantity. On the basis of a number of laser-based measurements, this value was taken as 299,792,458 m/s exactly, a rounded value in accord with the measurements of the several standards labs. This redefinition of 1983 took the form:

> *"the Meter is the length of the path traveled by light in vacuum during a time interval of 1/299,792,458 of a second.* The speed of light is
>
> $c = 299{,}792{,}458$ m/s , exactly.
>
> The *second* is determined to an uncertainty, $U = 1$ part in 10^{14} by the Cesium clock."

The General Conference also suggested several recommended radiations for realizing the meter at that time, eg.:

> "The wavelength of the iodine-stabilized Helium-Neon laser is
>
> $\lambda_{HeNe} = 632.99139822$ nm ,
>
> with an estimated relative standard uncertainty *(U)* of $\pm\, 2.5 \times 10^{-11}$."

In all of these changes in definition, the goal was not only to improve the precision of the definition, but also to change its actual length as little as possible. See [3]. With the speed of light defined, an optical frequency (linked to time) can thus serve as a length unit.

Fundamental Physics Issues in the Re-Definition of Length

At the times of these redefinitions, there were some concerns that we were switching the physical basis for the Metre definition. For example, if in the future we discover that some of the "constants of Physics" actually are slowly changing, one could worry that the new definition might impact or even limit our discovery process. In any case, we would be unaware of a global change that would conserve the physical relationships we have discovered. But could there be a differential effect that might be observable? Before 1960 we were accepting the spacing of some lattice planes in the Pt-Ir alloy of the Meter Bar as our measurement basis for length: this length certainly would fundamentally involve Quantum Mechanics, and Electricity and Magnetism. And, considering the thermal vibration of molecules in the somewhat-anharmonic interatomic potentials, we can suppose that the nuclear masses – and thus the Strong Interactions – will also play a role in length via the thermal expansion. With the 1960 redefinition of the Metre in terms of a Krypton atom's radiation's wavelength, perhaps we were opening some opportunity for confusion? Now Quantum Mechanics and Electricity and Magnetism are still fundamentally involved, but the atom's mass is involved only in a reduced-mass correction, rather than via thermal effects. Certainly a new "constant," the speed of light, is linearly serving as the dimensioned scale constant. Initially the 1983 redefinition appears to be still a different sort compared with the 1960 Kr definition, but really it just repeats the energy level difference idea (now it is Cs in defining the second rather than Kr defining an optical energy) followed by a conversion of dimensions. Who knows if there is some fun hidden in here?

Where we have come to is that the SI is now functioning with six, rather than seven, basic units. The Metre has been demoted to a derived unit, and the significance of Time and Frequency have been further elevated. This begins a long story, with the SI base units being challenged by spectacular advances "at the bottom of a Dewar" [4], giving us a Josephson-effect based voltage standard (Nobel Prize of 1973), while the von-Klitzing-effect defines a quantum resistance standard (Nobel Prize of 1985). Taken together as V^2/R, an electrical Watt unit is apparent, while an SI Watt – defined as a Joule per second – would be represented as ½ kg $(m/s)^2/s$. The relationship between these is established by a "Watt Balance" experiment [5]. Recently the Single Electron Transistor begins to enable digital counting of electron charges per second, contacting the SI Ampere, the unit of electric current. This interface between metrology and quantum physics is becoming a "Hot Topic" of our time [1, 6]. The remarkable advances in Metrology also allow – and advances in Cosmology and Astronomy strongly motivate – curiosity about the "exactness" and "time-invariance" of the various physics numbers used in our description of physical reality.

CLOCKS AND TIME

Time represents our most precisely measurable quantity and so it always has attracted certain kinds of devoted researchers. But also, now with various sensors and microprocessor control, many physical parameters can be read out by frequency measurements, and so we add a huge number of scientists in other fields who want to recover the finest details within their measurements. (Still, many really important research subjects are not yet so well developed that these frequency tools are useful: for example, world-changing decisions about air pollution management are being made even though we scarcely are sure about the *sign* of some effects.)

But for technology people, the improvement of time measurement precision grows as a field of intense interest and competition worldwide. In no small part this is because of the very advances singled out by this year's Prize: a capability jump by several decades is uncommon in any field, let alone *the* field where the precision of measurement was already at the highest level, and had already been driven to near its apparently basic limits.

Of course interest in time has been part of man's history from our beginnings, but only in the last several recent centuries have some lucky subsets of people been somewhat isolated from seasonal variations, with leisure to think about Nature, and so *time* as an experimental parameter began to emerge. Nowadays we can look from the scientific and experimental point at the question: why would one be interested in time? **For those who love precision, the clear reason is that *time is the most powerful metrological variable.***

Scaling of Precision Attainable when we are Measuring Time

The precision of time measurements can be increased essentially without limit, by increasing the measurement duration and simply counting the increased number of cycles of some regularly-spaced events. However a stronger information growth with measurement duration is possible if we have a nice source that has coherence from the beginning of the measurement until the end. (For the present purpose we may take this "coherence" to mean that if we know the oscillation cycle's phase early in the measurement, the *coherent* source is so steady that the oscillation phase could be predicted at later times near the measurement's end to a precision of 1 radian of phase.) In this case we can have a measurement precision which will grow with the measurement interval τ according to $\tau^{3/2}$. A simple way to explain this assertion is to suppose we divided the measurement duration into 3 equal sections, each with N/3 measurements. In the starting zone we compare the reference clock and the unknown clock, with a relative phase precision which scales as $(N/3)^{1/2}$. Next, in the middle section, we merely note the number of events, N/3. In the last section we again estimate the analog phase relationship between test and reference waves, with a relative imprecision which is again $(N/3)^{1/2}$. Subtracting the two analog phases increases the uncertainty of one measurement by a factor $2^{1/2}$ so, altogether, the relative precision increases as $(1/2^{1/2}) * (N/3)^{3/2}$. Thinking of a microwave frequency measurement, with a base frequency of 10^{10} Hz, in a 1 s measurement we have a factor of 10^5 potentially to win. Commercial counters already can register 12 digits in 1 s for a reasonable input signal. One can see there is just a huge gain in measurement precision if we can measure a coherent frequency source in a proper way: No wonder we have the situation where metrology scientists as well as philosophers, sailors, and farmers are interested in clocks and time and seasons [7]. Indeed our most powerful test of the existence of Einstein's predicted gravitational radiation comes from the observed shortening of the year of the Hulse-Taylor binary pulsar: orbital clock physics vs. quantum frequency standard physics on the earth. This marvelous work was celebrated by the Nobel Prize of 1993.

What makes a clock?

The three essentials of clocks are: a source of regular events, a counter/integrator to totalize the events, and a suitable readout mechanism to present the current result to an interested human or machine. In many ways the frequency source is the most interesting part since it is intrinsically an analog system, where the design goal is to diminish as little as possible the intrinsic stability of some physical oscillation, in the course of reading out its information. In this game, nuance and subtlety count for a lot. It is customary that the performance of clocks based on some well-known source of regular "clicks" will be improved several orders of magnitude by the work of many people over many years, with the ultimate fate of becoming suddenly obsolete due to the introduction of a better kind of stable oscillator. The new idea must be a serious advance, since it must be competitive at the start of its life with the previous technology which has been enhanced and improved in many stages. Still, some technologies have had a long lifetime – for example one can still buy a good wristwatch based on a torsional oscillator, even though this balance wheel concept was used by Ch. Huygens in 1675.

Keeping time has been of serious interest since man turned agrarian, but became of critical interest with the expansion of lucrative international trade: "inevitable" shipwrecks could be avoided by better knowledge of position (mainly longitude) at sea. Parliament's Longitude Prize of 1714 (above $10 M in current terms) attracted John Harrison's attention and some *40 years* of his inventive work. In 1761 his H-4 clock demonstrated 1/5 s/d , $\delta\nu/\nu \sim 2.5 \times 10^{-6}$ even while at sea. This was several-fold better that the requirement, but only half the Prize was initially paid: in part the controversy was about the Intellectual Property! A second problem was conflict of interest within the judging Committee. (This story is well-told in [7].) Present customers of precise timekeeping include TV Networks (for synchronization), cellular telephone companies, the GPS users who need the limiting performance, radio astronomers, NASA Deep Space Tracking, and various other branches of Science in which a physical variable has been read out by frequency methods.

Evolution of Frequency Sources: Distinguishing Precision and Accuracy

In discussing the performance of a mechanical clock, or the electronic oscillators based on vibrational modes of quartz crystals, it is clear that the basic frequency is set by mechanical dimensions. Such a device could be stable and have good precision, in that its readout could be determined with many digits, but there can be no claim to any particular fixed or natural frequency. Still the stability of any particular crystal device could be remarkable: a drift of $>10^{-6}$ /day gradually improved to the present $\lesssim 1 \times 10^{-10}$ /day, while the shift with acceleration remains near 10^{-9} per "g". The high frequency of electronic oscillators served well for convenient interpolation between "clicks" of the absolute standard, provided by zenith sightings of the daily motion of the Sun, as codified by the 1875 Metre Convention. (Later the earth rotation data series were based on telescopic observations of the lunar occulation starts of various stars and planets.) By the 1950's the electronic oscillators were refined enough that variability $\sim 10^{-8}$ was inferred in the earth's spin rate, and was associated with changes of the earth-atmosphere system's moment of inertia due to North-South ocean tides, and large storms. The community wished to eliminate the variability, but still needed an absolute and universal (rather than local artifact) standard. The new choice in 1960 adopted a stated number of seconds in the "Tropical Astronomical Year 1900". Perhaps this was good in its motivation, in that the rotation of the earth around the sun would have a lower level of perturbation. However, a clock/oscillator that has only a single click per year will be hard to enjoy at its full precision. As a metrology principle we rather would prefer the basic frequency source to be at a very high frequency so that the integer multiple of the standard's clicks will be a huge digital number in our measurement of some interesting phenomenon, and the unavoidable noise and uncertainty of the remaining analog subdivision of the unit will be as insignificant as possible.

Electronic Clocks based on Quantum Transitions

Based on Otto Stern's atomic beam method, which had resulted in his Nobel Prize of 1943, I. I. Rabi introduced atomic beam resonance methods which allowed probing internal (hyperfine) quantum energy states of atoms such as Cesium with greater precision. This work was recognized by the Nobel Prize of 1944. Using atoms in this way, the *independent realizability* and *universality* requirements for a Primary Standard could be well addressed. In addition, the transition frequencies were near the high-frequency-end of the usable rf spectrum, so the Metrology aspects were optimized as well. The first Atomic Beam Clock was developed at NBS in 1949 based on microwave transitions in Ammonia, and by 1955 Cs beam clocks were in operation at the NPL and NBS. The powerful Method of Separated Oscillating Fields was invented by N. F. Ramsey, reported in 1955, and later recognized by the Prize in 1989. In this dual-excitation concept, suitable atoms were excited once, and then left to evolve their internal phase (ideally) free of perturbation, until a second excitation pulse effectively completed the interferometric comparison of the phase evolution rates between the atomic and laboratory oscillating systems. Progress on the Cs beam atomic frequency standard was widespread and rapid, allowing redefinition in 1967 of the SI Second as 9 192 631 770 units of the Cs hyperfine oscillation period. Correspondingly, the Cs oscillation frequency is defined as (exactly) 9 192 631 770 Hertz (cycles of per second). The specialists involved in this redefinition of Time and Frequency wisely did not specify exact details of the measurement process, leaving room for considerable progress. For example when laser-based optical pumping of atoms between hyperfine states became feasible and popular in the early 1990's, NIST colleagues built a new atomic beam cesium standard, NIST-7, based on optically transferring most of the population from the 16 available hyperfine levels into the special ($|3,0\rangle$) lower state involved in the clock transition. Along with this factor, ~16x, improvements of the atom source itself, and better frequency source and readout electronics were helpful. Above all, computer-based signal processing and active control of measurement systematic offsets made it possible to reduce the inaccuracy of realizing the Cs second at NIST to ~5 x10^{-15} . But as usual in the art-form of Precision Measurement, this "tour de force" system was soon made obsolete in a single step by a qualitatively better technology.

As shown by Kasevich and Chu [8], laser cooling of the Cs atoms made it possible to successfully implement the "atomic fountain" concept for the realization of the Cs-based frequency definition. By shifting laser frequencies or powers, a slowly moving ball of atoms could be dispatched vertically upward through the excitation rf cavity, reaching apogee a good part of a meter above the cavity, and then beginning the return trip to pass through the excitation cavity a few 100 ms later. With such a long coherent interaction time, instantly the resonance linewidth dropped to ~ 1 Hz, down from ~300 Hz in the previous epoch of thermal beam of atoms. Optical probing of the atoms below (and temporally after) the cavity could yield the excitation-probability vs. probe-frequency-tuning curve needed to control the source oscillator's frequency. By using suitably-closed optical transitions for readout, one can have many photons emitted per atom so that, even after solid angle and detection inefficiencies are considered, the measurement noise is not much larger than the minimum associated with the finite number of atoms. Andre Clairon and his colleagues made the first real Cs Fountain Frequency Standard, in 1995 [9] at the Paris Institute now known as LNE-SYRTE [10]. Even without the contemporary schemes to break this atom-shot-noise limit, the fountain clocks at NIST and SYRTE now achieve accuracy levels below 1×10^{-15} when all the known measurement and perturbation issues are taken into account [11]. Of course with the resolution improvement one hopes for more potential accuracy, but will have beforehand an expanded list of small shifts and niggling concerns to consider. After all, even with the extended interaction time, fewer than 10^{10} oscillation cycles are counted, so the achieved inaccuracy of 1×10^{-15} already corresponds to 10 ppm splitting of the atomic fountain's resonance linewidth. Fountain Cs clocks are limited by two newly important effects, collisionally induced frequency shifts due to the hugely increased atom density [12], and shifts due to the effects of the ambient thermal radiation associated with the vacuum system's walls. Attempting to split lines further always brings a diverging list of new small problems, leading to an effective barrier.

An important observation is that for many types of Quantum Absorber samples the line broadening processes will be the same for both radio and optical frequency domains. For example, the atomic fountain apparatus could explore optical transitions, rather than microwave ones, with the same interaction time. Clearly we would prefer the higher base frequency of the optical world, since the resonance feature of interest will then display a relative sharpness increased by roughly the same huge factor of optical/microwave frequency ratio. With sharper line shapes we can expect more precise measurements that will let us better see the small effects of various experimental parameters, leading to better *independent reproducibility* which, with major investment of efforts, can often be parlayed into nearly a corresponding increase in measurement *accuracy capability* as we come more fully to characterize the offset processes. But before the Millennium Year of the Optical Comb, just how did you plan to measure the absolute optical frequency?

This *repeatability* idea seems weaker than the gold standard of *accuracy*, which additionally conveys our being able to connect the measured result with the base units of the Systeme International. But in fact we now know several optical clock systems that have 10-fold smaller uncertainty than the Cs standard. So before a redefinition is appropriate, their comparisons will be most interesting, especially as an entry point for one of the most interesting branches of Science, trying to figure out which physical "laws" are essentially exact, which ones are ignoring some details to have a tidy presentation, and which are in fact stating "facts" about Nature which are not exactly actually true. Celestial mechanics, ideal gas laws ignoring molecular volumes, and parity conservation in atomic physics could be my examples.

STARTING THE DREAM OF OPTICALLY-BASED CLOCKS

The Laser Arrives

The future of metrology was changed fundamentally on 12 December 1960 when a small team at Bell Labs, led by Ali Javan, eventually found the right conditions for their Optical Maser to generate self-sustained optical oscillations. Their specially crafted gas discharge tube had the improbable situation in which the populations in two particular Neon atomic levels were reversed from the thermal norm: by means of the discharge in the more-abundant He gas, collisional energy transfer set up a population inversion, whereby more atoms were in the higher energy state. It is impressive that these conditions were established on the basis of careful measurements and modeling of the discharge conditions! Having the populations inverted from the usual case reverses the sign of the absorption that experience teaches us is a universal property of (normal) matter. Accordingly, with an inverted population, rather than absorption, Javan's group had optical emission. The atoms would provide amplification of any resonant optical signal passing down the discharge cell. A few percent gain wouldn't be very exciting normally, except that the utilized multilayer mirrors were designed and fabricated to have reflection losses that could be even smaller, setting the stage for a buildup of power on every pass. So finally they did obtain a self-sustained continuous optical oscillation, and observed the collimated beam that was anticipated by Charles Townes and Arthur Schawlow in a classic paper of 1958. Similar ideas were also considered in the Former Soviet Union, leading to the Nobel Prize of 1964 being shared by N. Basov, A. Prokhorov, and Townes.

Connection to Glauber's ~~Quantum~~ Coherent States of Light

In planning a theoretical study of optical fields, perhaps one can understand starting with known results for single-photon fields, then adding a few photons cautiously to see what happens. Actually, for all of us following Professor Glauber's work it was surprising just how few photons were needed for the new photon density distribution functions to change fundamentally from the customary Poisson limit: with increasing number of photons in a mode the fields start showing the small fractional fluctuations that would characterize a classical field. On the experimental side, for Javan's very first laser, the output laser power was ~ 1 milliWatt, about 10^{16} photons per second! We can proceed to estimate the expected fractional variation of 1/Sqrt(N), but with such an incredibly large number of coherent photons in one mode, the result is an unphysically small variation. Thousands of merely technical processes would cause fluctuations larger than the predicted $1{:}10^{8}$! An equivalent statement is that these lasers were operating strongly, very far into the domain of classical fields, and quantum fluctuations would be very hard to observe. Indeed it was not until the end of the 1970's that people began again to appreciate how to study manifestly quantum fields with just a few photons in them. At this vastly-reduced intensity, quantum correlations are a bit challenging to observe, but they are very interesting, since they correspond to rather significant fractional effects. For example, H. J. Kimble's group used phase-dependent Squeezed Light to make a spectroscopic measurement with about 2-fold better Signal/Noise than the naïve shotnoise limit [13]. To observe strong Squeezed Light effects, it is essential to minimize optical losses, as they work to revert the statistics toward the thermal limit. Regrettably, noise from technical sources will grow linearly in the laser power, while the advantage due to squeezing will grow more slowly. It seems that getting a factor 10 amplitude S/N improvement appears to be incredibly difficult.

Coherence of the Laser Field Enables Frequency-Diagnostics

The Bell Labs laser design success had grown out a semiclassical view of how Optical Masers would operate. Yes, amplification would be provided by quantum mechanical atomic systems, rather than radio tubes or klystrons, and yes each atom could contribute just one photon to the field in each event. But still, considering how huge is the number of photons in the field, the discreteness probably will hardly matter. Almost immediately the Bell-Labs team was testing this understanding by combining two separate laser beams into a single coaxial beam, and shining this onto the sensitive surface of a high-speed photodetector. They already were thinking of each laser oscillation as being an essentially classical field, satisfying reflection boundary conditions at the two mirrors. So this stable-and-repeating bouncing specification would define the possible wavelength(s) of the generated laser light. By luck and design the discharge was wonderfully calm, so one could expect the gas' refractive index would be essentially constant. Thus the interferometric boundary conditions would essentially define the oscillation *frequency* and, accordingly, one would expect to see a sharp optical frequency come out of this device. With two lasers' sharp frequencies on the nonlinear detector's surface, one should expect the difference frequency to be generated, which it was. I can still remember hearing the audio beat whistle that Javan had recorded when his two lasers were tuned almost to the same optical frequency. It was a ~1 kHz difference between two sources at 260 THz!

Actually the linewidth of these beats was remarkably narrow. We already expected that based on the numbers noted above: a stream of $\sim 10^{16}$ photons/s would have random power fluctuation of $\sim 10^{-8}$ relative to the full power. So the optical phase could be extremely well defined. However, the laser's Schawlow-Townes linewidth calculation includes the role of optical loss, which actually limits the laser coherence, giving ~ milliHz linewidth expectation.

In principle then we have a radiation of incredible sharpness, and should be ready to seek interesting physical effects. The immediate disappointing truth is that this tiny predicted laser phase fluctuation will be completely masked by noise of technical origins. We already noted that the strongest definition of the oscillation's frequency is fixed by the interferometric standing-wave condition bouncing on the laser cavity mirrors. But the lab is a noisy place, seismically speaking, with a quiet lab having a ground noise of $\sim 3 \times 10^{-9}$ m/Sqrt(Hz) in the vibration frequency band say 1–30 Hz. A laser cavity is some fraction of a meter in length, so it will be difficult to make a system arbitrarily stiff. Rather, some important fraction of the ground noise will appear as cavity length variations, and therefore laser frequency variations. Suppose we say only 1% couples in to relative length changes. One can instantly see the scale of the problem: ~10^{-10} fractional frequency variations will be our apriori scale. Even temperature variations will be painful, since the 10^{-10} scale already corresponds to a few milliKelvin temperature change for low expansion materials like fused silica. We can make progress by locking the laser to a stable reference cavity [14]. Optimizing for vibrational integrity, we will use a stiff structure for mounting the reference cavity mirrors, and then mount the assembly with a horizontally soft suspension. By focusing on the vibration isolation, Bergquist has obtained [15] a record narrow laser linewidth ~0.16 Hz! Another approach seeks to minimize the cavity acceleration sensitivity. By use of a vertically-symmetric mounting [16] of the reference cavities, our group recently reported Hz-level laser linewidths.

Coherence of the Laser Beats Enables Frequency-Based Laser Control

Considering the small intrinsic phase noise of the laser source, and the rather high power ~ mW, heterodyne detection of the beat frequency between two laser sources yields an interestingly high Signal/Noise ratio. Even with very short averaging times, say 1 μs, we have generous S/N performance. Additionally, for such short times a well-engineered laser will scarcely respond to the "garbage effects" of real life in the lab (temperature variations, power-supply variations, vibrations ...) – within 1 μs these have not changed the system very much. The duration of the perturbations is too small for them to begin to wreak havoc with the stability of the frequency-defining cavity. So we actually can make useful measurements of the laser's phase in such a quick time frame that the problems are not yet apparent! One begins to see a strategy coming up: We will quickly *measure* what the laser actually is doing, compared with our desired behavior, and then use feedback onto suitable actuators to *control* the laser's frequency. If we can make the corrections quickly enough and accurately enough, then the controlled laser will very closely approximate the ideal frequency-stable laser we need.

Implementation of this servo-control feedback concept is a multi-nuanced thing, in the perfection of which this author has invested something over 40 years of active work. It has led to a lot of interesting and useful electro-optic tools and techniques.

The Relative High Power of Lasers Empowers Nonlinear Spectroscopy and Sharp Resonances

Let's begin with the first approach to observation of narrow atomic resonances, using Saturated Absorption Spectroscopy. These phenomena were studied first within a laser cavity by Bill Bennett using the dispersion effects associated with the active Neon laser gas. Owing to the Doppler effect, the Neon atom's natural resonance linewidth of ~10 MHz becomes masked and broadened to ~1500 MHz. Thus most of the gas atoms are detuned, and in a velocity-specific way. Some atoms have velocities near the special one giving the Doppler shift that will bring them into resonance with the intracavity laser field. Actually there are two such velocities to consider, since the laser beam goes both directions as it is bouncing back and forth between the mirrors. These resonant atoms will interact rather strongly with the field, leading to an increased decay rate for excited state atoms of that velocity – their inverted population gets converted into cavity photons! If we imagine a plot of the population difference (upper state minus lower state populations) we can expect to see a local and rather narrow dip around the velocity which is being converted from population inversion into light quanta. Actually there are the two mirror-symmetric dips as noted before. The interesting effects come when we let the laser frequency be tuned toward the atom's rest-frame frequency. Then the resonant atoms will have lower and lower Doppler velocities, until finally the selected velocity is zero. Now a new thing happens: when detuned, we had two groups of active atoms contributing their power to the laser output. When we reach the central tuning, both running-wave fields interact with a single atom velocity group. So with fewer atoms contributing, the laser power decreases conspicuously, but only at the central tuning. This feature in the power output with laser tuning could be used for locking the laser to this central tuning dip, which is called "Lamb's dip" after Willis Lamb whose early theoretical work made clear this origin of the experimentally-observed effect. (His Nobel Prize in 1955 was for his work on the new sub-hyperfine structure in the Hydrogen spectrum.) As it turns out, operating pressures for optimum laser operation were rather large (~3 Torr, 400 Pa), which led to substantial probability of atom-atom collisions, even during the few 10's of ns optical lifetimes. So the Lamb dips would be broader and less deep, and had to be observed against a somewhat-peaked Doppler profile representing the distribution of available atom velocities. In addition to reducing the Lamb-dip contrast, significant frequency shifts were generated [17]. One could not arbitrarily reduce the gas pressure since the discharge pumping mechanism actually populated a metastable He* level, and collisions were needed to transfer this excitation to the Neon atom component in the discharge. So even though the wavelength of the laser's characteristic coherent light was more-readily-measurable than the incoherent light from the krypton discharge lamp (the existing wavelength standard), in fact the lasers' pressure shifts were simply too large to accept. Particularly this was the case since the discharge technology of the day led to important change of the fill gas pressure and species ratio with operation, due to electrode sputter-pumping.

The clearly important idea of separating the amplifier and the reference gas cells' functions was soon introduced by Lee and Skolnick. More discussion of those interesting developments is available elsewhere [18, 19], but for our present purposes we do need to consider some of the essentials. Since the purpose was to have a sheltered life for our reference atoms, it was attractive to be thinking in terms of *absorption*, rather than amplification. Then we didn't need any discharge or optical pumping of the reference quantum

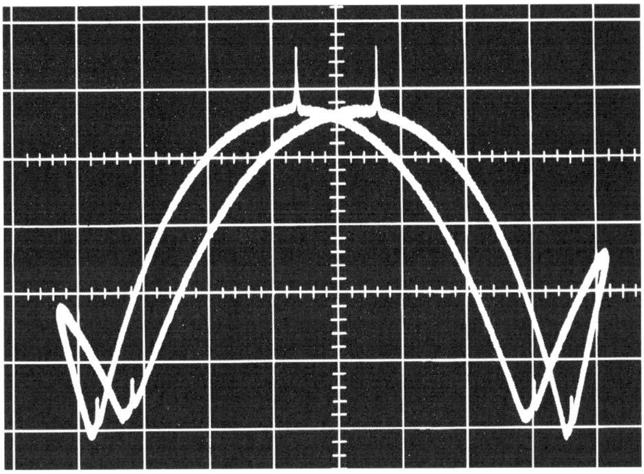

Figure 1. Saturated absorption peak in CH_4 molecules. HeNe laser at 3.39 μm is excited by rf discharge. CH_4 cell at 12 mTorr (16 mBar) is located inside laser cavity. Power output is 300 ~μW and peak contrast is ~12%. Peak width is ~270 kHz HWHM. At maximum power (~0.8 mW) contrast is ~15%. Cavity free spectral range is 250 MHz. Note cross-over resonances in two-mode region near cusps. Hysteresis of scan causes trace doubling.

resonators. Of course, to be able to use Lamb's nonlinear resonance for frequency stabilization, we certainly needed to be able to tune the laser to the reference cell's resonance frequency. Nowadays, this is no big problem, by just using tunable lasers. At that time, the best idea to get a wavelength coincidence would be to use molecules as the absorbers – then we would have zillions of absorption lines to choose from. The modern champion for this approach is molecular Iodine, with narrow useful absorption lines from the Near IR down to ~500 nm. For other molecules, utilizing transitions only between vibrational-rotational states, typical wavelengths are in the IR from ~2–10 micrometers range.

The first such dual-component optical frequency reference system, and still one of the better ones, uses a HeNe discharge cell to provide gain and laser oscillation at 3392 nm [18]. Also contained in the laser cavity is a cell containing CH_4 molecules, plain old tetrahedrally-symmetric methane, which has interesting lines that can be reached with the HeNe laser. To be brief, the necessary emitter/absorber spectral overlap is arranged by selection, based on good luck! The IR absorption band utilized, ν_3 , is a strong fundamental vibration band, providing 0.18 cm^{-1} absorption coefficient per Torr. Of course having the absorber gas inside the cavity means we don't need very much absorption to have an impact on the laser dynamics – just a few percent would be fine, since it would then be roughly ½ the loss associated with the output-coupling mirror. At 10 milliTorr, the associated pressure broadening of the CH_4 resonance would then be ~160 kHz, similar to the 130 kHz broadening associated with the molecular free-flight through the intracavity light beam, of 0.3 mm typical diameter. Importantly, the pressure-induced shift turns out to be *very* small for these transitions, only ~1 kHz under these conditions.

So we are talking about a system with a resonance in the power curve of ~0.6 MHz FWHM, with perhaps 5% relative contrast on the total laser output of say 200 μW. A little calculation leads one to a Signal/ShotNoise ratio ~10^6 in a 1 Hz measuring BW, while we're looking at the sub-MHz – wide peak produced at the central tuning, when both cavity running waves are bleaching the same absorbing molecules and thereby reducing the intracavity absorption losses. If this S/N were optimally used, the laser could be stabilized to have sub-Hz frequency deviations measured in 1 s intervals. In 1968 when this Saturated Absorption Optical Frequency Reference business began, our detectors and preamplifiers were not so good, and we didn't begin to approach the shot noise limit – that would have been a frequency (in)stability of ~2 $\times 10^{-14}$ at 1s. Early on, we did get $\delta\nu/\nu$ ~1 $\times 10^{-12}$, which was soon improved to 3 $\times 10^{-13}$ with better detectors and signal processing.

By locating the sample cell outside the laser resonator, the physical situation could be more-readily analyzed, and this arrangement was employed by Bordé, and by Hänsch, and by Chebotayev's group in early experiments. The interesting details are discussed in textbooks: see eg. Letokhov and Chebotayev [20], and Stenholm [21], Levenson and Kano [22]. Now we consider the transit-time linewidth issue.

Free-flying Molecules see a Light Pulse: two views of the Uncertainty Principle

For these transitions, the radiative lifetime (~ ms) was much larger than the transit time of the essentially free-flying molecules in crossing the laser beam. At low pressure the saturated absorption linewidth was not collisionally nor Doppler limited, so it could be immediately observed that the resonance linewidths could be reduced by increasing the field/molecule interaction time. Larger beams helped. So did liquid Nitrogen cooling of the glass cell. So a serious study began to really understand the lineshape in the free-flight regime. Chebotayev and his colleagues developed the theory analytically near the low-pressure, low optical power limit [23]. The JILA theory was based on computer integrations of the Density Matrix for absorbers making a free transit through the assumed Gaussian light beam mode [24]. Low intensity and weak interactions were assumed to simplify the calculations, but soon it became clear that most of the observed signal would be contributed by a very small number of slow molecules. The theoretical result is a logarithmic cusp at the exact line center. With long interaction times, even a "weak" power would lead to saturation and other strong-field effects.

We need low velocity in the longitudinal direction so that the molecule wouldn't cross wavefronts axially, and thereby begin to develop Doppler-related phase modulation. Effectively molecules should fly perpendicular to the axis, and leave the wavefront after the transit with only <1 radian geometrical phaseshift. We also need low transverse velocities, since a longer transit time will be directly imaged into a narrower line. We can see $\delta\nu \cdot \tau \approx 1$ will yield $\delta\nu = \beta\, v_{th}/w_0$, $\delta\nu$ is the HWHM of the observable resonance, v_{th}

is the thermal velocity, w_0 is the Gaussian beam radius, and β is a measured parameter. Experimentally we found $\beta\ v_{th}$ = 88 kHz mm for Methane at room temperature. Laser mode radius w_0 values from 56 μm to 9 cm were measured, with corresponding HWHM values from 1.6 MHz down to 940 Hz. (The interesting substructure will be addressed momentarily.) First it is useful to consider the transit-time broadening in the Fourier-dual domain: angular divergence. Corresponding to a Gaussian beam radius w_0 there is a minimum angular divergence of the collimated laser beam of $\delta\theta = \lambda / 2\pi w_0$. The k-vector spread, particularly the non axial components lead to a velocity-dependent Doppler shift of the *same* sign for both running waves, which will appear as broadening and shift of the resonance. Of course with a smaller mode diameter, the angular content is increased, and more broadening will appear spectrally.

While molecules typically do not have the "closed" optical transitions analogous to those needed for normal laser atom cooling, polar molecules *do* have a dipole moment. So with some electrical effort, one can arrange Sisyphus-like molecule slowing by switching the sign of the strong applied electric field, as shown by Meijer's group [25]. More recently Ye's group has achieved unprecedented high resolution microwave spectroscopy on Stark-slowed OH free radicals [26]. Certainly this will be an interesting frontier!

Other important directions are high sensitivity detection and improving the accuracy of locking to the molecular signals. For example some JILA work ("NICE-OHMS") shows a road to sensitivity increase by combining cavity enhancement and rf sideband techniques [27]. A fascinating physics avenue is the search for a parity-related frequency shift between suitable enantiomers [28]. Other important laser applications are considerred in Svanberg's book [29].

Momentum transfer from Light to Molecules – the Recoil Splitting

A full treatment of radiative interactions must include the field and molecular momenta, as well as the photon numbers and internal states of the quantum system. Such a treatment is essential for the case of pumping atoms with closed energy levels, which can allow the repeated interactions and deep velocity cooling celebrated by the 1997 Atom Cooling Prize of Phillips, Chu, and Cohen-Tannoudji. For the molecular sample of interest here, there are many decay channels, and likely even impact on the vacuum chamber walls before any particular molecule reappears in the laser fields: so a single interaction picture is reasonable. A clear observation of the transfer of momentum from field to atomic system is available with Saturated Absorption Spectroscopy, basically because it is a two-step process. Let's consider absorbers that initially have essentially zero velocity along the light beam. Then the left-running light beam can be tuned to $v = v_0\left(1 + \frac{hv}{2Mc^2}\right)$, the extra (recoil) energy being needed beyond the transition energy ν_0 to provide the kinetic energy associated with the recoil momentum the molecule will have after the transition occurs. The opposite-running beam will also deplete this zero-velocity group. So at this resonance tuning, the resulting nonlinear decrease of molecular opacity will lead to a peak in the transmission spectrum, and it is shifted slightly to the blue of the rest frequency. Another interesting case occurs when the molecules have a velocity $v = h/M\lambda$, ie., there is enough molecular momentum initially so that when the red-detuned laser interacts with this molecule, the photon and molecular momenta just cancel, and the original kinetic energy can make up for the photon's energy deficit. The result is an excited molecule with zero axial velocity. Now the laser beam in the other running direction will experience amplification from this particular tuning condition, again leading to a relative peak in the sample's transmission. With the molecule initially possessing some kinetic energy, the laser tuning for this upper-state resonant condition will be $v = v_0\left(1 - \frac{hv}{2Mc^2}\right)$. So considering photon recoil, the nonlinear interaction is associated with either the ground or excited state population being accessed by both beams for the same detuning, namely zero velocity in either one of the two states. For methane the splitting between the two peaks is 2.163 kHz and may be seen clearly in Figure 2 [30].

While the JILA and University of Paris Nord work exploited mainly the large diameter optical beams to gain a longer molecular interaction time, Chebotayev, Bagayev, and colleagues in the Novosibirsk group made good use also of another physical idea, namely the use of super-slow molecules to contribute the main part of the observed signal. In this way an additional 20-fold linewidth reduction to <50 Hz was achieved [31]. An important aspect of this approach is that the total 3-D effective molecular temperature is below 0.1 K, leading to a much-reduced second-order Doppler shift, of << 1 Hz. An average velocity 13x below thermal for slow C_2HD molecules was shown by Ye *et al.* [27], and was feasible only because of the very large sensitivity provided by the NICE-OHMS technique.

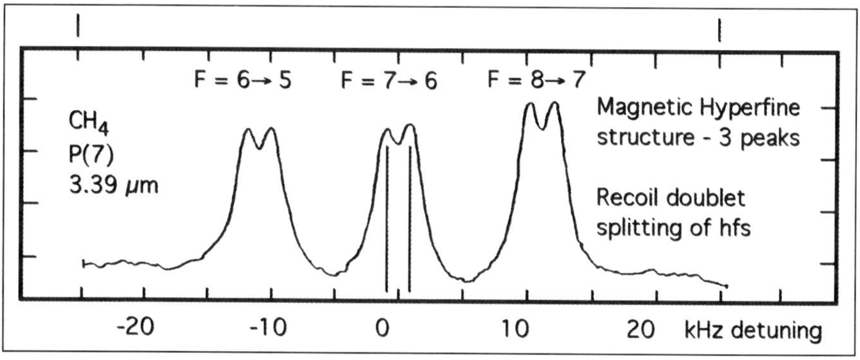

Figure 2. Recoil Splittings of Hyperfine-Structure Peaks in free-flight Methane Molecules. The vertical strokes indicate the positions of the two recoil components in one of the Hyperfine components.

Other Optical Frequency References Based on NonLinear Spectroscopy

Many research groups have been attracted to working with laser stabilization for Measurement Standards applications, such as interferometric calibration of gage blocks that serve to check reference standards used by industry. For this kind of application it is highly desirable that the reference laser beam be visible, as well as stable enough and reproducible enough. A huge success in this area is the 633 nm HeNe laser with an intracavity Iodine cell, and well developed systems of this type are even available commercially. This HeNe/I_2 system was the one whose frequency was measured by the NBS efforts in the early 1980's, with an uncertainty of 70 kHz. (Being the first measurement of such a visible system, it is perhaps understandable that several of the uncertainties were far from fundamental in their origins). Other labs joined in and over the next decade many labs gained experience and a few had frequency measurements confirming the NBS result, so that it became attractive to reconsider the definition of the International Unit of Length, the SI Metre.

As may be seen, the world of spectroscopy offers us an unending garden of fascinating details. Presumably Parity-Non-Conservation will lead to a next generation of fine structures in chiral molecules, particularly with the development of cold-molecule techniques. But enough about the "ticks" of the clock: now we must return to the main story, the development of frequency stabilization and cycle-counting measurement tools – **The inside Gear-Works of the Optical Clock!**

MEASURING OPTICAL FREQUENCIES WITH OPTICAL COMBS

The Metre redefinition of 1983 was not really a kindness to metrologists tasked with actually measuring some physical parts, because the practical methods for application to measurements were not yet spelled out. But it was a *boon* to the metrology researchers: it became their task to explore just which good stabilized laser system would have the optimal properties for precision interferometry, for outdoor surveying, for servo-loop guidance of milling

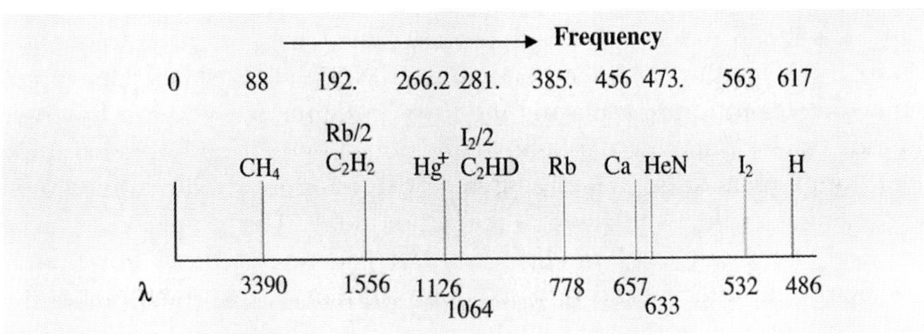

Figure 3. Stable Lasers based on NonLinear Doppler-free Resonances in Gases (1995). The frequency axis (above) is in THz units, the wavelength scale (below) is in nm.

machines, for … ? So within a dozen years after the redefinition there were at least 10 well-developed optical frequency standards, as illustrated in Figure 3.

As may be seen in Figure 3, there are stable frequency sources available from roughly 10 μm (30 THz) to ~280 nm (~1PHz), well beyond the visible range. It was striking that the difference between lines were surprisingly similar frequency intervals, ~88 THz, approximately the frequency of the CH_4 – stabilized laser. This led to schemes where doubled frequency of one laser would be compared with the sum of the two straddling lasers. Some "pocket change" of frequency, a few THz, could be synthesized as sidebands using a Kourogi comb, based on a microwave modulator in a cavity whose length provided resonance enhancement of all the generated sidebands [32]. In such a way we measured the 532 nm Iodine standard in terms of the difference of frequency between twice the HeNe Iodine system at 633 nm, and the Rb two photon line at 782 nm [33].

This was our introduction to the elegance of having an optical comb – a coherent ensemble of spectral lines whose frequencies are accurately represented by a simple formula. Our system covered just a few nm wavelength. How sweet it would be to cover the entire visible band, giving several million accurately known frequency reference lines all at once!

One way to broaden this comb's spectral width would be to provide intracavity gain, to compensate the modulator's optical losses, a scheme which was demonstrated by Diddams using an OPO crystal also inside the resonator. Oscillation and generation of hundreds of FM sidebands were easily observed [34]. For some tuning conditions the phase of the several spectral components led to pulse generation, rather than pure FM emission. In many ways this was just the hard way to do what the Ultra-Fast Laser scientists appreciated about the Ti:Sapphire self-mode-locked lasers: stable, self-organized, ultra-short high repetition rate pulse trains. Elsewhere our group's papers discuss the technical richness of these lasers and the comb business [35]. This is just one further note about the mutual coupling between "independent" research streams: we switched to Ti:Sapphire fs lasers and never looked back.

Coincidentally, in these final days of the last Millennium, this laser community received a fundamentally-important gift from the laser industry. There would probably be no widely-used frequency combs without it. This "gift" was the introduction of high-power visible lasers, based on frequency-doubling the output of a laser-diode-pumped Nd solid state laser. These were immediately put to use replacing the fussy and quite noisy Argon Ion laser in wide use for pumping the Solid State lasers. Competitive forces led these new pump lasers to be well engineered, with intensity stabilization to yield exceedingly low levels of residual amplitude noise. This property is crucial because of the way a self-mode-locked laser operates – these Ti:Sapphire lasers are self mode-locked by a self-induced optical lens which makes the cavity less lossy when the laser modes are all synchronized to form an "optical bullet" in the laser medium [36]. This temporary lens is formed by the radial index gradient, induced and present only if a light bullet is present.

So the laser cavity is originally set up to need this extra focusing to produce low-loss cavity modes, and after the laser is started in the pulse regime, stable self-mode-locking is maintained. Consider that the pulse lengths are only ~10 fs, while the repetition periods are ~10 ns. With ideal synchronization, the peak power/average power ratio is ~10^6 . A typical laser will emit ~0.5 W through an output mirror of 5% transmission. So we have 10 W average internal power, and 10 MW peak power, which is focused to a ~14 μm radius spot in the Ti:Sapphire laser crystal. This active area is only 3 x10^{-6} cm^2 , so with 10 MW peak power we have 3 TW/cm^2 ! The associated electric field is ~10% of the interatomic fields in the crystal, so it is not so surprising that a significant optically-induced increase of the index of refraction occurs (optical Kerr effect). The low amplitude noise of the pump laser is now seen to be critical: an intensity-dependent phase-shift though the laser crystal will produce amplitude → frequency conversion and thus unacceptable phase noise if the pump is noisy. In a good case the linewidth of laser comb-lines without frequency control is ~3 – 10 kHz due to this cause, before the servo is used. Details of the process have been studied [37].

So leaving the laser is a pulse train of ~500 kW peak power, much of which we will focus into the special nonlinear fibers that brought in the age of the Optical Comb. Because of the microstructure design of the fiber, full light guiding is possible even with fiber core sizes of 1.5–2 μm diameter. So now when we calculate the fiber's active area, it is roughly 200-fold smaller than the laser's, while the power level is ~20-fold lower. The 10-fold higher intensity produces a 3-fold higher electric field in the silica fiber, being now essentially comparable with interatomic field and setting the stage for SERIOUS NonLinear interactions. Forget Mr. Taylor's expansion here: this is strong signal NonLinear physics! All frequency components from the laser are mixed with each other, resulting in a drastic spectral broadening. By the fiber's optical design, a broad range of optical frequencies can travel through the fiber with little speed variation, which allows these frequency conversion processes to remain phase-matched and accumulate power into the newly created frequencies. Essentially, in a few cm of length, the input spectrum is converted to white light and covers an octave or more of optical bandwidth. Actually the light is not quite "white" since it still carries the basic heartbeat of the original fs laser, for example 100 MHz. As explained previously, this intrinsically generates a comb spectrum with component widths just connected to the spectral resolving power employed. Eventually, at the kHz level and below, the broadly-active phase modulation processes that affect all lasers will broaden these lines also (before the servo control is ON).

COMPLEMENTARITY, COOPERATION, AND COMPETITION

The Basics

The remarkable insights of Professor Hänsch's Stanford work [38] were published in ~1978, and already demonstrated using a repetitively-pulsing laser to generate an optical comb which could serve as a spectral ruler. However, the bandwidth of the covered spectrum was too small for general frequency measurements – only a GHz or two. Since these intervals could be spanned in other ways, the methods were not widely adopted. Basically there was not a technical growth path available at the time. Principle, yes; Tool, no.

The hard work, straight-ahead "government" approach to frequency measurement had been demonstrated at NBS in 1972 [39], following the pioneering work of Ali Javan's MIT frequency measurement group (See references in [40]). But this was a heroic effort and mainly only national standards laboratories took much interest. Laser after different laser had to be lined up and frequency-related to the doubled frequency of its predecessor, to step-by-step build up the frequency measurement chain. This kind of work required development of frequency- and phase-locking schemes now in wide use. We also got a "one-of-a-kind" physical result: a single laser frequency was measured by the cooperative and extended work of the NBS group [41]. But it was enough to get the Metre redefinition process started.

The Divide and Conquer Scheme

In a notable paper (1990), Professor Hänsch and his colleagues suggested an excellent way to simplify the frequency chains: one should use the *difference frequencies* between lasers as the entities that were harmonically marching up the spectrum [42]. In this way, the ensemble of lasers would all have nearly the same wavelength, and could be built essentially by duplication of a basic diode laser unit. Then with nonlinear crystals, fast photodetectors, and suitable phase-locking electronics one could progress from microwaves to optical frequencies. This system also felt rather elaborate and specialized, but was used with good results in Garching. A related strategy was developed at NRC [43], based on difference frequencies, using CO_2 lasers. Inspecting such a system, one came to see that the first 9 or 10 of the 14 stages served only to get the frequency up into the low THz range.

Then in 1994 came Kourogi and Ohtsu's multiply-resonant cavity approach, allowing one to reach a few THz in a single step [32]. Eventually the buildup of phase noise – according to the high harmonic of the original microwave source – would have been a problem in going into the visible range. But the fs laser Comb arrived and offers an easier and better way. See below.

A Brief History of the Optical Miracle of 1999–2000

Fibers for Spectral Broadening

By now the JILA group had accepted the fs laser as a great source of pulsed laser light. Ours had ~ 80 nm bandwidth at 800 nm. But the optical frequency standards we wanted to connect were at 1064 nm (fundamental of Iodine-stabilized Nd laser) and 778 nm (Rb two-photon-stabilized diode laser). An ordinary communication fiber was found to be just barely capable of spectral broadening the necessary amount – 104 THz. This paper was submitted at the end of September 1999 [44].

MicroStructure Fibers for Serious Nonlinearity

The Conference on Laser and Electro-Optics of June 1999 had a spectacular postdeadline presentation by a Bell Labs team [45], wherein a normal fs laser pulse evolved its color in a dramatic way in propagating through a few meters of a special fiber. Such a fiber *did* make collimated white light, in the form of stably-repeating pulses, just as Ted Hänsch had postulated for his (unpublished) frequency measurement proposal. Using that previously-unknown light source, most of the rest should be possible. (Seeing the repetitively-pulsed laser-like white light the fiber generated instantly convinced me that Ted's Concept actually could be a real and physical possibility! Without a repetitive white-light laser, there was no chance.) Lengthy appeals for scientific collaboration with the fiber owners' organization ultimately became irrelevant due to the miraculous appearance in JILA of a sample of this Magic Fiber. The concept of "band-gap" or "Photonic-Crystal" fibers was introduced in 1996 by Knight *et al.*, pointing out the possibility of controlling the spatial modes and effective group velocity dispersion by the mechanical design of the air holes [46]. Our first JILA experiments were made using microstructured fiber drawn from a preform prepared on September 10, 1997 by Robert S. Windeler of Bell Labs [47], using a construction technique of his own devising. A broad range of fiber designs was investigated in Bath, UK, by P. St. J. Russell and colleagues.

The Race is ON

Of course in JILA we didn't know that the Garching team had already gone from a plan to the first demonstration of a comb-based phase coherent link from microwaves to the visible, and had submitted their *Phys. Rev. Letter* in November 1999. Even before we got the Magic Fiber! They used a comb of somewhat limited bandwidth, 44 THz, but their divider stages could connect the optical frequency with the 28th harmonic of the difference between the comb's edges. It is a beautiful result, and appeared finally on 10 April 2000 [48]. In the meantime the JILA team was working hard with the Magic Fiber's white light output to implement and demonstrate our phase-coherent locking of the carrier-envelope offset frequency in terms of the laser's repetition rate. Our Disclosure of the scheme called this "Self-Referencing". The control electronics we built had a digital click switch so the phase could be set on any integer multiple of 1/16 of a phase-slip cycle per pulse. The JILA experimental demonstration was based on interferometrically determining the carrier-envelope phase difference between two optical pulses, separated by one intervening pulse. Finally the new electronics worked, the experimental data were clear and our report [49] appeared in *Science* on 28 April 2000. A *PRL* joint article celebrating the success of the combined Garching, Bell Labs, and JILA teams appeared on 29 May 2000 [50]. Within the next year there was an avalanche of absolute optical frequency measurements from labs all over the world. This was a glorious chapter in optical physics history, in no small part because of the high mutual respect of the two teams for each other, aided by the complete openness fostered by the frequent exchange of postdocs Scott Diddams and Thomas Udem between the two hotly-competing groups.

Some Frequency Measurement Results

Many laser frequency standards were being actively studied worldwide so that, when the Comb breakthrough came, there were many things to be accurately measured – many for the first time. A few of the world-wide results include the work shown in the following Table.

The comb technology spread explosively in 2000, bringing vast simplification of optical frequency measurements, along with a steady improvement in the accuracy. Very soon after the initial measurements, it has become the case that the comb's measurement precision can exceed that of the standards being measured. Recent tests at NIST, BIPM, and ECNU [51] confirm the earlier MPQ experiments [52] showing that the comb principle is strictly correct up to a measurement precision of more than 18 digits.

Ca	657 nm	Schnatz	PTB	PRL	1 Jan '96
Rb	780 nm	Ye	JILA	Opt. Lett.	August '96
C_2H_2	1500 nm	Nakagawa	NRLM	JOSA-B	Dec '96
I_2	532 nm	Hall	JILA	IEEE Instr,Meas	April '99
Sr^+	674 nm	Bernard	NRC	PRL	19 Apr '99
In^+	236 nm	v. Zanthier	MPQ	Opt.Comm.	Aug'99
H	243 nm	Reichert	MPQ	PRL	10 Apr '00
Rb	778 nm	D. Jones	JILA	Science	28 Apr '00
I_2	532 nm	Diddams	JILA	PRL	29 May '00
H	243 nm	Niering	MPQ	PRL	12 June '00
Yb^+	467 nm	Roberts	NPL	PRA	7 July '00
In^+	236 nm	v. Zanthier	MPQ	Opt. Lett.	1 Dec.'00
Ca	657 nm	Stenger	PTB	PRA	17 Jan '01
Hg^+	282 nm	Udem	NIST	PRL	28 May '01
Ca	657 nm	Udem	NIST	PRL	28 May '01
Yb^+	435 nm	Stenger	PTB	Opt. Lett.	5 Oct '01

Table I. **Measured Optical Frequencies.** The reference atom/molecule and its transition wavelength are indicated, followed by lead author and institution, the journal name and date. The first fs Comb measurement was Hydrogen, by Reichert *et al.* The first direct fs optical measurements were by the JILA team (Jones). Note the brevity of time between publications!

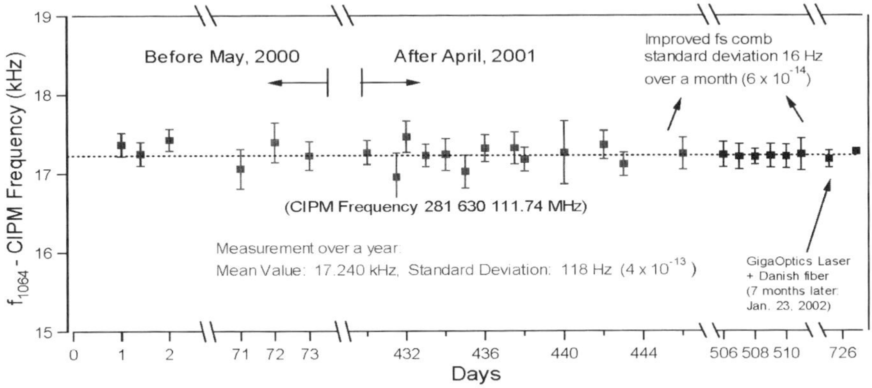

Figure 4. Long-term frequency stability of Iodine-based Optical Clock. This figure conveys the refinement and small frequency offset of this stable optical clock's frequency from previous, much less accurate measurements. With improved technology in 2002 the uncertainty was further reduced to ~6 x10^{-14}.

Molecular Iodine Optical Frequency Standard

The Iodine-stabilized Nd:YAG laser is a sweet spot in the stabilized laser domain, counting on its excellent performance and relative simplicity. One system was made in Japan that met airlines cabin baggage limitations and still delivered excellent performance [53]. Because of the Iodine's great atomic mass, the second-order Doppler correction for this system is only ~5 x10^{-13} and it is likely that independent reproducibility perhaps 5-fold superior to this can come with improved technical realizations. In particular, providing an offset-free modulation strategy is still a challenge. The advantage of this system is its compactness and potentially reasonable cost. Taken with an optical comb, one can have an attractive clock [54]. See Figure 4. The frequency (in-)stability of all the 1 million optical comb lines is ~ 4 x10^{-14} $/\sqrt{\tau}$.

Recently stable Yb:YAG single frequency lasers became available, with output tunable to 1029 nm. When frequency doubled, excellent stabilization performance should be possible with the I_2 transitions at 514.5 nm, considering that the linewidth is at least 5-fold smaller than for the 532 nm line [55]. Single frequency fiber systems can also offer this wavelength.

SO WHAT COMES NEXT ?

In addition to the simplification of optical frequency measurements, the resulting new capabilities are unbelievably rich in terms of the tools and capabilities that have been created, and these in turn are reinforcing progress in these contributing fields. This paper can't even attempt to present a myriad of delicious physical effects, which are normally understood as being in different fields, but which in their now-unified relationships can be seen as creating a truly remarkable and enabling advance of the research tools available in optical science. But let me still give a few examples.

After the frenzy of Generation I frequency measurements of Table I, some of the Generation II comb applications in Jun Ye's group include: low-jitter time synchronization (~fs) between ultrafast laser sources [56]; coherent stitching-together the spectrum of separate fs laser sources so as to spectrally broaden and temporally shorten the composite pulse [57]; precision measurement of optical nonlinearities using the phase measurement sensitivity of rf techniques [58] ; coherently storing a few hundred sequential pulses and then extracting their combined energy to generate correspondingly more intense pulses at a lower repetition rate [59]; and searching for a change in the physical constants by the Garching team [60]. Exciting topics of research for Generation III applications now include connecting optical frequency interim standards at the sub-Hz level (in spite of their different locations spectrally and physically), allowing precise remote synchronization of accelerator cavity fields, providing stable reference oscillators for Large Array Microwave Telescopes, and potentially reducing the relative phase-noise of the oscillator references used for deep space telescope arrays (NASA, VLBI ...) That's part of the first five years.

And the next projects? What about 14.4 keV comb-line harmonics to look at Mössbauer ^{57}Fe nuclear resonances? Another sharp line is in ^{181}Ta at 6.2 keV. How about parallel processing to determine biological activity of a candidate drug, by means of CARS using synchronized pulse lasers to excite specific ligand Raman resonances of a single molecule that was attracted to and stuck by a particular test protein patch on a surface?

In a larger framework, we now find ourselves at an almost unique point in the development of Science, where we have the remarkable ability to "under-stand" practically all phenomena, to compute accurate predictions from our equations, and to integrate a variety of details into our models. Consider for example the GPS system, in which different kinds of physics such as gravity and relativity are successfully merged with our sophisticated atomic clocks – not to forget satellite dynamics and radio engineering and computer software – so that in the total we have a coherent and highly useful practical tool. Remarkably, the system is simple for the end user to apply. We must count this GPS achievement as one of the all-time ultimate technical success levels ever achieved.

The work recognized by the 2005 Nobel Physics Prize represents entry of another dramatic, major and enabling advance, and one which we can expect to show some flavors of the same breadth and character just noted regarding GPS. But in these first moments after its birth, our opto-electronic technology is new and is barely illustrated, not much beyond the first cases of interest to frequency-standards people and metrologists. We know that the *accuracy* of optical frequency measurements is now limited to "just" 15 digits by the present microwave standard of frequency, but the "Comb" technology actually allows two optical frequencies to be compared with several orders of magnitude more precision. If the history of physics is any guide, we realistically can expect to find some nice surprises ahead as these capabilities become even more widespread, and are applied to ingenious fundamental measurements by a growing and imaginative community of "fundamental physics" scientists.

After considering all the known progress in Science, would *you* bet that we have already opened the Russian Matryoshka doll of Nature and already found the ultimate inside limit?

ACKNOWLEDGEMENTS

The joy of interacting with superb young scientists is clearly one of life's treasures. Among these many I must single out for special thanks Jim Bergquist, Leo Hollberg, Miao Zhu, and Jun Ye for their enthusiasm and unique contributions to the JILA program. The NIST management is enthusiastically thanked for accepting and sponsoring over the years a series of risky proposals in Laser Spectroscopy. Particularly Leo Hollberg, Steve Cundiff and I were glad they accepted the fs comb frequency synthesizer proposal in Spring 1999. Scott Diddams and David Jones were the excellent colleagues additionally involved in these experiments, and JILA's research force was hugely expanded when Jun Ye rejoined JILA in 1999 to start his own group. As always, whenever Professor Long Sheng Ma was visiting us from Shanghai our pace was strongly advanced. Visiting scientists such as Christian Bordé continue to be collaborators even three decades after their JILA time. I am pleased and grateful to acknowledge that the work has been supported in part by the NSF, ONR, AFOSR, and NASA, and for over 4 decades by the NIST. I benefited greatly from the knowledge and generous sharing of ideas and opportunities by my NBS mentor, Peter L. Bender. Above all I am indebted to my patient and insightful lifetime friend Lindy Hall for her understanding, her great efforts and contributions to this scientific work and, more importantly, to our joyous and fun life together. It has been wonderful in the course of these 45 years to see a progression of experiments and technical advances make possible this ultimate payoff in the optical comb. Now we're discussing if it's almost time to clean my office at JILA and pack up into a new camper unit to go out and further explore another domain in the world.

Appendix: The Full Comb Story in an Undergraduate's World

I'm glad you asked how to think about frequency combs. Suppose you have a sinewave voltage or field. Then a plot in time shows a smooth oscillation and a plot in frequency shows a single Fourier component, namely a sharp line. Now add a few harmonics onto this wave. The spectrum now has a few more lines at exact harmonic frequencies, while the time picture has a rather complicated shape. By adjusting the phases of the harmonics, we can begin to synthesize some disturbance in time that begins to remind one of a pulse, or more exactly, a series of identical pulses. Carry this a step forward by having a large number of harmonics. The more we add, the sharper is the pulse we can synthesize, and of course the richer is the spectrum of this wave. Going further in this direction of adding coherent harmonics, the spectrum now has zillions of spectral lines, all at the harmonics of our original sinewave. Carrying this concept to the visible will require a few million (10^6) harmonics for a source with 100 MHz basic repetition rate. With the proper phase adjustments, the time domain pulse can be 10^6 times sharper in time than the original sinewave. So we can expect really narrow temporal pulses, and really wide spectral bandwidths.

This situation fits well with what we would expect from Fourier analysis of a single pulse: such an impulse will have Fourier components at all frequencies, with their nearly-constant amplitudes gradually decreasing for frequencies above the reciprocal temporal pulsewidth. If we have a repetitive pulse train in time, but insist to ask about its spectrum, we will need an analyzer with a narrower passband compared with the repetition frequency, otherwise it couldn't resolve the harmonic structure. But a narrow spectral passband corresponds to a long temporal response time. So the output of the spectrometer at any particular wavelength or frequency setting will be the result of coherent addition of the contributions of many pulses. While an individual pulse has a broad and continuous spectrum, when we coherently add their spectral amplitudes we can expect to have interferences that will modulate the spectrum. Adding more pulses temporally (narrower spectral resolution) will give deeper modulation. Eventually we arrive at very sharp spectral lines, evenly arranged as Fourier harmonics. Until we encounter technical issues such as phase-noise of the repetition rate, the sharper an analysis resolution we apply to the waveform, the sharper will be the spectral lines we observe. So the spectrum does indeed remind one of a "comb." You can demonstrate these ideas safely at home for yourself easily in the electronics domain, but of course the optical and electronics worlds *should* work the same …

In fact, with the fs lasers used to generate these pulses, there is one more little item of interest. That is that the laser can oscillate in any one of its cavity modes, defined by having a repeating phase after going one loop around the cavity. All the many modes involved have their own longitudinal quantum numbers, essentially how many full optical cycles are contained in the closed loop. This calculation clearly involves the wavelength-dependent phase velocity, and some average of the propagation through the many optical components. Another reality is that the laser operates in a self-organized repetitive pulsing mode. Effectively the laser's optical losses can be made large enough to inhibit laser action, unless all the cavity modes can adjust their phases to synthesize a delta-function spatially. The critical thing is to have a short pulse when passing through the Ti:Sapphire crystal, since the short pulse will correspond to very high peak power, and that will interact with the laser rod's material in a quadratic way (optical Kerr effect) to produce a positive lens: a bigger index on the axis where the intensity is maximum. So the self-organized pulse situation is stable in which the laser's cavity has a high diffraction-loss (doesn't have quite enough positive lens power), but the losses are periodically remedied by a bullet of light which uses its self-action on the crystal to produce the needed extra refraction that makes the cavity losses be suitably less.

Now the pulse envelope that describes this light "bullet" results from superposition of many cavity modes, and the shape will evolve if there are temporal delay differences with wavelength. We are now just discussing the group velocity concept, whereby the shape of a disturbance will evolve unless all the frequencies have the same propagation speed. In the physical laser we must include some optical elements specifically to deal with the fact that blue light in the laser crystal will travel more slowly than red light. To get the shortest pulses the time delays around the loop need to be essentially the same, although you can see this becomes a little complicated in that the laser pulses themselves act to influence the time delays. In any case, the light which comes out of the laser's coupling mirror will be a regular time series of sharp pulses, and will display a comb-like structure under frequency analysis. However the underlying fast optical oscillations will in general have a different phase each time the pulse hits the mirror's surface. The fast oscillation's phase will shift a bit forward or backward from one pulse to the next, and so the optical frequency comb may be offset a bit from the strictly Fourier harmonic case we first imagined. The usual case is a constant phase shift for each pulse, and so a constant rate of accumulating a phase beyond the repetition-rate's harmonic. We have developed an electro-optic scheme called "self-referencing" in which this additional frequency, the Carrier-Envelope Offset Frequency, is stably locked to the repetition frequency in a digital ratio. For example one could choose zero for the setpoint ratio and thereby have a strictly harmonic comb. With the offset = 1/2, one generates a comb offset by 1/2 the basic repetition rate, which itself is of course the frequency comb's tooth spacing. See Ref [48–52].

REFERENCES

[1] BIPM, "http://www.bipm.org/en/si/si_constants.html."

[2] These are Mass, kg; Time, s; Length, m; Current, A; Temperature, K; Quantity of Matter, mol; Unity of Light Intensity (Candela), cd.

[3] NIST, "http://www.mel.nist.gov/div821/museum/timeline.htm."

[4] J. L. Flowers and B. W. Petley, "Progress in our knowledge of the fundamental constants of physics," *Reports on Progress in Physics*, vol. 64, pp. 1191–1246, 2001.

[5] R. Steiner, E. R. Williams, D. B. Newell, and R. Liu, "Towards an electronic kilogram: an improved measurement of the Planck constant and electron mass," *Metrologia*, vol. 42, pp. 431–441, 2005.

[6] C. J. Bordé, "Base units of the SI, fundamental constants and modern quantum physics," *Philosophical Transactions of the Royal Society a-Mathematical Physical and Engineering Sciences*, vol. 363, pp. 2177–2201, 2005.

[7] D. Sobel, *Longitude, the True Story of a Lone Genius Who Solved the Greatest Scientific Problem of His Time*. New York: Penguin Books, 1995.

[8] M. A. Kasevich, E. Riis, S. Chu, and R. G. Devoe, "Rf Spectroscopy in an Atomic Fountain," *Physical Review Letters*, vol. 63, pp. 612–616, 1989.

[9] A. Clairon, P. Laurent, G. Santarelli, S. Ghezali, S. N. Lea, and M. Bahoura, "A Cesium Fountain Frequency Standard – Preliminary-Results," *IEEE Transactions on Instrumentation and Measurement*, vol. 44, pp. 128–131, 1995.

[10] T. P. Heavner, S. R. Jefferts, E. A. Donley, J. H. Shirley, and T. E. Parker, "NIST-F1: recent improvements and accuracy evaluations," *Metrologia*, vol. 42, pp. 411–422, 2005.

[11] K. Gibble and S. Chu, "Laser-Cooled Cs Frequency Standard and a Measurement of the Frequency-Shift Due to Ultracold Collisions," *Physical Review Letters*, vol. 70, pp. 1771–1774, 1993.

[12] E. S. Polzik, J. Carri, and H. J. Kimble, "Atomic Spectroscopy With Squeezed Light For Sensitivity Beyond the Vacuum-State Limit," *Applied Physics B-Photophysics and Laser Chemistry*, vol. 55, pp. 279–290, 1992.

[13] R. W. P. Drever, J. L. Hall, F. V. Kowalski, J. Hough, G. M. Ford, A. J. Munley, and H. Ward, "Laser Phase and Frequency Stabilization Using an Optical-Resonator," *Applied Physics B-Photophysics and Laser Chemistry*, vol. 31, pp. 97–105, 1983.

[14] B. C. Young, F. C. Cruz, W. M. Itano, and J. C. Bergquist, "Visible lasers with subhertz linewidths," *Physical Review Lett.*, vol. 82, pp. 3799–3802, 1999.

[15] B. C. Young, R. J. Rafac, J. A. Beall, F. C. Cruz, W. M. Itano, D. J. Wineland, and J. C. Bergquist, "Hg+ OPTICAL FREQUENCY STANDARD: Recent Progress," in *ICOLS 2005*, vol. XVII, *LASER SPECTROSCOPY*. Cairngorm, Aviemore, Scotland: World Scientific, 2005, pp. 433.

[16] M. Notcutt, L. S. Ma, J. Ye, and J. L. Hall, "Simple and compact 1-Hz laser system via an improved mounting configuration of a reference cavity," *Optics Letters*, vol. 30, pp. 1815–1817, 2005.

[17] J. L. Hall, "The laser absolute wavelength standard problem," *IEEE Journal of Quantum Electronics*, vol. QE-4, pp. 638–41, 1968.

[18] R. L. Barger and J. L. Hall, "Pressure shift and broadening of methane line at 3.39 μm studied by laser-saturated molecular absorption," *Physical Review Letters*, vol. 22, pp. 4–8, 1969.

[19] J. L. Hall, "Optical frequency measurement: 40 years of technology revolutions," *IEEE Journal of Selected Topics in Quantum Electronics*, vol. 6, pp. 1136–44, 2000.

[20] V. S. Letokhov and V. P. Chebotayev, *NonLinear Laser Spectroscopy*. Berlin: Springer-Verlag, 1977.

[21] S. Stenholm, *Foundations of Laser Spectroscopy*. New York, Wiley, 1984.

[22] M. D. Levenson and S. S. Kano, *Introduction to NonLinear Laser Spectroscopy*. San Diego: Academic, 1988.

[23] S. N. Bagaev, L. S. Vasilenko, A. K. Dmitriev, M. N. Skvortsov, and V. P. Chebotaev, "Narrowing of Nonlinear Resonances in Low Pressure Gases," *JETP Letters*, vol. 23, pp. 360–363, 1976.

[24] C. J. Bordé, J. L. Hall, C. V. Kunasz, and D. G. Hummer, "Saturated absorption line shape: Calculation of the transit-time broadening by a perturbation approach," *Physical Review A (General Physics)*, vol. 14, pp. 236–63, 1976.

[25] H. L. Bethlem, G. Berden, and G. Meijer, "Decelerating neutral dipolar molecules," *Physical Review Letters*, vol. 83, pp. 1558–1561, 1999.

[26] E. R. Hudson, H. J. Lewandowski, B. C. Sawyer, and J. Ye, "Cold molecule spectroscopy for constraining the evolution of the fine structure constant," *Physical Review Letters*, vol. 96, 2006.

[27] L. S. Ma, J. Ye, P. Dubé, and J. L. Hall, "Ultrasensitive frequency-modulation spectroscopy enhanced by a high-finesse optical cavity: theory and application to overtone transitions of C2H2 and C2HD," *Journal of the Optical Society of America B-Optical Physics*, vol. 16, pp. 2255–2268, 1999.

[28] M. Ziskind, C. Daussy, T. Marrel, and C. Chardonnet, "Improved sensitivity in the search for a parity-violating energy difference in the vibrational spectrum of the enantiomers of CHFCIBr," *European Physical Journal D*, vol. 20, pp. 219–225, 2002.

[29] S. Svanberg, *Atomic and Molecular Spectroscopy, 2nd Edition*, Berlin, Springer-Velag, 1992.

[30] J. L. Hall, C. J. Bordé, and K. Uehara, "Direct Optical Resolution of Recoil Effect Using Saturated Absorption Spectroscopy," *Physical Review Letters*, vol. 37, pp. 1339–1342, 1976.

[31] S. N. Bagayev, V. P. Chebotayev, A. K. Dmitriyev, A. E. Om, Y. V. Nekrasov, and B. N. Skvortsov, "2nd-Order Doppler-Free Spectroscopy," *Applied Physics B-Photophysics and Laser Chemistry*, vol. 52, pp. 63–66, 1991.

[32] M. Kourogi, T. Enami, and M. Ohtsu, "A coupled-cavity monolithic optical frequency comb generator," *IEEE Photonics Technology Letters*, vol. 8, pp. 1698–1700, 1996.

[33] J. L. Hall, L.-S. Ma, M. Taubman, B. Tiemann, F.-L. Hong, O. Pfister, and J. Ye, "Stabilization and frequency measurement of the I/sub 2/-stabilized Nd:YAG laser," *IEEE Transactions on Instrumentation and Measurement*, vol. 48, pp. 583–6, 1999.

[34] S. A. Diddams, L.-S. Ma, J. Ye, and J. L. Hall, "Broadband optical frequency comb generation with a phase-modulated parametric oscillator," *Optics Letters*, vol. 24, pp. 1747–9, 1999.

[35] J. L. Hall, J. Ye, S. A. Diddams, L.-S. Ma, S. T. Cundiff, and D. J. Jones, "Ultrasensitive spectroscopy, the ultrastable lasers, the ultrafast lasers, and the seriously nonlinear fiber: a new alliance for physics and metrology," *IEEE Journal of Quantum Electronics*, vol. 37, pp. 1482–92, 2001.

[36] F. Salin, J. Squier, and M. Piche, "Mode locking of Ti:Sapphire lasers and self-focusing: a Gaussian approximation," *Opt. Lett.*, vol. 16, pp. 1674–1678, 1991.

[37] K. W. Holman, R. J. Jones, A. Marian, S. T. Cundiff, and J. Ye, "Detailed studies and control of intensity-related dynamics of femtosecond frequency combs from mode-locked Ti : Sapphire lasers," *IEEE Journal of Selected Topics in Quantum Electronics*, vol. 9, pp. 1018–1024, 2003.

[38] J. N. Eckstein, A. I. Ferguson, and T. W. Hänsch, " High-resolution 2-photon spectroscopy with picosecond light-pulses," *Physical Review Letters*, vol. 40, pp. 847–850, 1978.

[39] K. M. Evenson, J. S. Wells, F. R. Petersen, B. L. Danielson, G. W. Day, R. L. Barger, and J. L. Hall, "Speed of light from direct frequency and wavelength measurements of the methane-stabilized laser," *Physical Review Letters*, vol. 29, pp. 1346–9, 1972.

[40] A. Sanchez, C. F. Davis, K. C. Liu, and A. Javan, "Mom Tunneling Diode – Theoretical Estimate of Its Performance at Microwave and Infrared Frequencies," *Journal of Applied Physics*, vol. 49, pp. 5270–5278, 1978.

[41] D. A. Jennings, C. R. Pollock, F. R. Petersen, R. E. Drullinger, K. M. Evenson, J. S. Wells, J. L. Hall, and H. P. Layer, "Direct Frequency-Measurement of the I/sub 2/-Stabilized He-Ne 473-THz (633-nm) Laser," *Optics Letters*, vol. 8, pp. 136–138, 1983.

[42] H. R. Telle, D. Meschede, and T. W. Hänsch, "Realization of a New Concept for Visible Frequency-Division – Phase Locking of Harmonic and Sum Frequencies," *Optics Letters*, vol. 15, pp. 532–534, 1990.

[43] B. G. Whitford, "Measurement of the Absolute Frequencies of CO2-Laser Transitions by Multiplication of CO2-Laser Difference Frequencies," *IEEE Transactions on Instrumentation and Measurement*, vol. 29, pp. 168–176, 1980.

[44] S. A. Diddams, D. J. Jones, L. S. Ma, S. T. Cundiff, and J. L. Hall, "Optical frequency measurement across a 104-THz gap with a femtosecond laser frequency comb," *Optics Letters*, vol. 25, pp. 186–188, 2000.

[45] R. W. J. Ranka, and A. Stentz, "Visible continuum generation in air-silica microstructure optical fibers with anomalous dispersion at 800 nm," *Opt. Lett.*, vol. 25, p. 25, 2000.

[46] J. C. Knight, T. A. Birks, P. S. Russell, and D. M. Atkin, "All-silica single-mode optical fiber with photonic crystal cladding," *Optics Letters*, vol. 21, pp. 1547–1549, 1996.

[47] R. S. Windeler, private communication to J. L. Hall, May 2006.

[48] J. Reichert, M. Niering, R. Holzwarth, M. Weitz, T. Udem, and T. W. Hänsch, "Phase coherent vacuum-ultraviolet to radio frequency comparison with a mode-locked laser," *Physical Review Letters*, vol. 84, pp. 3232–3235, 2000.

[49] D. J. Jones, S. A. Diddams, J. K. Ranka, A. Stentz, R. S. Windeler, J. L. Hall, and S. T. Cundiff, "Carrier-envelope phase control of femtosecond mode-locked lasers and direct optical frequency synthesis," *Science*, vol. 288, pp. 635–639, 2000.

[50] S. A. Diddams, D. J. Jones, J. Ye, T. Cundiff, J. L. Hall, J. K. Ranka, R. S. Windeler, R. Holzwarth, T. Udem, and T. W. Hänsch, "Direct link between microwave and optical frequencies with a 300 THz femtosecond laser comb," *Physical Review Letters*, vol. 84, pp. 5102–5, 2000.

[51] L. S. Ma, Z. Y. Bi, A. Bartels, L. Robertsson, M. Zucco, R. S. Windeler, G. Wilpers, C. Oates, L. Hollberg, and S. A. Diddams, "Optical frequency synthesis and comparison with uncertainty at the 10(−19) level," *Science*, vol. 303, pp. 1843–1845, 2004.

[52] T. Udem, R. Holzwarth, and T. W. Hänsch, "Optical frequency metrology," *Nature*, vol. 416, pp. 233–237, 2002.

[53] F. L. Hong, J. Ishikawa, K. Sugiyama, A. Onae, H. Matsumoto, J. Ye, and J. L. Hall, "Comparison of independent optical frequency measurements using a portable iodine-stabilized Nd : YAG laser," *IEEE Transactions on Instrumentation and Measurement*, vol. 52, pp. 240–244, 2003.

[54] J. Ye, L. S. Ma, and J. L. Hall, "Molecular iodine clock," *Physical Review Letters*, vol. 87, pp. 270801–4, 2001.

[55] W. Y. Cheng, L. S. Chen, T. H. Yoon, J. L. Hall, and J. Ye, "Sub-Doppler molecular-iodine transitions near the dissociation limit (523–498 nm)," *Optics Letters*, vol. 27, pp. 571–573, 2002.

[56] R. K. Shelton, S. M. Foreman, L. S. Ma, J. L. Hall, H. C. Kapteyn, M. M. Murnane, M. Notcutt, and J. Ye, "Subfemtosecond timing jitter between two independent, actively synchronized, mode-locked lasers," *Optics Letters*, vol. 27, pp. 312–314, 2002.

[57] R. K. Shelton, L. S. Ma, H. C. Kapteyn, M. M. Murnane, J. L. Hall, and J. Ye, "Phase-coherent optical pulse synthesis from separate femtosecond lasers," *Science*, vol. 293, pp. 1286–9, 2001.

[58] T. M. Fortier, D. J. Jones, J. Ye, and S. T. Cundiff, "Highly phase stable mode-locked laser's," *IEEE Journal of Selected Topics in Quantum Electronics*, vol. 9, pp. 1002–1010, 2003.

[59] R. J. Jones and J. Ye, "High-repetition-rate coherent femtosecond pulse amplification with an external passive optical cavity," *Optics Letters*, vol. 29, pp. 2812–2814, 2004.

[60] M. Fischer, N. Kolachevsky, M. Zimmermann, R. Holzwarth, T. Udem, T. W. Hänsch, M. Abgrall, J. Grunert, I. Maksimovic, S. Bize, H. Marion, F. P. Dos Santos, P. Lemonde, G. Santarelli, P. Laurent, A. Clairon, C. Salomon, M. Haas, U. D. Jentschura, and C. H. Keitel, "New limits on the drift of fundamental constants from laboratory measurements," *Physical Review Letters*, vol. 92, 2004.

Portrait photo of John L. Hall by photographer Sears. P. Studio.

THEODOR W. HÄNSCH

I was born in Heidelberg, Germany, on October 30, 1941. My parents had moved there from their native Breslau a few years earlier. As far as I can tell, I am the only academic in our family. My father Karl Hänsch was a businessman engaged in the export of farming machinery, while my mother Marta raised her three children as a house wife. My younger brother Julius entered the book printing business, and my sister Lucia married a fellow physics graduate student and now helps run a small electronics engineering company.

Growing up during and after the second world war left some vivid memories. I can still see our family huddled together in the basement bomb shelter of our home in Heidelberg listening to the piercing sound of air raid sirens. After the war, our family had lost its estate in Breslau, and we had to share our small ground floor apartment in Heidelberg with some war refugees as subtenants. Without childhood diversions such as television, my brother and I gained a strong sense of independence and adventure by playing in the bombed ruins at the nearby railway station or exploring the many hiking trails on the slopes of the Gaisberg and Königstuhl. My father had long been disillusioned with the Nazi political leadership and raised us as rebels in spirit, distrusting any official authority.

My father also kindled my early interest in science. During the first world war, volunteering at a pharmacy, he became interested in medicine and chemistry. In Heidelberg we lived at Bunsenstrasse 10, in the house that had once belonged to the chemist Robert Bunsen. When I was about six years old, I asked my father what Bunsen had done to have a street named after him. On the next day he brought home a Bunsen burner which we connected to the gas stove in the kitchen. With a sprinkle of table salt, the blue flame turned to a bright yellow. My father explained that this is the characteristic color emitted by sodium atoms that are excited in the flame. It was obvious to me that I had to find out more about light and atoms. A little later, my father took me to visit the metallurgical laboratory of the Heinrich Lanz AG in Mannheim, where I was impressed by researchers in white lab coats who allowed me to look into their fancy microscopes. At a time when other boys dreamt about steering steam locomotives, I started to see myself as a future scientist.

In 1952, I entered the Helmholtz Gymnasium in Heidelberg, then located at the Kettengasse in the old town below the castle. Although the school emphasized modern languages and science, my father enrolled me in a rather small class with Latin as the first language to maintain my option of studying medicine. During the later years we enjoyed some remarkable teachers. Dr.

Mampel, our physics and chemistry teacher, gave me free reign of the school's collection of demonstration apparatus, and Dr. Biser, a Kaplan at the nearby Jesuitenkirche, who later became an eminent religious philosopher, turned the obligatory religious studies into a fascinating course on Western philosophy.

Early on, my interest in science dominated my activities outside school. I eagerly read popular science and science fiction books from the public library until I learned how to check out textbooks from the University library. I also liked doing experiments with my own hands. Intrigued by the world of chemistry, I started to spend my weekly allowance in pharmacies willing to sell substances like fuming nitric acid or white phosphorous to a young boy who stored his growing collection of chemicals in the bedroom of his parents. After an intimidating accident with bomb-making materials, my interests moved from chemistry to physics and electronics. Around 1957, I acquired an old cold cathode X-ray tube which I operated at home after winding a large Ruhmkorff-style induction coil. I also built a transistorized Geiger counter to perform experiments with a radioactive sample of 0.1 millicurie of Mesothorium which I had bought at a factory for radioactive luminous paint. To calibrate the Geiger counter, I went to the nuclear physics laboratory of Professor Otto Haxel at the University of Heidelberg, where an assistant was very kind and willing to introduce me to the real world of physics research. At that time, I set my sights on becoming a nuclear physicist and university professor.

STUDY AT THE UNIVERSITY OF HEIDELBERG

After the Abitur in 1961, I enrolled at the University of Heidelberg as a physics student. During the first two years most of my energy went to the study of mathematics. The lectures on physics and chemistry seemed like entertaining diversions by comparison. I was awed by the power and elegance of pure mathematical reasoning. But after a while I realized how much the complexity of an abstract formalism can sometimes distract from true physical insights. Since then I have acquired a compulsion to always try and construct the simplest possible intuitive model to "understand" a physics phenomenon. Such models have often helped me to perform quick order of magnitude estimates and to rapidly weed out half-baked ideas. Playing around with intuitive concepts I frequently arrive at interesting ideas only to find out that the results have been worked out long ago and are well known. But every once in a while I have experienced the immense joy that comes with some entirely new insight or invention.

After the Vordiplom in 1963, I enrolled in the Betatron laboratory of Professor Hans Kopfermann for the Grosspraktikum, an initial laboratory project of about six months. Unfortunately, Professor Kopfermann had died just before I could begin work on my assignment, the construction of a transistorized fast linear gate for a semiconductor detector of alpha particles which I quickly completed. In the spring of 1964, I attended my first meeting of the German Physical Society. Listening to different talks in a nuclear and particle physics session describing the work of large teams working at big machines, I lost some of my enthusiasm for this kind of research.

Instead, I became intrigued by the growing excitement about lasers which had been invented a few years earlier. In the neighboring Insitute of Applied Physics at Albert-Überle-Strasse, Professor Christoph Schmelzer had started to design a linear accelerator for heavy ions that was later realized at the GSI in Darmstadt. Since he felt that lasers might help to synchronize the phases of the individual resonators, he had hired Dr. Peter Toschek, a former student of Professor Wolfgang Paul in Bonn, as an assistant to set up a laser group in Heidelberg. Visiting this laboratory I was awed by the sight of a helium neon laser with its glowing discharge tube emitting an intense collimated beam of red laser light that produced an otherworldly speckle pattern. I sensed a large unexplored new world, and I instantly decided to switch fields. Fortunately Peter Toschek accepted me into his group so that I could pursue my two years of diploma research on gas lasers. Since commercial lasers were not yet available, we had to build everything ourselves, including the glass discharge tubes with their electrodes and Brewster windows, the gas filling stations, the high voltage power supplies, and even the dielectric mirrors and their adjustable mounts. In hindsight, this was excellent training for a budding experimentalist. My adviser was a scholar of high intellectual standards who made sure that we kept track of every single publication in the emerging field of lasers and quantum electronics. In my diploma research, I studied saturation effects in the gas laser medium by observing the light emitted spontaneously to the side. In the end I was able to determine a number of previously unknown radiative transition rates in the neon atom.

After receiving my physics diploma degree in 1966, I continued to study laser saturation phenomena in my thesis research. I had become intrigued by the sharp central Lamb dip, a drop in laser power, that Ali Javan had first observed when scanning the frequency of a single mode gas laser across the Doppler-broadened gain profile. The Lamb dip allowed a new kind of nonlinear Doppler-free high resolution spectroscopy, albeit limited to the study of laser transitions or to the "inverted Lamb dips" produced by molecular absorption lines in accidental coincidence.

In my own experiments, I studied the cross saturation of two coupled laser transitions in neon that share the same lower level. Soon, I observed strange line asymmetries that could not be understood within a hole burning model. I tentatively ascribed the observed phenomena to Raman-like two photon transitions and the dynamic Stark effect. After laboring for considerable time as a theorist I was able to explain the observations quantitatively with a semiclassical model that relied on the density matrix formalism to account for quantum interference effects in coupled three-level systems. This work, published in 1970 with Peter Toschek, is still cited frequently today, because it laid the groundwork for the understanding of phenomena such as lasers without inversion, electromagnetically induced transparency, and slow light. In January of 1969, I received my doctor degree from the University of

Heidelberg (Dr. rer. nat., "summa cum laude"), and I continued to work in Heidelberg for another year as an assistant of Professor Schmelzer. Aspects of coherence and quantum interference have remained a recurring theme in my later research, with intuitive insights from classical wave optics often guiding my thoughts and ideas.

AT STANFORD UNIVERSITY

In March 1970, I left Germany to join the laboratory of Professor Arthur L. Schawlow at Stanford University as a NATO postdoctoral fellow. I had first met Art Schawlow, co-inventor of the laser, at a summer school at Carberry Tower in Scotland in 1969, and I was immediately captivated by his warmth, his keen mind, and his contagious sense of humor. Fortunately, Art agreed to take me on as a postdoc.

On my way to California, I stopped at the east coast to visit Ali Javan at MIT and Bill Bennett at Yale University. I also arranged a vist to the famous Bell laboratories at Holmdel. There, Charles Shank and Herwig Kogelnik showed me a small pulsed dye laser, pumped by a nitrogen laser made by AVCO. At a repetition frequency of 100 Hz, the beam looked almost continuous to the eye, and the color could be changed by simply tilting the angle of a diffraction grating. This was a far cry from the few pulses per minute that I had obtained playing with a simple home-built flashlamp-pumped dye laser at Heidelberg, following the discovery of lasing in organic dye solutions by Peter Sorokin and Fritz Schäfer in 1966.

When I arrived at Stanford, I told Art Schawlow about the interesting experiments at Bell Laboratoies, and I proposed that I would like to try and make a nitrogen laser pumped dye laser so highly monochromatic that it could be used for Doppler-free saturation spectroscopy of gaseous absorption lines. When Art asked me how I would go about it, I explained that I would try holographic diffraction gratings, Lyot filters, etalons, or whatever else was necessary to restrict laser action to a single longitudinal mode. Art and I had already discovered that we shared a strong passion for clever gadgets. Art was sufficiently intrigued by my proposal to let me purchase an AVCO nitrogen laser, using funds from a post-Sputnik era Army contract. The nitrogen laser arrived in July of 1970 and immediately proved to be an irresistible toy. During the next six months we enjoyed some very entertaining experiments, ranging from edible lasers to dye laser image amplifiers.

Soon, I found the intellectual atmosphere at Stanford quite exhilarating. I was surrounded by legendary scientists such as Felix Bloch or Robert Hofstadter, and I could discuss laser science with some heroes of my graduate student years, including Tony Siegman, Steve Harris, and Robert Byer. At the heart of budding Silicon Valley, one could sense a "can do" atmosphere that seemed immensely liberating. Art kept pointing out that one did not have to know everything about a field in order to discover something new. If our German approach to research had resembled well-planned agriculture, the work at Stanford could be compared to game hunting. With my instinctive

Arthur L. Schawlow (right) and me at Stanford University.

aversion against organized planning, I enjoyed this atmosphere tremendously. At least, we did not have to be afraid of research results that made all planning obsolete. We soon found ourselves at the heart of a revolution in laser spectroscopy that brought plenty of such results.

Towards the end of 1970, I began to focus my efforts increasingly on the goal of making a widely tunable dye laser highly monochromatic. Like other experimenters before me, I was working with a rather small beam diameter inside the dye laser cavity. Suddenly, I realized that the spectral resolving power must be limited if only a small number of grating lines is illuminated. I happened to carry a small Zeiss monocular telescope in my pocket, which I often used to read the small print of slides or transparencies from the back of a lecture room. Quickly I mounted this telescope as a beam expander inside the cavity to fill the grating area more efficiently, and instantly I observed a dramatic improvement in the laser line width. With a larger beam expanding telescope and an additional etalon inside the cavity, the spectral width of the pulsed dye laser could be reduced to an unprecedented 0.0004 nm, and an additional external filter etalon soon permitted the first experiments on Doppler-free saturation spectroscopy of atomic resonance lines. To this end, I devised a scheme for saturation spectroscopy outside the laser cavity that was highly immune to the intensity fluctuations of our still primitive dye lasers. The technique became later known as Hänsch-Bordé method, since Christian Bordé in Paris had independently pursued similar ideas.

When Art Schawlow saw the first Doppler-free spectra of the sodium D lines which I had left on his desk after an exhilarating night, he suggested that we should do the same with the red Balmer-alpha line of atomic hydrogen. This line had been at the center of attention of atomic spectroscopists in the 1930s, because of suspected discrepancies from the predictions of the relativistic Dirac theory. With Issah S. Shahin, a graduate student from Jordan, we quickly set up an old-fashioned Wood-style hydrogen gas discharge tube, and soon we were able to resolve single fine structure components of the red Balmer line for the first time so that we could observe the 2S Lamb shift directly in the optical spectrum. A few years later, the first laser measurement of the Rydberg constant improved the accuracy of this important fundamental constant by an order of magnitude. This was the beginning of a long quest for ever higher resolution and measurement precision in optical spectroscopy of the simple hydrogen atom which permits unique confrontations between experiment and fundamental theory. This pursuit has culminated in the invention of the femtosecond laser frequency comb, a tool that is revolutionizing precision measurements of time and frequency, as recounted in my Nobel lecture.

When word about the new tunable laser and its powers spread, an unending series of visitors began to file through our unpretentious little laboratory, and an article describing the dye laser soon became a "citation classic." This experience taught me that a simple and imperfect proof-of-principle experiment can sometimes find a much wider resonance than a complex experiment of intimidating perfection. In 1973, Art Schawlow and I were named "California Scientists of the Year" by the California Museum of Science and Industry in Los Angeles for this work. At the same ceremony, William Hewlett and David Packard were honored as "California Industrialists of the Year". With such recognition, it became easy for me to clinch tenure as Associate Professor at Stanford. Soon, I received offers of full professorships from the University of Heidelberg, Yale University, and Harvard University. In the end, I decided to remain at Stanford, accepting a promotion to full professor in 1975, and I continued to work close to Art Schawlow for another 11 years before returning to my native Germany in 1986. Our early work with hydrogen was prominently cited when Art Schawlow received the Nobel Prize for laser spectroscopy in 1981.

Many of my other highly cited papers from the Stanford years describe relatively simple experiments such as ultrasensitive fluorescence spectroscopy with the power to detect the light from single atoms, sensitive intracavity absorption spectroscopy with a multimode dye laser, the first demonstration of continuous wave Doppler-fee two-photon spectroscopy, and the experiments with my student Carl Wieman on Doppler-free polarization spectroscopy. Even the roots of the laser frequency comb can be traced to the exhilarating seventies at Stanford. With my student Jim Eckstein and visiting Lindeman Fellow Allister Fergusson we used the comb of regularly spaced longitudinal modes of a mode-locked sub-picosecond dye laser to measure some fine structure intervals of atomic sodium.

Discussing possible ways to increase the interaction time of hydrogen atoms with a laser beam, Art Schawlow and I came up with the idea of laser cooling of atomic gases in early 1974. Vladilen Letokhov in Troisk was one of the first to start experiments with one-dimensional radiation pressure cooling of a sodium atomic beam. It took ten more years until Steve Chu and his team at Bell Laboratories realized "optical molasses", demonstrating 3-dimensional Doppler cooling as envisioned in our original proposal. Considering the dramatic subsequent developments, Art and I have often regretted that we did not immediately follow up our proposal with our own experiments. However, we did not know how to laser cool hydrogen atoms, and there were many other interesting things to do that seemed much easier. Art still remembered some unhappy experiences with atomic beam machines during his thesis research at the University of Toronto and advised against any experiments involving serious vacuum. In 1978, my former thesis adviser Peter Toschek and his group were the first to demonstrate the related method of laser cooling of trapped ions that had been proposed in 1975 by Hans Dehmelt and David Wineland.

At Stanford in the mid-seventies, one could smell the beginning microcomputer revolution, ignited by the early microprocessors introduced by the nearby Intel Corporation. After soldering together my own IMSAI computer from a kit and advancing from binary programming by flipping switches to assembly language, I became an avid visitor of the weekly meetings of the Stanford Hombrew Computer Club, where Bill Gates sold rolls of punched paper tape with 4k and 8k versions of ALTAIR BASIC. Steve Jobs, the later founder of Apple Computer came to my undergraduate classes on electricity and magnetism. In 1976, Art Schawlow and I bought one of the first Apple I Computers, a bare printed circuit board, at the Mountain View Byte Shop. Both Art and I succumbed to the microcomputer fever, spending a forbidding amount of personal money on a growing collection of computers and peripherals, so that our offices started to look like space mission control centers. Art sometimes joked about our role as early adopters: "The pioneers are the ones with the arrows in their backs." Around 1980, Art and I even started a small mail-order software business, selling a little graphics program Autoplot written in BASIC for the Radio Shack TRS80 computer. Sales were brisk until the hardware became obsolete.

My long exposure to the challenging intellectual climate at the Stanford Physics Department played an important role in shaping my own academic values. I witnessed an Assistant Professor who ruined his chance for tenure by writing a book reviewing science rather than pursuing original research. I could easily live with such priorities because doing research is what I love most. For the same reason, I have always resisted trading academic freedom for the power that comes with administrative and management responsibilities, even though I am grateful to colleagues who shoulder such burdens.

Karl Ludwig Kompa, Herbert Walther, Siegbert Witkowsky and me at the new Max-Planck-Institut of Quantum Optics in Garching in 1986.

AT MUNICH AND GARCHING

In 1978, Professor Herbert Walther at the University of Munich invited me to spend a sabbatical in Germany with an Alexander von Humboldt Senior US Scientist Award. Herbert had played a key role in promoting laser science in Germany. Since 1976, he was directing a Project Group for Laser Science in Garching that became the Max-Planck Institute of Quantum Optics in 1981. A few years later he helped to organize a very tempting lure to make me return to my native Germany. In March 1986, after agonizing over the decision for almost two years, I accepted an offer to join the Ludwig-Maximilians University of Munich as a Professor of Experimental Physics and to build a new Division of Laser Spectroscopy at the Max-Planck-Institute that was just about to move into an attractive new building at the southern end of the Garching research campus. The University laboratories which I inherited from my predecessor, Professor Josef Brandmüller, where located downtown, at Schellingstrasse 4, in the Max-Vorstadt, surrounded by bookstores and small restaurants. Since I felt that the downtown location would make it easier to attract graduate students, I decided to set up laboratories at both locations.

With considerably more space, positions for assistants and graduate students, and ample start-up funds at my disposal, it was obvious that I could pursue many more projects than before, but I could no longer maintain my Stanford style of closely working with a small group of graduate students, getting intimately involved in every detail of one or two hot experiments. At first, this adjustment felt rather painful. I had to rely much more on the initiative and judgement of my students and postdocs, resigning to the expectation that they would make expensive mistakes. Fortunately, I have been able to attract some exceptionally gifted young coworkers, sometimes with an energy, patience, and discipline that far exceeds my own. The best young people blossom if they feel free to follow their own ideas. Therefore, I try to guide my students in rather subtle ways, letting them arrive at their own insights, goals, and research plans during our discussions, while I show enthusiasm and excitement when they move in a promising direction. Once the fire is lit it tends to become self-sustaining, and I am rather proud that more than thirty of my former students and postdocs are now Professors at universities around the world, running their own laboratories. Since I did not want to give up my own hands-on experimental work, I established my own small "toy" laboratory at the University, equipped to set up improvised laser and optics experiments quickly, to explore crazy new ideas and to stay abreast of ever advancing technologies.

Shortly before moving to Munich in April 1986, I had met Gerd Binnig, the co-inventor of the scanning tunneling microscope, at the laboratory of Cal Quate at Stanford. Gerd told me that he was planning to move to Munich to set up an IBM physics group as an advance guard for an envisioned new IBM research laboratory. Since I was intrigued by the possibility of combining tunneling microscopy and laser spectroscopy, I offered Gerd to set up shop for a few years at our downtown university laboratory. In October of 1986, we could help celebrate the Physics Nobel Prize to Gerd Binnig and Heinrich Rohrer. Even though IBM soon abandoned the plans for a Munich laboratory, Gerd and his group stayed with us for the next ten years. During this time we pursued some serious research on scanning microscopy ourselves, including studies of biomolecules on graphite surfaces with Wolfgang Heckl, and a proof-of-principle demonstration of an aperturte-less scanning optical near field microscope with Johannes Pedarnig. Unfortunately, the combination of tunneling microscopy and laser spectroscopy did not advance as easily as hoped. When Gerd Binnig returned to the IBM Rüschlikon laboratory in 1996, scanning microscopy had evolved into a large worldwide enterprise and we would have had to concentrate considerable resources to compete effectively. Instead, we decided to abandon our microscopy research.

At that time the quantum physics of ultracold atoms had become an important focus of our research. With Andreas Hemmerich we were the first to demonstrate and explore two- and three-dimensional optical lattices bound by light in the early 1990s. With Tilman Esslinger, we explored new tricks for laser cooling below the recoil limit. After Immanuel Bloch joined us as a graduate student in 1997, we realized Bose-Einstein condensation (BEC) of rubidium atoms in a novel magnetic QUIC trap in February of 1998, as only the second group outside the USA. We later exploited the high magnetic stability of our small trap in the first continuous wave atom laser that dominated German science news in 1999.

In the same year, Markus Greiner joined our group as a diploma student. He constructed a new BEC apparatus transporting cold atoms magnetically from a magneto-optical trap into a high vacuum glass cell where BEC can be achieved without the need for any further laser cooling. Continuing his work as a Ph.D. student, Markus Greiner used the unobscured optical access to load the Bose-Einstein condensate into a three-dimensional optical dipole force lattice potential formed at the intersection of three orthogonal far-detuned standing wave laser fields. In 2001, he and Immanuel Bloch were the first to demonstrate a reversible quantum phase transition from a wave-like superfluid atomic state to a particle-like Mott insulator crystal, by simply adjusting the height of the lattice potential wells, as theoretically predicted in 1998 by Dieter Jaksch and Peter Zoller. This realization of a strongly correlated quantum gas has triggered an avalanche of work at the interface between atomic physics and condensed matter physics, with the prospect of quantum simulators for elusive phenomena ranging from antiferromagnetism to high Tc superconductivity.

Starting around 1994, we have also been exploring microscopic magnetic traps for the manipulation of cold paramagnetic atoms. Such traps can produce large field gradients and field curvatures without the need for strong currents. From the beginning I was intrigued by the prospect of tayloring complex magnetic potentials with lithographically fabricated circuit patterns, combining traps, wave guides, and other atom optical elements to create a quantum laboratory on a chip. Discussing such ideas with some of my coworkers in a Schwabing restaurant, I sketched a proposal for a conveyor belt for cold atoms onto a napkin. Graduate student Wolfgang Hänsel soon modeled the magnetic fields on his computer and concluded that such a device might work. In 1996, Dr. Jakob Reichel had joined our laboratory after receiving his doctorate with Christoph Salomon and Claude Cohen-Tannoudji at the ENS in Paris. Together with Wolfgang Hänsel, and later joined by Peter Hommelhoff, he set out to realize such an atom motor chip. With the mirror-MOT we found an effective method for loading cold atoms into microscopic traps, and soon we could demonstrate a working conveyer belt for atoms. In June 2001, we were the first to achieve Bose-Einstein condensation entirely on a microfabricated atom chip. More recently, Philip Treutlein has demonstrated long atomic coherence times close to the chip surface, raising the prospects for accurate atomic clocks and for quantum information processing on a chip.

One fascinating line of research at the MPQ until today has been pursuit of precision laser spectroscopy of the simple hydrogen atom. Numerous graduate students have advanced the state of the art during their thesis research in Garching, starting in the late 1980s with Reinald Kallenbach, Claus Zimmermann, Ferdinand Schmidt-Kaler, and Martin Weitz. With a beam of cold hydrogen atoms and a highly stabilized continuous laser source of 243 nm, we could greatly improve the resolution of the sharp 1S-2S two-photon resonance. Thanks to the efforts of Thomas Udem we later learned to build optical frequency interval dividers which made it possible to measure the optical frequency of the ultraviolet 1S-2S resonance against the infrared frequency of a transportable methane-stabilized He-Ne laser. This intermediate reference was shuttled to the PTB in Braunschweig many times for calibration with an elaborate harmonic laser frequency chain. These experiments led to a new Rydberg constant, tests of QED, such as a precise measurement of the Lamb shift of the 1S ground state, and accurate determinations of the rms charge radius of the proton and the structure radius of the deuteron. Krzysztof Pachucky, Savely Karshenboim and Ulrich Jentschura have provided essential theoretical support.

Starting in 1997, our efforts to measure the frequency of laser light led to the vastly simplified approach of the femtosecond laser frequency comb, as recounted in my Nobel lecture. In a proof-of-principle experiment in the fall of 1998, we used a commercial mode-locked femtosecond laser with a comb spanning 70 THz to compare the frequency of a blue dye laser directly with the microwave frequency of a commercial cesium atomic clock in our own laboratory. In June of 1999, we reached a precision of 14 decimal digits in a comparison of the 1S-2S frequency with the transportable cesium fountain clock built at BNM SYRTE in Paris. A second such measurement in February 2003, took advantage of an even simpler octave spanning laser frequency comb synthesizer as first demonstrated in late 1999 in Boulder, CO and at Garching. Frequency comparisons of this kind allow sensitive searches for possible slow variations of fundamental constants.

Our laboratory has long been a partner of the ATRAP collaboration, one of two international teams working at CERN with the goal of applying precision laser spectroscopy to anti-hydrogen, searching for conceivable differences between matter and antimatter.

Over the years, I have enjoyed the hospitality of many universities around the world hosting me as a visiting professor or lecturer. I am particularly grateful to Professor Salvatore Califano and Massimo Inguscio at the University of Florence, Italy, who gave me the opportunity to participate in the creation of the European Laboratory for Nonlinear Spectroscopy (LENS) and to teach a course on laser spectroscopy to Ph.D. students at the University of Florence in the enchanting hills of Arcetri. A serendipitous experiment with Dr. Marco Bellini at LENS in February of 1997, on the coherence of white light continuum pulses produced with an amplified femtosecond laser was actually the crucial step that convinced me that an octave-spanning optical frequency comb synthesizer could be realized.

During my Stanford years, I had enjoyed the liberating climate of entrepreneurship that was omnipresent in the heart of Silicon Valley. Fortunately the old aversion between academia and industry is now waning in Germany, and in 2001, my former students Ronald Holzwarth and Michael Mei have taken the risk of starting a spin-off company, Menlo Systems GmbH, to develop commercial frequency comb synthesizers. The name has been inspired by Menlo Park in New Jersey where Thomas Alva Edison invented the light bulb.

Sometimes people comment on my growing collection of prizes and awards, wondering if I am pursuing research in order to win still more prizes. The honest answer is that I like prizes to reassure our sponsors that their money is being spent well. Prizes are also important as recognition and source of pride and motivation for our team. However, the most important reward for me has always been the joy that comes from new insights, discoveries, and inventions.

PASSION FOR PRECISION

Nobel Lecture, December 8, 2005

by

THEODOR W. HÄNSCH

Max-Planck-Institut für Quantenoptik, Garching, and
Department of Physics, Ludwig-Maximilians-Universität München, Germany.

ABSTRACT

Optical frequency combs from mode-locked femtosecond lasers have revolutionized the art of counting the frequency of light. They can link optical and microwave frequencies in a single step, and they provide the long missing clockwork for optical atomic clocks. By extending the limits of time and frequency metrology, they enable new tests of fundamental physics laws. Precise comparisons of optical resonance frequencies of atomic hydrogen and other atoms with the microwave frequency of a cesium atomic clock are establishing sensitive limits for possible slow variations of fundamental constants. Optical high harmonic generation is extending frequency comb techniques into the extreme ultraviolet, opening a new spectral territory to precision laser spectroscopy. Frequency comb techniques are also providing a key to attosecond science by offering control of the electric field of ultrafast laser pulses. In our laboratories at Stanford and Garching, the development of new instruments and techniques for precision laser spectroscopy has long been motivated by the goal of ever higher resolution and measurement accuracy in optical spectroscopy of the simple hydrogen atom which permits unique confrontations between experiment and fundamental theory. This lecture recounts these adventures and the evolution of laser frequency comb techniques from my personal perspective.

INTRODUCTION

In our highly complex and ever changing world it is reassuring to know that certain physical quantities can be measured and predicted with very high precision. Precision measurements have always appealed to me as one of the most beautiful aspects of physics. With better measuring tools, one can look where no one has looked before. More than once, seemingly minute differences between measurement and theory have led to major advances in fundamental knowledge. The birth of modern science itself is intimately linked to the art of accurate measurements.

Since Galileo Galilei and Christiaan Huygens invented the pendulum clock, time and frequency have been the quantities that we can measure with the highest precision. Today, it is often a good strategy to transform other quantities such as length or voltage into a frequency in order to make accurate measurement. This is what my friend and mentor Arthur Schawlow at Stanford University had in mind when he advised his students: "Never measure anything but frequency!" Measuring a frequency, i.e. counting the number of cycles during a given time interval, is intrinsically a digital procedure that is immune to many sources of noise. Electronic counters that work up to microwave frequencies have long been available. In 1967, the Conference Generale des Poids et Mesures (CGPM) has defined the second, our unit of time, as the period during which a cesium-133 atom oscillates 9 192 631 770 times on a hyperfine clock transition in the atomic ground state. Today, after 50 years of continuous refinement, microwave cesium atomic clocks reach a precision of 15 decimal digits [1].

Even much higher precision is expected from future optical atomic clocks which use atoms or ions oscillating at the frequency of light as the "pendulum". By slicing time into a hundred thousand times finer intervals, such clocks will greatly extend the frontiers of time and frequency metrology. The long missing clockwork mechanism can now be realized with a femtosecond laser frequency comb, an ultra-precise measuring tool that can link and compare optical frequencies and microwave frequencies phase coherently in a single step. Laser frequency combs provide powerful tools for new tests of fundamental physics laws. Precise comparisons of optical resonance frequencies of atomic hydrogen and other atoms with the microwave frequency of a cesium atomic clock are already establishing sensitive limits for possible slow variations of fundamental constants. Optical high harmonic generation is extending frequency comb techniques into the extreme ultraviolet, opening a new spectral territory to precision laser spectroscopy. Frequency comb techniques are also providing a key to attosecond science by offering control of the electric field of ultrafast laser pulses.

Femtosecond laser frequency combs have been highlighted in the citation for the 2005 Nobel Prize in Physics. Although perfected only about seven years ago, they have already become standard tools for precision spectroscopy and optical frequency metrology in laboratories around the world. Commercial instruments have quickly moved to the market, and extensive review articles and books have been written on frequency comb techniques [2 - 4]. In this lecture I will try to give my personal perspective on the evolution of these intriguing measuring tools for time and frequency. Far from attempting a comprehensive review, I have selected references that helped guide my own insights along a winding path.

THE DAWN OF DOPPLER-FREE LASER SPECTROSCOPY

High resolution laser spectroscopy and precise spectroscopic measurements have appealed to me since I was a graduate student at the University of Heidelberg. For my diploma and thesis research I worked with helium-neon gas lasers in the group of Peter Toschek at the Institute of Applied Physics, headed by Christoph Schmelzer. I was intrigued by the central narrow Lamb dip that Abraham Szöke and Ali Javan had first observed while scanning the frequency of a single-mode gas laser across the Doppler-broadened gain profile [5]. Such a dip had been predicted by Willis Lamb in his semiclassical laser theory [6]. Bill Bennett was the first to give a simple explanation in terms of saturation and spectral hole burning the two counter propagating waves inside the standing wave laser cavity [7]. Other researchers such as John Hall, Veniamin Chebotaev, or Christian Bordé soon explored "inverted Lamb dips" by placing some absorbing molecular gas inside the laser cavity [8]. With resonances of unprecedented spectral resolution, one could almost smell the revolution in laser spectroscopy that would unfold within the next few years. At that time, however, such Doppler free spectroscopy remained limited to the study of gas laser transitions or of a few molecular absorption lines in accidental coincidence. In my own work with Peter Toschek, I studied quantum interference effects in coupled atomic three-level systems [9, 10], demonstrating phenomena which have recently been recognized as important, such as lasing without inversion or electromagnetically induced transparency. They are also essential to understand slow light.

In 1970, I joined Arthur L. Schawlow at Stanford University as a postdoc. Collaborating in separate experiments with Peter Smith, then at Berkeley [11], and with Marc Levenson at Stanford [12], I perfected a new method of Doppler-free saturation spectroscopy that did not require the sample to be placed inside a laser cavity. Soon afterwards, I succeeded in making a nitrogen-laser-pumped widely tunable pulsed dye laser so highly monochromatic that we could apply Doppler-free saturation spectroscopy to arbitrarily chosen atomic resonance lines [13, 14]. Broadly tunable laser action in liquid solutions of organic dyes had been discovered in 1966, independently by Fritz Schäfer [15] and Peter Sorokin [16].

LASER SPECTROSCOPY OF ATOMIC HYDROGEN

Arthur Schawlow at Stanford suggested to apply our technique to the red Balmer-α line of atomic hydrogen that had been at the center of attention of atomic spectroscopists in the 1930s because of suspected discrepancies between the observed line profile and the predictions of Dirac's relativistic quantum theory [17]. In those days, spectroscopists could only observe a blend of unresolved fine structure components because Doppler broadening is particularly large for the light hydrogen atoms. Spectroscopy of the simple hydrogen atom has long played a central role in the history of atomic physics. The visible Balmer spectrum was the Rosetta stone that allowed us to decipher the laws of quantum physics. It has inspired the path breaking discoveries of Niels Bohr, Arnold Sommerfeld, Louis De Broglie, Erwin Schrödinger, Paul Dirac, and even Willis Lamb at the origin of modern quantum electrodynamics.

In 1972, graduate student Issa Shahin and myself were proud to present to Arthur Schawlow a Doppler-free saturation spectrum of the red hydrogen

Balmer-α line, recorded with our pulsed tunable dye laser [18]. The 2S Lamb shift, i.e. the splitting between the $2S_{1/2}$ and $2P_{1/2}$ states that should be degenerate according to the Dirac theory, appeared clearly resolved in the optical spectrum. This was the beginning of a long adventure in precision spectroscopy of the simple hydrogen atom, which permits unique confrontations between experiment and theory. This quest continues today. It has inspired many advances in spectroscopic techniques, including the first proposal for laser cooling of atomic gases [19], and, most recently, the femtosecond laser frequency comb.

Fig. 1 illustrates how the accuracy of optical spectroscopy of atomic hydrogen has advanced over time [20]. Classical spectroscopists remained limited to about six or seven digits of precision by the large Doppler broadening of hydrogen spectral lines. In 1971, our group at Stanford overcame this barrier by nonlinear spectroscopy with a tunable dye laser. Other groups, notably in New Haven, Oxford, and Paris, soon joined in to improve the accuracy by three orders of magnitude over the next two decades. Around 1990, a new barrier appeared: the limits of optical wavelength metrology due to unavoidable geometric wave front errors. Progress beyond a few parts in 10^{10} has been achieved only because we have learned increasingly well how to measure the frequency of light rather than its wavelength. In 2003, the accuracy has reached 1.4 parts in 10^{14} [21]. Further progress is becoming difficult, because we are again approaching a barrier: the limits of how well we know our unit of time, the second. Cesium atomic clocks have been continually refined over the past 50 years [1], as shown by the dashed line in Fig. 1, but the potential for further improvements seems almost exhausted. However, our optical frequency counting techniques make it now feasible to develop optical atomic clocks, based on sharp optical resonances in laser-cooled trapped ions, neutral atoms or molecules. With such clocks future spectroscopic measurements may reach accuracies of parts in 10^{18} and beyond.

In atomic hydrogen, the highest resolution can be achieved on the ultraviolet 1S-2S two-photon resonance with a natural line width of only 1 Hz. First order Doppler shifts cancel if this transition is excited with two counter-propagating laser waves, as was first pointed out by Veniamin Chebotaev [22]. The first Doppler-free spectra have been recorded in our laboratory at Stanford in 1975 [23]. At Garching, we observe this resonance by collinear excitation of a cold hydrogen atomic beam [21]. Starting in 1986, many generations of graduate students and postdocs have made important contributions to advance the state of the art.

Today, the hydrogen atoms are produced by microwave dissociation of molecules and cooled to a temperature of about 6 K by collisions with the walls of a nozzle mounted to a helium cryostat. A collinear standing wave field at 243 nm for Doppler-free two-photon excitation is produced by coupling the frequency-doubled output of a dye laser into a buildup cavity inside the vacuum chamber. Atoms excited to the 2S metastable state after traveling along a path of about 10 cm are detected by applying a quenching electric field and counting the emitted vacuum ultraviolet Lyman-α photons. The

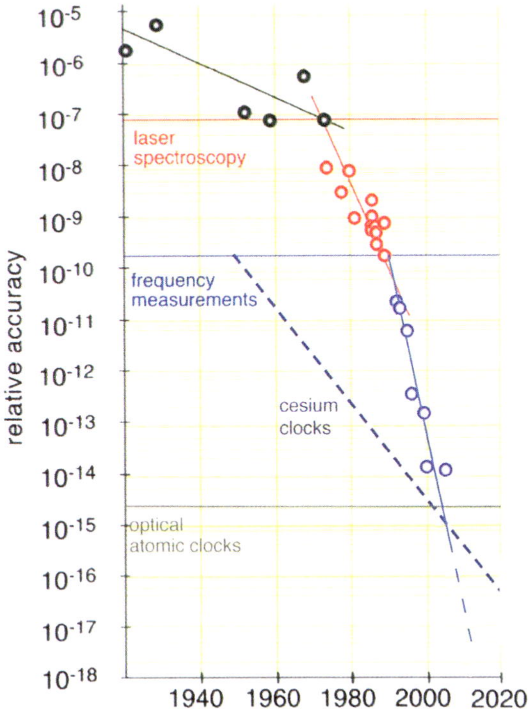

Figure 1. The relative accuracy in optical spectroscopy of atomic hydrogen is charted over eight decades. Major barriers have been overcome in the early 1970s with the advent of Doppler-free laser spectroscopy and in the early 1990s with the introduction of optical frequency measurements. The accuracy of such measurement will soon be limited by the performance of cesium atomic clocks. Dramatic future advances are expected from the development optical atomic clocks.

laser light is periodically blocked by a chopper, and the photon counts are sorted into bins corresponding to different delay times. With slow atoms selected by a delay time of 1.3 ms, the line width is now reduced to about 530 Hz at 243 nm corresponding to a resolution of 4.3 parts in 10^{13}. We would have to reach an accuracy of 5 parts in 10^{15} in order to measure the line position to 1% of this width.

MEASURING OPTICAL FREQUENCIES

The observation of sharp optical resonances by nonlinear laser spectroscopy with a resolution much beyond the measurement limits of wavelength interferometry had long created a strong need for methods to measure the frequency rather than the wavelength of light. The quest for an optical frequency counter is almost as old as the laser itself. Ali Javan, the co-inventor of the helium-neon laser, was the first to superimpose the beams from two different lasers with a beam splitter on a photo detector to observe a beat note, similar to the interference of the sound waves from two tuning forks [24]. This was an extraordinary result, because it proved that laser light can behave like a classical radio wave. A coherent laser wave can have a well defined phase and amplitude, so that it must be possible to count the ripples of such a light wave. However, at a frequency near 500 000 billion oscillations per second, there are no electronic detectors and circuits fast enough to build an optical frequency counter.

At MIT in the early 1960s, Ali Javan started a research project, aimed at extending microwave frequency counting techniques into the optical spectral region. He experimented with whisker-like metal-insulator-metal point contacts as antennas, detectors and mixers for infrared laser waves. Such elements were later used by John Hall and Ken Evenson at NBS (now NIST) in Boulder to realize the first harmonic laser frequency chain, that was used to determine the speed of light by measuring both the wavelength and the frequency of a methane-stabilized 3.39 μm helium-neon gas lasers [25]. Harmonic laser frequency chains were highly complex systems, engineered to measure just one particular optical frequency, and only a handful of these chains have ever been constructed at a number of well-equipped national metrology laboratories. In the early 1980s, a chain at NBS in Boulder had been perfected so that it could measure the frequencies of some iodine-stabilized visible helium-neon lasers to 10 decimal digits. This demonstration led the Conference Generale des Poids et Mesures in 1983, to redefine the meter by defining the speed of light in vacuum c as exactly 299 792 458 meters per second. One meter is then the distance traveled by light during the time of 1/299 792 458 seconds. From now on, one could determine the precise wavelength of a laser in vacuum, λ, by simply measuring the frequency f, since $f \cdot \lambda = c$.

Unfortunately, the complex NBS frequency chain had to be abandoned soon after this definition was in the books, and for the next decade there was not a single laboratory in the U.S. that could have followed this prescription. A number of European laboratories did better, notably the Observatoire in Paris (now BNM SYRTE) and the Physikalisch-Technische Bundesanstalt (PTB) in Braunschweig. In an article published in early 1996 [26], a team from the PTB laid claim to the first phase coherent frequency measurement of visible radiation. An elaborate frequency chain filling three large laboratories spread over two separate buildings was assembled to compare the frequency of the red intercombination line of atomic calcium with the microwave frequency of a cesium atomic clock. To reach sufficient phase stability, the clock frequency was first reduced to the 100 MHz of a stable quartz oscillator. From here, the chain traversed the entire electromagnetic spectrum in discrete steps, always generating some harmonic frequency in a suitable nonlinear element and producing enough power for the next step with a phase-locked transfer oscillator. A tricky puzzle had to be solved to reach the desired final frequency with the help of several auxiliary oscillators.

It was obvious that we could not afford to assemble such a harmonic laser frequency chain for our hydrogen experiments at Garching. As a simpler alternative, I proposed a frequency interval divider chain in 1988, that worked with frequency differences rather than the frequencies themselves so that one could stay in a convenient region of the electromagnetic spectrum such as the near infrared where compact diode laser sources are available [27]. The basic building block is an interval divider stage with a laser that is servo-controlled to oscillate at the precise mid-point of two input frequencies. To this end, the second harmonic frequency of the central laser is compared to the sum of the two input frequencies, as generated in a nonlinear optical crystal. With a cascaded chain of n such interval dividers, a large frequency interval can be divided by 2^n. To measure an absolute laser frequency f, one could start with the interval between f and its second harmonic 2f, which is just equal to the frequency f. After repeatedly cutting this interval in half with perhaps 15 stages, the remaining frequency gap is small enough that it can be observed as a beat note with a fast photo-detector and measured with a microwave frequency counter. With Harald Telle, who had joined us from the PTB, and Dieter Meschede, we demonstrated the first working interval divider in 1990 [28].

We never assembled a complete optical frequency counter, but we constructed a chain of four interval dividers to measure a frequency interval of 1 THz that we encountered when comparing our hydrogen 1S-2S frequency with the infrared frequency of a methane-stabilized infrared helium-neon laser at 3.39 μm as the starting point of our own short harmonic laser frequency chain [29]. This intermediate frequency reference had to be repeatedly shuttled to Braunschweig to be calibrated with the PTB frequency chain against a cesium clock. In 1997, we established a new record in optical frequency metrology [29] by determining the ultraviolet 1S-2S frequency to within 3.7 parts in 10^{13}. From this and other spectroscopic measurement in hydrogen, we were able to derive a new value of the Rydberg constant, the scaling factor for any spectroscopic transition, and the most precisely known of the fundamental constants. We could also derive the Lamb shift of the 1S ground state accurately enough to provide a stringent new test of bound-state QED. Assuming that QED is correct we could, moreover, determine new values for the rms charge radius of the proton and the structure radius of the deuteron [29, 30]. We were rather proud that the accuracy achieved with our table-top experiments exceeded that of electron scattering experiments with large accelerators by an order of magnitude.

Soon, a number of metrology laboratories set out to build optical frequency counters based on optical interval division. At Garching, we were also experimenting with electro-optic frequency comb generators kindly provided by Motonobu Kourogi, that could produce an evenly spaced comb of modulation sidebands extending over several THz [31]. A frequency counter could have been realized with only six or seven interval divider stages, if the final frequency gap was bridged with such an electro-optic comb generator. During an extended visit to Garching, Motonobu Kourogi showed us how to observe even feeble comb lines by heterodyne detection, improving the signal to noise ratio with optical balanced receivers and variable beam splitters. We soon verified the accuracy of his frequency comb generator and our frequency interval divider chain in a direct comparison [32].

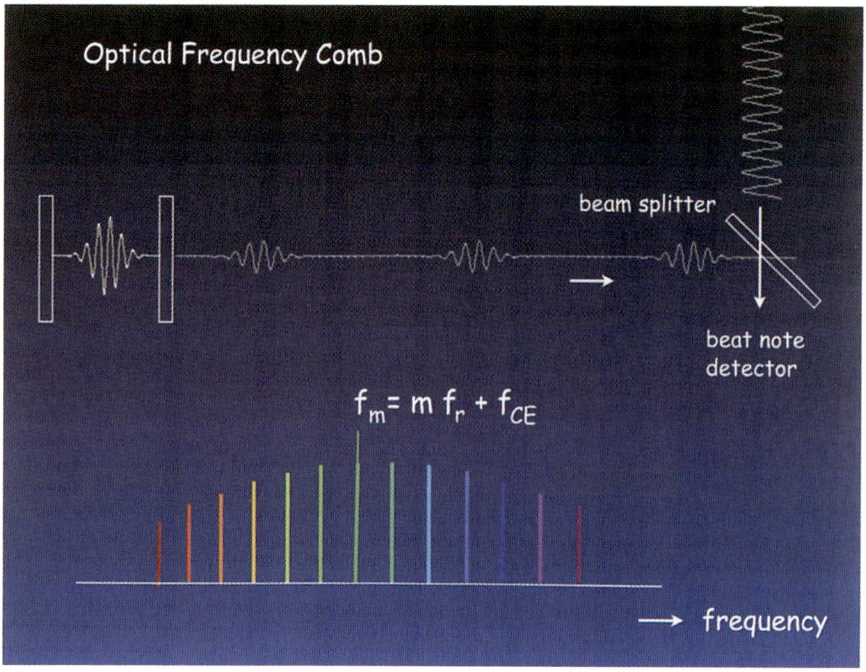

Figure 2. Scheme of a femtosecond laser frequency comb synthesizer.

Many other alternatives have been explored in the long quest for precise optical frequency measurements, including interferometry with modulated laser waves [33, 34] or frequency division with phase-locked optical parametric oscillators [35]. David Wineland has proposed to synchronize the cyclotron motion of a single electron to a laser wave [36]. In the meantime, all these approaches have become obsolete. Since 1998, optical frequency measurements have been enormously simplified with the advent of femtosecond laser optical frequency comb synthesizers [2,3].

FEMTOSECOND LASER OPTICAL FREQUENCY COMBS

The scheme of a frequency comb synthesizer is rather simple, as illustrated in Fig. 2. At the heart is a single mode-locked femtosecond laser which maintains a soliton-like short pulse circulating inside the optical cavity. This laser can be compared to Einstein's Gedanken light clock. With each roundtrip, an attenuated copy of the light pulse escapes so that the laser emits a regular train of ultrashort pulses. To measure the unknown frequency of a laser wave, the beam and the pulse train are superimposed with a beam splitter, and a photo detector registers an interference signal. In the idealized case of a perfectly periodic pulse train, we would expect a low frequency beat note whenever the laser frequency comes close to a value where an integer number of oscillations fits in the time interval between two pulses. To give an example, if we know that the laser emits precisely one billion pulses per second and if we can be sure that the laser wave oscillates precisely 500 000 times during the pulse repetition period, then we know that the optical frequency must be 500 000 billion cycles per second.

In the frequency domain, we can argue that the coupled longitudinal modes of the pulsed laser form an evenly spaced comb of spectral lines. A low frequency beat note is expected whenever the unknown laser frequency approaches one of these comb lines. The origin of the comb spectrum is well explained by Antony E. Siegman in his classic textbook [37]. Consider an arbitrary optical waveform circulating inside an optical cavity. During each roundtrip, an attenuated copy escapes through a partly transmitting mirror. A single copy will have a broad and more or less complicated spectrum. However, two identical copies end-to-end will produce interference fringes in the spectrum, somewhat reminiscent of Young's double slit experiment. Three copies produce a spectrum that resembles the interference pattern of a triple-slit, and an infinite series of copies produces sharp lines which can be identified with the modes of the cavity. Mathematically, an ideal periodic pulse train can be described in terms of a Fourier series, and the comb lines correspond to the elements of this series.

The separation between two modes or comb lines is just equal to the repetition frequency f_r. This remains true even if the pulses are not identical replicas but if we allow for a (reproducible) slip of the phase of the electromagnetic "carrier"-wave relative to the pulse envelope from pulse to pulse [38, 2, 3]. Such phase slips are unavoidable in a real laser because of dispersion in the cavity. The entire comb will then be shifted relative to the integer harmonics of the repetition frequency f_r by a carrier-envelope offset frequency f_{CE}, that equals the net phase slip modulo 2π per pulse interval. The frequency of a comb line with integer mode number m is then given by

$$f_m = m\, f_r + f_{CE}.$$

Such a comb acts like a ruler in frequency space that can be used to measure a large separation between two different optical frequencies in terms of the pulse repetition rate f_r. If these two frequencies are known multiples or fractions of the same laser frequency f, such a measurement reveals the optical frequency f itself. With a known repetition frequency f_r, the beat signal between a known optical frequency f and the nearest comb line reveals the previously unknown offset frequency f_{CE}. The frequency of any comb line can be calculated from the two radio-frequencies f_r and f_{CE} together with the integer mode number.

It has been surprising to most experts how far this frequency comb approach can be pushed. The frequency spectrum of a femtosecond laser oscillator can be broadened in a nonlinear optical medium to span more than an optical octave without destroying the integrity of the comb lines. In a now common implementation, the pulse train from a Kerr-lens mode-locked Ti:sapphire laser is sent through a micro structured silica fiber, with a small solid fiber core surrounded by air-filled holes [39, 40]. The large change in refractive index at the silica-air interface permits guiding by total internal reflection even if the incoming beam is tightly focused to a high intensity. Since part of the light travels as an evanescent wave in air, an additional engineering parameter is available in such a fiber to reduce the spreading of an injected pulse due to group velocity dispersion. Inside the fiber, the pulse spectrum is broadened by self-phase modulation due to the intensity dependent refractive index, soliton splitting, shock wave formation, and other nonlinear optical processes. The emerging white light can be dispersed with a grating to form a rainbow of colors. However, this is not ordinary white light. Remarkably, the processes generating the white light can be so highly reproducible that successive pulses are still correlated in their phases and can interfere in the spectrum to form a comb of several hundred thousand sharp spectral lines.

By now, it has been confirmed in many experiments that the line spacing is very precisely equal to the repetition frequency f_r. With a comb spanning more than an octave, it is particularly simple to measure the carrier envelope offset frequency f_{CE}. One can simply select a few thousand comb lines from the red end of the spectrum and send the light pulses through a frequency doubling crystal, so that new comb lines are generated which are now displaced by twice the offset frequency f_{CE}. A collective beat note with the corresponding original comb lines near the blue end of the spectrum directly reveals the shift f_{CE}. Once this frequency can be measured, it can be controlled, for instance by adjusting the dispersion in the laser cavity or simply by changing the pump power. One can even set f_{CE} to zero so that the frequencies of the comb lines become precise integer multiples of the laser repetition frequency f_r.

So far, we have treated all light waves as classical electromagnetic waves. The quantum optical aspects of frequency combs, i.e. expected correlations in the noise due to photons and their entanglement, have not yet been explored. Such studies may lead to a rich new field of research.

A laser frequency comb provides a direct link between optical frequencies and microwave frequencies. This link can be used in either direction. We can measure or control the repetition frequency f_r with a cesium atomic clock to synthesize several hundred thousand sharp optical reference frequencies which are precisely known in terms of the primary standard of time. Any unknown frequency can then be determined by first making a wavelength measurement with a conventional wave meter that is sufficiently accurate to determine the integer order number m of the nearest comb line. The precise distance from this reference line is then measured by feeding a beat signal to a microwave counter. In the reverse direction, we can start with a sharp optical reference line in some cold trapped ion, cold atoms, or slow molecules, and lock a nearby comb line to this optical reference. All the other comb line frequencies are then rational multiples of the optical reference frequency, and the repetition frequency becomes a precisely known fraction.

Frequency comb synthesizers act as if we had several hundred thousand ultra-stable and precisely tuned lasers operating at once. With the help of nonlinear sum and difference frequency generation, they can precisely measure any frequency from radio waves to the near ultraviolet. They provide the long-missing clockwork for optical atomic clocks. They can even generate microwaves with extreme phase stability [41]. As sources of phase-stabilized femtosecond pulses they have even given us a key to the intriguing field of attosecond science [42]. With their electronic servo controls, frequency-comb synthesizers can be relatively simple, robust, and increasingly user-friendly devices.

THIS IS A SIMPLE IDEA! WHAT TOOK SO LONG?

In hindsight, the ideas behind the optical frequency comb look rather simple and almost obvious. Why did all the experts, including our own laboratory, struggle for so long with much more cumbersome harmonic laser frequency chains?

The main reason may be that nobody believed seriously that such an approach could actually work. There were good arguments why it should be impossible to bridge the gap between radio frequencies and optical frequencies in a single step. The phase noise of even the best quartz oscillator is so large that any comb structure would be completely washed out if one could somehow multiply its frequency up into the visible region. In a harmonic frequency chain, the intermediate transfer oscillators act as phase noise filters and electromagnetic "flywheels" to overcome this "coherence collapse" [43].

Another reason may be that two separate scientific communities have evolved since the early days of laser science. People interested in precise high resolution spectroscopy used their ingenuity to perfect the frequency stability of continuous wave lasers. On the other side, there were people who invented clever techniques to produce ever shorter pulses with mode-locked lasers. They applied their spectrally broad light flashes to the study of ultrafast phenomena in semiconductors, liquids, or in chemical reaction dynamics or to generate ever higher peak intensities for experiments in plasma physics. These two communities went to their own separate conferences, and neither side felt a strong need to keep track of the other frontier.

For our own work, I cannot hide behind this latter excuse. I knew since the early experiments with multi-mode helium-neon lasers [44] that the longitudinal modes of a laser are well defined and their phases can be coupled so as to produce a short light pulse circulating inside the cavity [45, 46]. Much shorter pulses were produced some years later with broadband dye lasers by locking their axial modes with the help of a saturable absorber or by synchronous pumping with a modulated argon laser [47]. At Stanford in the mid-seventies, I became intrigued by the idea of high resolution spectroscopy of atomic resonance lines by Ramsey-like excitation with a coherent train of multiple light pulses [48]. Resonant excitation with separated light pulses has also been explored at the time by Michael Salour at MIT [49] and by Veniamin Chebotaev at Novosibirsk [50]. After initial encouraging experiments with a dye laser pulse injected into a passive cavity [48], our group at Stanford with graduate student Jim Eckstein and visiting Lindemann Fellow Allister Ferguson demonstrated that a synchronously pumped mode-locked picosecond dye laser could produce a stable phase coherent pulse train which we used for Doppler-free two-photon excitation of atomic sodium [51]. The comb lines served as a frequency ruler to measure some atomic fine structure intervals. To improve the precision, we replaced the original radio frequency driver for the modulator of our argon pump laser with a high quality frequency synthesizer. Much to our delight, the performance of the dye laser improved so much that we were the first to generate sub-picosecond pulses directly from a synchronously pumped dye laser [52]. At that time we should have learned an important lesson: what is good for frequency stability is also good for the generation of ultrashort light pulses! We later explored Doppler-free polarization spectroscopy with our frequency comb [53], as well as two-photon spectroscopy with the frequency comb of an FM mode-locked laser where the phases of the modes adjust so that the intensity remains constant but the frequency is sweeping back and forth periodically [54].

During the Stanford experiments, we were painfully aware that we could not know the absolute positions of our comb lines because the dispersion inside the laser resonator would lead to unknown phase slips of the carrier wave relative to the pulse envelope. Such phase slips shift the entire comb spectrum by an unknown amount f_{CE}, as worked out in considerable detail in the 1978 Stanford Ph.D. thesis of Jim Eckstein [38]. With a comb spectrum spanning only 800 GHz, we had no means to observe and measure the offset frequency f_{CE}. Therefore, we did not know how to measure absolute optical frequencies with our laser frequency combs in the late seventies.

The idea of somehow generating much broader frequency combs surfaced again in my mind after I had moved back to Germany in 1986. In 1990, I published a proposal for a synthesizer of sub-femtosecond pulses that would superimpose the frequencies from separate phase-locked cw laser oscillators to form a very wide comb [55]. In the early 1990s, the technology of ultrafast lasers advanced dramatically with the discovery of Kerr-lens mode locking by Wilson Sibbett at the University of St. Andrews [56]. Soon, commercial Ti:sapphire femtosecond lasers became available, that made the generation of ultrashort light pulses much easier. Intrigued by these new sources I discussed with Peter Lambropoulos at Garching the possibility of finding some highly nonlinear effect, such as above-threshold-ionization (ATI) , that would depend on the phase of the electric field relative to the pulse envelope and could reveal the offset frequency f_{CE} of the laser comb lines. Calculations soon showed that such effects would be observable only for pulses lasting at most a few optical cycles [57]. Today, such sources have become available, and the phase dependence of ATI has since been demonstrated by Gerhard Paulus and Herbert Walther [58]. In 1994, I also discussed the problem of the carrier-envelope phase slips with Ferenc Krausz at the Technical University of Vienna. His group was the first to observe such pulse-to-pulse phase slips in an interferometric correlation experiment in 1996 [59].

I remember a trade show in 1994, when I was captivated by an exhibit of a

(Coherent Mira) mode-locked Ti:sapphire femtosecond laser with regenerative amplifier. The laser beam was focused into a glass slide to produce a white light continuum which a prism dispersed into a rainbow of colors. Such white light pulses are produced by a combination of self-focusing, self phase modulation, and other nonlinear processes, and they have been used for a long time in ultrafast pump-probe experiments [60]. A striking feature was the laser-like speckle pattern in the rainbow colors which indicated a high degree of spatial coherence. It occurred to me that such a system might produce an octave-spanning frequency comb if the phases of successive pulses were sufficiently correlated. Such a wide comb could then be used as a ruler to measure the large interval between a laser frequency and its second harmonic, which must be equal to the laser frequency itself. Even though the pulse repetition frequency of a few hundred kHz remained inconveniently low for frequency comb experiments, I felt sufficiently intrigued to acquire such a system for our frequency metrology laboratory at Garching in 1994. In the back of my mind I had the hope that it might somehow be possible to produce white light directly with the pulses from the laser oscillator, without the regenerative amplifier with its much lower pulse rate, by sending the pulse train into a small waveguide made from some highly nonlinear optical material, so that it would not be necessary to reach the threshold power for self-focusing.

We did not pursue the femtosecond laser approach seriously right away, because we had come quite far in perfecting our alternative scheme of optical interval division. An accurate measurement of the 1S-2S frequency seemed almost within reach. We also felt that we would need an independent tool to verify any measurement with a femtosecond laser frequency comb, since the frequency metrology community would otherwise distrust our results. The hydrogen measurements were finally completed in 1997 [29, 30].

In February 1997, I visited the European Laboratory for Nonlinear Spectroscopy, LENS, in Florence, Italy. There, Marco Bellini was working with an amplified Ti:sapphire femtosecond laser producing pulses of 1 mJ energy at a rate of 1 kHz. As is common in many ultrafast laboratories, he produced a white light continuum for pump-probe experiments by focusing part of the laser beam into a thin plate of CaF_2. I asked what would happen if we split the laser beam in two parts and focus these beams at two spatially separate spots. Would the two white light pulses interfere?

In an earlier joint experiment at the Lund Laser Center, we had investigated the same question for the generation of high harmonic radiation in a gas jet [61]. From that time, Marco Bellini still had a Michelson interferometer on his shelve that we could quickly place into the laser beam, somewhat misaligned so that two beams would escape in two slightly different directions. By adjusting the length of one arm we could make sure that the two focused pulses arrived on the CaF_2 plate at precisely the same time. I felt electrified when we observed stable interference fringes of high contrast for all the colors that I could record with my handheld camcorder electronic notebook, as shown in Fig. 3 [62]. The white light pulses had to be phase-locked to the

Figure 3. Interference fringes between two white light pulses [Ref. 62].

driving laser field! No matter how complicated the process of white light continuum generation might be, the process was reproducible. If such pulses were separated in time rather than in space, they would interfere in the spectrum to produce a very broad frequency comb.

By March 30, 1997, I had written a confidential 6-page proposal for a universal optical frequency comb synthesizer "*...which produces a wide comb of absolutely known equidistant marker frequencies throughout the infrared, visible, and ultraviolet spectral range. To this end, a white light continuum with a pulse repetition rate* f_r *is produced by focusing the output of a mode-locked femtosecond laser into an optical fiber or bulk medium with a third order nonlinear susceptibility. The rate of phase slippage of the laser carrier relative to the pulse envelope,* f_{CE}*, is monitored by observing a beat signal between the white light continuum and the second harmonic of the laser.*" The envisioned self-referencing scheme could find the carrier-envelope offset frequency f_{CE} without any auxiliary laser. I asked Thomas Udem and Martin Weitz in our laboratory to witness and sign every page on April 4, 1997, since this might become important for later patent applications.

At Garching, we soon started a serious experimental effort towards optical frequency measurements with femtosecond laser frequency combs. Even if we did not yet know how to broaden the spectrum of our Mira laser oscillator to more than an optical octave, we could always follow the 1988 proposal [27] and measure the frequency of the dye laser in our hydrogen spectrometer by implementing a short chain of two or three interval divider stages, using small semiconductor lasers, to arrive at a frequency gap that could be bridged with our laser comb. Propelled by such visions, Thomas Udem and Jörg Reichert investigated the frequency comb spectrum of the Mira femtosecond laser. They were later joined by Ronald Holzwarth. By that time, hundreds of such lasers were in use in laboratories around the world, but they were mostly used to study ultrafast phenomena. Nobody had ever looked for any comb lines, as far as we could tell. With a repetition frequency of 76.5 MHz, the comb spectrum of our femtosecond laser was so densely spaced that no spectrometer in our laboratory could resolve the comb lines. Therefore, we resorted to heterodyne detection, employing a cw diode laser as a local oscillator. The diode laser beam and the pulse train were superimposed with a beam splitter, and a beat signal was detected with an avalanche photodiode after some spectral filtering. After paying attention to the mechanical stability of the femtosecond laser, we could observe stable comb lines. Next, we investigated the spacing of these lines. We phase-locked two diode lasers to two arbitrarily chosen comb lines and used an optical interval divider stage to produce a new frequency precisely at the center. A beat note with the nearest comb line confirmed, much to our delight, that the comb lines were perfectly evenly spaced, way out into the wings of the emission spectrum, within a few parts in 10^{17} [63].

It was now certain that the frequency comb of such a mode-locked femtosecond laser did not suffer from "coherence collapse" and could serve as a ruler in frequency space to measure large optical frequency intervals. In a first demonstration of an optical frequency measurement with a femtosecond laser comb, we determined the frequency interval between the cesium D1 resonance line and the fourth harmonic of a transportable CH_4-stabilized 3.39 μm He-Ne-laser, which had been calibrated with a harmonic laser frequency chain at the PTB Braunschweig [64]. The optical cesium frequency was needed for a determination of the fine structure constant α from the atomic recoil energy as measured by atom interferometry in the group of Steve Chu at Stanford. Soon these experiments found a considerable resonance in newspapers and journals. They demonstrated to the optical frequency metrology community that femtosecond laser frequency combs could be powerful tools to measure the frequency of light.

We next turned to the more ambitious goal of measuring the absolute optical frequency of our 486 nm dye laser in the hydrogen 1S-2S spectrometer. At that time, we had broadened the frequency comb of our mode-locked laser oscillator by self-phase-modulation in a short length of ordinary optical fiber to span 60 or 70 THz. Rather than building a few new interval divider stages, we recognized a more expedient approach for a first proof-of-principle experiment. Our hydrogen spectrometer [29] required only minor modifications to produce two different fractional subharmonics, 4/7 and 1/2, of the dye laser frequency, which we could bridge with our femtosecond laser comb [65, 66]. The CH_4-stabilized helium neon laser served now as part of an interval divider and no longer as an intermediate reference standard. The primary reference for our first absolute frequency measurements was a commercial HP cesium atomic beam clock which we used to determine the pulse repetition rate f_r and the carrier-envelope offset frequency f_{CE} [65]. Once the optical frequency gap was measured, we knew the absolute frequency of the dye laser as well as the absolute frequencies of all the comb lines. To control the position of the comb lines, we learned how to change the frequency f_{CE} of our Mira Ti:sapphire laser by tilting an end mirror where the spectrum is slightly dispersed by a prism pair inside the cavity. In this way, we were the first to generate femtosecond laser pulses with controlled slips of the carrier-envelope phase.

In October 1998 we proudly showed our experiment to Norman Ramsey, who had come to the Max-Planck-Institute as a member of the Scientific Advisory Board. For the first time, we could compare the hydrogen 1S-2S frequency directly with a cesium atomic clock in our own laboratory, without involving any large harmonic laser frequency chain. Later in the same year, we demonstrated our experiment to John Hall who had come to Munich to attend a meeting commemorating our mutual friend Veniamin Chebotaev. John soon became an ardent evangelist for "this goofy technique, that makes everything obsolete that we have worked on for so long." He started to assemble a powerful professional team in Boulder to advance research on femtosecond laser frequency combs, and he persuaded his colleague Steve Cundiff at JILA, an expert on femtosecond lasers from Bell Laboratories, to visit our laboratory in the spring of 1999. An increasingly heated competition did much to accelerate the development of the new tools in the coming months and to ignite a firework of novel applications in the years to follow [3].

Until the summer of 1999, we had felt as if the new femto-comb playing ground belonged just to us. Thomas Udem had reported on our experiments at a conference in Perth, Australia, in late 1998 [43]. But we had delayed publications such as a first article on "measuring the frequency of light with mode-locked lasers" [67] to appear after the filing of our first patent application in March 1999 because, according to German law, an invention can no longer be patented once it has been published.

In June 1999, we could directly compare the hydrogen frequency with a highly accurate transportable cesium fountain clock (PHARAO), built at the LPTF (now BNM SYRTE) in Paris [66]. This measurement yielded a new value of the hydrogen 1S-2S frequency accurate to 1.8 parts in 10^{14}, surpassing all earlier optical frequency measurements by more than an order of magnitude. By now, the compelling advantages of laser frequency combs had been clearly demonstrated. A number of different possible approaches for carrier-envelope offset phase control were soon proposed [68].

As a next step in our own work, we wanted to drastically simplify our setup. Around that time, a new tool had appeared on the horizon that made it likely that we would no longer need any optical interval dividers. At the CLEO conference at Baltimore, MD, held in May 1999, researchers from Bell Laboratories had reported on a novel micro-structured "rainbow fiber" that could broaden the spectrum of pulses of a Ti:sapphire femtosecond laser oscillator without further amplification to a rainbow of colors [40]. After the white light interference experiments in Florence [62], I felt rather confident that this magic fiber would preserve the phase coherence of successive pulses and produce comb lines with a desirable large frequency spacing.

In June 1999, John Hall came to Germany to participate in the annual retreat of our research group at the Ringberg Castle near the Tegernsee south of Munich. Together, we phoned many of our old friends at Bell Laboratories, to try and obtain a sample of the magic fiber. We were hoping that we could demonstrate an octave-spanning frequency comb while John Hall was still in Germany. Unfortunately, this plan was foiled by the lawyers at Lucent Technologies who did not allow the fiber to leave Bell Laboratories. Ronald Holzwarth traveled to Bell Laboratories at Holmdel, NJ, during his 1999 summer vacation, but he had to leave without a piece of the fiber. John Hall's team in Boulder experienced similar difficulties at first, but in October 1999, they could demonstrate the first octave-spanning self-referencing laser frequency comb after finally securing some of the holey fiber [69, 70]. At Garching we realized a similar comb system a few weeks later [71], after we had received some "photonic crystal fiber" from the group of Philip Russell at the University of Bath in the UK. We had found out too late that these British researchers had actually pioneered micro-structured silica fibers some years earlier [39]. Both laboratories submitted their first short publications on octave-spanning frequency combs on the same day (Nov. 12, 1999) to the CLEO/QELS 2000 conference in San Francisco.

Similar to the Boulder experiments, we used a commercial small Ti:sapphire ring laser for our first octave-spanning frequency comb, producing pulses of about 25 fs duration at a repetition frequency of 625 MHz. Launching about 170 mW into a 30 cm length of photonic crystal fiber, we immediately produced a frequency comb spanning more than an octave. The spectrum showed a complicated structure, with valleys and peaks, but it offered useable comb lines everywhere. Together with a nonlinear interferometer for control of the offset frequency f_{CE}, the entire optical setup did easily fit on a single breadboard. While the traditional harmonic frequency chains with their factory halls full of lasers could measure just one single optical frequency, our new system was ready to measure any frequency throughout the visible and near infrared.

Since then, Ti:sapphire femtosecond lasers have been developed that produce an octave-spanning spectrum directly from the oscillator, without any need for external spectral broadening [72]. Octave spanning combs can also be generated with erbium-doped fiber lasers [73], pumped by very reliable and robust laser diodes developed for telecommunications. As turn-key instruments, such fiber comb generators can run for months without human attention.

In a first stringent test, Ronald Holzwarth has compared an octave spanning frequency comb synthesizer with the more complex frequency synthesizer used in the 1999 hydrogen frequency measurement [71]. By starting with a common 10 MHz radiofrequency reference and comparing comb lines near 350 THz, he could verify agreement within a few parts in 10^{16}, limited by Doppler shifts due to air pressure changes or thermal expansion of the optical tables. In 2002, a group at the PTB in Braunschweig demonstrated how a femtosecond laser frequency comb generator can be used as a transfer oscillator to precisely measure optical frequency ratios [74]. As a test case, they measured the frequency ratio between the second harmonic of a Nd:YAG laser and the fundamental frequency, verifying the expected value of 2 with an uncertainty of 7 parts in 10^{19}. More recently, Marcus Zimmermann in our laboratory has pushed a related experiment to an uncertainty of 6 parts in 10^{21} [75]. In 2004, researchers in Boulder compared four different frequency combs from different laboratories, finding agreement between neighboring comb lines at an uncertainty level of 10^{-19} [76]. So far, no systematic error has been identified which would limit the potential accuracy of future precision spectroscopy or optical atomic clocks.

NEW FREQUENCY MEASUREMENT OF HYDROGEN 1S-2S IN 2003: ARE THE FUNDAMENTAL CONSTANTS CONSTANT?

In February 2003, we used an octave spanning comb synthesizer in a new measurement of the hydrogen 1S-2S transition frequency [21]. Marc Fischer and Nikolai Kolachevsky had implemented many improvements in the hydrogen spectrometer. Light from the dye laser was sent through a fiber into the frequency metrology laboratory, where an octave-spanning Ti:sapphire femtosecond laser frequency comb synthesizer was used to compare the optical frequency to the radio frequency of the Paris PHARAO atomic fountain clock, which had again been brought to Garching.

With such immediate absolute frequency calibration, the hydrogen spectroscopy could be performed with much confidence. Compared to the 1999 measurements, the statistical error of the data recorded on a given day was much reduced. Nonetheless, the day-to-day fluctuations of our measurements remained of similar magnitude as before. They indicate some uncontrolled systematic errors. After careful further experiments probing for possible causes of systematic line shifts and after a thorough statistical analysis of all recorded data, we believe that the fluctuations are caused by some residual small first order Doppler shifts. Such shifts are expected if the two counter-propagating wave fronts of the exciting 243 nm radiation do not match perfectly. A mismatch may be caused by imperfect mode-matching of the frequency-doubled dye laser beam that enters the build-up cavity inside the atomic beam apparatus through a 2% input coupling mirror. Another cause may be the accumulation of frozen molecular hydrogen on the walls of the cold copper nozzle for the hydrogen atomic beam, which can grow until it distorts the optical wave fronts by vignetting and diffraction. A second cause of systematic line shifts in two-photon spectroscopy with frequency-doubled laser light may be unwanted correlations between amplitude noise and phase noise that could be caused by imperfect servo-locks of the enhancement cavities. For the future, we are preparing an ultraviolet build-up cavity of higher finesse, and we are working towards an all-solid-state laser source with a line width of only a few Hz.

From the 2003 measurement, we find a frequency of 2 466 061 102 474 851 ± 34 Hz for the F=1 to F'=1 hyperfine component of the hydrogen 1S-2S frequency, with a relative uncertainty of 1.4 parts in 10^{14}. The new results agrees within the error limits with the 1999 measurement of 2 466 061 102 474 880 ± 46 Hz. A difference of 29 ± 57 Hz in 44 months corresponds to a relative drift of the 1S-2S transition frequency of $(3.2 \pm 6.3) \times 10^{-15}$ per year, i.e. it is compatible with zero drift.

This experiment has attracted some attention because it can be considered as a test for a possible slow variation of the electromagnetic fine structure constant α. During the 2003 experiment, theorist Harald Fritzsch called frequently to get some preliminary results, because he had predicted an observable drift of the microwave frequency of the cesium clock relative to the hydrogen frequency [77]. Starting point for his arguments were astronomical observations of spectral lines in the light of distant quasars performed at the Keck Observatory [78]. Differential red shifts seemed to suggest that the electromagnetic fine structure constant α in the early universe was somewhat smaller than today. Making the simplest assumption of a linear drift, the data would indicate a drift of $(6.4 \pm 1.35) \times 10^{-16}$ per year, too small to be observable in our laboratory experiment. However, Fritzsch had argued with ideas from grand unification and quantum chromodynamics that α cannot change simply by itself. If all known forces are to remain unified at very high energies, other coupling constants must change as well. As a result, the masses and magnetic moments of hadrons (in units of the Bohr magneton) should change relative to those of the electron. Fritzsch pointed out a possible magnifying effect, that could change the hyperfine transition frequency of the cesium atomic clock about 20 times faster than the optical hydrogen frequency. So far, we have not found any evidence for such a drift. There are also other more recent observations of quasar spectra that do not support the evidence for a changing fine structure constant [79].

Regardless of such speculations, we have to admit that the hydrogen measurements of 1999 and 2003 do not strictly rule out a changing fine structure constant α. It is conceivable that the magnetic moment of the cesium nucleus also changes at just the right rate to give a null result in our experiment. Fortunately, such measurements are not limited to hydrogen. One can also measure transition frequencies in heavier atoms with stronger relativistic effects that respond in a different way to changes in α. One such candidate is the clock transition in a single cold Hg^+ ion, that has been compared by Jim Bergquist and his team at Boulder with a cesium atomic clock in 2000 and 2002, also using a laser frequency comb [80]. In addition, Eckhard Peik and his team at the PTB have made two separate measurements of the clock transition in a single Yb^+ ion [81]. Together, these laboratory experiments give now some upper limits for the possible rates of change of the fine structure constant α and the cesium nuclear magnetic moment μ_{Cs} on the order of $(-0.3 \pm 2.0) \times 10^{-15}$ per year and $(2.4 \pm 6.8) \times 10^{-15}$ per year, respectively.

Within the next few years we can expect much more stringent experimental limits on possible variations of physical constants from such laboratory experiments. If we found such changes it would not have any consequences for our everyday lives, but it would give reason for fascinating speculations about the nature of the universe.

placed a xenon gas jet at an intra-cavity focus. The high harmonic radiation is coupled out by external reflection from a thin sapphire Brewster plate, that has a refractive index smaller than 1 in the extreme ultraviolet. Similar experiments have also been reported by Jun Ye in Boulder [89].

In a future ambitious project, we plan to apply frequency combs in the XUV directly to precision spectroscopy of sharp resonances in laser-cooled trapped ions. The hydrogen-like helium ion with a 1S-2S two-photon transition near 60 nm is a particularly interesting candidate. In one envisioned scenario, helium ions will be sympathetically cooled by laser-cooled magnesium ions in the same trap, and the signal might be detected via the production of doubly charged helium ions due to photo ionization.

CONCLUSIONS

Spectroscopy of the simple hydrogen atom has sparked off the cross fertilization of two seemingly unrelated frontiers, precise optical spectroscopy and the study of ultrafast phenomena. Femtosecond frequency combs are revolutionizing precision measurements of time and frequency. Future optical atomic clocks will find important applications in many areas of science and technology. Ultraprecise optical spectroscopy can be harnessed for new tests of fundamental physics laws. However, many other spectroscopic applications of laser frequency combs can be envisioned, such as massively parallel ultra-sensitive cavity ring-down spectroscopy [90] or broadband spectral interferometry. At the same time, frequency comb techniques are also offering powerful new tools for ultrafast physics. By controlling the phase of the electric field of intense light pulses lasting for only a few cycles, they make it possible to study ultrafast electronic processes in light matter interactions, such as the production of single sub-femtosecond pulses of soft X-rays in high harmonic generation [42]. Only the future can show what we will discover with such exquisite new instruments.

ACKNOWLEDGEMENT

More than ninety students, postdocs, visiting scholars, and senior colleagues have made essential contributions to our research on precision laser spectroscopy during more than 4 decades. I owe particular gratitude to Thomas Udem and Ronald Holzwarth who played a key role in making the frequency comb synthesizer become reality. I am also very grateful to John L. Hall, who has long shared his precious insights on how to make lasers unbelievably stable.

REFERENCES

[1] C. Audoin, and G. Bernard, *The Measurement of Time: Time, Frequency, and the Atomic Clock*, Cambridge: Cambridge University Press, 2001

[2] Th. Udem, R. Holzwarth, and T.W. Hänsch, Nature, **416,** 233 (2002)

[3] Jun Ye, and S. Cundiff, eds., *Femtosecond Optical Frequency Comb: Principle, Operation and Applications*, Springer Verlag, New York, 2005

[4] P. Hannaford, ed., *Femtosecond Laser Spectroscopy*, Springer Verlag, New York, 2005

[5] A. Szöke, and A. Javan, Phys. Rev. Lett. **10**, 521 (1963)

[6] W.E. Lamb, Phys. Rev. **134**, 1429 (1964)

[7] W.R. Bennett, Phys. Rev. **126,** 580 (1962)

[8] V.S. Letokhov, and V.P. Chebotaev, *Nonlinear Laser Spectroscopy*, Springer Series in Optical Sciences, Vol. 4, Springer Verlag, New York, 1977

[9] T.W. Hänsch, R. Keil, A. Schabert, Ch. Schmelzer, and P. Toschek, Z. Physik, **226**, 293 (1969)

[10] T.W. Hänsch, and P. Toschek, Z. Physik **236**, 213 (1970)

[11] P.W. Smith, and T.W. Hänsch, Phys. Rev. Lett. **26,** 740 (1971)

[12] T.W. Hänsch, M.D. Levenson, and A.L. Schawlow, Phys. Rev. Lett. **26**, 949 (1971)

[13] T.W. Hänsch, Appl. Opt. **11,** 895 (1972)

[14] T.W. Hänsch, I.S. Shahin, and A.L. Schawlow, Phys. Rev. Lett. **27,** 707 (1971)

[15] P.P. Sorokin, and J.R. Lankard, IBM J. Res. Dev. **10**, 162 (1966)

[16] F.P. Schäfer, W. Schmidt, and J. Volze, Appl. Phys. Lett. **9**, 306 (1966)

[17] G.W. Series, *Spectrum of Atomic Hydrogen*, Oxford University Press, Oxford, 1957

[18] T.W. Hänsch, I.S. Shahin, and A.L. Schawlow, Nature **235**, 63 (1972)

[19] T.W. Hänsch, and A.L. Schawlow, Opt. Comm. **13**, 68 (1975)

[20] S.G. Karshenboim, F.S. Pavone, F. Bassani, M. Inguscio, and T.W. Hänsch, eds., *The Hydrogen Atom, Precision Physics of Simple Atomic Systems*, Springer Verlag, Lecture Notes in Physics, New York, 2001

[21] M. Fischer, N. Kolachevsky, M. Zimmermann, R. Holzwarth, Th. Udem, T.W. Hänsch, M. Abgrall, J. Grünert, I. Maksimovic, S. Bize, H. Marion, F. Pereira Dos Santos, P. Lemonde, G .Santarelli, P. Laurent, A. Clairon, and C. Salomon, M. Haas, U. D. Jentschura, and C. H. Keitel, Phys. Rev. Lett. **92**, 230802 (2004)

[22] E.V. Baklanov, and V.P. Chebotaev, Opt. Comm. **12**, 312 (1974)

[23] T.W. Hänsch, S.A. Lee, R. Wallenstein, and C. Wieman, Phys. Rev. Lett. **34**, 307 (1975)

[24] A. Javan, E.A. Ballik, and W.L. Bond, J. Opt. Soc. Am. 7, 553 (1962)

[25] K. M. Evenson, J. S. Wells, F. R. Petersen, B. L. Danielson, G. W. Day, R. L. Barger, and J. L. Hall, Phys. Rev. Lett. **29**, 1346 (1972)

[26] H. Schnatz, B. Lipphardt, J. Helmcke, F. Riehle, and G. Zinner, Phys. Rev. Lett. **76**, 18, (1996)

[27] T.W. Hänsch, in *The Hydrogen Atom, Proceedings of the Symposium held in Pisa, Italy, June 1988*, G.F. Bassani, M. Inguscio, and T.W. Hänsch, eds., Springer Verlag, New York, 1989, pp. 93

[28] H.R. Telle, D. Meschede, and T.W. Hänsch, Opt. Lett. **15,** 532 (1990)

[29] T. Udem, A. Huber, B. Gross, J. Reichert, M. Prevedelli, M. Weitz, and T.W. Hänsch, Phys. Rev. Lett. **79**, 2646 (1997)

[30] A. Huber, T. Udem, B. Gross, J. Reichert, M. Kourogi, K. Pachucki, M. Weitz, and T.W. Hänsch, Phys. Rev. Lett. **80**, 468 (1998)

[31] M. Kourogi, K. Nakagawa, and M. Ohtsu, IEEE J. Quant. El. 29, 2693 (1993)

[32] T. Udem, J. Reichert, M. Kourogi, and T.W. Hänsch, Optics Letters, **23**, 1387 (1998)

[33] Z. Bay, G.G. Luther, and J.A. White, Phys. Rev. Lett. **29**, 189 (1972)

[34] R. G. DeVoe, C. Fabre, K. Jungmann, J. Hoffnagle, and R. G. Brewer, Phys. Rev. A 37, 1802 (1988)

[35] N.C. Wong, Opt. Lett. **17**, 13 (1992)

[36] D.J. Wineland, J. Appl. Phys. **50**, 2528 (1979)

[37] A.E. Siegmann, *Lasers*, University Science Books, Mill Valley, 1986
[38] J.N. Eckstein, *Ph.D. Thesis*, Stanford University, 1978
[39] T.A. Birks, P.J. Roberts, P. St. J. Russell, D.M. Atkin, and T.J. Shepherd, Electron. Lett. **31**, 1941 (1995)
[40] J.K. Ranka, R.S. Windeler, and A.J. Stentz, Opt. Lett. 25, 25 (2000)
[41] J.J. McFerran, E.N. Ivanov, A. Bartels, G. Wilpers, C.W. Oates, S.A. Diddams, and L. Hollberg, Electron. Lett. **41**, 650 (2005)
[42] A. Baltuska, Th. Udem, M. Uiberacker, M. Hentschel, E. Goulielmakis, Ch. Gohle, R. Holzwarth, V.S. Yakovlev, A. Scrinzi, T.W. Hänsch, and F. Krausz, Nature **421**, 611 (2003)
[43] A.N. Luiten, ed., *Frequency Measurement and Control: Advanced Techniques and Future Trends,* Topics in Applied Physics Vol. 79, Springer Verlag, New York, 2001
[44] L.E. Hargrove, R.L. Fork, and M.A. Pollack, Appl. Phys. Lett. 5, 4 (1964)
[45] A. Yariv, J. Appl. Phys. 36, 388 (1965)
[46] O.P. McDuff and S.E. Harris, IEEE J. Quantum Electron. **3**, 101 (1967)
[47] C.V. Shank, and E.P. Ippen, in *Dye Lasers*, F.P. Schäfer. Ed., Topics in Applied Physics, Vol. 1, Springer Verlag, New York, 3rd edition, 1990, pp 139
[48] R. Teets, J. Eckstein, and T.W. Hänsch, Phys. Rev. Lett. **38**, 760 (1977)
[49] M.M. Salour, and C. Cohen-Tannoudji, Phys. Rev. Lett. **38**, 757 (1977)
[50] E.V. Baklanov and V.P. Chebotaev, Kvantovaya Elektronika **4**, 2189 (1977)
[51] J.N. Eckstein, A.I. Fergusons, and T.W. Hänsch, Phys. Rev. Lett. **40**, 847 (1978)
[52] A.I. Ferguson, J.N. Eckstein, and T.W. Hänsch, J. Appl. Phys. **49**, 5389 (1978)
[53] A.I. Ferguson, J.N. Eckstein, and T.W. Hänsch, Appl. Phys. **18**, 257 (1979)
[54] T.W. Hänsch, and N.C. Wong, Metrologia **16**, 101 (1980)
[55] T.W. Hänsch, Opt. Comm. **80**, 71 (1990)
[56] D.E. Spencer, P.N. Kean, and W. Sibbett, Opt. Lett. **16**, 42 (1991)
[57] T. Nakajima and P. Lambropoulos, Phys. Rev. A **50**, 595 (1994)
[58] G.G. Paulus, F. Lindner, H. Walther, A. Baltuska, E. Gouielmakis, M. Lezius, and F. Krausz, Phys. Rev. Lett. 91, 253004 (2003)
[59] L. Xu, C. Spielmann, A. Poppe, T. Brabec, F. Krausz, and T.W. Hänsch, Opt. Lett. **21**, 2008 (1996)
[60] R.L. Fork, C.V. Shank, C. Hirlimann, R. Yen, and W.J. Tomlison, Opt. Lett. **8**, 1 (1983)
[61] R. Zerne, C. Altucci, M.B. Gaarde, A. L'Huillier, C. Lynga, C.-G. Wahlström, M. Bellini, and T.W. Hänsch, Phys. Rev. Lett., **79**, 1006 (1997)
[62] M. Bellini and T.W. Hänsch, Opt. Lett. **25**, 1049 (2000)
[63] Th. Udem, J. Reichert, R. Holzwarth, and T.W. Hänsch, Opt. Lett. **24**, 881 (1999)
[64] Th. Udem, J. Reichert, R. Holzwarth, and T.W. Hänsch, Phys. Rev. Lett. **82**, 3568 (1999)
[65] J. Reichert, M. Niering, R. Holzwarth, M. Weitz, Th. Udem, and T. W. Hänsch, Phys. Rev. Lett. **84**, 3232 (2000)
[66] M. Niering, R. Holzwarth, J. Reichert, P. Pokasov, Th. Udem, M. Weitz, T. W. Hänsch, P. Lemonde, G. Santarelli, M. Abgrall, P. Laurent, C. Salomon, and A. Clairon, Phys. Rev. Lett. **84**, 5496 (2000)
[67] J. Reichert, R. Holzwarth, Th. Udem, and T. W. Hänsch, Opt. Comm., **172**, 59 (1999)
[68] H.R. Telle, G. Steinmeyer. A.E. Dunlop, J. Stenger, D.H. Sutter, and U. Keller, Appl. Phys. B **69**, 327 (1999)
[69] S. A. Diddams, D. J. Jones, J. Ye, S. Cundiff, J.L. Hall, J. K. Ranka, R. Windeler, R. Holzwarth, Th. Udem, and T. W. Hänsch, Phys. Rev. Lett. **84**, 5102 (2000)
[70] D.J. Jones, S.A. Diddams, J.K. Ranka, A. Stentz, R.S. Windeler, J.L. Hall, and S. Cundiff, Science **288**, 635 (2000)
[71] R. Holzwarth, Th. Udem, T.W. Hänsch, J. C. Knight, W. J. Wadsworth, and P.St.J. Russell, Phys. Rev. Lett. **85**, 2264 (2000)

[72] L. Matos, D. Kleppner, O. Kuzucu, T. R. Schibli, J. Kim, E. P. Ippen, and F.X. Kaertner, Opt. Lett. **29**, 1683 (2004)
[73] B.R. Washburn, S.A. Diddams, N.R. Newbury, J.W. Nicholson, M.F. Yan, and C.G. Jorgensen, Opt. Lett. **29**, 250 (2004)
[74] J. Stenger, H. Schnatz, C. Tamm, and H.R. Telle, Phys. Rev. Lett. **88**, 073601 (2002)
[75] M. Zimmermann, Ch. Gohle, R. Holzwarth, Th. Udem, and T.W. Hänsch, Opt. Lett. **29**, 310-312 (2004)
[76] L.S. Ma, B. Zhiyi, A. Bartels, L. Robertsson, M. Zucco, R.S. Windeler, G. Wilpers, C. Oates, L. Hollberg, and S.A. Diddams, Science **303**, 1843 (2004)
[77] X. Calmet and H. Fritzsch, Phys. Lett. B **540**, 173 (2002)
[78] M.T. Murphy, J.K. Webb, and V.V. Flambaum, MNRAS 345, 609 (2003)
[79] S. Srianand, H. Chand, P. Petitjean, and B. Aracil, Phys. Rev. Lett. **92**, 121302 (2004)
[80] S. Bize, S.A. Diddams, U. Tanaka, C.E. Tanner, W.H. Oskay, R.E. Drullinger, T.E. Parker, T.P. Heavner, S.R. Jefferts, L. Hollberg, W.M. Itano, and J.C. Bergquist, Phys. Rev. Lett. 90, 150802 (2003)
[81] E. Peik, B. Lipphardt, H. Schnatz, T. Schneider, C. Tamm, and S.G. Karshenboim, Phys. Rev. Lett. **93**, 170801 (2004)
[82] L. Hollberg, C.W. Oates, G. Wilpers, C.W. Hoyt, Z.W. Barber, S.A. Diddams, W.H. Oskay, and J.C. Bergquist, J. Phys. B **38**, S469 (2005)
[83] M. Takamoto, F.L. Hong, R. Higashi, and H. Katori, Nature **435**, 321 (2005)
[84] A. McPherson, G. Gibson, H. Jara, U. Johann, T.S. Luk, I. McIntyre, K. Boyer, and C.H. Rhodes, J. Opt. Soc. Am, B 4, 595 (1987)
[85] M. Ferray, A. L'Huiller, X.F. Li, A. Lompré, G. Mainfray, and C. Manus, J. Phys. B **21**, L31 (1988)
[86] P. Corkum, Phys. Rev. Lett. **71**, 1994 (1993)
[87] M. Bellini, C. Lynga, A. Tozzi, M.B. Gaarde, A. L'Huillier, C.-G. Wahlström, and T.W. Hänsch, Phys. Rev. Lett., **81**, 297 (1998)
[88] Ch. Gohle, Th. Udem, J. Rauschenberger, R. Holzwarth, M. Herrmann, H.A. Schüssler, F. Krausz, and T.W. Hänsch, Nature, **436**, 234 (2005)
[89] R. J. Jones, K. D. Moll, M. J. Thorpe, and J. Ye, Phys. Rev. Lett. **94**, 193201 (2005)
[90] Jun Ye, private communication

Portrait photo of Theodor W. Hänsch by photographer F. M. Schmidt.